A Series In

The History of Modern Physics, 1800-1950

The History of Modern Physics, 1800-1950

The History of Modern Physics, 1800-1950

TITLES IN SERIES

VOLUME I
Alsos
by Samuel A. Goudsmit

VOLUME II
Project Y: The Los Alamos Story
Part I: *Toward Trinity*
by David Hawkins.
Part II: *Beyond Trinity*
by Edith C. Truslow and Ralph Carlisle Smith

VOLUME III
American Physics in Transition: A History of Conceptual Change in the Late Nineteenth Century
by Albert E. Moyer

VOLUME IV
The Question of the Atom: From the Karlsruhe Congress to the First Solvay Conference, 1860-1911
by Mary Jo Nye

INTRODUCTORY NOTE

The Tomash series in the History of Modern Physics offers the opportunity to follow the evolution of physics from its classical period in the nineteenth century when it emerged as a distinct discipline, through the early decades of the twentieth century when its modern roots were established, into the middle years of this century when physicists continued to develop extraordinary theories and techniques. The one hundred and fifty years covered by the series, 1800 to 1950, were crucial to all mankind not only because profound evolutionary advances occurred but also because some of these led to such applications as the release of nuclear energy. Our primary intent has been to choose a collection of historically important literature which would make this most significant period readily accessible.

We believe that the history of physics is more than just the narrative of the development of theoretical concepts and experimental results: it is also about the physicists individually and as a group—how they pursued their separate tasks, their means of support and avenues of communication, and how they interacted with other elements of their contemporary society. To express these interwoven themes we have identified and selected four types of works: reprints of "classics" no longer readily available; original monographs and works of primary scholarship, some previously only privately circulated, which warrant wider distribution; anthologies of important articles here collected in one place; and dissertations, recently written, revised, and enhanced. Each book is prefaced by an introductory essay written by an acknowledged scholar, which, by placing the material in its historical context, makes the volume more valuable as a reference work.

The books in the series are all noteworthy additions to the literature of the history of physics. They have been selected for their merit, distinction, and uniqueness. We believe that they will be of interest not only to the advanced scholar in the history of physics, but to a much broader, less specialized group of readers who may wish to understand a science that has become a central force in society and an integral part of our twentieth-century culture. Taken in its entirety, the series will bring to the reader a comprehensive picture of this major discipline not readily achieved in any one work. Taken individually, the works selected will surely be enjoyed and valued in themselves.

A Series In

The History of Modern Physics, 1800-1950

VOLUME IV

THE Question OF THE Atom

THE

Question

OF THE

Atom

FROM THE KARLSRUHE CONGRESS
TO THE FIRST SOLVAY CONFERENCE,
1860–1911

A COMPILATION OF PRIMARY SOURCES
SELECTED AND INTRODUCED BY

Mary Jo Nye

Tomash Publishers

LOS ANGELES / SAN FRANCISCO

SECOND PRINTING, 1986

This printing is part of a copublishing agreement between Tomash Publishers and the American Institute of Physics.

Printed in the United States of America.

Library of Congress Cataloging in Publication Data
Main entry under title:

The Question of the atom.

(History of modern physics, 1800–1950 ; v. 4)
Includes bibliographical references.
1. Atomic theory. I. Nye, Mary Jo. II. Series.
QC173.Q45 1983 539.7 82-50748
ISBN 0-938228-07-2

ACKNOWLEDGMENTS

The author and publisher wish to thank the following publishers and societies for permission to reproduce facsimile copies of original material, the bibliographic details of which can be found in the Primary Source section of the Appendix: Taylor and Francis, Verlag Chemie GMBH, Open Court Publishing Company, Princeton University Press, Pergamon Press, the Royal Society of Chemistry (Journal of the Chemical Society) and the Royal Society of Edinburgh (Alembic Club Reprints). Most of the reproductions were made from materials located at the libraries of the University of California at Los Angeles, and we gratefully acknowledge the assistance of the librarians who helped locate the volumes. The Alembic Club Reprints were obtained courtesy of the Niels Bohr Library of the American Institute of Physics.

For their help in procuring the photographs which were reproduced in the book, we should also like to express our thanks to Spencer R. Weart, Eleanor A. Maass, Aaron J. Ihde, and Donald G. Miller. Permission to reproduce the photographs of Adolphe Wurtz, Stanislao Cannizzaro, Lord Kelvin, A.W. Williamson, Ernst Mach, J. Norman Lockyer, James Clerk Maxwell, Charles Marignac, Marcellin Berthelot, Hermann Von Helmholtz, Svante Arrhenius, Dmitri Mendeleev, J.J. Thomson, Henri Poincaré, Frederick Soddy, Albert Einstein, Jean Perrin, and Ernest Rutherford was given by the ACS Center for The History of Chemistry, University of Pennsylvania (Edgar Fahs Smith Collection); those of Ludwig Boltzmann and of the Solvay Conference, 1911, by the AIP Niels Bohr Library. Donald G. Miller gave us the likeness of Pierre Duhem. To these individuals and organizations we are most thankful.

The brief biographies in this anthology rely heavily on biographical entries in the *Dictionary of Scientific Biography* edited by Charles C. Gillispie (New York: Charles Scribner's Sons, 1970–1980). Invaluable as sources, too, are J.R. Partington, *A History of Chemistry*, Volume IV (London: Macmillan, 1972) and Aaron J. Ihde, *The Development of Modern Chemistry*, (New York: Harper and Row, 1964).

I am grateful to Adele Clark at Tomash Publishers and to those associated with the History of Modern Physics series editorial board for their advice and comments on this anthology. I have also benefited from the reading of my introductory essay by Alan Rocke, David Knight, and Robert Nye, and from the typing and retyping of the essay and biographical sketches by Valli Powell. In making selections of primary sources for the anthology, an important consideration was the availability of English translations. I am particularly pleased that this anthology offers the first published translation, by John Greenberg and William Clark, of Adolphe Wurtz's account of the Karlsruhe Congress proceedings.

EPIGRAPH

Well, it you took a piece of lead, and halved it, and halved the half, and went on like that, where do you think you'd come to in the end? Do you think it would be lead for ever? . . . If you went on long enough, you'd come to an atom of lead, an atom, do you hear, an atom, and if you split that up, you wouldn't have lead anymore. What do you think you would have? . . . pieces of positive and negative electricity. Just that . . . . That's all you are. Just positive and negative electricity—and, of course an immortal soul.

The Schoolmaster Luard to his class.
C. P. Snow, *The Search* (New York:
Charles Scribner's Sons, 1934, 1958), p. 10.

Contents

Introduction

ONE HUNDRED AND FORTY CHEMISTS GATHERED, IN THE EARLY AUTUMN OF 1860, at a meeting hall at Karlsruhe, in the German Rhineland, for the first international scientific congress. The extraordinary meeting was the idea of August Kekulé, a professor of chemistry at the University of Ghent and the recent author of the controversial notion of carbon chains. Discussions at the Congress focused on chemists' definitions of "atoms," "molecules," and "equivalents" in an attempt to bring coherence and consistency into chemical notation. The only formal recommendation made by the Congress was that chemists be encouraged to use a barred notation whenever they employed the convention denoting the atomic weight of oxygen to be 16. Thus the symbol for oxygen would be Θ in the formula H^2O, when the atomic weight of oxygen is taken to be 16 rather than 8 (Carlsruhe "Compte Rendu," p. 688).[1]

Decades later, in the autumn of 1911, another international congress convened, this time at Brussels, sponsored by the successful industrialist Ernest Solvay. In opening this first Solvay Congress of Physics, the Dutch physicist H. A. Lorentz indicated that the questions under discussion were to be the fundamental principles of mechanics and the ultimate properties of matter. Walter Nernst, the Berlin specialist in thermodynamics who organized the Congress sessions, recalled to the twenty-one assembled physicists the earlier Karlsruhe meeting of 1860: "That Congress was likewise called in order to study the one fundamental question of atomistics."[2]

Over a period of some fifty years, from the Karlsruhe Congress to the Brussels meeting, physical scientists continued to debate the efficacy and validity of the atomic theory. What was at issue was not the *nature* of the atom, but the *very existence* of both the atom and the molecule. Were they real? The problem of the "atom" was one which perplexed both chemists and physicists, directing each group's attention to the work of the other. The Italian physicist Stanislao Cannizzaro anticipated the coming synthesis of chemistry and physics when he denied, at the Karlsruhe meeting, that there was any meaningful distinction between the "chemical atom" and the "physical atom" (Carlsruhe "Compte Rendu," p. 676). But the Zurich physicist Rudolf Clausius showed little awareness of chemistry when he proposed a few years earlier, in 1857, the notion of a diatomic molecule, a notion which chemists had been discussing since Amedeo Avogadro and André Marie Ampère independently proposed the gas law of "equal volumes-equal numbers of particles" in 1811 and 1814.[3] Later on, J. Norman Lockyer, who was noted for his work in spectroscopy, still claimed, in 1874, that "the chemist never thinks about encounters, and the physicist is careless as to atomic weight" (Lockyer, p. 117).

While Cannizzaro's identity principle was to become a fundamental assumption of both chemists and physicists by the time of the 1911 Solvay conference, we can see that this hardly happened overnight. Many chemists, recalcitrant in employing Cannizzaro's atomic weights, were in fact resisting the subordination of chemical laws and chemical science to general *physical* principles. This physical reductionism was exemplified in the claim, made in 1867, by the Scottish physicist William Thomson (Lord Kelvin) that his *physical* vortex atom could explain *chemical* affinities (Kelvin, pp. 16–17). Similarly, the Austrian physicist Ludwig Boltzmann praised the kinetic-atomic theory as a key to understanding chemical combination and chemical isomerism whereby substances differing in their properties might have the same chemical composition (Boltzmann, pp. 73–74). In contrast to Kelvin and Boltzmann's claims, August Kekulé spoke for many chemists at Karlsruhe when he said that physical considerations should be invoked only as a standard of control for molecular ideas deduced from chemical laws (Carlsruhe "Compte Rendu," p. 676); and the Parisian chemist Marcellin Berthelot, who opposed the substitution of atomic weights for equivalent combining weights, introduced physical properties into his chemistry courses at the Collège de France "only in a subordinate way" to give greater precision to chemical analogies (Berthelot, p. 185).

The idea that chemistry was a different kind of discipline from physics was defended, too, by Pierre Duhem, professor of theoretical physics at the University of Bordeaux. He claimed that chemical analogies and chemical substitution are a system of explanation "in the manner of ideas employed by the naturalists," and that "chemistry is to physics as morphology is to physiology." Duhem cited the 1867 *Leçons sur l'affinité* of his Parisian colleague Henri Sainte-Claire Deville as a classic statement of this view of the nature of chemical knowledge: that chemistry is a natural science and its theories are methods of classification (Duhem, pp. 28–29, 125).[4]

But the history of the "chemical atom" from 1860–1911 is the history principally of its *physical* confirmation and elucidation. This was clearly evident in the focus placed on atomism at the 1911 Solvay Congress of Physics. The wave of the future lay in J.J. Thomson's 1897 attempt, while working at Cambridge University's Cavendish Laboratory, to deduce Mendeleev's periodic chemical properties from physical configurations of particles within a composed chemical atom (Thomson, p. 313). When Henri Poincaré heralded the birth of physical chemistry in 1909 at the International Physics Congress in Paris, he was heralding an event long fought by some members of the chemical community: the reduction of chemistry to physics through the identification of the chemical with the physical atom.

Both chemists and physicists in the nineteenth century sought to agree on *invariable* molecular properties defined by fundamental laws. Wilhelm Ostwald, the leader of the Leipzig school of physical chemistry, expressed this goal at the 1895 Lübeck meeting of the Society of German Scientists and Physicians when he said, "What we seek in scientific concepts are invariants" (Ostwald, p. 422). But in 1860 the chemists had no invariants. What they had were two systems for deriving sets of numerical values for combining weights of the elements: using relative atomic weights on the one hand, and equivalent combining proportions on the other. The values for atomic weights depended upon an arbitrary weight chosen for oxygen or hydrogen when combining to form water, as well as upon hypothetical chemical formulas chosen for water and other compound substances, according to rules of simplicity and analogy. Similarly, the values for equivalent combining proportions (or gas volumes) depended upon a set of conventional assumptions or rules, as well as upon the experimentally observed function of each element in particular chemical reactions. That one element might have different equivalent values, depending upon the chemical reaction considered, was noted, in 1869, by A. W. Williamson, who taught chemistry at University College, London. For example, nitrogen combines with hydrogen in the weight proportions $1:4\frac{2}{3}$ in ammonia; but $1:2\frac{4}{5}$ in sal ammoniac. Therefore nitrogen has at least two equivalent weights, each the direct result of experiment and each a submultiple of the atomists' single value of 14 for the atomic weight of nitrogen (Williamson, pp. 332–333).[5]

As the search for invariants proceeded, experimental arguments regarding physical constants in the atomic theory centered fundamentally around the questions of the difference between an atom and a molecule, the relative weights of atoms and molecules, the arrangements of atoms in molecules, and the divisibility of the atom itself. The most important experimental work bearing upon these problems had to do with vapor densities, specific heats, and spectral lines. Most antiatomists were unwilling to entertain the idea that elementary atoms, if they really exist, are indivisible; and some atomists, such as the Swiss chemist Charles Marignac, agreed with antiatomist Marcellin Berthelot that the indivisible atom could not exist (Marignac, p. 95). Other atomists, Williamson, for example, hedged the issue by saying that "atom . . . denotes the fact that [the elements] do not undergo decomposition under any conditions known to us" (Williamson, pp. 343, 365). Chemists insisted that a fundamental difference between chemical change and physical change was in the discontinuous nature of chemical transformations: there are sudden jumps in the masses of the elements, claimed Russian chemist Dmitri Mendeleev, and "not . . . a continuous evolution" (Mendeleev, p. 167).

Results of spectroscopic analysis suggested that chemical elements, whether atoms or molecules, have a capacity for internal vibrations. The Viennese physicist Ernst Mach was influenced in his antiatomism by the failure of physicists to successfully explain spectra by atomic modelling with vortex atoms, for example (Mach, p. 87). James Clerk Maxwell concluded from spectroscopy that a molecule "cannot . . . be a mere material point, but [must be] a system capable of changing its form" (Maxwell, p. 504). Norman Lockyer's comparison of stellar spectra to each other and to terrestrial spectra led him to the controversial conclusion that great heat and strong electrical action might dissociate terrestrial elements into simpler particles (Lockyer, p. 126). This speculation renewed interest in the old Proutian hydrogen or protyle hypothesis, the idea proposed by William Prout around 1815 that hydrogen atoms are the material from which all other elements are formed. Through the work of Lockyer, supported especially by his British colleague William Crookes, the idea of the "dissociation" of matter became more and more popular, and J. J. Thomson founded his 1897 interpretation of electrons on the view that "atoms of the different chemical elements are different aggregations of atoms of the same kind." The cathode-ray "corpuscles," Thomson said, are the product of the dissociation" of the molecules of gas in the very intense electric field near the cathode (Thomson, p. 311).

The word "dissociation" had most frequently been used by chemists in connection with studies of vapor densities at high temperatures. Like spectroscopic studies, the investigation of vapor densities and specific heats was crucial in the search among atomists and antiatomists for physical constants. From the time that John Dalton first calculated the relative atomic weights of elementary substances, chemists used vapor densities to estimate atomic and molecular weights of elementary and compound gases and vapors. To do this, either one must assume that gases or vapors under identical physical conditions contain the same number of particles, or one must adopt rules for deciding ratios of numbers of particles in different gases.

The first two volumes of Dalton's influential treatise on atomism, the *New System of Chemical Philosophy*, appeared in 1808 and 1810. Jöns Jacob Berzelius in Sweden and Jean-Baptiste Dumas in France were among the most influential European chemists who devised precise methods to calculate relative atomic weights. But Dumas, assuming all elementary gas particles are split into identical halves in chemical reactions, ran into some difficulty in deciding upon atomic weight values for sulfur, phosphorus, and mercury vapors. Later, Henri Sainte-Claire Deville demonstrated the break-up of vapors at high temperatures (Cannizzaro, pp. 6–7, Carlsruhe, "Compte Rendu," pp. 684–686); and A. A. Kundt and Emil Warburg in 1875 demonstrated that mercury vapor has exactly the ratio of specific heats (c_P/c_V) predicted from kinetic theory for single, billiard-ball gas particles. These results helped establish that all elementary gas molecules need not contain the same number of atoms. Mercury vapor is made up of single-atom molecules, whereas sulfur vapor ordinarily exists as S_6 and phosphorus vapor as P_4. Working out these relations removed some objections to calculations for atomic weights.[6]

Still, spectroscopic studies suggested that the billiard-ball atom vindicated by Kundt and Warburg must yet have an internal capacity for vibration; yet no consistent representation of internal vibrations, external translations, and specific heats could be found. Maxwell regarded the problems concerning specific heat as the "greatest difficulty which the molecular theory has yet encountered" (Maxwell, pp. 502–504). While the specific heats of metals such as lead, mercury, and iron had provided a guide for atomists to enable them to double the equivalentists' values for weight (Cannizzaro, p. 27; and Williamson, pp. 357–358), Berthelot and antiatomists emphasized the unreliability of specific heats, and, in the case of solids, their dependence upon temperature (Berthelot, p. 186). A crucial breakthrough in the controversy over vapor densities, specific heats, and atomic and molecular weights was the demonstration by J. H. Van't Hoff, Svante Arrhenius, François Raoult, J. D. Van der Waals, and others, that the gas laws could be applied reliably to solutions. Molecular weights could now be calculated from osmotic pressures, freezing-point depressions, boiling-point elevations, and ionization activity coefficients, independently corroborating the values favored by those using specific heats and confirming the atomist system based on H = 1, O = 16 (Arrhenius, p. 53).

Direct measurements of the absolute weights, sizes, and numbers of atoms or molecules in a volume of gas or liquid seemed far-fetched for many years after Karlsruhe. Critics of both chemical and physical atomism demanded, in the meantime, that the atomic theory must be useful, even if it could not be proved true. This does not mean that chemists and physicists abandoned debate about the reality of atoms and molecules, but rather that the grounds of debate often shifted to the topic of utility. Striking exceptions to this rule were remarks made during the 1869 discussion of atomism at the London Chemical Society and also during the 1877 debates at the Paris Academy of Sciences. On the former occasion, Williamson was understood by the audience to defend a realist atomic theory, and George Carey Foster, a professor of physics at University College, London, insisted that the "question is not whether the atomic theory is useful or useless, but whether it is true or false" (Williamson, p. 437).[7] In 1877 Berthelot voiced his strong opposition to the realist implications of any atomic hypothesis when he demanded of the Academy, "Who has ever seen, I repeat, a gaseous molecule or an atom?"[8]

There were two criteria for the usefulness of the atomic-molecular hypothesis: first, a hypothesis must be fruitful and suggestive as an instrument of discovery; and second, it must be innocuous in the sense that it does not lead to false conclusions. Practically every chemist and physicist paid lip service to the heuristic value in scientific discovery of carefully chosen hypotheses; some critics of atomism compared the heuristic hypothesis to disposable scaffolding which is discarded once a building or structure is complete. Why not treat atomism in the same manner? Others stressed the possible harm caused by hypotheses not capable of swift and sure verification through direct experience. This objection was voiced by those influenced by the ideas of the French philosopher Auguste Comte who had taught, in the 1830s, that hypotheses are to be used only if "susceptible, by their very nature, of a positive verification, sooner or later."[9] Comte's words are echoed almost exactly in Poincaré's view that a hypothesis should, as soon and as often as possible, be submitted to verification (Poincaré, p. 523).

In the empiricist, positivist tradition of Comte, Ernst Mach warned that we must not turn facts, such as the conservation of weight, into metaphysical conclusions, such as the conservation of matter. Imagining a three-dimensional spatial structure for elementary atoms to explain spectral lines or chemical properties is similarly unwarranted. "We are not justified in ascribing spatial properties to things which are not perceived by the senses" (Mach, pp. 48–49, 86–87). Hypotheses based on extrapolations of our sensations may lead to false theories if we seek an identity where only a loose analogy is valid. "Are we to understand," asked

Poincaré, "that God, in contemplating his work, feels the same sensations as we in the presence of a billiard match? . . . Hypotheses of this nature have only a symbolic sense" . . . the sense of "metaphor" (Poincaré, p. 531).

The *kind* of useful hypothesis allowed in scientific theory was a crucial issue among antiatomists. Both Mach, an antiatomist, and Boltzmann, an atomist, seemed to agree that in the end, hypotheses (or "unintelligibilities" as Mach termed them) are a "matter of taste" (Mach, p. 55; and Boltzmann, p. 65). Certain kinds of hypotheses, namely physical models, appealed very little to those critical of the atomic-molecular theory. Both Ostwald and Mach identified the molecular theory with the mechanical tradition of masses-and-forces, a physical view which Mach termed "an evil" for its allegedly nonempirical foundations (Ostwald, p. 434; and Mach, p. 57). Duhem turned his ire especially against what he called the English school of physical modelling, the engineer's universe. In a famous critique of English mechanical physics, Duhem complained of it:

> In it there are nothing but strings which move around pulleys, which roll around drums, which go through pearl beads, which carry weights; and tubes which pump water while others swirl and contract; toothed wheels which are geared to one another and engage hooks. We thought we were entering the tranquil and neatly ordered abode of reason, but we find ourselves in a factory.[10]

It became clear that there was special resistance among antiatomists to fanciful models proposed to meet problems of spectral lines, specific heats, and chemical affinities. Further, resistance emerged to the new, three-dimensional carbon-atom models proposed independently in 1874 by the Dutch physical chemist J. H. Van't Hoff and the French chemist J. A. Le Bel. Dismay over absurdities in mechanical representation prompted Ostwald to criticize the "naive drawings" of a contemporary chemist who "visualized" the catalytic effect of pounded glass on certain gases by drawing the sharp edges of glass splinters cutting gas molecules into atoms which could then combine freely.[11]

If the atomic-molecular theory had been the only serious contender in explaining physical and chemical properties of substances, its critics might have been silenced. In fact, there were at least two promising alternative strategies. One of these strategies took shape in Maxwell's electromagnetic theory, which relates electricity, magnetism, and light to one another by the same set of fundamental equations. From this, some theorists inferred that the only physical realities are electromagnetic ether and electric particles which themselves may be reducible to properties of ether. Lord Kelvin worked out his ideas about the dynamic conception of matter inherent in the electromagnetic program in an influential paper of 1867, where he presented the atom as a kind of compromise between Newton's hard, impenetrable particles and Descartes' vortex continuum. The paper concluded with predictions of spectral lines and the development of general electromagnetic ideas as a natural outgrowth of properties inherent in the structure of matter (Kelvin, esp. pp. 17–24).

While the British physicists William Thomson (Lord Kelvin), Joseph Larmor, Oliver Lodge, and J. J. Thomson characteristically used mechanical analogies or models to draw out the implications of Maxwell's theory, some continental physicists attempted to eliminate mechanical conceptions and even ether itself from that theory. Wilhelm Wien, in Germany, successfully formulated Newton's first and second laws as special cases of the more general laws of electromagnetism in 1900; and Lorentz, in 1904, offered two proposals: that the *masses of all material particles vary with motion as do electron masses*, and that all mass may be electromagnetic in origin.[12] It appeared, as Helmholtz had argued in his 1881 Faraday Lecture, that the electromagnetic theory of Maxwell might restore to natural philosophy the old nonmaterial definition of "substance" (Helmholtz, p. 138).

The relevance of electrical theory to chemical science had hardly been disputed since the invention of the Voltaic pile and the pioneering work of Humphry Davy and Michael Faraday early in the nineteenth century. Cannizzaro treated the "capacity for saturation" of one substance reacting chemically with another partially as an electrical problem, as did J.J. Berzelius and J. B. Dumas before him. He divided elementary and compound reactants into electropositive and electronegative "radicals" (Cannizzaro, p. 36). By the end of the century, the term "valence" replaced "capacity" (or "atomicity," the term used by the Parisian chemist Adolphe Wurtz). Using the notion of electrochemical equivalents developed decades earlier by Faraday, the twentieth-century definition of equivalent weight was to become the ratio of atomic weight to valence. In the late nineteenth century, perhaps the most influential exposition of the necessary relationship between electricity and chemical affinity was Helmholtz's 1881 Faraday Lecture given before the London Chemical Society. While declining to speculate on the "real nature" of electricity, Helmholtz argued that "the very mightiest among the chemical forces are of electrical origin." Faraday had often relinquished the atom theory whenever he focused on electrical and magnetic forces. But, Helmholtz postulated, if we accept the hypothesis that substances are composed of atoms, we cannot avoid the conclusion that electricity itself is divided into definite elementary portions which behave like atoms of electricity (Helmholtz, pp. 155–158, 145).

While electromagnetic theories had their proponents, a second alternative strategy to atomism was the new thermodynamics based upon the general laws of the conservation of energy and the increase of entropy in closed systems. Just as Lorentz had attempted to give physical meaning to the formal equations of electromagnetism, so Ostwald and his colleague Georg Helm attemped to give physical meaning to the formalism of thermodynamics. Their views received much attention at the 1895 Lübeck meeting of the Society of German Scientists and Physicians where they argued that the underlying reality of the universe was neither matter nor ether, but energy.[13] Matter, suggested Ostwald, is a group of energy forms coordinated in space (Ostwald, p. 432).

An advantage of thermodynamics over atomism lay in its empirical foundations in measurements of heat gain and loss; and most antiatomists avoided Ostwald's tendency to read physical significance into the concept of energy. Poincaré, for example, described the principle of conservation of energy as the empirical principle that "there is something that remains constant" (Poincaré, p. 532). Ostwald championed empirical thermodynamics, or what Berthelot called "molecular mechanics," as a particularly promising instrument for developing the fundamental laws of chemical dynamics (Ostwald, p. 420; Berthelot, p. 187). Mach argued that heat of combustion gives a clearer idea of chemical stability and combination than any pictorial, molecular representations. The most rational way of writing chemical combinations, he suggested, might be to write the components in a circle, drawing a line connecting the components, and writing out the respective heats of combination. This avoided any particular view about the atomic or molecular constitution of bodies (Mach, pp. 53–54).

The thermodynamic strategy had an advantage over atomism which was not shared by electromagnetism. The second law of thermodynamics is a law of irreversible physical processes, whereas the classical laws of both atomism and electromagnetism in no way preclude the reversibility of processes in the natural world. Critics of atomism often argued that the "mechanical world view," which describes all natural phenomena as the result of matter in motion, cannot explain the distinction between past and present, that is, the irreversibility of life. As Ostwald expressed the argument, atomism cannot tell us why the tree does not return to the sapling, nor the old man to the babe (Ostwald, p. 428).

Following the example of Maxwell, atomists began to deal statistically with the problem of deducing irreversible results from molecular collisions. One of the most successful demonstrations of statistical and probabilistic laws was that of Brownian motion which described the random motions of microscopic particles in a colloid. Jean Perrin confirmed these laws in a brilliant series of experiments from 1908 to 1911. Poincaré anticipated the importance of this application of kinetic theory when he referred, in his 1900 address at the International Congress of Physics, to the recent work of the Lyon physicist Georges Gouy on Brownian motion (Poincaré, p. 541); and so, by the time of the Solvay Congress in 1911, Perrin had provided, experimentally, a convincing riposte to the antiatomists' demands for a demonstration of Carnot's thermodynamic law of irreversibility, beginning with the random motions of molecular processes (Perrin, pp. 1–74).[14]

Another difficulty for atomism was its alliance with the general program of mechanical reductionism, the aim of which was to explain both biological and psychological processes from the laws of molecular physics. Those who resisted a purely mechanistic approach to *all kinds of natural phenomena* often criticized atomism as the backbone of the program, and physical atomism in chemical explanation was viewed as an avenue to biological reductionism. That physical atomists did not balk at explaining reason and sensation in terms of atomic processes is indicated in Boltzmann's *pointilliste* speculations about vision:

> Our sensations of sight correspond to the excitation of a finite number of nerve fibers, and are thence doubtless better represented by a mosaic than by a continuous surface. The same holds true of other sensations. Is it not more probable, therefore, that the models representing complexes of sensations are more fitly composed of discrete parts? (Boltzmann, p. 69).

That the mechanical formulas of the great Napoleonic astronomer Pierre-Sadi Laplace might allow the prediction of future positions of all particles in the universe, was an idea, claimed Mach, which led inevitably to the mechanics of "the atoms of the brain."[15]

The goal of a *unified science* was no small concern in the nineteenth century, as is illustrated by the popularity of Auguste Comte's positivism. This philosophy required an empirical and nonmetaphysical methodology to establish true knowledge of things and processes. While uniting the individual scientific disciplines logically in a hierarchical structure, Comte's positivism also described the necessary order of the historical evolution of the sciences. Critics of atomism often argued that atoms could not (and should not) provide a fundamental unifying theme for all the sciences. But arguments regarding the proper *methods* of a unified science frequently failed to distinguish methods of *discovery* from methods of *demonstration* or exposition. Indeed, so prevalent was this confusion, that one wonders if some scientists, especially the antiatomists, might really have thought that there was nothing new to discover.[16]

Cannizzaro's 1858 pamphlet "Sunto di un Corso di Filosofia Chimica," which was distributed to participants in the Karlsruhe Congress, was a defense of atomism on the grounds that it simplified the teaching of chemistry. The conviction that chemical notation must be reformed in the interest of both scientific education and industrial progress was voiced at Karlsruhe by both the distinguished Parisian chemist J.B. Dumas and by Carl Weltzien, who taught at Karlsruhe's technical school (Carlsruhe "Compte Rendu," pp. 674 and 680).[17]

Many antiatomists, including Ernst Mach, J. B. Stallo, and Edouard LeRoy argued explicitly that the atom was a logically unintelligible basis for the teaching or demonstration of scientific knowledge. Bad teaching damaged the integrity of what must be a unified scientific enterprise, and inconsistencies and confusions in scientific theory might mislead the public into thinking the whole scientific enterprise bankrupt (Poincaré, p. 529).[18] J. B. Stallo in America echoed his concern:

> . . . the atom can not be a cube or oblate spheroid for physical, and a sphere for chemical purposes. And a group of constant atoms can not be an aggregate of extended and absolutely inert and impenetrable masses in a crucible or retort, and a system of mere centres of force as part of a magnet or of a Clamond batter . . . . If physics *as a science* are [sic] not to fall into utter disrepute, it is time to evoke some order from the confusion which prevails among the very first principles, theories, and definitions of theoretical physics.[19]

Clearly, the expository or demonstrative integrity of the atomic theory was a central issue in the atomic debates. Indeed the stubborn resistance of Sainte-Claire Deville and Berthelot to atomism may be explained by their stress upon the teaching or exposition of scientific theory. The primary function of the French University in the nineteenth century was teaching, not research. French professors trained their students for examinations and not for independent thinking. Perhaps the most powerful objection to the molecular-kinetic theory in France was that it did not meet the needs of the educational system. Eleuthère Mascart, who held the chair of physics at the Collège de France from 1872–1908, wrote that the atomic doctrine might be useful to chemists themselves, but "the needs of education are more exacting: the precision of hypotheses and principles there is *de rigeur*; consequences ought to be deduced without compromise. . . ."[20]

Following upon the 1877 atomic debates at the Paris Academy, Charles Marignac, who had studied in Paris and Giessen before teaching chemistry in Geneva, suggested that atomism was especially thwarted in France because of the centralized structure of French education and the influence of antiatomists on national curriculum, textbooks, and examinations. Berthelot denied this: the pressure for conformity on examination candidates, he claimed, came from the atomists and not the equivalentists (Marignac, p. 91; Berthelot, p. 184). But Marignac had hit upon an important point. Victor Grignard, French Nobel Laureate in chemistry in 1912, later recalled that as late as the 1890s: ". . . we had not yet emerged from the period in which the influence of Berthelot, exerting a despotic influence on secondary training, prevented the atomic theory from replacing that of equivalents."[21]

For a period of some fifty years, from 1860 to 1911, the alternative strategies of thermodynamics and electrodynamics competed with atomism. During that time, atomism made its surest strides in English science, perhaps because of the indigenous strength of Newtonian corpuscularism and mechanics. Although chemical atomism was the subject of debate at the Chemical Society of London in 1869, and Oxford University professor Benjamin Brodie proposed a "calculus of chemical operations" as an alternative to the atomic theory,[22] the *notation* of atoms had already almost wholly replaced that of equivalents in influential chemistry textbooks in Great Britain and Germany. By contrast, in France, chemists still were debating, in 1877, the introduction of the modern chemical notation into textbooks of chemistry. As atomic notation was finally introduced into French textbooks in the 1890s, Pierre Duhem made the startling suggestion, in 1892, that "atomic weights" should be *rechristened* "chemical equivalents" since their real meaning was a conventional one.[23] Duhem, like Berthelot, succumbed to the *convention* that the weight of oxygen is to be taken as 16 and the formula of water as H_2O. Their capitulation was not to a realistic atomic theory, however. They did not accept the notion of the existence of indivisible chunks of matter corresponding to the chemical elements.

By the time of the 1895 Lübeck meeting of the Society of German Scientists and Physicians, Ostwald had begun his campaign for "energetics" as a substitution for "matter" theory. But the battle was soon to be over. The grounds of argument shifted from decades-long debate about methodology, epistemology, chemical notation, and interpretations of gas laws, specific heats, vapor densities, and spectral lines. What moved most chemists and physicists to active enthusiasm for the atomic theory was the discovery of a set of new, exciting properties and problems for the atom's long-sought substructure. Divided into pieces in cathode-ray tubes and in the products of radioactive decay, the physico-chemical atom became more useful and more provocative than energy or ether in both exposition and promotion of further discoveries. The atom's moving electrified substructure now explained chemical properties, as well as electromagnetic radiations and other forms of energy associated with matter. The new atom of J. J. Thomson and Ernest Rutherford became the principle of unification for chemistry and physics.

As early as 1901, Jean Perrin proposed a solar-system model for the atom, and the European-trained Japanese physicist Nagoaka Hantarō mathematically worked out a similar hypothesis in 1904.[24] But there were good objections to the solar-system model, both on mechanical and electromagnetic grounds, and alternatives were offered. The mathematician Ferdinand Lindemann, for example, attempted, in 1905, to explain chemical and physical properties of the atom and molecule using classical dynamical conceptions of ether. His approach registered resistance to "the widely circulated notion that the individual atoms of a molecule describe definite paths around particular centers according to which each molecule would represent a miniature planetary system."[25]

But Rutherford and Frederick Soddy moved the physico-chemical community inexorably toward the solar-system model when they explained radioactivity and the chemical changes associated with it in terms of subatomic transformations (Rutherford and Soddy, p. 371). By the 1911 Solvay Conference, Rutherford published his evidence that "the atom contains a central charge distributed through a very small volume . . . that the value of the central charge for different atoms is approximately proportional to their atomic weights'' (Rutherford, pp. 687–688). The distribution of the negative electrons was not so simple. A crucial insight into the complexity of the new atom with its moving electrons lay in Albert Einstein's 1906 paper ''On a Heuristic Point of View about the Creation and Conversion of Light,'' in which Einstein applied Boltzmann's kinetic theory of gases to vibrating electrons. Here Einstein pointed the way to unity in the laws of matter, electricity, and heat through Planck's new quantum theory, a tack taken up in later years by Niels Bohr and a new school of quantum physicists (Einstein, esp. pp. 93–96, 98, 100–101).[26] At the 1911 Solvay Congress, the quantum theory and atomism went hand-in-hand, with physicists thoroughly committed now to solving both physical and chemical problems by the same means.

Not only was the new atom a principle of unity, but it also was *real*. Tracked down in electrolytes, radioactive matter, gases, and colloids, atoms and molecules became measurable in absolute and invariable parameters. The atom's diameter and the number of particles in a liter of gas could be calculated. The laws of the kinetic theory were shown to explain phenomena as disparate as the viscosity of gases, the diffusion of dissolved substances, the emission of alpha-particles, and the energy of the infra-red spectrum (Perrin, p. 90). The atom's fundamental unit of charge, e, joined the speed of light, c, and Planck's constant, h, as fundamental, invariable parameters unifying the natural world.

All this was summed up masterfully by Jean Perrin at the Solvay conference of 1911. Perrin believed that atoms were not only *useful* entities, but that they were *facts*: atoms were *real*.[27] However, the atom Perrin described in 1911 was neither the old-fashioned elemental atom of the proatomist chemists nor the discrete mass *cum* vortex ring of the proatomist physicists. Rather, the 1911 atom incorporated the critical ideas and requirements of nineteenth-century antiatomists, both chemists and physicists. Perrin's atomic-molecular theory drew upon the electromagnetic and thermodynamic traditions of the nineteenth century, as surely as it did upon Newton's mechanical world view. The new theory was a true synthesis, living up to Henri Poincaré's requirement that the goal of science be ''unity,'' not merely "mechanics."[28] The elucidation of the physico-chemical atom was an achievement within the framework of classical science. As the First World War marked the demise of the nineteenth-century world order in Europe, so the Solvay conference marked the last expression of a truly classical atomic theory.

MARY JO NYE
The University of Oklahoma

NOTES

1. Short titles and page numbers indicated in parentheses refer to primary sources reprinted in this volume. On the Karlsruhe Congress, see Clara de Milt, "Carl Weltzein [sic] and the Congress at Karlsruhe," *Chymia,* 1 (1948), 153-170. Superscripts rather than subscripts were commonly used in molecular formulae in 1860.

2. In Maurice de Broglie, *Les Premiers Congrès de Physique Solvay et l'Orientation de la Physique depuis 1911* (Paris: Michel, 1951), 16 and 21–22. On the Solvay Congress, see also the second part of Russell McCormmach's "Henri Poincaré and the Quantum Theory," *Isis*, 58 (1967), 37–55.

3. It was the "equal volumes-equal numbers" rule that Marcellin Berthelot especially vilified: "The point at which atomists go beyond the domain of immediate experience is in the supposition of Avogadro (1811) and Ampère (1814) that equal volumes of all gases in the same physical conditions contain exactly the same number of molecules." In M. Berthelot. "La Théorie atomique," *Revue Scientifique*, 16 (1875), 442–447 (on 443). On the reception of Avogadro's ideas, see John Hedley Brooke, "Avogadro's Hypothesis and Its Fate: A Case-Study in the Failure of Case-Studies," *History of Science*, 19 (1981), 235–273; and N. W. Fisher, "Avogadro, the Chemists, and Historians of Chemistry. Part I," *History of Science*, 20 (1982), 77–102. On Clausius, see Edward E. Daub, "Rudolf Clausius," in *Dictionary of Scientific Biography* III (New York: Charles Scribner's Sons, 1971), 303–311 (on 307).

4. See Pierre Duhem, "Notation atomique et hypothèses atomiques," *Revue des Questions Scientifiques*, 31 (1892), 391–454 (on 392 and 453–454).

5. Using modern notation and values for atomic weights, the ratio in ammonia [NH_3] of the combining weight of nitrogen [14] to hydrogen [1] is $^{14}/_{(3\times1)} = 4.67$. In sal ammoniac [$NH_4Cl$], where 1 gram of hydrogen is chemically equivalent to 35.5 grams of chlorine [Cl], the ratio of the combining weight of nitrogen [14] to hydrogen [1] is $^{14}/_{[(4\times1)+1]} = 14/5 = 2.8$. See the fine discussions by Alan J. Rocke, "Atoms and Equivalents: The Early Development of the Chemical Atomic Theory," *Historical Studies in the Physical Sciences,* IX (Baltimore, 1978), 225–263; and *Chemical Atomism in the Nineteenth Century: From Dalton to Cannizzaro*, manuscript. Also, the essays in D. S. L. Cardwell, ed. *John Dalton and the Progress of Science* (New York: Manchester University Press, 1968).

6. On the theory of dissociation and other experimental problems, see Peter Clark, "Atomism vs. Thermodynamics," in Colin Howson, ed., *Method and Appraisal in the Physical Sciences* (Cambridge, England: Cambridge University Press, 1976), 41–105; and Stephen G. Brush, *The Kind of Motion We Call Heat. A History of the Kinetic Theory of Gases in the 19th Century*, 2 vols., (New York: American Elsevier, 1976). Also Mary Jo Nye, "The Nineteenth-Century Atomic Debates and the Dilemma of an 'Indifferent Hypothesis,' " *Studies in the History and Philosophy of Science,* 7 (1976), 245–268.

7. On these debates, see W. H. Brock, ed., *The Atomic Debates. Brodie and the Rejection of the Atomic Theory* (Leicester, England: Leicester University Press, 1967), and D. M. Knight, *Atoms and Elements. A Study of Theories of Matter in England in the Nineteenth Century* (London: Hutchinson, 1967).

8. Marcellin Berthelot, "Réponse à la note de M. Wurtz," *Comptes Rendus de l'Académie des Sciences*, 84 (28 May 1877), 1189–1195 (on 1194). On the debates at the Academy, see P. Colmant, "Querelle à l'Institut entre équivalentistes et atomistes," *Revue des Questions Scientifiques,* 143 (1972), 493–519.

9. See Laurens Laudan, "Towards a Reassessment of Comte's 'Méthode Positive,' " *Philosophy of Science*, 38 (1971), 35–53.

10. Pierre Duhem, *The Aim and Structure of Physical Theory*, trans. Philip P. Wiener (New York: Atheneum, 1974), 70–71. On resistance to the "mechanical view of nature" see Martin J. Klein, "Mechanical Explanation at the End of the Nineteenth Century," *Centaurus*, 17 (1972), 58–82.

11. Wilhelm Ostwald, "On Catalysis," in *Nobel Lectures. Chemistry 1909–1921* (New York: American Elsevier, 1966), 150–169 (on 62). On Van't Hoff and Le Bel, see G. M. Richardson, ed., *The Foundations of Stereochemistry. Memoirs by Pasteur, Van't Hoff, LeBel, and Wislicenus* (New York, 1901).

12. On this result, see Russell McCormmach, "H. A. Lorentz and the Electromagnetic View of Nature," *Isis*, 61 (Winter, 1970), 459–497.

13. See the accounts of Lübeck in Erwin N. Hiebert, "The Conception of Thermodynamics in the Scientific Thought of Mach and Planck," *Ernst Mach Institut. Wissenschaftlicher Bericht*, 5 (1968), 106 pp.; and H. R. Post, "Atomism 1900," *Physics Education*, 3 (1968), 1–12.

14. On the problem of statistical laws, irreversibility, and Brownian motion, see Mary Jo Nye, *Molecular Reality. A Perspective on the Scientific Work of Jean Perrin* (London: Macdonald, and New York: American Elsevier, 1972).

15. Ernst Mach, "The Economical Nature of Physical Inquiry," (1882), in *Popular Scientific Lectures*, trans. Philip E. Jourdain (Chicago: Open Court, 1911), 186–213 (on 188).

16. On the conclusion that nineteenth-century science had made all the most important discoveries, see Lawrence Badash, "The Completeness of Nineteenth-Century Science," *Isis*, 63 (1972), 48–58.

17. So effective was Cannizzaro's 1858 exposition of atomic theory that his younger German colleague Lothar Meyer, upon reading it, felt ". . . as though the scales fell from my eyes, doubt disappeared, and the feeling of tranquil certainty stepped into its place." In Lothar Meyer's "Anmerkungen" to Cannizzaro's paper reprinted in Ostwald's *Klassiker*, no. 30, (Leipzig: Wilhelm Engelmann), 51–61 (on 59).

18. On the issue of the "bankruptcy of science," see Harry W. Paul, "The Debate over the Bankruptcy of Science in 1895," *French Historical Studies,* 5 (1968), 299-327.

19. J. B. Stallo, *The Concepts and Theories of Modern Physics*, ed. Percy W. Bridgman (Cambridge, Mass.: Harvard University Press, 1960), 18, 47.

20. Eleuthère Mascart, "Les Théories chimiques dans l'enseignement," *Revue Scientifique*, 11 (1873), 958-970 (on 969-970).

21. Heinrich Rheinboldt, "Fifty Years of the Gringard Reaction," *Journal of Chemical Education*, 27 (1950), 476-488 (on 477).

22. W. H. Brock, ed., *The Atomic Debates* (cited in note 7).

23. On this history, see M. M. Pattison Muir, *A History of Chemical Theories and Laws* (New York: John Wiley and Sons, 1906), 141-142; and Colmant (cited in note 8). In 1894, L. J. Troost's 26th edition of the *Précis de Chimie* carried the new subtitle "Atomic Notation." For Duhem's suggestion, see "Notation atomique," pp. 403-405 (cited in note 4).

24. Jean Perrin, "Les Hypothèses moléculaires," *Revue Scientifique*, 15 (1901), 449-461 (on 460); and Nagaoka Hantarō, "Kinetics of a System of Particles Illustrating the Line and the Band Spectrum and the Phenomena of Radioactivity," *Philosophical Magazine*, [6], 7 (1904), 445-455.

25. Ferdinand Lindemann, "On the Form and Spectrum of Atoms," trans. Lydia G. Robinson, *The Monist*, 16 (January, 1906), 1-16 (on 13-15).

26. Niels Bohr, "On the Constitution of Atoms and Molecules [Part I]," *Philosophical Magazine,* [6], 26 (1913), 1-25.

27. Jean Perrin, "Les Preuves de la réalite moléculaire," in *La Théorie du Rayonnement et les Quanta* (Paris: Gauthier-Villars, 1912), 153-250.

28. See Henri Poincaré, "The Theories of Modern Physics," in *Foundations of Science*, trans. G. B. Halsted (New York: Science Press, 1913), 140-154 (on p.150).

THE Question OF THE Atom

WURTZ, CHARLES-ADOLPHE. (b. Wolfisheim, near Strasbourg, France, 26 November 1817; d. Paris, France, 12 May 1884.)

When Adolphe Wurtz wrote the "Compte Rendu" of the Karlsruhe Congress, he was professor of organic chemistry at the Paris Faculty of Medicine. A Lutheran of Alsatian origin, Wurtz took a doctorate in medicine in 1843 at Strasbourg, writing his thesis on a chemical topic concerning fibrin and albumin. He studied briefly with Justus Liebig at Giessen and moved to Paris in 1844 to work in Jean-Baptiste Dumas's laboratory. Wurtz succeeded Dumas at the Paris Medical Faculty in 1849, and he transferred to a chair of organic chemistry at the Sorbonne in 1874. Among his most important experimental work was the synthesis of primary amines in 1849, demonstrating that they are organic compounds of the ammonia type.

Entering into chemists' debates about the possible existence of free radicals, an idea favored by dualists of the Berzelius tradition, Wurtz proved that Hermann Kolbe and Edward Frankland had not isolated methyl and ethyl free radicals, as they thought they had done. In 1854, using a method that proved generally useful for synthesizing alkanes, Wurtz applied sodium to a mixture of alkyl iodides, obtaining the mixed radical gases ethyl butyl and butyl amyl. This demonstrated that the gases Frankland and Kolbe had isolated contained two equivalents of the radicals and not the free radical itself.

In France, Wurtz was an important supporter of the ideas of Auguste Laurent and Charles Gerhardt on the classification of organic compounds into four types of molecules which react with other substances by substitution or double decomposition. Wurtz was, as well, a staunch advocate of the distinction between atoms and molecules and the use of Avogadro's hypothesis. He developed the notion of the multiple combining power ("atomicity") of the elements, i.e., the view that an element might exhibit more than one valence. His 1879 book, *La Théorie Atomique*, was an important primer for the new chemistry, and it was in his laboratory that many leading chemists studied, including Couper, Butlerov, LeBel, Van't Hoff, Friedel, and Lecoq de Boisbaudran. Fluent in French, German, and English, Wurtz had international interests and connections. He regularly translated German papers into French for the *Annales de Chimie*, beginning with Liebig's work in 1844. It was entirely appropriate that Wurtz should draw up an account of the first international chemistry congress of 1860.

ACCOUNT OF THE SESSIONS OF THE INTERNATIONAL CONGRESS OF CHEMISTS IN KARLSRUHE, ON 3, 4, AND 5 SEPTEMBER 1860[1]*

It was Mr. Kekulé's idea to bring about an international meeting of chemists. During the fall of 1859 he had an opportunity to make the initial overtures in this regard—first to Mr. Weltzien, then to Mr. Wurtz. At the end of March 1860, these three scientists, all in Paris at that moment, devised the initial steps to be taken in order to carry out the plan in question. An initial circular was composed, which had as its aim winning the support of the most outstanding men of the science. It noted, in general terms, the differences that had arisen between the theoretical views of chemists and the urgency of putting an end to these differences by a common agreement, at least where certain questions were concerned.

The first appeal having been favorably received, an understanding was reached on the time and place of the meeting, and printing of a second circular addressed to all European chemists, which explained the objectives and goals of an international congress in the following terms, was agreed upon:[2]

Paris, 15 June 1860

[1]I am indebted to my colleague, Prof. Eugen Gaufinez in Bonn, for examining the French text.

*Compte rendu des séances du Congrès international des chimistes réuni à Carlsruhe le 3, 4 et 5 septembre 1860. See Appendix ii for original text.

[2]The circular was sent in German, French, and English. The German text is dated: "Carlsruhe, July 10, 1860"; the English text: "London, July 1, 1860." With the French version of the proceedings of the Congress, I have incorporated the French text of the circular, which, compared with the German and English, contains the name "Regnault" amongst the undersigned; the English text lacks the name "Mitscherlich," as well.

Dear Distinguished Colleague,

The great development that has taken place in chemistry in recent years, and the differences in theoretical opinions that have emerged, make a Congress, whose goal is the discussion of some important questions as seen from the standpoint of the future progress of the science, both timely and useful.

The undersigned invite to this meeting all chemists authorized by their work or position to express an opinion in a scientific discussion.

Such an assembly cannot deliberate on behalf of everyone, nor can it pass resolutions by which everyone must abide, but by means of a free and thorough discussion, certain misunderstandings could be eliminated, and a common agreement facilitated on some of the following points: the definition of important chemical notions, such as those expressed by the words atom, molecule, equivalent, atomic, basic; the examination of the question of equivalents and of chemical formulae; the institution of a notation and a uniform nomenclature.

Knowing that the assembly's deliberations would not be of a nature such as to reconcile all opinions and eliminate all disagreements immediately, the undersigned believe, nevertheless, that such works could pave the way for a much desired agreement between chemists in the future, at least where the most important questions are concerned. A commission could be charged to continue the investigation of these questions and to interest in them learned academies or societies with the necessary material means for resolving them.

The Congress will convene in Karlsruhe on 3 September 1860.

Our colleague, Mr. Weltzien, Professor at the Polytechnic School in this city, wishes to take on the duties of General Commissioner. In this capacity, he will be in charge of registering prospective members for the Congress and will open the assembly at nine o'clock in the morning on the day indicated.

In conclusion, and with the aim of avoiding any unfortunate omissions, the undersigned request that the individuals to whom this circular will be sent please communicate it to their scientist friends who are duly authorized to attend the planned conference.

Babo de, Freiburg
Balard, Paris
Bekétoff, Kasan
Boussingault, Paris
Brodie, Oxford
Bunsen, Heidelberg
Bussy, Paris
Cahours, Paris
Cannizzaro, Genoa
Deville, H., Paris
Dumas, Paris
Engelhardt, St. Petersburg
Erdmann, O.L., Leipzig
Fehling de, Stuttgart
Frankland, London
Fremy, Paris
Fritzsche, St. Petersburg
Hofmann, A.W., London
Kekulé, Ghent
Kopp, H. Giessen
Hlasiwetz, Innsbruck
Liebig, J. de, Munich
Malaguti, Rennes
Marignac, Geneva
Mitscherlich, Berlin
Odling, London
Pasteur, Paris
Payen, Paris
Pebal, Vienna
Peligot, Paris
Pelouze, Paris
Piria, Turin
Regnault, V., Paris
Roscoe, Manchester
Schroetter, A., Vienna
Socoloff, St. Petersburg
Staedler, Zurich
Stas, Brussels
Strecker, Tübingen
Weltzien, C., Karlsruhe
Will, H., Giessen
Williamson, W., London
Wöhler, F., Göttingen
Wurtz, Ad., Paris
Zinin, St. Petersburg

Nota Bene: You can sign up for the conference either directly with Mr. Weltzien, Polytechnic School, Karlsruhe, or with Mr. A. Kekulé, Professor of Chemistry at the University of Ghent, who will pass it on to Mr. Weltzien.

The number of people who wanted to participate was considerable, and on 3 September 1860, 140 chemists [3]met together in the meeting room of the second Chamber of State, which was made available by the Archduke of Baden.

The list of the chemists in attendance follows:[4]

[3]The printed list of members, supplemented by handwritten additions, contains 126 names. (A.)

[4]I have arranged the participants by country and by the cities in which they worked at that time. (A.)

I. Belgium. *Brussels:* Stas; *Ghent:* Donny, A. Kekulé.

II. Germany. *Berlin:* Ad. Baeyer, G. Quinke; *Bonn:* Landolt; *Breslau:* Lothar Meyer; *Kassel:* Guckelberger; *Klausthal:* Streng; *Darmstadt:* E. Winkler; *Erlangen:* v. Gorup-Besanez; *Freiburg* i.B.: v. Babo, Schneyder; *Giessen:* Boeckmann, H. Kopp, H. Will; *Göttingen:* F. Beilstein; *Halle a.S.:* W. Heintz; *Hanover:* Heeren; *Heidelberg:* Becker, O. Braun, R. Bunsen, L. Carius, E. Erlenmeyer, O. Mendius, Schiel; *Jena:* Lehmann, H. Ludwig; *Karlsruhe:* A. Klemm, R. Muller, J. Nessler, Petersen, K. Seubert, Weltzien; *Leipzig:* O. L. Erdmann, Hirzel, Knop, Kuhn; *Mannheim:* Gundelach, Schroeder; *Marburg* a.L.: R. Schmidt, Zwenger; *Munich:* Geiger; *Nuremberg:* v. Bibra; *Offenbach:* Grimm; *Rappenau:* Finck; *Schönberg:* R. Hoffmann; *Speyer:* Keller, Mühlhaüser; *Stuttgart:* v. Fehling, W. Hallwachs; *Tübingen:* Finckh, A. Naumann, A. Strecker; *Wiesbaden:* Kasselmann, R. Fresenius, C. Neubauer; *Würzburg:* Scherer, v.Schwarzenbach.

III. England. *Dublin:* Apjohn; *Edinburgh:* Al. Crum Brown, Wanklyn, F. Guthrie; *Glasgow:* Anderson; *London:* B. J. Duppa, G. C. Foster, Gladstone, Müller, Noad, A. Normandy, Odling; *Manchester:* Roscoë; *Oxford:* Daubeny, G. Griffeth, F. Schickendantz; *Woolwich:* Abel.

IV. France. *Montpellier:* A. Béchamp, A. Gautier, C.G. Reischauer; *Mülhousen* i.E.: Th. Schneider; *Nancy:* J. Nicklès; *Paris:* Boussingault, Dumas, C. Friedel, L. Grandeau, Le Canu, Persoz, Alf. Riche, P. Thénard, Verdét, Wurtz; *Strasbourg* i.E.: Jacquemin, Oppermann, F. Schlagdenhaussen, Schützenberger; *Tann:* Ch. Kestner, Scheurer-Kestner.

V. Italy. *Genoa:* Cannizarro; *Pavia:* Pavesi.

VI. Mexico. Posselt.

VII. Austria. *Innsbruck:* Hlasiwetz; *Lemberg:* Pebal; *Pesth:* Th. Wertheim; *Vienna:* V.v.Lang, A. Lieben, Folwarezny, F. Schneider.

VIII. Portugal. *Coïmbra:* Mide Carvalho.

IX. Russia. *Kharkov:* Sawitsch; *St. Petersburg:* Borodin, Mendelyeev; L. Schischkoff, Zinin; *Warsaw:* T. Lesinski, J. Natanson.

X. Sweden. *Harpenden:* J. H. Gilbert; *Lund:* Berlin, C. W. Blomstrand. *Stockholm:* Bahr.

XI. Switzerland. *Bern:* C. Brunner, H. Schiff; *Genf:* C. Marignac; *Lausanne:* Bischoff; *Reichenau bei Chur:* A.v. Planta; *Zurich: J. Wislicenus.*

XII. Spain. *Madrid:* R. de Suna.

First Session of the Congress

Mr. Weltzien, General Commissioner, opened the first session with the following speech:

As provisional chairman, I have the honor to inaugurate a Congress which has no precedent for its kind, the nature of which has never before met. To be sure, German Natural Scientists and Physicians, upon the instigation of Oken, and emulating their Swiss Colleagues, have assembled almost annually for scientific conferences in various cities of their Fatherland since 1822. Following the lead of such congresses, English, French, and during the past several years, also Scandinavian Natural Scientists have convened for a similar purpose. Devotees of the different branches of Natural Science and Medicine were regularly present, although all participants were invariably of the same nationality. The business of these congresses was for the most part characterized by reports, the topics of which were not integrated into any previously arranged program, but rather, as presentations of work in progess, left to the discretion of each individual. A lively and amicable intercourse, flavored by a sequence of festivities, united the ethnically and linguistically related Natural Scientists and Physicians for several days.

Not so for the Congress convened here today. For the first time, the representatives of a single, and indeed the newest Natural Science have assembled. These representatives belong, however, to nearly every nationality. We may be of differing ethnic origin and speak different languages, but we are related by professional specialty, are bound by scientific interest, and are united by the same design. We are assembled for the specific goal of attempting to initiate unification around points of vital concern for our beautiful science. Due to the extraordinarily swift development of Chemistry, and especially because of the massive accumulation of factual materials, the theoretical standpoints of researchers and the means of expression, both in words and symbols, have begun to diverge more than is expedient for mutual understanding, and, especially, more than is suitable for instruction. Considering the importance of Chemistry for other Natural Sciences and its indispensability for technology, it seems exceedingly desirable and advisable to cast our science in a more rigorous form, so that it will be possible to communicate it in a relatively more concise manner.

In order to achieve this, we should not be constrained to only review various viewpoints and writing conventions, the variety of which offers little of importance; and we should not be burdened with a nomenclature, which in view of a plethora of unnecessary symbols lacks any rational basis, and which, making matters worse, is derived, for the most part, from a theory whose validity can hardly be maintained today. The ample attendance at this Congress is surely a clear indication that these nuisances are universally recognized and that their removal from the path toward unification appears desirable. The achievement of this end is a prize of such beauty that it is well worth the effort to undertake the task here.

The original notion of a Chemistry-Congress was communicated to me some time ago by our colleague Kekulé. Early this year, I initiated the first steps toward its realization. The ripeness of this undertaking was manifoldly acknowledged; I obtained unsolicited support from all quarters. Because of this, I do not doubt that this Congress will be called upon to lay the foundations for a not unimportant epoch in the history of our science.

The city of Karlsruhe, which two years ago had the great fortune to host one of the most splendid congresses of German Natural Scientists and Physicians, has now the honor of seeing the first international Chemistry-Convention within its city walls. Karlsruhe is the capital of a small but blessed province, in which, under the auspices of a noble prince and a liberal government, the Arts and Sciences flourish, and where its devotees, esteemed and sustained, can follow their calling with devotion and good cheer. Since it is my pleasure to bid you a hearty welcome to this city, I expect that the same good cheer will permeate our Congress, and hope that our science will one day look back with satisfaction upon our assembly.

After this speech the general commissioner first asks Mr. Bunsen to preside, but the latter refuses and asks the assembly to encourage Mr. Weltzien to direct the deliberations during the first session. Mr. Weltzien is named President. Messrs. Wurtz, Strecker, Kekulé, Odling, Roscoe, and Schischkoff are appointed to serve as Secretaries.

At the suggestion of Mr. Kekulé, the assembly decides that a commission will be put in charge of drawing up a list of questions that will be submitted to the Congress for deliberation.

Mr. Kekulé takes the floor in order to explain the schedule of questions. (Then an analysis of Mr. Kekulé's speech follows.[5])

Mr. Erdmann emphasizes the necessity of directing the assembly's deliberations and resolutions towards questions of form, rather than doctrinal points.

A discussion begins over the question of whether the Commission will hold its sessions behind closed doors or during plenary sessions of the entire assembly.

After some words among Messrs. Fresenius, Kekulé, Wurtz, Boussingault, and H. Kopp, the assembly decides, upon the motion made by the last individual named, that the sessions will take place behind closed doors.

First Session of the Commission

The commission met on 3 September at 11 A.M., Mr. H. Kopp presiding.

The chairman suggests that the discussion begin with the notions of molecule and atom, and he asks Mr. Kekulé and Mr. Cannizzaro, whose studies have especially encompassed this issue, to take the floor.

Mr. Kekulé emphasizes the need to distinguish between the molecule and atom, and, in principle at least, the physical molecule and the chemical molecule.

Mr. Cannizzaro is unable to conceive of the notion of the chemical molecule. For him there are only physical molecules, and the Ampère-Avogadro law is the basis for considerations relating to the chemical molecule. The latter is nothing other than the gaseous molecule.

Mr. Kekulé thinks, on the contrary, that the chemical facts must serve as the basis for the definition and determination of the

[5]Cf. Appendix 9. (A.)

(chemical) molecule and that physical considerations should only be invoked as a check.

Mr. Strecker points out that in certain cases the atom and molecule are identical, as in the case of ethylene.

Mr. Wurtz says that a certain difficulty can be sensed in defining the chemical molecule of oxygen and the diatomic elements in general, which are comparable to ethylene. The view of these as molecules formed of two atoms derived from physical considerations [6] but until now no chemical fact appears to militate in favor of this doubling.

Mr. H. Kopp, summarizing the discussion, says that the need to separate the idea of the molecule from that of the atom appears to be established; that the notion of the molecule can be fixed with the help of purely chemical consisderations; that the definition does not have to involve density alone; and, finally, that it appears natural to call the largest quantity the molecule, and the smallest quantity the atom. In concluding, the speaker formulates the first question to be put to the assembly. This question is as follows: "Is it appropriate to establish a distinction between the terms molecule and atom, and to call molecules, which are comparable as far as physical properties go, the smallest quantities of bodies which enter into or come out of a reaction, and to call atoms the smallest quantities of bodies which are contained in these molecules?"

Mr. Fresenius calls attention to the expression *compound atom* and said that these two words entail a contradiction. Mr. Fresenius's remark motivates the drawing up of a second question to be submitted to the assembly, which is as follows: "Can the expression *compound atom* be eliminated and replaced by the expressions *radical* or *residue*?"

Mr. Kopp goes back to the program explained by Mr. Kekulé, and he calls attention to the definition of the word *equivalent*. It seems to him that the notion of equivalent is perfectly clear and is sharply distinguished from the notion of molecule and that of atom. Consequently, the commission adopts, without discussion, the third proposition to be submitted to the assembly, which is

[6]In Kekulé's manuscript, he has added here the following marginal note: "Not always!" (A.)

as follows: "The notion of equivalents is empirical and independent of the idea of molecule and that of atom."

The session continues, with Mr. Erdmann presiding.[7] The discussion on *notation* gets underway: Mr. Kekulé points out that molecular and atomic notation, or the notation of equivalents, can be employed, but that as in any system it is necessary to stick rigorously to the particular notation, whichever it be, once adopted.

The meaning of the word "equivalent" is the object of several remarks. Mr. Béchamps says that equivalence can only be assumed in cases where the functions of bodies are identical.

Mr. Schischkoff is not of the same opinion, He thinks that the notation for equivalence and equivalent quantities is independent of chemical functions. Everyone assumes an equivalence between chlorine and hydrogen. After a few observations presented by other members on the same subject, the session was adjourned.

Second Session of the Congress

Mr. Boussingault *presiding*

In taking the chair, Mr. Boussingault thanks the Congress for having bestowed the honor of presiding upon a scientist whose studies have had matters of applied chemistry as their object, rather than points of abstract theory. The chairman sees in this choice that the Congress wanted to testify to the unity of the so-called old and new chemistries. He protests against these terms, and remarks that it is not chemistry that grows old, but chemists.

The chairman announces that the work of the Commission is not ready, but that it has agreed upon the drawing up of three questions to be submitted to the assembly for deliberation. He asks one of the Secretaries to make these known to the assembly.

Mr. Strecker takes the floor and reads to the assembly the questions drafted by the commission and indicated above.

Mr. Kekulé enlarges upon the points specified in the first question.

[7]In Kekulé's manuscript, he has added: "Kekulé and Will declined."

Concerning the fundamental hypothesis which can be made about the nature of matter, the speaker wonders if it is necessary to adopt the atomic hypothesis or if a dynamical hypothesis is enough. The first alternative seems preferable to him. Dalton's hypothesis was verified by everything known about the nature of gases. One is authorized to assume small units or small components in gases, and when the same body can affect the gaseous state, the solid state, and the crystalline state, it is possible that the crystalline molecules are precisely the small gaseous components in question, or that these are a fraction of others. But the nature of these relations cannot be specified. What is certain is that in chemical reactions there exists a quantity that enters into or comes out of reaction in the smallest proportion, and never as a fraction of this proportion. These quantities are the smallest that can exist in a free state. These are the molecules defined chemically. But these quantities are not indivisible; chemical reactions succeed in cutting them and resolving them into absolutely indivisible particles. These particles are atoms. The elements themselves, when they are free, consist of molecules formed of atoms. [8] Thus the free chlorine molecule is formed from two atoms. This leads one to assume different molecular and atomic units:

1. physical molecules
2. chemical molecules
3. atoms

The gaseous physical molecules have not been shown to be identical with the physical molecules of solids and liquids. Secondly, the chemical molecules have not been shown to be identical with gaseous molecules. Thus it is not established if the smallest quantity of a substance that enters into a reaction is also the smallest quantity of this substance that plays a role in heat phenomena.

It must be said, however, that the chemical molecule is normally identical to the physical molecule. It has even been maintained that the first never represents anything more than the

[8]In Kekulé's manuscript, there is the following marginal note: "A molecule is 1, at most 2 atoms." (A.)

second. For the speaker it is not like that. The chemical molecule has an independent existence, and in order to allow the distinction in question to be assumed, it is enough to demonstrate its reality in a few cases. But that is easy. Has it not been shown that, for the density of sulphur vapor, the chemical molecules do not always completely separate from one another, but remain fused together in certain conditions (at 500°) to form physical molecules?

The speaker adds that the existence and magnitude of chemical molecules can and must be determined by chemical demonstrations and that the physical facts are not enough to achieve this result.[9] With the help of physical considerations, how could it be shown that hydrochloric acid is formed from a single hydrogen atom and a single chlorine atom? Would it not be enough to multiply the formula HCl by a certain coefficient, and to do the same for all other formulae, in order to establish a perfect agreement between physical properties?

Mr. Cannizzaro takes the floor in order to point out that the distinction between physical and chemical molecules appears to him neither to be necessary nor clearly established.

Mr. Wurtz expresses the opinion that this is a secondary point and can be reserved. It seems to him, on the contrary, that the question relative to establishing the distinction between the terms "molecule" and "atom" has nearly been concluded, and that everyone seems to recognize the utility of such a distinction. It is a matter of clarifying the sense of words generally in common use; it is purely a matter of a definition, and the speaker thinks that in a question of this type, there would perhaps be propriety and usefulness in the expression, by the assembly, of an opinion following the discussion. This opinion, moreover, would not bind anyone, and there would be nothing obligatory about it.

Discussion of the second question relative to the words "compound radical" begins. Mr. Miller thinks that scientific language could not do without the words "compound atom." There are atoms of simple substances; there are atoms of compound substances.

[9]The following note is in Kekulé's manuscript: "Striking example: NH_4Cl, $SO_3 \cdot OH_2$." (A.)

Messrs. Kekulé, Natanson, Strecker, Ramon de Luna, Nicklès, Béchamps and other members present varied observations in one direction or another, but the discussion of this question, like that of the preceding one, leads to no resolution from the assembly.

SECOND SESSION OF THE COMMISSION

Mr. H. Kopp *presiding*

Mr. Kekulé explains his ideas on chemical notation. He points out that either an atomic-molecular notation or a notation in equivalents can be employed. In the first case, the chemical formula represents the molecule; in the second, it represents equivalence. The following examples illustrate this distinction:

atomic molecular notation	notation in equivalents
H Cl	H Cl
H^2O	HO
H^3Az	H az[10]

The important thing is not to mix these notations and get them confused, as is often done. They get mixed up if water is written HO = 9 and ammonia, H^3Az = 17, etc.

Mr. Cannizzaro stresses the importance of considerations relating to volumes in the question of notation. The arguments elaborated by the speaker are reproduced in extenso in the report of the third session of the Congress (see ahead).

The chairman draws attention to the overly detailed course of the discussion, and he points out that, within the Commission, the questions should be indicated rather than investigated. He also deems that the discussion relative to notation in equivalents, such as had just been formulated by Mr. Kekulé, can be left aside. No one is served by it. The speaker believes that it would be proper not to get too attached to theoretical matters concerning the content of things, and to stick to questions of form. Several members express a similar opinion, and Mr. Erdmann, in particular, draws attention to the urgency of adopting a notation whose symbols always represent one and the same given value.

Summarizing the discussion, the chairman acknowledges that

[10] H = 1. C = 8. Θ = 16. Az = 14. az = $^{14}/_{3}$

given the recent advances in the science, it is likely that certain atomic weights ought to be doubled, but that it would be useful to take into consideration the notation that has, until now, generally been employed in introducing notation to represent these double weights and not to adhere too rigorously to the symbols in the latest notation representing different values. As a transitional measure and to avoid confusion, he thinks it convenient to adopt certain signs to indicate the differences in question. Consequently, the chairman approves of the habit of some chemists, that of barring [writing a horizontal bar through the alphabetical letters symbolizing the element] double atomic weights. In concluding, he formulates the question to be submitted before the Congress in the following manner:

"Is it desirable to harmonize chemical notation with recent advances in the science by doubling a certain number of atomic weights?"

Third Session of the Commission

Mr. Dumas *presiding*

Mr. Kekulé summarizes the discussion of the preceding session and repeats the question, announced by Mr. Kopp, in a slightly mitigated form. According to the speaker, this question should be posed in the following way:

"Do the recent advances in science warrant a change in notation?"

Mr. Strecker proposes that atomic notation be adopted in principle.

The chairman stresses forcefully the disadvantages that result from current confusion. He points out that this state of affairs, were it to continue, would be such as to undermine not only the proper direction of teaching and advances in science, but the reliability of industrial work as well. Let us think back, says the chairman, to what we remember from twenty years earlier. Berzelius's table of atomic weights was both the underlying support for the whole science of chemistry and the infallible guide to industrial operations. There is nothing today to replace this universally acknowledged authority, and we have to be careful that chemistry does not fall from the high rank that it has enjoyed among the sciences until now.

Mr. Wurtz is pleased to acknowledge that Mr. Dumas has gotten to the core of the issue, and he thinks it necessary to return to the principles of atomic weights and to Berzelius's notation. According to the speaker, marginal changes in the interpretation of some facts would suffice to bring the principles and this notation into harmony with the requirements of modern science. The notation suitable for adoption today is not exactly Gerhardt's. Gerhardt rendered enormous services to the science. Today he is dead, and his name, the speaker says, should only be spoken with respect. However, it seems that this chemist made two mistakes. One concerns form alone, the other is inherent in the root of things.

First, instead of presenting his notation as founded upon new principles, he more moderately linked it to Berzelius's principles, thus sheltering his innovation under the authority of this great name. Secondly, it seems that Gerhardt made a mistake in likening all of the oxides of inorganic chemistry to silver oxide and to anhydrous potassium oxide, and in attributing to them, as in the case of the latter, the formula $\left.\begin{matrix}R\\R\end{matrix}\right\}\Theta$. In organic chemistry there should be oxides corresponding to ethylene oxide, just as there are representatives of ethyl oxide and of others that correspond to glycerol oxide, and if potassium hydrate, for example, can be compared to alcohol, then other hydrates should be compared with glycol and with glycerine. It is understandable that these considerations are such as to prompt and justify some changes in Gerhardt's notation and in the atomic weights that he attributed to certain metals.

After a discussion in which Messrs. Cannizzaro, Wurtz, and Kekulé take part, Mr. Kekulé opines that the question is well enough prepared to submit to the Congress for deliberation, and he asks that the writing of this question be entrusted to the Secretaries.

This proposition is adopted by the Commission.

Third Session of the Congress

Mr. Dumas *presiding*

In taking the chair, the chairman addresses some words of thanks to the assembly and expresses the hope that a common agreement on some of the questions aired before the Congress will be reached.

Next, the Chairman suggests to the assembly that two Vice-chairmen be designated. Messrs. Will and Miller are appointed to these positions. Mr. Odling replaces Mr. Roscoe, who had to leave, as Secretary. Next, the Secretaries read the questions, whose writing had been entrusted to them, and which have been worked out by the Commission. These questions are conceived as follows:

"Is it desirable to harmonize chemical notation with advances in the science?"

"Is it appropriate to adopt the principles of Berzelius again, where notation is concerned, in bringing about some modifications of these principles?"

"Is it desirable to distinguish new chemical symbols from those which were generally in use fifteen years ago with the help of particular signs?"

Mr. Cannizzaro takes the floor in order oppose the second proposition. It scarcely appears fitting or logical to him to move science back to the time of Berzelius, so as to make chemistry again cover the path that it has already taken. In effect, Berzelius's system has already undergone successive modifications, and these modifications have led to Gerhardt's system of formulae. And these changes were not at all introduced abruptly or without transitions into the system; they were the result of successive advances. If Gerhardt had not proposed them, Mr. Williamson or Mr. Odling, or another chemist who had taken part in the evolution of science, would have done so.

"The source that Gerhardt's system goes back to is the Avogadro-Ampère theory of the uniform constitution of substances in the gaseous state. This theory leads us to view the molecules of certain simple substances as susceptible to division in the future. Mr. Dumas understood the importance of Avogadro's theory and all of its consequences. He posed this question: Is there agreement between the results of Avogadro's theory and the results deduced by means of other methods used to determine the relative weights of molecules? Realizing that science was still short of experimental results of this kind, he wanted to gather together the largest possible number before *risking any general conclusion on this subject*. Thus he got down to work, and with the help of the method which he applied to determine vapor densities, he furnished science with valuable results. However, it appears that he never got far enough along with the method to be able to infer from the results acquired the general conclusion for which he aimed. Be that as it may concerning this reserve he thought necessary, it can be said that it was he who put chemists on the path to Avogadro's theory, because he, more than anyone else, was responsible for introducing the habit of choosing formulae for volatile substances corresponding to the same volume as that taken up by hydrochloric acid and ammonia.

The most evident display of this influence of Mr. Dumas's school appears in a paper by one of his students, Mr. Gaudin. Mr. Gaudin accepted Avogadro's theory without reservation. He established a clear-cut distinction between the words *atom* and *molecule*, by means of which he was able to reconcile all facts with theory. This distinction had already been made by Mr. Dumas, who had called the molecule *the physical atom* in his lessons on chemical philosophy. It is certainly a mainspring of Gerhardt's system.

Sticking more closely to Avogadro's theory than Gerhardt did later, and taking advantage of new experimental data on vapor densities, Mr. Gaudin established that atoms are not always the same fraction of the molecules of simple bodies—that is to say, these molecules do not always result from the same number of atoms; while the molecules of oxygen, hydrogen, and other halogens are formed from two atoms, the molecule of mercury is made of a single atom. He went so far as to compare the composition of equal volumes of alcohol and ether in order to deduce the relative composition of their molecules. But his mind did not seize upon all of the results of this comparison, and chemists have forgotten the idea that he had. And yet this comparison was one of the starting points for Gerhardt's proposed reform.

Other chemists, Proust among them, also accepted Avogadro's theory and arrived at the same general conclusions as Mr. Gaudin.

What did Gerhardt do in this state of the science?

He accepted Avogadro's theory and the consequence that atoms of simple bodies are divisible, and he applied this theory to deduce the relative make-up of the molecules of hydrogen, oxygen, chlorine, nitrogen, hydrochloric acid, water, and ammonia. If he had stopped there, he would not have gotten ahead of Avogadro and Mr. Dumas. But he then subjected all of the formulae of organic chemistry to a general investigation, and he realized that all of these formulae corresponding to equal volumes of hydrochloric acid and ammonia were confirmed by all reactions and by all chemical analogies. Thus he contemplated modifying formulae that were the exception to the rule introduced by Mr. Dumas. He tried to show that the reasons for the violation of the equal volumes rule were unfounded. To reduce the formulae of all volatile substances of organic chemistry to equal volumes had been the starting point for Gerhardt's proposed reforms. The modifications of atomic weights of certain simple sustances, the discovery of the relations that the hydrates, whether acidic or basic, have with water, had been the consequence of this first step. What happened next? The unforgettable experiments of Mr. Williamson on etherification, on mixed ethers, on acetones, those of Gerhardt on anhydrous acids, those of Mr. Wurtz on alcoholic radicals, etc., successively confirmed what Gerhardt had predicted as a consequence of his system. Thus there occurred in chemistry something analogous to what happened in optics when the undulatory theory was introduced. This theory predicted with wonderful accuracy the facts that experiments later confirmed. Gerhardt's system in chemistry was not less fruitful in exact predictions. It is intimately mixed up with and tied to all of the works of chemistry which had preceded it and to all of the advances that followed it in the history of the science. It is not an abrupt leap, an isolated event. It is a regular step forward, small in appearance, but large in results. From now on, this system cannot be effaced from the history of science. It can and must be discussed and modified. But it is the system that must be taken as the starting point, when it is a matter of introducing into chemical science a system of formulae in accord with the actual state of our knowledge. Some chemists will perhaps be tempted to say: the difference between Gerhardt's formulae and those of Berzelius is very small, because the formula for water, for example, is the same in the two systems. But we must be careful. The difference is very small in appearance, but it is large at bottom. Berzelius was under the influence of Dalton's ideas. The idea of a difference between the atom and the molecule of substances never entered his mind. In all of his arguments he assumed implicitly that atoms of simple substances, are, vis-à-vis physical forces, units of the same order, compound atoms. For this reason he began by assuming that equal volumes contain the same number of atoms. Soon he realized that this rule could only be applied to simple substances, and throughout the whole of his scientific career, he attributed no value to atoms of compound substances in choosing formulae. He was even forced to restrict the rule for equal numbers of atoms in equal volumes to a very small number of simple substances—that is, to those that are permanent gases—thereby introducing into the makeup of gases and vapors a difference that no physicist was ever able to admit. Berzelius did not assume that molecules of simple substances could divide in combining. On the contrary, he assumed that two molecules often form the quantity that enters wholly into the combination. This is what he called *double atoms*. Thus he assumed that water and hydrochloric acid contain the same quantity of hydrogen—a quantity equal to two physical molecules joined together.

So you see, gentlemen, what a profound difference exists between the ideas of Berzelius and those of Avogadro, Ampère, Mr. Dumas, and Gerhardt.

I am surprised that Mr. Kekulé, who said in his book that Gerhardt is the first and only one who completely understood the atomic theory, has accepted the commission's proposition.

I believe that I have shown, Mr. Cannizzaro continued, that a discussion of formulae must take as starting points, the formulas of Gerhardt, but I do not maintain that all of them must be accepted in the form that he proposed them. Far from it; I tried, some years ago, to introduce certain modifications into them, in such a way as to avoid the inconsistencies which appeared to me to exist in Gerhardt's system. In effect, it is strange to see how this chemist renounced Avogadro's theory after having used it as the basis for his reforms. Here is how he put it himself: 'There are molecules in 1, 2 and 4 volumes, as there are in ½, in ⅓, in ¼ of a volume.' (*Comptes rendus des travaux de Chimie*, 1851, p. 146). And he continued as follows (p. 147): 'It is perhaps surprising to see me defend this thesis, when I have recommended and still recommend every day that a regular notation in organic chemistry be followed, in representing all volatile substances by the same number of volumes, by 2 or by 4. The chemists who see 2 contradictory assertions in that forget that I never acknowledged the preceding principle as a molecular truth, but as a condition to be satisfied in order to arrive at the knowledge of certain laws or certain relations that would be allowed to escape the observer's attention in an arbitrary notation, or one suitable for special cases.'

There certainly were facts that forced Gerhardt to renounce Avogadro's theory, but there were also unwarranted hypotheses. The facts were the densities of monohydrate sulfuric acid vapor, of sal ammoniac, and of phosphorus perchloride.

You already know, gentlemen, that on the occasion of the publication of the paper of Mr. Deville on the dissociation of certain compounds by heat, I was the first to try to interpret the occurrence of these abnormal densities, by supposing that the bodies in question are split in two, and that in reality a mixture of vapors is weighed in the determination of these densities. After me, Mr. H. Kopp proposed the same interpretation in his own way.

I will not repeat here the arguments that we invoked in favor of this interpretation. I will only add that one of the members of this Congress just told me that the boiling point of sulphuric acid is almost constant at very different pressures—a fact that shows that it is not a matter of a boiling point here, but of a decomposition point. I am convinced that other facts will confirm the interpretation that we have given to abnormal densities, and, as a result, will dispel the doubts that some scientists still appear to harbor concerning Avogadro's theory.

But independently of the facts that I have just cited, there were also unwarranted hypotheses that led Gerhardt away from Avogadro's theory. I am going to show that this is the case.

Gerhardt took as a demonstrated truth that all metallic compounds have formulae analogous to those of the corresponding hydrogen compounds. From that it follows that the formulae for the mercury chlorides are HgCl, Hg^2Cl, in assuming that the free mercury molecule is formed of two atoms, like that of hydrogen. Let us observe that the vapor densities lead to a different result. In effect, in order to represent the composition of equal volumes of the following five bodies: hydrogen, hydrochloric acid, mercury, mercurous chloride, and mercuric chloride, we will have the following formulae:

$$H^2, HCl, Hg^2, Hg^2Cl, Hg^2Cl^2$$

The comparison of these formulae show that in the molecules of free mercury and of its two chlorides, there exists the same amount of mercury, expressed by Hg^2, and that mercurous chloride is analogous to hydrochloric acid, while the mercuric chloride contains twice the amount of chlorine in its molecule.

As a result, the same reason that directed us to double the carbon atom also commits us to double the mercury atom. This comes down to saying that the amount of mercury expressed by Hg^2 in the preceding formulae represents a single atom. In this case it is seen that the atom is equal to the molecule of the free body; and that in mercurous salts this atom is the equivalent of a single hydrogen atom, while in the mercuric salts it is the equivalent of two hydrogen atoms. In other terms, to employ the language generally in common use today, the mercury is monoatomic in the mercurous salts, but in the mercuric salts it is diatomic like the radicals of Mr. Wurtz's glycols.

It is important to point out now that in doubling the atomic weight of mercury, as was done with the atomic weight of sulphur, we arrive at numbers that accord with the law of specific heats.

But if the atomic weight of mercury is doubled, one is led by analogy to double those of copper, zinc, lead, tin, etc.—in a word, one ends up back in Mr. Regnault's system of atomic weights that agree with specific heats, with isomorphism, and with chemical analogies.

That Gerhardt's system conflicted with the law of specific heats, as well as with isomorphism, was the truly unfortunate thing. This clash has produced two different chemistries—one, which dealt with inorganic substances, and accorded great value to isomorphism; and the other, which investigated organic substances, that took no account of this. Therefore, the same substance could not have the same formula in one chemistry as in the other. The clash that I have just indicated stemmed from the fact that Gerhardt's system was not entirely consistent; it disappears as soon as the inconsistencies are done away with.

Vapor densities provide a means of determining the weight of molecules of substances, whether simple or compound. Specific heats are used to check the weights of atoms and not those of molecules. Isomorphism reveals analogies in molecular constitution.

In support of the modification of atomic weights of certain metals which I have just suggested, I will cite the following facts: all of the volatile compounds of mercury, zinc, tin, and lead contain amounts of metal represented in ordinary notation by Hg^2, Zn^2, Sn^2, Pb^2. This fact alone is enough to indicate to us that these quantities represent the true atoms of the metals in question. The fact that there exist three oxalates of potassium and ammonium (monoatomic radicals), while there exist only two oxalates of barium and calcium (diatomic radicals), could also be cited. But for the moment, I do not stress this point, and I cannot deny, on the other hand, that there is one case where the atomic weight deduced from the comparison of molecular compositions is in conflict with the one deduced from specific heat. This case is relative to carbon. But it could be that the law of atomic heats remained masked by other causes that intervene in specific heat.

In summary, gentlemen, I propose that Gerhardt's system be

accepted, taking into consideration the modifications of the atomic weights of certain metals and the formulae for their salts which I suggest be brought about.

And if we are unable to reach a complete agreement upon which to accept the basis for the new system, let us at least avoid issuing a contrary opinion that would serve no purpose, you can be sure. In effect, we can only obstruct Gerhardt's system from gaining advocates every day. It is already accepted by the majority of young chemists today who take the most active part in advances in science.

In this case let us restrict ourselves to establishing some conventions for avoiding the confusion that results from using identical symbols that stand for different values. Generalizing already established custom, it is thus that we can adopt barred letters to represent the doubled atomic weights."

Mr. Strecker offers some clarifications concerning the drafting of the second proposition submitted to the Congress. This draft originally mentioned Gerhardt's name, but the majority of the committee had wanted to substitute Berzelius's name. The speaker did not share the opinion of the majority. It did not seem to him that there was good reason to go back to Berzelius, who could perhaps be criticized for a logical flaw on the question of atoms and equivalents. The useful and urgent thing is to improve what exists by taking into account the advances of science since Berzelius. Mr. Strecker adds that the doctrines expressed in "Gerhardt's system" offer real advantages. As for himself, he will henceforth adopt the new atomic weights in his papers, but he does not think that the time has come to introduce them into teaching and into elementary books.

Mr. Kekulé shares all of Mr. Cannizzaro's opinions. It appears useful to him, however, to have reservations about one point of detail. Mr. Cannizzaro considers mercurous chloride as containing HgCl (Hg = 200). It appears more rational to Mr. Kekulé to envision it as a combination analogous to "Dutch liquid" [oil of the Dutch chemists (clorinated ethylene, i.e., $C_2H_4Cl_2$)], that is, as containing Hg^2Cl^2 (Hg = 200) and to assume that at the moment of vaporization the molecule Hg^2Cl^2 splits.[11]

[11]Cf. Kekulé's Ann. (1857), *104*, 132n.

Mr. Will does not wish to enter into the details of the questions submitted before the Congress. He confines himself to pointing out that the Congress must proceed directly to its goal. This goal is to find a clear, logical notation, incapable of generating confusion in the minds of those uninitiated in the formulae, and suitable for not only expressing the long-accepted facts in science, but those that modern discoveries add to it each day as well.

Mr. Erdmann suggests that the first two questions be dropped and that discussion be confined to the last one. It appears difficult to him to reach an agreement concerning questions of principle, and especially to impose a notation by vote, as it were.

Mr. Wurtz points out that it was not anyone's idea to impose some opinion or other. One is faced with two kinds of questions—those that concern the very root of things, and others that are questions of form. If there were not yet good grounds for resolving the first by vote, because they were not yet ripe enough, nothing prevents agreement on, and even voting on, the purely formal questions.

Mr. Hermann Kopp notes that, on many theoretical points, the opinions of chemists are divided. These differences of opinions are caused in part by misunderstandings and are reflected in the notation itself. A discussion could be very useful for terminating the misunderstandings.

Mr. Erlenmeyer suggests that barred symbols always be used to express atomic weights that represent the old double equivalents.[12]

Mr. L. Meyer points out that this point seemed settled, because no one has raised an objection in this respect.

A discussion among several members on the suitability of casting a vote began.

Mr. Cannizzaro's opinion is that it is pointless to vote on the third question.

Mr. Boussingault draws attention to the possible difficulty in misunderstanding the meaning of the votes the Congress can cast concerning the questions submitted before it. Voting is an expres-

[12]Originally suggested by Williamson. (A.)

sion of the wishes of the Congress and the Congress is not intending to impose the majority opinion on anyone.

Mr. Will aligns himself with the same opinion.

Mr. Normandy points out that the scientists who suggest the introduction of certain reforms concerning notation into science are those who principally cultivate organic chemistry. Now, it can be noted that the scientists do not even agree among themselves on some points. Thus it appears premature to apply principles which are still under discussion to inorganic chemistry.

Mr. Odling speaks about the question of barred symbols. He remembers that Berzelius introduced them into the science to express double atoms. The bar, he says, is thus the sign of divisibility, and it appears contrary to logic to bar symbols expressing indivisible atoms of oxygen and carbon.

Agreeing completely that Berzelius's double atoms had a different meaning from the indivisible atoms, the barring of whose symbols had been proposed, Mr. Kekulé points out that these barred symbols must express not the divisibility of atoms, but the divisibility of the value represented by these symbols, which is twice what they had been taken to be in the past.

In reply to the observations made by Messrs. Erdmann and Normandy, Mr. Kekulé adds that it is not enough to impose a theoretical opinion or a notation by vote, but that a discussion of such subjects is necessary and useful and will not fail to bear fruit.

The Congress consulted by the chairman expresses the wish that the use of barred symbols, representing atomic weights twice those that have been assumed in the past, be introduced into science.

Mr. Dumas adjourns the third and last session of the Congress, after paying respects to the Grand Duke of Baden and thanking him for his hospitality.

CANNIZZARO, STANISLAO. (b. Palermo, Sicily, 13 July 1826; d. Rome, Italy, 10 May 1910)

Cannizzaro studied chemistry at the University of Pisa with Raffaele Piria, before participating in the 1847–1849 Sicilian uprising against the Bourbons. After a brief stay in M. E. Chevruel's laboratory in Paris, Cannizzaro returned to Italy, where he taught chemistry in Alessandria (1851–1855), Genoa (1855–1861), Palermo (1861–1871), and Rome (1871–1909). His experimental work, especially the study of aromatic alcohols, was in organic chemistry. In 1853 he discovered the chemical process still called the "Cannizzaro reaction" wherein benzaldehyde combines with potassium hydroxide to produce both benzoic acid and benzyl alcohol. According to historian of chemistry Henry M. Leicester, Cannizzaro was the first to propose the name "hydroxyl" for the OH radical.

Cannizzaro's reputation in the history of chemistry rests upon the letter he wrote in 1858 to his friend Sabastiano de Luca, who taught chemistry at the University of Pisa. In this letter, Cannizzaro described the lecture course in chemistry which he had devised at Genoa. The letter was published in the journal *Nuova Cimento*, established at Pisa by Cannizzaro's mentor Piria, and it was reprinted as a pamphlet in 1859. In September 1869 Cannizzaro took copies of the pamphlet with him to the Karlsruhe Congress, and the copies were distributed at the conclusion of the three-day meeting by his friend Angelo Pavesi, professor of chemistry at the University of Pavia. The pamphlet presents arguments for distinguishing atoms from molecules and for establishing atomic weights and molecular formulas. It became central to the chemistry of the late nineteenth century and especially to the revival of interest in Avogadro's hypothesis of the relationship between gas volumes and numbers of gas molecules.

Alembic Club Reprints—No. 18

SKETCH OF A COURSE

OF

CHEMICAL PHILOSOPHY

BY

STANISLAO CANNIZZARO

(1858)

Edinburgh:
Published by THE ALEMBIC CLUB

Edinburgh Agent:
JAMES THIN, 54, 55, and 56 South Bridge

London Agents:
SIMPKIN, MARSHALL, HAMILTON, KENT, & CO., LTD.

1910

PREFACE

THE value of the hypothesis of the Italian physicist Avogadro* as a systematising principle in chemistry was practically unrecognised for forty years after its publication. It had been, it is true, considered and in part applied by Dumas, Gerhardt, and others, but the young Italian chemist Cannizzaro was the first to show its consistent applicability to the selection of atomic weights, and to harmonise with it the results of other methods directed towards the same end.

The eminence of Cannizzaro as a teacher is plain in every page of the summary of his lecture course on chemical philosophy which is here translated. The facts are marshalled and their bearing explained with absolute mastery of pedagogic method, and one is impelled to the conclusion that Cannizzaro's students of 1858 must have had clearer conceptions of chemical theory than most of his scientific colleagues of a much later date.

Permission to publish this translation was received from the venerable chemist a few days before his death on 10th May 1910.

J. W.

* Alembic Club Reprint, No. 4, p. 28.

LETTER OF

PROFESSOR STANISLAO CANNIZZARO

TO

PROFESSOR S. DE LUCA:

SKETCH OF A COURSE OF CHEMICAL PHILOSOPHY

*Given in the Royal University of Genoa.**

I BELIEVE that the progress of science made in these last years has confirmed the hypothesis of Avogadro, of Ampère, and of Dumas on the similar constitution of substances in the gaseous state; that is, that equal volumes of these substances, whether simple or compound, contain an equal number of molecules: not however an equal number of atoms, since the molecules of the different substances, or those of the same substance in its different states, may contain a different number of atoms, whether of the same or of diverse nature.

In order to lead my students to the conviction which I have reached myself, I wish to place them on the same path as that by which I have arrived at it—the path, that is, of the historical examination of chemical theories.

I commence, then, in the first lecture by showing how, from the examination of the physical properties

* From *Il Nuovo Cimento*, vol. vii. (1858), pp. 321-366.

of gaseous bodies, and from the law of Gay-Lussac on the volume relations between components and compounds, there arose almost spontaneously the hypothesis alluded to above, which was first of all enunciated by Avogadro, and shortly afterwards by Ampère. Analysing the conception of these two physicists, I show that it contains nothing contradictory to known facts, provided that we distinguish, as they did, molecules from atoms ; provided that we do not confuse the criteria by which the number and the weight of the former are compared, with the criteria which serve to deduce the weight of the latter ; provided that, finally, we have not fixed in our minds the prejudice that whilst the molecules of compound substances may consist of different numbers of atoms, the molecules of the various simple substances must all contain either one atom, or at least an equal number of atoms.

In the second lecture I set myself the task of investigating the reasons why this hypothesis of Avogadro and Ampère was not immediately accepted by the majority of chemists. I therefore expound rapidly the work and the ideas of those who examined the relationships of the reacting quantities of substances without concerning themselves with the volumes which these substances occupy in the gaseous state; and I pause to explain the ideas of Berzelius, by the influence of which the hypothesis above cited appeared to chemists out of harmony with the facts.

I examine the order of the ideas of Berzelius, and show how on the one hand he developed and completed the dualistic theory of Lavoisier by his own electro-chemical hypothesis, and how on the other hand, influenced by the atomic theory of Dalton (which had been confirmed by the experiments of Wollaston), he applied this theory and took it for his guide in his later researches, bringing it into agreement with the dualistic electro-chemical theory, whilst at the same time he extended the laws of Richter and tried to harmonise them with the results of Proust. I bring out clearly the reason why he was led to assume that the atoms, whilst separate in simple bodies, should unite to form the atoms of a compound of the first order, and these in turn, uniting in simple proportions, should form composite atoms of the second order, and why (since he could not admit that when two substances give a single compound, a molecule of the one and a molecule of the other, instead of uniting to form a single molecule, should change into two molecules of the same nature) he could not accept the hypothesis of Avogadro and of Ampère, which in many cases leads to the conclusion just indicated.

I then show how Berzelius, being unable to escape from his own dualistic ideas, and yet wishing to explain the simple relations discovered by Gay-Lussac between the volumes of gaseous compounds and their gaseous components, was led to formulate a hypothesis very different from that of Avogadro and of Ampère, namely, that equal volumes of simple substances in the gaseous state contain the same number of atoms, which in combination unite intact; how, later, the vapour densities of many simple substances having been determined, he had to restrict this hypothesis by saying that only simple substances which are permanent gases obey this law; how, not believing that composite atoms even of the same order could be equidistant in the gaseous state under the same conditions, he was led to suppose that in the molecules of hydrochloric, hydriodic, and hydrobromic acids, and in those of water and sulphuretted hydrogen, there was contained the same quantity of hydrogen, although the different behaviour of these compounds confirmed the deductions from the hypothesis of Avogadro and of Ampère.

I conclude this lecture by showing that we have only to distinguish atoms from molecules in order to reconcile all the experimental results known to Berzelius, and have no need to assume any difference in constitution between permanent and coercible, or between simple and compound gases, in contradiction to the physical properties of all elastic fluids.

In the third lecture I pass in review the various researches of physicists on gaseous bodies, and show that all the new researches from Gay-Lussac to Clausius confirm the hypothesis of Avogadro and of Ampère that the distances between the molecules, so long as they remain in the gaseous state, do not depend on their nature, nor on their mass, nor on the number of atoms they contain, but only on their temperature and on the pressure to which they are subjected.

In the fourth lecture I pass under review the chemical theories since Berzelius: I pause to examine how Dumas, inclining to the idea of Ampère, had habituated chemists who busied themselves with *organic substances* to apply this idea in determining the molecular weights of compounds; and what were the reasons which had stopped him half way in the application of this theory. I then expound, in continuation of this, two different methods—the one due to Berzelius, the other to Ampère and Dumas—which were used to determine formulæ in inorganic and in organic chemistry respectively until Laurent and Gerhardt sought to bring both parts of the science into harmony. I explain clearly how the discoveries made by Gerhardt, Williamson, Hofmann, Wurtz, Berthelot, Frankland, and others, on the constitution of organic compounds confirm the hypothesis of Avogadro and Ampère, and how that part of Gerhardt's theory which corresponds best with the facts and best explains their connection, is nothing but the extension of Ampère's theory, that is, its complete application, already begun by Dumas.

I draw attention, however, to the fact that Gerhardt did not always consistently follow the theory which had given him such fertile results; since he assumed that equal volumes of gaseous bodies contain the same number of molecules, only in the majority of cases, but not always.

I show how he was constrained by a prejudice, the reverse of that of Berzelius, frequently to distort the facts. Whilst Berzelius, on the one hand, did not admit that the molecules of simple substances could be divided in the act of combination, Gerhardt supposes that all the molecules of simple substances are divisible in chemical action. This prejudice forces him to suppose that the molecule of mercury and of all the metals consists of two atoms, like that of hydrogen, and therefore that the compounds of all the metals are of the same type as those of hydrogen. This error even yet persists in the minds of chemists, and has prevented them from discovering amongst the metals the existence of biatomic radicals perfectly analogous to those lately discovered by Wurtz in organic chemistry.

From the historical examination of chemical theories, as well as from physical researches, I draw the conclusion that to bring into harmony all the branches of chemistry we must have recourse to the complete application of the theory of Avogadro and Ampère in order to compare the weights and the numbers of the molecules; and I propose in the sequel to show that the conclusions drawn from it are invariably in accordance with all physical and chemical laws hitherto discovered.

I begin in the fifth lecture by applying the hypothesis of Avogadro and Ampère to determine the weights of molecules even before their composition is known.

On the basis of the hypothesis cited above, the weights of the molecules are proportional to the densities of the substances in the gaseous state. If we wish the densities of vapours to express the weights of the molecules, it is expedient to refer them all to the density of a simple gas taken as unity, rather than to the weight of a mixture of two gases such as air.

Hydrogen being the lightest gas, we may take it as the unit to which we refer the densities of other gaseous bodies, which in such a case express the weights of the molecules compared to the weight of the molecule of hydrogen = 1.

Since I prefer to take as common unit for the weights of the molecules and for their fractions, the weight of a half and not of a whole molecule of hydrogen, I therefore refer the densities of the various gaseous bodies to that of hydrogen = 2. If the densities are referred to air = 1, it is sufficient to multiply by 14.438 to change them to those referred to that of hydrogen = 1; and by 28·87 to refer them to the density of hydrogen = 2.

I write the two series of numbers, expressing these weights in the following manner:—

Names of Substances.	Densities or weights of one volume, the volume of Hydrogen being made = 1, *i.e.*, weights of the molecules referred to the weight of a whole molecule of Hydrogen taken as unity.	Densities referred to that of Hydrogen = 2, *i.e.*, weights of the molecules referred to the weight of half a molecule of Hydrogen taken as unity.
Hydrogen . . .	1	2
Oxygen, ordinary. .	16	32
Oxygen, electrised .	64	128
Sulphur below 1000° .	96	192
Sulphur* above 1000° .	32	64
Chlorine . . .	35·5	71
Bromine . . .	80	160
Arsenic . . .	150	300
Mercury . . .	100	200
Water	9	18
Hydrochloric Acid .	18·25	36·50†
Acetic Acid . . .	30	60

* This determination was made by Bineau, but I believe it requires confirmation.

† The numbers expressing the densities are approximate: we arrive at a closer approximation by comparing them with those derived from chemical data, and bringing the two into harmony.

Whoever wishes to refer the densities to hydrogen = 1 and the weights of the molecules to the weight of half a molecule of hydrogen, can say that the weights of the molecules are all represented by the weight of two volumes.

I myself, however, for simplicity of exposition, prefer to refer the densities to that of hydrogen = 2, and so the weights of the molecules are all represented by the weight of one volume.

From the few examples contained in the table, I show that the same substance in its different allotropic states can have different molecular weights, without concealing the fact that the experimental data on which this conclusion is founded still require confirmation.

I assume that the study of the various compounds has been begun by determining the weights of the molecules, *i.e.*, their densities in the gaseous state, without enquiring if they are simple or compound.

I then come to the examination of the composition of these molecules. If the substance is undecomposable, we are forced to admit that its molecule is entirely made up by the weight of one and the same kind of matter. If the body is composite, its elementary analysis is made, and thus we discover the constant relations between the weights of its components: then the weight of the molecule is divided into parts proportional to the numbers expressing the relative weights of the components, and thus we obtain the quantities of these components contained in the molecule of the compound, referred to the same unit as that to which we refer the weights of all the molecules. By this method I have constructed the following table :—

[TABLE

Name of Substance.	Weight of one volume, *i.e.*, weight of the molecule referred to the weight of half a molecule of Hydrogen = 1.	Component weights of one volume, *i.e.*, component weights of the molecule, all referred to the weight of half a molecule of Hydrogen = 1.	
Hydrogen	2	2 Hydrogen	
Oxygen, ordinary	32	32 Oxygen	
„ electrised	128	128 „	
Sulphur below 1000°	192	192 Sulphur	
„ above 1000° (?)	64	64 „	
Phosphorus	124	124 Phosphorus	
Chlorine	71	71 Chlorine	
Bromine	160	160 Bromine	
Iodine	254	254 Iodine	
Nitrogen	28	28 Nitrogen	
Arsenic	300	300 Arsenic	
Mercury	200	200 Mercury	
Hydrochloric Acid	36·5	35·5 Chlorine	1 Hydrogen
Hydrobromic Acid	81	80 Bromine	1 „
Hydriodic Acid	128	127 Iodine	1 „
Water	18	16 Oxygen	2 „
Ammonia	17	14 Nitrogen	3 „
Arseniuretted Hyd.	78	75 Arsenic	3 „
Phosphuretted Hyd.	35	32 Phosphorus	3 „
Calomel	235·5	35·5 Chlorine	200 Mercury
Corrosive Sublimate	271	71 „	200 „
Arsenic Trichloride	181·5	106·5 „	75 Arsenic
Protochloride of Phosphorus	138·5	106·5 „	32 Phosphorus
Perchloride of Iron	325	213 „	112 Iron
Protoxide of Nitrogen	44	16 Oxygen	28 Nitrogen
Binoxide of Nitrogen	30	16 „	14 „
Carbonic Oxide	28	16 „	12 Carbon
„ Acid	44	32 „	12 „
Ethylene	28	4 Hydrogen	24 „
Propylene	42	6 „	36 „
Acetic Acid, hydrated	60	4 „ 32 Oxygen 24 Carbon	
„ anhydrous	102	6 Hydrogen 48 Oxygen 48 Carbon	
Alcohol	46	6 Hydrogen 16 Oxygen 24 Carbon	
Ether	74	10 Hydrogen 16 Oxygen 48 Carbon	

All the numbers contained in the preceding table are comparable amongst themselves, being referred to the same unit. And to fix this well in the minds of my pupils, I have recourse to a very simple artifice: I say to them, namely, "Suppose it to be shown that the half molecule of hydrogen weighs a millionth of a milligram, then all the numbers of the preceding table become concrete numbers, expressing in millionths of a milligram the concrete weights of the molecules and of their components: the same thing would follow if the common unit had any other concrete value," and so I lead them to gain a clear conception of the comparability of these numbers, whatever be the concrete value of the common unit.

Once this artifice has served its purpose, I hasten to destroy it by explaining how it is not possible in reality to know the concrete value of this unit; but the clear ideas remain in the minds of my pupils whatever may be their degree of mathematical knowledge. I proceed pretty much as engineers do when they destroy the wooden scaffolding which has served them to construct their bridges, as soon as these can support themselves. But I fear that you will say, "Is it worth the trouble and the waste of time and ink to tell me of this very common artifice?" I am, however, constrained to tell you that I have paused to do so because I have become attached to this pedagogic expedient, having had such great success with it amongst my pupils, and thus I recommend it to all those who, like myself, must teach chemistry to youths not well accustomed to the comparison of quantities.

Once my students have become familar with the importance of the numbers as they are exhibited in the preceding table, it is easy to lead them to discover the law which results from their comparison. "Compare," I say to them, "the various quantities of the same element contained in the molecule of the free substance and in those of all its different compounds, and you will not be able to escape the following law: *The different quantities of the same element contained in different molecules are all whole multiples of one and the same quantity, which, always being entire, has the right to be called an atom.*"

Thus:—

One molecule	of free hydrogen .	contains	2	of hydrogen	= 2 × 1
,,	of hydrochloric acid .	,,	1	,,	= 1 × 1
,,	of hydrobromic acid	,,	1	,,	= 1 × 1
,,	of hydriodic acid .	,,	1	,,	= 1 × 1
,,	of hydrocyanic acid .	,,	1	,,	= 1 × 1
,,	of water . . .	,,	2	,,	= 2 × 1
,,	of sulphuretted hydrogen . .	,,	2	,,	= 2 × 1
,,	of formic acid . .	,,	2	,,	= 2 × 1
,,	of ammonia . .	,,	3	,,	= 3 × 1
,,	of gaseous phosphuretted hydrogen .	,,	3	,,	= 3 × 1
,,	of acetic acid . .	,,	4	,,	= 4 × 1
,,	of ethylene . .	,,	4	,,	= 4 × 1
,,	of alcohol . .	,,	6	,,	= 6 × 1
,,	of ether . . .	,,	10	,,	= 10 × 1

Thus all the various weights of hydrogen contained in the different molecules are integral multiples of the weight contained in the molecule of hydrochloric acid, which justifies our having taken it as common unit of the weights of the atoms and of the molecules. The atom of hydrogen is contained twice in the molecule of free hydrogen.

In the same way it is shown that the various quantities of chlorine existing in different molecules are all whole multiples of the quantity contained in the molecule of hydrochloric acid, that is, of 35.5; and that the quantities of oxygen existing in the different molecules are all whole multiples of the quantity contained in the molecule of water, that is, of 16, which quantity is half of that contained in the molecule of free oxygen, and an eighth part of that contained in the molecule of electrised oxygen (ozone).

Thus :—

One molecule	of free oxygen .	contains	32	of oxygen	= 2 × 16
"	of ozone . .	"	128	"	= 8 × 16
"	of water . .	"	16	"	= 1 × 16
"	of ether . . .	"	16	"	= 1 × 16
"	of acetic acid .	"	32	"	= 2 × 16
	etc. etc.				

One molecule	of free chlorine .	contains	71	of chlorine	= 2 × 35·5
"	of hydrochloric acid	"	35·5	"	= 1 × 35·5
"	of corrosive sublimate	"	71	"	= 2 × 35·5
"	of chloride of arsenic	"	106·5	"	= 3 × 35·5
"	of chloride of tin .	"	142	"	= 4 × 35·5
	etc. etc.				

In a similar way may be found the smallest quantity of each element which enters as a whole into the molecules which contain it, and to which may be given with reason the name of atom. In order, then, to find the atomic weight of each element, it is necessary first of all to know the weights of all or of the greater part of the molecules in which it is contained and their composition.

If it should appear to any one that this method of finding the weights of the molecules is too hypothetical, then let him compare the composition of equal volumes of substances in the gaseous state under the same conditions. He will not be able to escape the following law: *The various quantities of the same element contained in equal volumes either of the free element or of its compounds are all whole multiples of one and the same quantity ;* that is, each element has a special numerical value by means of which and of integral coefficients the composition by weight of equal volumes of the different substances in which it is contained may be expressed. Now, since all chemical reactions take place between equal volumes, or integral multiples of them, it is possible to express all chemical reactions by means of the same numerical values and integral coefficients. The law enunciated in the form just indicated is a direct deduction from the facts: but who is not led to assume from this same law that the weights of equal volumes represent the molecular weights, although other proofs are wanting? I thus prefer to substitute in the expression of the law the word molecule instead of volume. This is advantageous for teaching, because, when the vapour densities cannot be determined, recourse is had to other means for deducing the weights of the molecules of compounds. The whole substance of my course consists in this: to prove the exactness of these latter methods by showing that they lead to the same results as the vapour density when both kinds of method can be adopted at the same time for determining molecular weights.

The law above enunciated, called by me the law of atoms, contains in itself that of multiple proportions and that of simple relations between the volumes; which I demonstrate amply in my lecture. After this I easily succeed in explaining how, expressing by symbols the different atomic weights of the various elements, it is possible to express by means of formulæ the composition of their molecules and of those of their compounds, and I pause a little to make my pupils familiar with the passage from gaseous volume to molecule, the first directly expressing the fact and the second interpreting it. Above all, I study to implant in their minds thoroughly the difference between molecule and atom. It is possible indeed to know the atomic weight of an element without knowing its molecular weight; this is seen in the case of carbon. A great number of the compounds of this substance being volatile, the weights of the molecules and their composition may be compared, and it is seen that the quantities of carbon which they contain are all integral multiples of 12, which quantity is thus the atom of carbon and expressed by the symbol C; but since we cannot determine the vapour density of free carbon we have no means of knowing the weight of its molecule, and thus we cannot know how many times the atom is contained in it. Analogy does not in any way help us, because we observe that the molecules of the most closely analogous substances (such as sulphur and oxygen), and even the molecules of the same substance in its allotropic states, are composed of different numbers of atoms. We have no means of predicting the vapour density of carbon; the only thing that we can say is that it will be either 12 or an integral multiple of 12 (in my system of numbers). The number which is given in different treatises on chemistry as the theoretical density of carbon is quite arbitrary, and a useless datum in chemical calculations; it is useless for calculating and verifying the weights of the molecules of the various compounds of carbon, because the weight of the molecule of free carbon may be ignored if we know the weights of the molecules of all its compounds; it is useless for determining the weight of the atom of carbon, because this is deduced by comparing the composition of a certain number of molecules containing carbon, and the knowledge of the weight of the molecule of this last would scarcely add a datum more to those which are already sufficient for the solution of the problem. Any one will easily convince himself of this by placing in the following manner the numbers expressing the molecular weights derived from the densities and the weights of the components contained in them:—

Names of Compounds of Carbon.	Weights of the molecules referred to the atom of Hydrogen.	Weights of the components of the molecules referred to the weight of the atom of Hydrogen taken as unity.	Formulæ, making H = 1 C = 12 O = 16 S = 32
Carbonic Oxide .	28	12 Carbon 16 Oxygen	CO
,, Acid .	44	12 ,, 32 ,,	CO^2
Sulphide of Carbon	76	12 ,, 64 Sulphur	CS^2
Marsh Gas . .	16	12 ,, 4 Hydrogen	CH^4
Ethylene . .	28	24 ,, 4 ,,	C^2H^4
Propylene . .	42	36 ,, 6 ,,	C^3H^6
Ether . . .	74	{48 ,, 10 ,, 16 Oxygen}	$C^4H^{10}O$
etc.	etc.	etc.	etc.

In the list of molecules containing carbon there might be placed also that of free carbon if the weight of it were known; but this would not have any greater utility than what we would derive by writing in the list one more compound of carbon; that is, it would do nothing but verify once more that the quantity of carbon contained in any molecule, whether of the element itself or of its compounds, is 12 or $n \times 12 = C^n$, n being an integral number.

I then discuss whether it is better to express the composition of the molecules of compounds as a function of the molecules of the components, or if, on the other hand, it is better, as I commenced by doing, to express the composition of both in terms of those constant quantities which always enter by whole numbers into both, that is, by means of the atoms.

Thus, for example, is it better to indicate in the formula that one molecule of hydrochloric acid contains the weight of half a molecule of hydrogen and half a molecule of chlorine, or that it contains an atom of one and an atom of the other, pointing out at the same time that the molecules of both of these substances consist of two atoms?

Should we adopt the formulæ made with symbols indicating the molecules of the elements, then many coefficients of these symbols would be fractional, and the formula of a compound would indicate directly the ratio of the volumes occupied by the components and by the compounds in the gaseous state. This was proposed by Dumas in his classical memoir, *Sur quelques points de la Théorie atomique* (Annales de Chimie et de Physique, tom. 33, 1826).

To discuss the question proposed, I give to the molecules of the elements symbols of a different kind from those employed to represent the atoms, and in this way I compare the formulæ made with the two kinds of symbols.

Atoms or Molecules.	Symbols of the molecules of the Elements and formulæ made with these symbols.	Symbols of the atoms of the Elements and formulæ made with these symbols.	Nos. expressing their weights.
Atom of Hydrogen . . .	$\mathfrak{H}^{\frac{1}{2}}$	$= H =$	1
Molecule of Hydrogen . .	$\mathfrak{H}$	$= H^2 =$	2
Atom of Oxygen . . .	$\mathfrak{O}^{\frac{1}{2}} = \mathfrak{O}\mathfrak{z}^{\frac{1}{8}}$	$= O =$	16
Molecule of ordinary Oxygen .	$\mathfrak{O}$	$= O^2 =$	32
Molecule of electrised Oxygen (Ozone)	$\mathfrak{O}\mathfrak{z}$	$= O^8 =$	128
Atom of Sulphur . . .	$\mathfrak{S}^{\frac{1}{2}} = \mathfrak{S}\mathfrak{a}^{\frac{1}{6}}$	$= S =$	32
Molecule of Sulphur above 1000° (Bineau)	$\mathfrak{S}$	$= S^2 =$	64
Molecule of Sulphur below 1000°	$\mathfrak{S}\mathfrak{a}$	$= S^6 =$	192
" Water . . .	$\mathfrak{H}\mathfrak{O}^{\frac{1}{2}} = \mathfrak{H}\mathfrak{O}\mathfrak{z}^{\frac{1}{8}}$	$= H^2O =$	18
" Sulphuretted Hydrogen	$\mathfrak{H}\mathfrak{S}^{\frac{1}{2}} = \mathfrak{H}\mathfrak{S}\mathfrak{a}^{\frac{1}{6}}$	$= H^2S =$	34

These few examples are sufficient to demonstrate the inconveniences associated with the formulæ indicating the composition of compound molecules as a function of the entire component molecules, which may be summed up as follows:—

1°. It is not possible to determine the weight of the molecules of many elements the density of which in the gaseous state cannot be ascertained.

2°. If it is true that oxygen and sulphur have different densities in their different allotropic states, that is, if they have different molecular weights, then their compounds would have two or more formulæ according as the quantities of their components were referred to the molecules of one or the other allotropic state.

3°. The molecules of analogous substances (such as sulphur and oxygen) being composed of different numbers of atoms, the formulæ of analogous compounds would be dissimilar. If we indicate, instead, the composition of the molecules by means of the atoms, it is seen that analogous compounds contain in their molecules an equal number of atoms.

It is true that when we employ in the formulæ the symbols expressing the weights of the molecules, *i.e.*, of equal volumes, the relationship between the volumes of the components and those of the compounds follows directly; but this relationship is also indicated in the formulæ expressing the number of atoms; it is sufficient to bear in mind that the atom represented by a symbol is either the entire molecule of the free substance or a fraction of it, that is, it is sufficient to know the atomic formula of the free molecule. Thus, to take an example, it is sufficient to know that the atom of oxygen, O, is one-half of the molecule of ordinary oxygen and an eighth part of the molecule of electrised oxygen—to know that the weight of the atom of oxygen is represented by $\frac{1}{2}$ volume of free oxygen and $\frac{1}{8}$ of electrised oxygen. In short, it is easy to accustom students to consider the weights of the atoms as being represented either by a whole volume or by a fraction of a volume, according as the atom is equal to the whole molecule or to a fraction of it. In this system of formulæ, those which represent the weights and the composition of the molecules, whether of elements or of compounds, represent the weights and the composition of equal gaseous volumes under the same conditions. The atom of each element is represented by that quantity of it which constantly enters as a whole into equal volumes of the free substance or of its compounds ; it may be either the entire quantity contained in one volume of the free substance or a simple sub-multiple of this quantity.

This foundation of the atomic theory having been laid, I begin in the following lecture—the sixth—to examine the constitution of the molecules of the chlorides, bromides, and iodides. Since the greater part of these are volatile, and since we know their densities in the gaseous state, there cannot remain any doubt as to the approximate weights of the molecules, and so of the quantities of chlorine, bromine, and iodine contained in them. These quantities being always integral multiples of the weights of chlorine, bromine, and iodine contained in hydrochloric, hydrobromic, and hydriodic acids, *i.e.*, of the weights of the half molecules, there can remain no doubt as to the atomic weights of these substances, and thus as to the number of atoms existing in the molecules of their compounds, whose weights and composition are known.

A difficulty sometimes appears in deciding whether the quantity of the other element combined with one atom of these halogens is 1, 2, 3, or *n* atoms in the molecule ; to decide this, it is necessary to compare the composition of all the other molecules containing the same element and find out the weight of this element which constantly enters as a whole. When we cannot determine the vapour densities of the other compounds of the element whose atomic weight we wish to determine, it is necessary then to have recourse to other criteria to know the weights of their molecules and to deduce the weight of the atom of the element. What I am to expound in the sequel serves to teach my pupils the method of employing these other criteria to verify or to determine atomic weights and the composition of molecules. I begin by making them study the following table of some chlorides, bromides, and iodides whose vapour densities are known ; I write their formulæ, certain of justifying later the value assigned to the atomic weights of some elements existing in the compounds indicated. I do not omit to draw their attention once more to the atomic weights of hydrogen, chlorine, bromine, and iodine being all equal to the weights of half a molecule, and represented by the weight of half a volume, which I indicate in the following table :—

	Symbol.	Weight.
Weight of the atom of Hydrogen or half a molecule represented by the weight of ½ volume .	H	1
Weight of the atom of Chlorine or half a molecule represented by the weight of ½ volume .	Cl	35·5
Weight of the atom of Bromine or half a molecule represented by the weight of ½ volume .	Br	80
Weight of the atom of Iodine or half a molecule represented by the weight of ½ volume .	I	127

These data being given, there follows the table of some compounds of the halogens :—

Names of the Chlorides.	Weights of equal volumes in the gaseous state, under the same conditions, referred to the weight of ½ volume of Hydrogen = 1; *i.e.*, weights of the molecules referred to the weight of the atom of Hydrogen = 1.	Composition of equal volumes in the gaseous state, under the same conditions, *i.e.*, composition of the molecules, the weights of the components being all referred to the weight of the atom of Hydrogen taken as unity, *i.e.*, the common unit adopted for the weights of atoms and of molecules.			Formulæ expressing the composition of the molecules or of equal volumes in the gaseous state under the same conditions.
Free Chlorine . . .	71	71 of Chlorine			Cl^2
Hydrochloric Acid . . .	36·5	35·5 ,,	1 of Hydrogen . . .		HCl
Protochloride of Mercury or Calomel	235·5	35·5 ,,	200 of Mercury . . .		$HgCl$
Bichloride of Mercury or Corrosive Sublimate . .	271	71 ,,	200 ,, . . .		$HgCl^2$
Chloride of Ethyl . . .	64·5	35·5 ,,	5 of Hydrogen	24 of Carbon	C^2H^5Cl
,, Acetyl . . .	78·5	35·5 ,, 24 of Carbon	3 ,, 16 of Oxygen . . .		C^2H^3OCl
,, Ethylene . .	99	71 of Chlorine	4 of Hydrogen	24 of Carbon	$C^2H^4Cl^2$
,, Arsenic . .	181·5	106·5 ,,	75 of Arsenic . . .		$AsCl^3$
Protochloride of Phosphorus .	138·5	106·5 ,,	32 of Phosphorus . . .		PCl^3
Chloride of Boron . . .	117·5	106·5 ,,	11 of Boron		BCl^3
Bichloride of Tin . . .	259·6	142 ,,	117·6 of Tin		$SnCl^4$
,, Titanium . .	198	142 ,,	56 of Titanium . . .		$TiCl^4$
Chloride of Silicon . . .	170	142 ,,	28 of Silicon		$SiCl^4$
,, Zirconium . .	231	142 ,,	89 of Zirconium . . .		$ZrCl^4$
,, Aluminium . .	267	213 ,,	54 of Aluminium . . .		Al^2Cl^6
Perchloride of Iron . . .	325	213 ,,	112 of Iron		Fe^2Cl^6
Sesquichloride of Chromium .	319	213 ,,	106 of Chromium . . .		Cr^2Cl^6

I stop to examine the composition of the molecules of the two chlorides and the two iodides of mercury. There can remain no doubt that the protochloride contains in its molecule the same quantity of chlorine as hydrochloric acid, that the bichloride contains twice as much, and that the quantity of mercury contained in the molecules of both is the same. The supposition made by some chemists that the quantities of chlorine contained in the two molecules are equal, and on the other hand that the quantities of mercury are different, is supported by no valid reason. The vapour densities of the two chlorides having been determined, and it having been observed that equal volumes of them contain the same quantity of mercury, and that the quantity of chlorine contained in one volume of the vapour of calomel is equal to that contained in the same volume of hydrochloric acid gas under the same conditions, whilst the quantity of chlorine contained in one volume of corrosive sublimate is twice that contained in an equal volume of calomel or of hydrochloric acid gas, the relative molecular composition of the two chlorides cannot be doubtful. The same may be said of the two iodides. Does the constant quantity of mercury existing in the molecules of these compounds, and represented by the number 200, correspond to one or more atoms? The observation that in these compounds the same quantity of mercury is combined with one or two atoms of chlorine or of iodine, would itself incline us to believe that this quantity is that which enters always as a whole into all the molecules containing mercury, namely, the atom; whence $Hg = 200$.

To verify this, it would be necessary to compare the various quantities of mercury contained in all the molecules of its compounds whose weights and composition are known with certainty. Few other compounds of mercury besides those indicated above lend themselves to this; still there are some in organic chemistry the formulæ of which express well the molecular composition; in these formulæ we always find $Hg^2 = 200$, chemists having made $Hg = 100$ and $H = 1$. This is a confirmation that the atom of mercury is 200 and not 100, no compound of mercury existing whose molecule contains less than this quantity of it. For verification I refer to the law of the specific heats of elements and of compounds.

I call the quantity of heat consumed by the atoms or the molecules the product of their weights into their specific heats. I compare the heat consumed by the atom of mercury with that consumed by the atoms of iodine and of bromine in the same physical state, and find them almost equal, which confirms the accuracy of the relation between the atomic weight of mercury and that of each of the two halogens, and thus also, indirectly, between the atomic weight of mercury and that of hydrogen, whose specific heats cannot be directly compared.

Thus we have—

Name of Substance.	Atomic weight.	Specific heat, *i.e.*, heat required to heat unit weight 1°.	Products of specific heats by atomic weights, *i.e.*, heat required to heat the atom 1°.
Solid Bromine .	80	0·08432	6·74560
Iodine . .	127	0·05412	6·87324
Solid Mercury .	200	0·03241	6·48200

The same thing is shown by comparing the specific heats of the different compounds of mercury. Woestyn and Garnier have shown that the state of combination does not notably change the calorific capacity of the atoms ; and since this is almost equal in the various elements, the molecules would require, to heat them 1°, quantities of heat proportional to the number of atoms which they contain. If Hg = 200, that is, if the formulæ of the two chlorides and iodides of mercury are HgCl, HgI, $HgCl^2$, HgI^2, it will be necessary that the molecules of the first pair should consume twice as much heat as each separate atom, and those of the second pair three times as much ; and this is so in fact, as may be seen in the following table :—

Formulæ of the compounds of Mercury.	Weights of their molecules $=p$.	Specific heats of unit weight $=c$.	Specific heats of the molecules $=p \times c$.	Number of atoms in the molecules $=n$.	Specific heats of each atom $=\frac{p \times c}{n}$.
HgCl .	235·5	0·05205	12·257745	2	6·128872
HgI .	327	0·03949	12·91323	2	6·45661
$HgCl^2$.	271	0·06889	18·66919	3	6·22306
HgI^2 .	454	0·04197	19·05438	3	6·35146

Thus the weight 200 of mercury, whether as an element or in its compounds, requires to heat it 1° the same quantity of heat as 127 of iodine, 80 of bromine, and almost certainly as 35.5 of chlorine and 1 of hydrogen, if it were possible to compare these two last substances in the same physical state as that in which the specific heats of the above-named substances have been compared.

But the atoms of hydrogen, iodine, and bromine are half their respective molecules: thus it is natural to ask if the weight 200 of mercury also corresponds to half a molecule of free mercury. It is sufficient to look at the table of numbers expressing the molecular weights to perceive that if 2 is the molecular weight of hydrogen, the weight of the molecule of mercury is

200, *i.e.*, equal to the weight of the atom. In other words, one volume of vapour, whether of protochloride or protoiodide, whether of bichloride or of biniodide, contains an equal volume of mercury vapour ; so that each molecule of these compounds contains an entire molecule of mercury, which, entering as a whole into all the molecules, is the atom of this substance. This is confirmed by observing that the complete molecule of mercury requires for heating it 1°, the same quantity of heat as half a molecule of iodine, or half a molecule of bromine. It appears to me, then, that I can sustain that what enters into chemical actions is the half molecule of hydrogen and the whole molecule of mercury : both of these quantities are indivisible, at least *in the sphere of chemical actions actually known*. You will perceive that with this last expression I avoid the question if it is possible to divide this quantity further. I do not fail to apprise you that all those who faithfully applied the theory of Avogadro and of Ampère, have arrived at this same result. First Dumas and afterwards Gaudin showed that the molecule of mercury, differing from that of hydrogen, always entered as a whole into compounds. On this account Gaudin called the molecule of mercury monatomic, and that of hydrogen biatomic. However, I wish to avoid the use of these adjectives in this special sense, because to-day they are employed as you know in a very different sense, that is, to indicate the different capacity for saturation of the radicals.

The formulæ of the two chlorides of mercury having been demonstrated, I next compare them with that of hydrochloric acid. The atomic formulæ indicate that the constitution of the protochloride is similar to that of hydrochloric acid, if we consider the number of atoms existing in the molecules of the two ; if, however, we compare the quantities of the components with those which exist in their free molecules, then a difference is perceived. To make this evident I bring the atomic formulæ of the various molecules under examination into comparison with the formulæ made with the symbols expressing the weights of the entire molecules, placing them in the manner which you see below :—

	Symbols of the molecules of the elements and formulæ of their compounds made with these symbols, *i.e.*, symbols and formulæ representing the weights of equal volumes in the gaseous state.	Symbols of the atoms of the elements, and formulæ of their compounds made with these symbols.	Numbers expressing the corresponding weights.
Atom of Hydrogen . . .	$\mathfrak{H}^{\frac{1}{2}}$ =	H =	1
Molecule of Hydrogen . .	$\mathfrak{H}$ =	H^2 =	2
Atom of Chlorine . . .	$\mathfrak{Cl}^{\frac{1}{2}}$ =	Cl =	35·5
Molecule of Chlorine . .	$\mathfrak{Cl}$ =	Cl^2 =	71
Atom of Bromine . . .	$\mathfrak{Br}^{\frac{1}{2}}$ =	Br =	80
Molecule of Bromine . .	$\mathfrak{Br}$ =	Br^2 =	160
Atom of Iodine . . .	$\mathfrak{I}^{\frac{1}{2}}$ =	I =	127
Molecule of Iodine . . .	$\mathfrak{I}$ =	I^2 =	254
Atom of Mercury . . .	$\mathfrak{Hg}$ =	Hg =	200
Molecule of Mercury . .	$\mathfrak{Hg}$ =	Hg =	200
,, Hydrochloric Acid	$\mathfrak{H}^{\frac{1}{2}}\mathfrak{Cl}^{\frac{1}{2}}$ =	HCl =	36·5
,, Hydrobromic Acid	$\mathfrak{H}^{\frac{1}{2}}\mathfrak{Br}^{\frac{1}{2}}$ =	HBr =	81
,, Hydriodic Acid .	$\mathfrak{H}^{\frac{1}{2}}\mathfrak{I}^{\frac{1}{2}}$ =	HI =	128
Mol. of protochloride of Mercury	$\mathfrak{Hg}\mathfrak{Cl}^{\frac{1}{2}}$ =	HgCl =	235·5
,, protobromide of Mercury	$\mathfrak{Hg}\mathfrak{Br}^{\frac{1}{2}}$ =	HgBr =	280
,, protoiodide of Mercury .	$\mathfrak{Hg}\mathfrak{I}^{\frac{1}{2}}$ =	HgI =	327
,, bichloride of Mercury .	$\mathfrak{Hg}\mathfrak{Cl}$ =	$HgCl^2$ =	271
,, bibromide of Mercury .	$\mathfrak{Hg}\mathfrak{Br}$ =	$HgBr^2$ =	360
,, biniodide of Mercury .	$\mathfrak{Hg}\mathfrak{I}$ =	HgI_2 =	454

The comparison of these formulæ confirms still more the preference which we must give to the atomic formulæ, which indicate also clearly the relations between the gaseous bodies. It is sufficient to recall that whilst the atoms of chlorine, bromine, iodine, and hydrogen are represented by the weight of $\frac{1}{2}$ volume, the atom of mercury is represented by the weight of a whole volume.

I then come to the examination of the two chlorides of copper. The analogy with those of mercury forces us to admit that they have a similar atomic constitution, but we cannot verify this directly by determining and comparing the weights and the compositions of the molecules, as we do not know the vapour densities of these two compounds.

The specific heats of free copper and of its compounds confirm the atomic constitution of the two chlorides of copper deduced from the analogy with those of mercury. Indeed the composition of the two chlorides leads us to conclude that if they have the formulæ CuCl, $CuCl^2$, the atomic weight of copper indicated by Cu is equal to 63, which may be seen from the following proportions :—

	Ratio between the components expressed by numbers whose sum = 100.	Ratio between the components expressed by atomic weights.
Protochloride of Copper .	36·04 : 63·96 Chlorine. Copper.	35·5 : 63 Cl. Cu.
Bichloride of Copper .	52·98 : 47·02 Chlorine. Copper.	71 : 63 Cl^2. Cu.

Now 63 multiplied by the specific heat of copper gives a product practically equal to that given by the atomic weight of iodine or of mercury into their respective specific heats. Thus:

63	×	0·09515	=	6
Atomic weight of copper.		Specific heat of copper.		

The same quantity of heat is required to heat the weight of 63 of copper in its compounds through 1°. Thus :—

Formulæ of the compounds of Copper.	Weights of their molecules $=p$.	Specific heats of unit weights $=c$.	Specific heats of the molecules $=p\times c$.	Number of atoms in the molecules $=n$.	Specific heat of each atom $=\frac{p\times c}{n}$.
CuCl .	98·5	0·13827	13·619595	2	6·809797
CuI .	190	0·06869	14·0511	2	7·0255

After this comes the question, whether this quantity of copper which enters as a whole into the compounds, the calorific capacity of the atoms being maintained, is an entire molecule or a sub-multiple of it. The analogy of the compounds of copper with those of mercury would make us inclined to believe that the atom of copper is a complete molecule. But having no other proof to confirm this, I prefer to declare that there is no means of knowing the molecular weight of free copper until the vapour density of this substance can be determined.

I then go on to examine the constitution of the chlorides, bromides, and iodides of potassium, sodium, lithium, and silver. Each of these metals makes with each of the halogens only one well characterised and definite compound ; of none of these compounds is the vapour density known ; we are therefore in want of the direct means of discovering if in their molecules there are one, two, or more atoms of the halogens. But their analogies with the protochloride of mercury, HgCl, and with the protochloride of copper, CuCl, and the specific heats of the free metals and of their compounds make us assume that in the molecules of each of these compounds there is one atom of metal and one of halogen. According to this supposition, the atomic weight of potassium K = 39, that of sodium Na = 23, that of silver Ag = 108. These numbers multiplied by the respective specific heats give the same product as the atomic weights of the substances previously examined.

Name of Substance.	Atomic weight $=p$.	Specific heats of unit weight $=s$.	Specific heats of the atoms $=p \times c$.
Solid Bromine .	80	0·08432	6·74560
Iodine . .	127	0·05412	6·87324
Solid Mercury .	200	0·03241	6·48200
Copper . .	63	0·09515	6
Potassium . .	39	0·169556	6·612684
Sodium . .	23	0·2934	6·7482
Silver . .	108	0·05701	6·15708

Besides this, the specific heats of the chlorides, bromides, and iodides of these metals confirm the view that their molecules contain the same number of atoms of the two components. Thus :—

Formulæ and Names of the compounds.	Weights of their molecules $=p$.	Specific heats of unit weight $=c$.	Specific heats of the molecules $=p \times c$.	No. of atoms in the molecules $=n$	Specific heat of each atom $=\frac{p \times c}{n}$
KCl . Chl. of Potassium.	74·5	0·17295	12·884775	2	6·442387
$NaCl$. Chl. of Sodium.	58·5	0·21401	12·519585	2	6·259792
$AgCl$. Chl. of Silver.	143·5	0·09109	13·071415	2	6·535707
KBr . Brom. of Potassium	119	0·11321	13·47318	2	6·73659
$NaBr$. Brom. of Sodium.	103	0·13842	14·25726	2	7·12863
$AgBr$. Brom. of Silver.	188	0·07391	13·89508	2	6·94754
KI . . Iod. of Potassium.	166	0·08191	13·59706	2	6·79853
NaI . . Iodide of Sodium.	150	0.08684	13·0260	2	6·5130
AgI . . Iodide of Silver.	235	0·06159	14·47365	2	7·23682

Are the atoms of potassium, sodium, lithium, and silver equal to ½ molecule, like that of hydrogen, or equal to a whole molecule, like that of mercury? As the vapour densities of these elements are wanting, we cannot answer the question directly; I will give you later some reasons which incline me to believe that the molecules of these elements, like that of hydrogen, are composed of two atoms.

Gold makes with each of the halogens two compounds. I show that the first chloride is analogous to calomel, *i.e.*, that it has AuCl as it formula. The atomic weight of gold deduced from the composition of the protochloride to which this formula is given corresponds to the law of specific heats, as may be seen from what follows:

196·32	×	0·03244	=	6·3696208
Au		Specific heat of Gold.		

I show in the sequel that the first or only chlorides of the following metals have a constitution similar to the bichloride of mercury and of that of copper, that is, for each atom of metal they contain two atoms of chlorine.

Not knowing the density in the gaseous state of these lower or only chlorides, we cannot show directly the quantity of chlorine existing in their molecules, yet the specific heats of these free metals and of their compounds show what I have said above. I write the quantities of these different elements combined with the weight of two atoms of chlorine in the lower or only chlorides, and confirm in these quantities the properties of the other atoms; I write the formulæ of the lower chlorides, bromides, and iodides all as MCl^2, and verify that they correspond to the laws of specific heats of compound substances.

Names of Substances.	Symbols and weights of the atoms.	Specific heats of unit weight.	Specific heats of the atoms.
Iodine . .	I = 127	0·05412	6·87324
Solid Mercury	Hg = 200	0·03241	6·48200
Copper . .	Cu = 63	0·09515	6
Zinc . .	Zn = 66	0·09555	6·30630
Lead . .	Pb = 207	0·0314	6·4998
Iron . .	Fe = 56	0·11379	6·37224
Manganese .	Mn = 55	0·1181	6·4955
Tin . .	Sn = 117·6	0·05623	6·612648
Platinum .	Pt = 197	0·03243	6·38871
Calcium . .	Ca = 40		
Magnesium .	Mg = 24		
Barium . .	Ba = 137		

Formulæ of the compounds.	Weights of their molecules $=p$.	Specific heats of unit weight $=c$.	Specific heats of the molecules $=p\times c$.	No. of atoms in the molecules $=n$.	Specific heat of each atom $=\frac{p\times c}{n}$.
$HgCl^2$.	271	0·06889	18·66919	3	6·22306
$ZnCl^2$.	134	0·13618	18·65666	3	6·21888
$SnCl^2$.	188·6	0·10161	19·163646	3	6·387882
$MnCl^2$.	126	0·14255	17·96130	3	5·98710
$PbCl^2$.	278	0·06641	18·46198	3	6·15399
$MgCl^2$.	95	0·1946	18·4870	3	6·1623
$CaCl^2$.	111	0·1642	18·2262	3	6·0754
$BaCl^2$.	208	0·08957	18·63056	3	6·21018
HgI^2 .	454	0·04197	19·05438	3	6·35146
PbI^2 .	461	0·04267	19·67087	3	6·55695

Some of the metals indicated above make other compounds with chlorine, bromine, and iodine, whose molecular weights may be determined and compositions compared ; in such cases the values found for the atomic weights are confirmed. Thus, for example, a molecule of perchloride of tin weighs 259.6, and contains 117.6 of tin (= Sn) and 142 of chlorine (= Cl^4). A molecule of perchloride of iron weighs 325, and contains 112 of iron (= Fe^2) and 213 of chlorine (= Cl^6).

For zinc there are some volatile compounds which confirm the atomic weight fixed by me. Chemists believing chloride of zinc to be of the same type as hydrochloric acid, made the atom of zinc Zn = 33, that is half of that adopted by me; having then prepared some compounds of zinc with the alcohol radicals, they were astonished that, expressing the composition by formulæ corresponding to gaseous volumes equal to those of other well-known compounds, it was necessary to express the quantity of zinc contained in the molecule by Zn^2. This is a necessary consequence of the quantity of zinc represented by other chemists by Zn^2 being only a single atom, which is equivalent in its saturation capacity to two atoms of hydrogen. Since in the sequel of my lectures I return to this argument, you will therefore find it spoken of later in this abstract.

Are the atoms of all these metals equal to their molecules or to a simple sub-multiple of them? I gave you above the reasons which make me think it probable that the molecules of these metals are similar to that of mercury; but I warn you now that I do not believe my reasons to be of such value as to lead to that certainty which their vapour densities would give us if we only knew them.

Reviewing what I show in the lecture of which I have given you an abstract, we find it amounts to the following:—Not all the lower chlorides corresponding to the oxide with one atom of oxygen have the same constitution; some of them contain a single atom of chlorine, others two, as may be seen in the following list :—

HCl	$HgCl$	$CuCl$	KCl	$NaCl$	$LiCl$	$AgCl$	$AuCl$
Hydrochloric acid.	Protochloride of mercury.	Protochloride of copper.	Chloride of potassium.	Chloride of sodium.	Chloride of lithium.	Chloride of silver.	Protochloride of gold.

$HgCl^2$	$CuCl^2$	$ZnCl^2$	$PbCl^2$	$CaCl^2$	$SnCl^2$	$PtCl^2$	etc. etc.
Bichloride of mercury.	Bichloride of copper.	Chloride of zinc.	Chloride of lead.	Chloride of calcium.	Protochloride of tin.	Protochloride of platinum.	

Regnault, having determined the specific heats of the metals and of many of their compounds, had observed that it was necessary to modify the atomic weights attributed to them, namely, to divide by 2 those of potassium, sodium, and silver, leaving the others unaltered ; or, *vice versa*, to multiply these latter by 2, leaving unaltered those of potassium, sodium, silver, and hydrogen. From this he drew the conclusion that the chlorides of potassium, sodium, and silver, are analogous to calomel (protochloride of mercury) and to protochloride of copper : on the other hand, that those of zinc, lead, calcium, etc., etc., are analogous to corrosive sublimate and to bichloride of copper ; but he supposed that the molecules of calomel and of the analogous chlorides all contained 2 atoms of metal and 2 of chlorine, whilst the molecules of corrosive sublimate and the other analogous chlorides contained 1 atom of metal and 2 of chlorine. Here follows the list of the formulæ proposed by Regnault.

H^2Cl^2	Hg^2Cl^2	Cu^2Cl^2	K^2Cl^2	Na^2Cl^2	Li^2Cl^2	Ag^2Cl^2	Au^2Cl^2
Hydrochloric acid.	Protochloride of mercury.	Protochloride of copper.	Chloride of potassium.	Chloride of sodium.	Chloride of lithium.	Chloride of silver.	Protochloride of gold.

$HgCl^2$	$CuCl^2$	$ZnCl^2$	$PbCl^2$	$CaCl^2$	etc. etc.
Bichloride of mercury.	Bichloride of copper.	Chloride of zinc.	Chloride of lead.	Chloride of calcium.	

In truth, using the data for specific heats alone, it is not possible to decide whether the molecules of the chlorides written in the first horizontal line are MCl or M^2Cl^2; the only thing that can be said is that they contain the same number of atoms of metal and of chlorine. But knowing the densities in the gaseous state of hydrochloric acid and of the two chlorides of mercury, and thus the weights of their molecules, we can compare their composition and decide the question; and I have already explained to you how I show to my pupils that the molecules of the two chlorides of mercury contain the same weight of mercury, and that the molecule of one of them contains the same quantity of chlorine as hydrochloric acid, *i.e.*, $\frac{1}{2}$ molecule of free chlorine, whilst the molecule of the other chloride contains twice as much. This shows with certainty that the two formulæ Hg^2Cl^2, $HgCl^2$ are inexact, because they indicate that in the molecules of the two chlorides there is the same quantity of chlorine and different quantities of mercury, which is precisely the opposite of what is shown by the vapour densities. The formulæ proposed by me harmonise the results furnished by the specific heats and by the gaseous densities.

Now I wish to direct your attention to an inconsistency of Gerhardt. From the theory of Avogadro, Ampère, and Dumas, that is, from the comparison of the gaseous densities as representing the molecular weights, Gerhardt drew arguments in support of the view that the atoms of hydrogen, of chlorine, and of oxygen are half molecules; that the molecule of water contains twice as much hydrogen as that of hydrochloric acid; that in the molecule of ether there is twice as much of the radical ethyl as in that of alcohol; and that to form one molecule of anhydrous monobasic acid two

C

molecules of hydrated acid must come together: and yet Gerhardt did not extend to the whole of chemistry the theory of Ampère, but arbitrarily, in opposition to its precepts, assumed that the molecules of chloride of potassium, of bichloride of mercury, in fact of all the chlorides corresponding to the protoxides, had the same atomic constitution as hydrochloric acid, and that the atoms of all the metals were, like that of hydrogen, a simple sub-multiple of the molecule.

I have already explained to you the reasons which show the contrary.

After having demonstrated the constitution of the chlorides corresponding to the oxides containing one atom of oxygen, I postpone the study of the other chlorides to another lecture, and now define what I mean by capacity for saturation of the various metallic radicals.

If we compare the constitution of the two kinds of chlorides, we observe that one atom of metal is now combined with one atom of chlorine, now with two; I express this by saying that in the first case the atom of metal is equivalent to 1 of hydrogen, in the second case to 2. Thus, for example, the atom of mercury, as it is in calomel, is equivalent to 1 of hydrogen, whereas in corrosive sublimate it is equivalent to 2; the atoms of potassium, sodium, and silver are equivalent to 1 of hydrogen: the atoms of zinc, lead, magnesium, calcium, etc., to 2. Now it is seen from the study of all chemical actions that the number of atoms of the various substances which combine with one and the same quantity of chlorine combine also with one and the same quantity of oxygen, of sulphur, or of any other substance, and *vice versa*. Thus, for example, if the same quantity of chlorine which combines with a single atom of zinc, or lead, or calcium combines with 2 atoms of hydrogen, of potassium, or of sodium, then the same quantity of oxygen or of any other substance which combines with a single atom of the first will combine with two of the second. This shows that the property possessed by the first atoms of being equivalent to 2 of the second depends on some cause inherent either in their own nature or in the state in which they are placed before combining. We express this constant equivalence by saying that each atom of the first has a saturation capacity twice that of each of the second. These expressions are not new to science, and we now only extend them from compounds of the second order to those of the first order.

For the same reasons given by chemists when they say that phosphoric acid assumes various saturation capacities without changing in composition, it may also be said that the atom of mercury and that of copper assume different saturation capacities according as they are found in the protochlorides or in the bichlorides. Thus, I express the fact that the atoms of these two metals being equivalent to 1 atom of hydrogen in the protochlorides, tend, in double decompositions, to take the place of a single atom of hydrogen, whilst in the bichlorides they tend to take the place of 2 atoms of hydrogen. For the same reason that we say there are three different modifications of phosphoric acid combined with various bases, we may also say that there are two different modifications of the same radical mercury or copper. I call the radicals of the protochlorides and of the corresponding salts, mercurous and cuprous; those of the bichlorides and of the corresponding salts are called mercuric and cupric radicals.

To express the various saturation capacities of the

different radicals, I compare them to that of hydrogen or of the halogens, according as they are electro-positive or electro-negative. An atom of hydrogen is saturated by one of a halogen, and *vice versa.* I express this by saying that the first is a monatomic electro-positive radical, and the second a monatomic electro - negative radical: thus, potassium, sodium, lithium, silver, and the mercurous and cuprous radicals are monatomic electro-positive radicals. The biatomic radicals are those which, not being divisible, are equivalent to 2 of hydrogen or to 2 of chlorine; among the electro - positive radicals there are the metallic radicals of the mercuric and cupric salts, of the salts of zinc, lead, magnesium, calcium, etc., and amongst the electro-negative we have oxygen, sulphur, selenium, and tellurium, *i.e.*, the amphidic substances. There are, besides, radicals which are equivalent to three or more atoms of hydrogen or of chlorine, but I postpone the study of these until later.

Before finishing the lecture I take care to make clear that the law of equivalents must be considered as a law distinct from the law of atoms.

The latter in fact only says that the quantities of the same element contained in different molecules must be integral multiples of one and the same quantity, but it does not predict, for example, that an atom of zinc is equivalent to 2 of hydrogen not only in its compounds with chlorine, but in all other compounds in which they may replace each other. These constant relations between the numbers of atoms of various substances which displace one another, whatever may be the nature and the number of the other components, is a law which restricts the number of possible combinations, and sums up with greater definiteness all the cases of double decomposition.

I occupy the whole of the seventh lecture in studying some monatomic and biatomic radicals, namely, cyanogen and the alcohol radicals.

I have already told you the method which I faithfully follow for ascertaining the weights and numbers of the molecules of the various substances whose vapour densities can be determined. This method, applied to all the substances which contain alcohol radicals, permits us, so to speak, to follow the path from one molecule to another. To discover the saturation capacity of a radical, it is expedient to begin with the examination of a molecule in which it is combined with a monatomic radical: thus for electro-negative radicals I begin by examining the compounds with hydrogen or with any other monatomic electro-positive radical; and conversely, for the electro-positive radicals, I examine their compounds with chlorine, bromine, and iodine. Those electro-negative radicals which form a molecule with a single atom of hydrogen are monatomic; those which combine with 2 of hydrogen are biatomic, and so on. Conversely, the electro-positive radicals are monatomic if they combine with a single atom of halogen, biatomic if they combine with 2.

With these rules I establish—

1°. That cyanogen, CN, is a monatomic electro-negative radical, and that the molecule of free cyanogen contains twice the quantity of carbon and nitrogen contained in the molecule of the monocyanides; and that in this way cyanogen, CN, behaves in all respects like an atom of chlorine, Cl;

2°. That cacodyle, C^2H^6As, methyl, CH^3, ethyl, C^2H^5, and the other homologous and isologous radicals, are, like the atom of hydrogen, monatomic, and like it cannot form a molecule alone, but must associate themselves with another monatomic radical, simple or compound, whether of the same or of a different kind;

3°. That ethylene, C^2H^4, propylene, C^3H^6, are biatomic radicals analogous to the radicals of mercuric and cupric salts, and to those of the salts of zinc, lead, calcium, magnesium, etc.; and that these radicals, like the atom of mercury, can form a molecule by themselves.

The analogy between the mercuric salts and those of ethylene and propylene has not been noted, so far as I know, by any other chemist. All that I have expounded previously shows it with such clearness that it appears useless to stop and discuss it with you at length. In fact, just as 1 volume of the vapour of mercury, combining with an equal volume of chlorine, makes 1 volume of vapour of mercury bichloride, so 1 volume of ethylene combined with an equal volume of chlorine makes a single volume of vapour of chloride of ethylene—(oil of Dutch chemists). If the formula of this last is $C^2H^4Cl^2$, that of bichloride of mercury should be $HgCl^2$; and if this is the formula of the bichloride of mercury, the chlorides of zinc, lead, calcium, etc., must also be MCl^2; that is, the atoms of all these metals are, like ethylene and propylene, biatomic radicals. Observing that all the electro-positive monatomic radicals which can be weighed free in the gaseous state, behave like hydrogen, that is, cannot of themselves form molecules, it appears to me very probable that a capacity of saturation equal to that of hydrogen in atoms, or groups which can act as their substitutes, constantly coincides with the fact of their not being able to exist in the isolated state. This is the reason why, until there is proof to the contrary, I believe that the molecules of potassium, sodium, lithium, and silver in the free state are formed of two atoms, that is, are represented by the formulæ K^2, Na^2, Li^2, Ag^2.

Conversely, observing that if the atom of mercury (which tends to form a biatomic rather than a monatomic radical) like ethylene and propylene can exist in the free state, forming a distinct molecule by itself, it appears to me probable that the atoms of zinc, lead, and calcium should be endowed also with this property, that is, that the molecules of these metals should consist of a single atom. If this correspondence between the number of atoms contained in the molecule and the capacity of the saturation of the atom, or of the group which takes its place, is verified, we may sum up as follows: *the metallic radicals whose molecules enter as a whole into compounds are biatomic, those whose atom is half a molecule are monatomic.* You already perceive the importance of this correlation, which forces us to conclude that one molecule of mercury (in mercuric salts), or of zinc, or ethylene, or propylene, etc., is equivalent to a molecule of hydrogen, of potassium, or of silver; thus the former as well as the latter combines with an entire molecule of chlorine, yet with this important difference that the former, not being capable of division, forms a single molecule with two atoms of chlorine, whilst the latter, being divisible, makes with the two atoms of chlorine two distinct molecules. But before drawing a general conclusion of such importance, it is necessary to demonstrate somewhat better the accuracy of the data on which it is founded.

In the eighth lecture I begin to compare the mode of behaviour in some reactions of monatomic and biatomic radicals. The compound radicals indicated in the preceding lecture, since they form volatile compounds, frequently afford the means of explaining by analogy what holds good for metallic compounds, the molecular weights of which cannot often be determined directly, since few of them are volatile. This is the great benefit which the study of organic chemistry has rendered to chemistry in general.

In the use of formulæ I adhere to the following rules, which I state before representing by means of equations the various types of reaction :—

1°. I use the coefficients of the symbols in the position of the exponents only when I wish to express that the number of atoms indicated is contained in one and the same molecule ; in other cases I place the coefficient before the symbols. Thus, when I wish to indicate two atoms of free hydrogen as they are contained in a single molecule, I write H^2. If, however, I wish to indicate four atoms as they are contained in two molecules, I do not write H^4 but $2H^2$; for the same reason I indicate *n* atoms of free mercury by the formula nHg.

2°. Sometimes I repeat in the same formula more than once the same symbol to indicate some difference between one part and another of the same element. Thus I write acetic acid $C^2H^3HO^2$, to indicate that one of the four atoms of hydrogen contained in the molecule is in a state different from the other three, it alone being replaceable by metals. Occasionally I write the same symbol several times to indicate several atoms of the same element, only to place better in relief what occurs in some reactions.

3°. For this last reason I often write the various atoms of the same component or the residues of various equal molecules in vertical lines. Thus, for example, I indicate the molecule of bichloride of mercury, $HgCl^2$, as follows:—$Hg\begin{cases}Cl\\Cl\end{cases}$; the molecule of acetate of mercury, $C^4H^6HgO^4$, as follows: $Hg\begin{cases}C^2H^3O^2\\C^2H^3O^2\end{cases}$; to indicate that the two atoms of chlorine or the two residues of acetic acid come from two distinct molecules of hydrochloric acid and of hydrated acetic acid.

4°. I indicate by the symbol R^i_m any monatomic metallic radical whether simple or compound; and with the symbol R^{ii}_m any biatomic metallic radical. If in the same formula or in the same equation I wish to indicate in general two or more monatomic radicals, the one different from the other, I add to the symbol the small letters *a*, *b*, *c*, etc., thus R^i_{ma} R^i_{mb} indicates a single molecule formed of two different monatomic radicals; such are the so-called mixed radicals.

The molecules of the monatomic metallic radicals are represented by the formula $(R^i_m)^2$; those of the biatomic radicals by the same symbol as for the radical existing in its compounds, since it is the character of these radicals to have the molecule formed of a single atom or of a single group which takes its place. You understand that in speaking of metallic radicals I include all those which can replace metals in saline compounds.

5°. Since all compounds containing in their molecule a single atom of hydrogen replaceable by metals behave similarly when they act on metals or on their compounds, it is convenient to adopt a general formula, and I shall use the following. In HX, X indicates all that there is in the molecule except metallic hydrogen; thus, for example, in the case of acetic acid, $X = C^2H^3O^2$, these being the components which together with H make up the molecule of hydrated acetic acid. Since there are compounds, also called acids, whose molecules contain two atoms of hydrogen replaceable by metals, and since owing to this last fact they behave in a similar manner towards molecules containing metals, I adopt for them the general formula H^2Y, indicating by Y all that there is in the molecules except the two atoms of hydrogen. I hasten to mention that when I indicate by X and by Y the things which in the molecules of acids are combined with H and H^2, I do not intend to affirm that X and H, or Y and H^2, are detached within the molecule as its two immediate components; but without touching the question of the disposition of the atoms within the molecule of acids, I only wish to indicate distinctly the part which is not changed in the transformation of the acid into its corresponding salts.

Before treating and discussing the various reactions, I remind my pupils once more that all the formulæ used by me correspond to equal gaseous volumes, the theory of Avogadro and Ampère being constantly the guiding thread which leads me in the study of chemical reactions.

This done, I now give very rapidly an abstract of what I explain in this lecture concerning some reactions of the monatomic and biatomic radicals. I always write the reaction of the molecule containing a monatomic radical alongside a corresponding one of a molecule containing a biatomic radical, in order that the comparison may be easier.

Direct Combination	
Of the Monatomic Metallic Radicals with the Halogens.	Of the Biatomic Metallic Radicals with the Halogens.
* $H^2 + Cl^2 = 2\,HCl$ 1 molecule of hydrogen. · 1 molecule of chlorine. · 2 molecules of hydrochloric acid.	$Hg + Cl^2 = HgCl^2$ 1 molecule of mercury. · 1 molecule of chlorine. · 1 molecule of bichloride of mercury.
$K^2 + Cl^2 = 2\,KCl$ 1 molecule of potassium. · 1 molecule of chlorine. · 2 molecules of chloride of potassium.	$Zn + Cl^2 = ZnCl^2$ 1 molecule of zinc. · 1 molecule of chlorine. · 1 molecule of chloride of zinc.
† $(CH^3)^2 + Cl^2 = 2\,CH^3Cl$ 1 molecule of methyl. · 1 molecule of chlorine. · 2 molecules of chloride of methyl.	$C^2H^4 + Cl^2 = C^2H^4Cl^2$ 1 molecule of ethylene. · 1 molecule of chlorine. · 1 molecule of chloride of ethylene.
$(R^{i}_{m})^2 + Cl^2 = 2\,R^{i}_{m}Cl$	$R^{ii}_{m} + Cl^2 = R^{ii}_{m}Cl^2$
Apparent direct combination, in reality molecular double decomposition, in virtue of which two molecules of different kinds give two of the same kind.	True direct combination or union of two different entire molecules into a single molecule.

* The direct combination of hydrogen and chlorine is expressed by some as $H + Cl = HCl$; in the equations used by me I always employ molecules.

† It appears that in practice this direct combination succeeds with difficulty, the chlorine having an action on the hydrogen of the radical; it has been indicated merely for comparison with that of ethylene.

From what precedes it may be observed that a complete molecule of chlorine, and thus of any halogen, always reacts with a complete molecule of a metallic radical; if the latter is monatomic it makes two molecules, if it is biatomic it forms only one.

Substitutions in Chlorides, Bromides, and Iodides

Of one Monatomic Metallic Radical for another.

K^2	+	HCl HCl	=	H^2	+	KCl KCl
1 molecule of potassium.		2 molecules of hydrochloric acid.		1 molecule of hydrogen.		2 molecules of chloride of potassium.
H^2	+	$AgCl$ $AgCl$	=	Ag^2	+	HCl HCl
1 molecule of hydrogen.		2 molecules of chloride of silver.		1 molecule of silver.		2 molecules of hydrochloric acid.
Ag^2	+	HI HI	=	H^2	+	AgI AgI
1 molecule of silver.		2 molecules of hydriodic acid.		1 molecule of hydrogen.		2 molecules of iodide of silver.
$(R^{i}_{ma})^2$	+	$R^{i}_{mb}Cl$ $R^{i}_{mb}Cl$	=	$(R^{i}_{mb})^2$	+	$R^{i}_{ma}Cl$ $R^{i}_{ma}Cl$

Substitution without change in the number of molecules.

Of one Biatomic Metallic Radical for a Monatomic.

Zn	+	HCl HCl	=	H^2	+	$Zn\begin{cases}Cl\\Cl\end{cases}$
1 molecule of zinc.		2 molecules of hydrochloric acid.		1 molecule of hydrogen.		1 molecule of chloride of zinc.
Zn	+	$AgCl$ $AgCl$	=	Ag^2	+	$Zn\begin{cases}Cl\\Cl\end{cases}$
1 molecule of zinc.		2 molecules of chloride of silver.		1 molecule of silver.		1 molecule of chloride of zinc.
Hg	+	HI HI	=	H^2	+	$Hg\begin{cases}I\\I\end{cases}$
1 molecule of mercury.		2 molecules of hydriodic acid.		1 molecule of hydrogen.		1 molecule of biniodide of mercury.
R^{ii}_{m}	+	$R^{i}_{m}Cl$ $R^{i}_{m}Cl$	=	$(R^{i}_{m})^2$	+	$R^{ii}_{m}\begin{cases}Cl\\Cl\end{cases}$

Substitution with change in the number of molecules; 3 become 2.

From what is written in this table it is seen that two molecules of hydrochloric acid or of another analogous monochloride always react with a single molecule of metallic radical; if this is monatomic, they change into two molecules of monochloride, if it is biatomic into a single molecule of bichloride. The cause of the last difference consists in this: that the molecule of the monatomic radical is divisible into two; that of the biatomic radical, not being capable of division, collects into a single molecule the residues of two molecules of monochloride or monoiodide.

The biatomic radicals behave similarly to the acids containing 1 atom of monatomic metallic radicals (H, Ag, K); collecting into a single molecule the residues of two molecules of acids or of salts, as may be seen in the following comparative table.

[TABLE

SUBSTITUTION IN THE ACIDS HX, AND IN GENERAL IN THE SALTS $R^{i}_{m}X$,

Of a Monatomic Metallic Radical, R^{i}_{ma}, for another, R^{i}_{mb}.

K^2	+	$HAzO^3$ $HAzO^3$	=	H^2	+	$KAzO^3$ $KAzO^3$
1 molecule of potassium.		2 molecules of hydrated nitric acid.		1 molecule of hydrogen.		2 molecules of nitrate of potassium.
Na^2	+	$HC^2H^3O^2$ $HC^2H^3O^2$	=	H^2	+	$NaC^2H^3O^2$ $NaC^2H^3O^2$
1 molecule of sodium.		2 molecules of hydrated acetic acid.		1 molecule of hydrogen.		2 molecules of acetate of sodium.
$(R^{i}_{ma})^2$	+	$R^{i}_{mb}X$ $R^{i}_{mb}X$	=	$(R^{i}_{mb})^2$	+	$R^{i}_{ma}X$ $R^{i}_{ma}X$

Of a Biatomic Metallic Radical, R^{ii}_{m}, for a Monatomic, R^{i}_{m}.

Zn	+	$HAzO^3$ $HAzO^3$	=	H^2	+	$Zn\begin{cases}AzO^3\\AzO^3\end{cases}$
1 molecule of zinc.		2 molecules of hydrated nitric acid.		1 molecule of hydrogen.		1 molecule of nitrate of zinc.
Zn	+	$HC^2H^3O^2$ $HC^2H^3O^2$	=	H^2	+	$Zn\begin{cases}C^2H^3O^2\\C^2H^3O^2\end{cases}$
1 molecule of zinc.		2 molecules of hydrated acetic acid.		1 molecule of hydrogen.		1 molecule of acetate of zinc.
R^{ii}_{m}	+	$R^{i}_{m}X$ $R^{i}_{m}X$	=	$(R^{i}_{m})^2$	+	$R^{ii}_{m}\begin{cases}X\\X\end{cases}$

These examples are sufficient to show that the compounds containing a monatomic metallic radical behave like the monochlorides: two molecules of these react with a single molecule of metallic radical, changing into two molecules if the latter is monatomic, into a single molecule if it is. biatomic. We can prove more easily that the biatomic metallic radicals bind in a single molecule the residues X of two molecules R_m^iX, by comparing the double decompositions or mutual substitutions of the chlorides of the monatomic and biatomic radicals with the compound R_m^iX.

I write in the following table some examples of these double decompositions.

[Table

Mutual Substitution of the Compounds containing a Monatomic Radical, $R^{i}_{m}X$,

With the Chlorides of the Monatomic Metallic Radicals $R^{i}_{m}Cl$.

KCl	+	$HAzO^3$	=	HCl	+	$KAzO^3$
1 molecule of chloride of potassium.		1 molecule of hydrated nitric acid.		1 molecule of hydrochloric acid.		1 molecule of nitrate of potassium.
KCl	+	$AgC^2H^3O^2$	=	$AgCl$	+	$KC^2H^3O^2$
1 molecule of chloride of potassium.		1 molecule of acetate of silver.		1 molecule of chloride of silver.		1 molecule of acetate of potassium.
C^2H^5, Cl	+	$AgC^2H^3O^2$	=	$AgCl$	+	$C^2H^5, C^2H^3O^2$
1 molecule of chloride of ethyl.		1 molecule of acetate of silver.		1 molecule of chloride of silver.		1 molecule of acetate of ethyl.
R^{i}_{ma}, Cl	+	R^{i}_{mb}, X	=	R^{i}_{mb}, Cl	+	R^{i}_{ma}, X

With the Chlorides of the Biatomic Metallic Radicals $R^{ii}_{m}Cl^2 = R^{ii}_{m}\left\{\begin{matrix}Cl\\Cl\end{matrix}\right.$

$Hg\left\{\begin{matrix}Cl\\Cl\end{matrix}\right.$	+	$\begin{matrix}HAzO^3\\HAzO^3\end{matrix}$	=	$\begin{matrix}HCl\\HCl\end{matrix}$	+	$Hg\left\{\begin{matrix}AzO^3\\AzO^3\end{matrix}\right.$
1 molecule of bichloride of mercury.		2 molecules of hydrated nitric acid.		2 molecules of hydrochloric acid.		1 molecule of mercuric nitrate.
$Hg\left\{\begin{matrix}Cl\\Cl\end{matrix}\right.$	+	$\begin{matrix}AgC^2H^3O^2\\AgC^2H^3O^2\end{matrix}$	=	$\begin{matrix}AgCl\\AgCl\end{matrix}$	+	$Hg\left\{\begin{matrix}C^2H^3O^2\\C^2H^3O^2\end{matrix}\right.$
1 molecule of bichloride of mercury.		2 molecules of acetate of silver.		2 molecules of chloride of silver.		1 molecule of mercuric acetate.
$C^2H^4\left\{\begin{matrix}Cl\\Cl\end{matrix}\right.$	+	$\begin{matrix}AgC^2H^3O^2\\AgC^2H^3O^2\end{matrix}$	=	$\begin{matrix}AgCl\\AgCl\end{matrix}$	+	$C^2H^4\left\{\begin{matrix}C^2H^3O^2\\C^2H^3O^2\end{matrix}\right.$
1 molecule of chloride of ethylene.		2 molecules of acetate of silver.		2 molecules of chloride of silver.		1 molecule of acetate of ethylene.
$R^{ii}_{m}\left\{\begin{matrix}Cl\\Cl\end{matrix}\right.$	+	$\begin{matrix}R^{i}_{m}, X\\R^{i}_{m}, X\end{matrix}$	=	$\begin{matrix}R^{i}_{m}Cl\\R^{i}_{m}Cl\end{matrix}$	+	$R^{ii}_{m}\left\{\begin{matrix}X\\X\end{matrix}\right.$

All the reactions indicated in this table may be summed up as follows :—Whatever is combined with one atom of hydrogen or any other equivalent radical =(X) replaces one atom of chlorine, and conversely is replaceable by the latter ; if an indivisible radical in the double decompositions is found combined in a single molecule with two atoms of chlorine, it will, if the chlorine is exchanged for X, remain combined in a single molecule with 2X.

That ethylene is combined with two atoms of chlorine in choride of ethylene, and that the acetate of ethylene contains in one molecule twice $C^2H^3O^2$, is shown by the comparison of the gaseous densities of these substances. From the vapour density and from the specific heats, it is further demonstrated that the molecule of corrosive sublimate, like that of chloride of ethylene, contains two atoms of chlorine. Hence the mercuric salts are constituted in a similar manner to those of ethylene, whilst the salts of potassium, sodium, and silver are formed like those of ethyl.

Having proved, then, as I think I have already sufficiently indicated, that the lower or only chlorides of iron, manganese, zinc, magnesium, calcium, barium, etc., are constituted like corrosive sublimate, that is, have the formula MCl^2, there can remain no further doubt that the salts which are obtained by means of these chlorides and of the monobasic acids, or of their salts, are all similar to those of ethylene, propylene, etc. These important conclusions may be summed up as follows :—

1°. Amongst the salts of monobasic acids only those of hydrogen, potassium, sodium, lithium, silver, together with mercurous and cuprous salts, are similar to those of methyl and ethyl, that is, to compounds

D

of the alcohols containing a monatomic radical; all the other salts, of the so-called protoxides, are similar to those of ethylene and propylene, that is, to the compound ethers of the alcohols with biatomic radicals.

2°. A single molecule of the first is not sufficient to form the anhydrous acid and the metallic oxide; two molecules instead are required; but a single molecule of the second contains the components of the molecule of the anhydrous acid and of that of the protoxide. This becomes clear by bringing the following equations into comparison :—

$\begin{matrix} AgC^2H^3O^2 \\ AgC^2H^3O^2 \end{matrix} = \left.\begin{matrix} Ag \\ Ag \end{matrix}\right\} O + C^4H^6O^3$	$Hg\left\{\begin{matrix} C^2H^3O^2 \\ C^2H^3O^2 \end{matrix}\right. = HgO + C^4H^6O^3$
2 molecules of acetate of silver. — 1 molecule of oxide of silver. — 1 molecule of anhydrous acetic acid.	1 molecule of mercuric acetate. — 1 molecule of oxide of mercury. — 1 molecule of anhydrous acetic acid.
$\begin{matrix} C^2H^5, C^2H^3O^2 \\ C^2H^5, C^2H^3O^2 \end{matrix} = \left.\begin{matrix} C^2H^5 \\ C^2H^5 \end{matrix}\right\} O + C^4H^6O^3$	$C^2H^4\left\{\begin{matrix} C^2H^3O^2 \\ C^2H^3O^2 \end{matrix}\right. = C^2H^4O + C^4H^6O^3$
2 molecules of acetate of ethyl. — 1 molecule of oxide of ethyl. — 1 molecule of anhydrous acetic acid.	1 molecule of acetate of ethylene. — 1 molecule of oxide of ethylene. — 1 molecule of anhydrous acetic acid.
$\begin{matrix} R^{i}_{m}X \\ R^{i}_{m}X \end{matrix} = \left.\begin{matrix} R^{i}_{m} \\ R^{i}_{m} \end{matrix}\right\} O + (2X - O)$	$R^{ii}_{m}\left\{\begin{matrix} X \\ X \end{matrix}\right. = R^{ii}_{m}O + (2X - O)$

The mercuric salts and the salts of zinc, etc., being similar to those of ethylene, it is probable that salts of this type exist containing the residues of two different monobasic acids. I indicate by what reactions they might be generated :—

$Hg\left\{\begin{matrix}Cl\\Cl\end{matrix}\right.$	+	$\begin{matrix}AgC^2H^3O^2\\AgC^2H^3O^2\end{matrix}$	=	$\begin{matrix}AgCl\\AgCl\end{matrix}$	+	$Hg\left\{\begin{matrix}C^2H^3O^2\\C^2H^3O^2\end{matrix}\right.$
1 molecule of bichloride of mercury.		2 molecules of acetate of silver.		2 molecules of chloride of silver.		1 molecule of acetate of mercury.
$Hg\left\{\begin{matrix}Cl\\Cl\end{matrix}\right.$	+	$\begin{matrix}AgC^2H^3O^2\\AgC^7H^5O^2\end{matrix}$	=	$\begin{matrix}AgCl\\AgCl\end{matrix}$	+	$Hg\left\{\begin{matrix}C^2H^3O^2\\C^7H^5O^2\end{matrix}\right.$
1 molecule of bichloride of mercury.		1 molecule of acetate and 1 of benzoate of silver.		2 molecules of chloride of silver.		1 molecule of benzacetate of mercury.
$C^2H^4\left\{\begin{matrix}Cl\\Cl\end{matrix}\right.$	+	$\begin{matrix}AgC^2H^3O^2\\AgC^7H^5O^2\end{matrix}$	=	$\begin{matrix}AgCl\\AgCl\end{matrix}$	+	$C^2H^4\left\{\begin{matrix}C^2H^3O^2\\C^7H^5O^2\end{matrix}\right.$
1 molecule of chloride of ethylene.		1 molecule of acetate and 1 of benzoate of silver.		2 molecules of chloride of silver.		1 molecule of benzacetate of ethylene.

Just as acetates are produced from anhydrous acetic acid and the oxides of biatomic metallic radicals, so from anhydrous benzacetic acid the benzacetates will be formed, as I indicate in the following equation :—

$$C^4H^6O^3 + R^{ii}_{m}O = R^{ii}_{m}C^4H^6O^4 = R^{ii}_{m}\left\{\begin{matrix}C^2H^3O^2\\C^2H^3O^2\end{matrix}\right.$$

$$C^9H^8O^3 + R^{ii}_{m}O = R^{ii}_{m}C^9H^8O^4 = R^{ii}_{m}\left\{\begin{matrix}C^2H^3O^2\\C^7H^5O^2\end{matrix}\right.$$

Having already proved that zinc is a biatomic radical, and that in consequence its atomic weight should be doubled, I stop to examine the reactions and the mode of formation of zinc ethyl, zinc methyl, etc. I show you by means of equations the method by which I interpret these reactions.

The vapour densities demonstrate the accuracy of the following formulæ corresponding to equal volumes : — C^2H^5Cl (chloride of ethyl) C^2H^5, H (hydride of ethyl) C^2H^5, C^2H^5 (free ethyl) C^2H^5, CH^3 (methyl ethyl), $Zn(C^2H^5)^2 = Zn\left\{\begin{matrix}C^2H^5\\C^2H^5\end{matrix}\right.$ (zinc ethyl).

$$C^2H^5Cl + H^2 = C^2H^5, H + HCl$$

$$\begin{matrix} C^2H^5Cl \\ C^2H^5Cl \end{matrix} + Zn = (C^2H^5)^2 + Zn\left\{\begin{matrix} Cl \\ Cl \end{matrix}\right.$$

$$\begin{matrix} C^2H^5Cl \\ C^2H^5Cl \end{matrix} + 2Zn = \left.\begin{matrix} C^2H^5 \\ C^2H^5 \end{matrix}\right\}Zn + Zn\left\{\begin{matrix} Cl \\ Cl \end{matrix}\right.$$

$$\begin{matrix} C^2H^5Cl \\ C^2H^5Cl \end{matrix} + \left.\begin{matrix} C^2H^5 \\ C^2H^5 \end{matrix}\right\}Zn = 2(C^2H^5)^2 + Zn\left\{\begin{matrix} Cl \\ Cl \end{matrix}\right.$$

$$\begin{matrix} C^2H^5Cl \\ CH^3Cl \end{matrix} + Zn = C^2H^5, CH^3 + Zn\left\{\begin{matrix} Cl \\ Cl \end{matrix}\right.$$

$$\begin{matrix} C^2H^5Cl \\ CH^3Cl \end{matrix} + 2Zn = \left.\begin{matrix} C^2H^5 \\ CH^3 \end{matrix}\right\}Zn + Zn\left\{\begin{matrix} Cl \\ Cl \end{matrix}\right.$$

No one has yet demonstrated, as far as I know, the existence of the type of compound indicated in the last equation. But it being proved from the density of zinc ethyl vapour, and from its specific heat, that the complete molecule of zinc ethyl contains a single atom of zinc combined with two ethyl radicals, that is, with the molecule of the free radical, no one can deny that there will be prepared compounds containing a single atom of zinc combined with two different monatomic radicals. It may also be predicted that ethylene and propylene will form compounds in whose molecules an atom of zinc is combined with the biatomic radical.

I will give you later an account of some of my experiments directed to show the existence of the compounds just mentioned.

After having spoken of the mode of behaviour of the compounds containing monatomic or biatomic radicals with regard to monobasic acids, I examine the mode of behaviour with regard to those compounds which contain in each molecule two atoms of hydrogen, or, as they are called, the bibasic acids, to which I have given the general formula H^2Y.

To predict the reactions, it is sufficient to bear in mind what follows :—

1°. The two atoms of hydrogen are united in a single molecule by the forces of all the other components which together we call Y, hence what is equivalent to H^2 can enter into a single molecule with Y.

2°. What is combined with H^2 is equivalent to two atoms of chlorine Cl^2; hence in double decomposition H^2Y will act either on a single molecule of a bichloride ($=R^{\text{ii}}_{\text{m}}Cl^2$) or on two molecules of a monochloride; what is combined with two atoms of chlorine, whether in one or in two molecules, will combine with Y; and H^2 combining with Cl^2 will always form two molecules of hydrochloric acid.

The examples of double decomposition which follow clearly show what I have just indicated.

Double Decompositions of Hydrated Sulphuric Acid, H^2SO^4,	
With the Monochlorides $R^{\text{ii}}_{\text{m}}Cl$.	With the Bichlorides $R^{\text{ii}}_{\text{m}}Cl^2$.
$\begin{matrix}NaCl\\NaCl\end{matrix} + H^2SO^4 = \begin{matrix}HCl\\HCl\end{matrix} + Na^2SO^4$	$Hg\begin{cases}Cl\\Cl\end{cases} + H^2SO^4 = \begin{matrix}HCl\\HCl\end{matrix} + HgSO^4$
$\begin{matrix}NaCl\\NaCl\end{matrix} + Ag^2SO^4 = \begin{matrix}AgCl\\AgCl\end{matrix} + Na^2SO^4$	$Hg\begin{cases}Cl\\Cl\end{cases} + Ag^2SO^4 = \begin{matrix}AgCl\\AgCl\end{matrix} + HgSO^4$
$\begin{matrix}C^2H^5Cl\\C^2H^5Cl\end{matrix} + Ag^2SO^4 = \begin{matrix}AgCl\\AgCl\end{matrix} + (C^2H^5)^2SO^4$	$C^2H^4\begin{cases}Cl\\Cl\end{cases} + Ag^2SO^4 = \begin{matrix}AgCl\\AgCl\end{matrix} + C^2H^4SO^4$

In connection with this point I compare the formulæ of the oxy-salts proposed by me with those of Berzelius and of Gerhardt, and discuss the causes of the differences and of the coincidences, which may be summed up as follows :—

1°. All the formulæ given by Berzelius to the oxy-salts of the biatomic metallic radicals are the same as those proposed by me, whether the acid is monobasic or bibasic ; all these oxy-salts contain in each molecule the elements of a complete molecule of oxide and of a complete molecule of anhydrous acid.

2°. There correspond also to the formulæ proposed by me all those of Berzelius for sulphates and analogous salts, if we introduce the modification by Regnault, *i.e.*, if we consider the quantity of metal contained in the molecules of potassic, argentic, mercurous, and cuprous sulphates equal to 2 atoms, and those on the other hand of metal contained in the molecules of mercuric, cupric, plumbic, zincic, calcic, baric, etc., sulphates, equal to a single atom.

3°. The formulæ proposed by me for the oxy-salts of potassium, sodium, silver, hydrogen, methyl, and all the other analogous monatomic radicals with a monobasic acid, are equal to half the formulæ proposed by Berzelius and modified by Regnault, *i.e.*, each molecule of them contains the components of half a molecule of anhydrous acid and half a molecule of metallic oxide.

4°. The formulæ of Gerhardt coincide with those proposed by me only for the salts of potassium, sodium, silver, hydrogen, methyl, and all the other monatomic radicals, but not for those of zinc, lead, calcium, barium, and the other metallic protoxides ; Gerhardt having wished to consider all the metals analogous to hydrogen, which I have shown to be erroneous.

In the succeeding lectures I speak of the oxides with monatomic and biatomic radicals, afterwards I treat of the other classes of polyatomic radicals, examining comparatively the chlorides and the oxides ; lastly, I discuss the constitution of acids and of salts, returning with new proofs to demonstrate what I have just indicated.

But of all this I will give you an abstract in another letter.

GENOA, *12th March* 1858.

THOMSON, SIR WILLIAM (BARON KELVIN OF LARGS). (b. Belfast, Ireland, 26 June 1824; d. Netherhall, near Largs, Ayrshire, Scotland, 17 December 1907)

Thomson's initial university training began at Glasgow when he was ten years old. At this time the Glasgow chemistry chair was held by Thomas Thomson, who first brought widespread attention to John Dalton's ideas about atoms. William Thomson studied at Cambridge from 1841 to 1845, and, from 1846 until 1879, he was professor of natural history at the University of Glasgow, where he established the first teaching laboratory in

Great Britain. Thomson's *Treatise on Natural Philosophy* (1867), coauthored with Peter Guthrie Tait, was among the most influential and best-known nineteenth-century physics texts. In addition, Thomson received public attention for his debates with Charles Lyell on the age of the earth, where Thomson calculated the earth's age on the basis of physical laws applied to the cooling of the sun. He was knighted in 1866 for his role in the successful construction of a transatlantic cable: Thomson designed a mirror galvanometer that would respond to the tiny voltages necessary for the successful rapid transmission of signals across the transatlantic cable length.

As a young man Thomson spent a brief period in Regnault's laboratory in Paris. Influenced both by the French treatment of the motive power of heat in the Carnot-Clapeyron tradition, and by James Prescott Joule's ideas on the dynamical origin of heat, Thomson presented a paper to the Royal Society of Edinburgh in 1851 on the dynamical theory of heat, reconciling Carnot's work with the conclusions of the British scientists Rumford, Davy, and Joule. This work gave Thomson a role as a founder of the second law of thermodynamics, denying the possibility of a perpetuum mobile via the passing of heat from a cooler to a warmer body.

Thomson's 1867 paper "On Vortex Atoms" is similarly a synthetic one, originating this time from his long-standing interest in analogies among heat, electricity, and magnetism. This interest can be seen in a series of papers in the 1840s which attempt to clarify the differences between Faraday's "curved lines" of electromotive force and Coulomb's and Poisson's mathematical theory of Newtonian, action-at-a-distance forces. Thomson's analogies between the laws of heat transfer and electrical action, including his papers on the mechanical properties of space, influenced James Clerk Maxwell in the development of his electromagnetic theory. Thomson's 1867 paper on vortex atoms is a powerful attempt to unify the particulate atoms of the dynamical theory of heat with the perpetually circulating vortices of Faraday's curved space.

II. *On Vortex Atoms.*
By Professor Sir William Thomson, *F.R.S.**

AFTER noticing Helmholtz's admirable discovery of the law of vortex motion in a perfect liquid—that is, in a fluid perfectly destitute of viscosity (or fluid friction)—the author said that this discovery inevitably suggests the idea that Helmholtz's rings are the only true atoms. For the only pretext seeming to justify the monstrous assumption of infinitely strong and infinitely rigid pieces of matter, the existence of which is asserted as a probable hypothesis by some of the greatest modern chemists in their rashly-worded introductory statements, is that urged by Lucretius and adopted by Newton—that it seems necessary to account for the unalterable distinguishing qualities of different kinds of matter. But Helmholtz has proved an absolutely unalterable quality in the motion of any portion of a perfect liquid in which the peculiar motion which he calls "Wirbelbewegung" has been once created. Thus any portion of a perfect liquid which has "Wirbelbewegung" has one recommendation of Lucretius's atoms—infinitely perennial specific quality. To generate or to destroy "Wirbelbewegung" in a perfect fluid can only be an act of creative power. Lucretius's atom does not explain

* Communicated by the Author, having been read before the Royal Society of Edinburgh.

any of the properties of matter without attributing them to the atom itself. Thus the "clash of atoms," as it has been well called, has been invoked by his modern followers to account for the elasticity of gases. Every other property of matter has similarly required an assumption of specific forces pertaining to the atom. It is as easy (and as improbable—not more so) to assume whatever specific forces may be required in any portion of matter which possesses the "Wirbelbewegung," as in a solid indivisible piece of matter; and hence the Lucretius atom has no *prima facie* advantage over the Helmholtz atom. A magnificent display of smoke-rings, which he recently had the pleasure of witnessing in Professor Tait's lecture-room, diminished by one the number of assumptions required to explain the properties of matter on the hypothesis that all bodies are composed of vortex atoms in a perfect homogeneous liquid. Two smoke-rings were frequently seen to bound obliquely from one another, shaking violently from the effects of the shock. The result was very similar to that observable in two large india-rubber rings striking one another in the air. The elasticity of each smoke-ring seemed no further from perfection than might be expected in a solid india-rubber ring of the same shape, from what we know of the viscosity of india-rubber. Of course this kinetic elasticity of form is perfect elasticity for vortex rings in a perfect liquid. It is at least as good a beginning as the "clash of atoms" to account for the elasticity of gases. Probably the beautiful investigations of D. Bernoulli, Herapath, Joule, Krönig, Clausius, and Maxwell on the various thermodynamic properties of gases, may have all the positive assumptions they have been obliged to make, as to mutual forces between two atoms and kinetic energy acquired by individual atoms or molecules, satisfied by vortex rings, without requiring any other property in the matter whose motion composes them than inertia and incompressible occupation of space. A full mathematical investigation of the mutual action between two vortex rings of any given magnitudes and velocities passing one another in any two lines, so directed that they never come nearer one another than a large multiple of the diameter of either, is a perfectly solvable mathematical problem; and the novelty of the circumstances contemplated presents difficulties of an exciting character. Its solution will become the foundation of the proposed new kinetic theory of gases. The possibility of founding a theory of elastic solids and liquids on the dynamics of more closely-packed vortex atoms may be reasonably anticipated. It may be remarked in connexion with this anticipation, that the mere title of Rankine's paper on "Molecular Vortices," communicated to the Royal Society of Edinburgh in 1849 and 1850, was a most suggestive step in physical theory.

Diagrams and wire models were shown to the Society to illustrate knotted or knitted vortex atoms, the endless variety of which is infinitely more than sufficient to explain the varieties and allotropies of known simple bodies and their mutual affinities. It is to be remarked that two ring atoms linked together, or one knotted in any manner with its ends meeting, constitute a system which, however it may be altered in shape, can never deviate from its own peculiarity of multiple continuity, it being impossible for the matter in any line of vortex motion to go through the line of any other matter in such motion or any other part of its own line. In fact, a closed line of vortex core is literally indivisible by any action resulting from vortex motion.

The author called attention to a very important property of the vortex atom, with reference to the now celebrated spectrum-analysis practically established by the discoveries and labours of Kirchhoff and Bunsen. The dynamical theory of this subject, which Professor Stokes had taught to the author of the present paper before September 1852, and which he has taught in his lectures in the University of Glasgow from that time forward, required that the ultimate constitution of simple bodies should have one or more fundamental periods of vibration, as has a stringed instrument of one or more strings, or an elastic solid consisting of one or more tuning-forks rigidly connected. To assume such a property in the Lucretius atom, is at once to give it that very flexibility and elasticity for the explanation of which, as exhibited in aggregate bodies, the atomic constitution was originally assumed. If, then, the hypothesis of atoms and vacuum imagined by Lucretius and his followers to be necessary to account for the flexibility and compressibility of tangible solids and fluids were really necessary, it would be necessary that the molecule of sodium, for instance, should be not an atom, but a group of atoms with void space between them. Such a molecule could not be strong and durable, and thus it loses the one recommendation which has given it the degree of acceptance it has had among philosophers; but, as the experiments shown to the Society illustrate, the vortex atom has perfectly definite fundamental modes of vibration, depending solely on that motion the existence of which constitutes it. The discovery of these fundamental modes forms an intensely interesting problem of pure mathematics. Even for a simple Helmholtz ring, the analytical difficulties which it presents are of a very formidable character, but certainly far from insuperable in the present state of mathematical science. The author of the present communication had not attempted, hitherto, to work it out except for an infinitely long, straight, cylindrical vortex. For this case he was working out solutions corresponding to every possible description of infinitesimal vibration, and intended to include them in a mathematical paper which he hoped soon to be able to communicate to the Royal Society. One very simple result which he could now state is the following. Let such a vortex be given with its section differing from exact circular figure by an infinitesimal harmonic deviation of order i. This *form* will travel as waves round the axis of the cylinder in the same direction as the vortex rotation, with an angular velocity equal to $\frac{i-1}{i}$ of the angular velocity of this rotation. Hence, as the number of crests in a whole circumference is equal to i, for an harmonic deviation of order i there are $i-1$ periods of vibration in the period of revolution of the vortex. For the case $i=1$ there is no vibration, and the solution expresses merely an infinitesimally displaced vortex with its circular form unchanged. The case $i=2$ corresponds to elliptic deformation of the circular section; and for it the period of vibration is, therefore, simply the period of revolution. These results are, of course, applicable to the Helmholtz ring when the diameter of the approximately circular section is small in comparison with the diameter of the ring, as it is in the smoke-rings exhibited to the Society. The lowest fundamental modes of the two kinds of transverse vibrations of a ring, such as the vibrations that were seen in the experiments, must be much graver than the elliptic vibration of section. It is probable that the vibrations which constitute the incandescence of sodium-vapour are analogous to those which the smoke-rings had exhibited; and it is therefore probable that the period of each vortex rotation of the atoms of sodium-vapour is much less than $\frac{1}{525}$ of the millionth of the millionth of a second, this being approximately the period of vibration of the yellow sodium light. Further, inasmuch as this light consists of two sets of vibrations coexistent in slightly different periods, equal approximately to the time just stated, and of as nearly as can be perceived equal intensities, the sodium atom must have two fundamental modes of vibration, having those for their respective periods, and being about equally excitable by such forces as the atom experiences in the incandescent vapour. This last condition renders it probable that the two fundamental modes concerned are approximately similar (and not merely different orders of different series chancing to concur very nearly in their periods of vibration). In an approximately circular and uniform disk of elastic solid the fundamental modes of transverse vibration, with nodal division into quadrants, fulfils both the conditions. In an approximately circular and uniform ring of elastic solid these conditions are fulfilled for the flexural vibrations in its plane, and also in its transverse vibrations perpendicular to its own plane. But the circular vortex ring, if created with one part somewhat thicker than another, would not remain so, but would experience longitudinal vibrations round its own circumference, and could not possibly have two fundamental modes of vibration similar in character and approximately equal in period. The same assertion may, it is probable*, be practically extended to any atom consisting of a single vortex ring, however involved, as illustrated by those of the models shown to the Society, which consisted of only a single wire knotted in various ways. It seems, therefore, probable that the sodium atom may not consist of a single vortex line; but it may very probably consist of two approximately equal vortex rings passing through one another like two links of a chain. It is, however, quite certain that a vapour consisting of such atoms, with proper volumes and angular velocities in the two rings of each atom, would act precisely as incandescent sodium-vapour acts—that is to say, would fulfil the "spectrum test" for sodium.

The possible effect of change of temperature on the fundamental modes cannot be pronounced upon without mathematical investigation not hitherto executed; and therefore we cannot say that the dynamical explanation now suggested is mathematically demonstrated so far as to include the very approximate identity of the periods of the vibrating particles of the incandescent vapour with those of their corresponding fundamental modes at the lower temperature at which the vapour exhibits its remarkable absorbing-power for the sodium light.

A very remarkable discovery made by Helmholtz regarding the simple vortex ring is that it always moves, relatively to the distant parts of the fluid, in a direction perpendicular to its plane, towards the side towards which the rotatory motion carries the inner parts of the ring. The determination of the velocity of this motion, even approximately, for rings of which the sectional radius is small in comparison with the radius of the circular axis, has presented mathematical difficulties which have not yet been overcome†. In the smoke-rings which have been actually observed, it seems to be always something smaller than the velocity of the fluid along the straight axis through the centre of the ring; for the observer standing beside the line of motion of the ring sees,

* [*Note*, April 26, 1867.—The author has seen reason for believing that the sodium characteristic might be realized by a certain configuration of a single line of vortex core, to be described in the mathematical paper which he intends to communicate to the Society.]

† See, however, note added to Professor Tait's translation of Helmholtz's paper (Phil. Mag. 1867, vol. xxxiii. Suppl.), where the result of a mathematical investigation which the author of the present communication has recently succeeded in executing, is given.

as its plane passes through the position of his eye, a convex* outline of an atmosphere of smoke in front of the ring. This

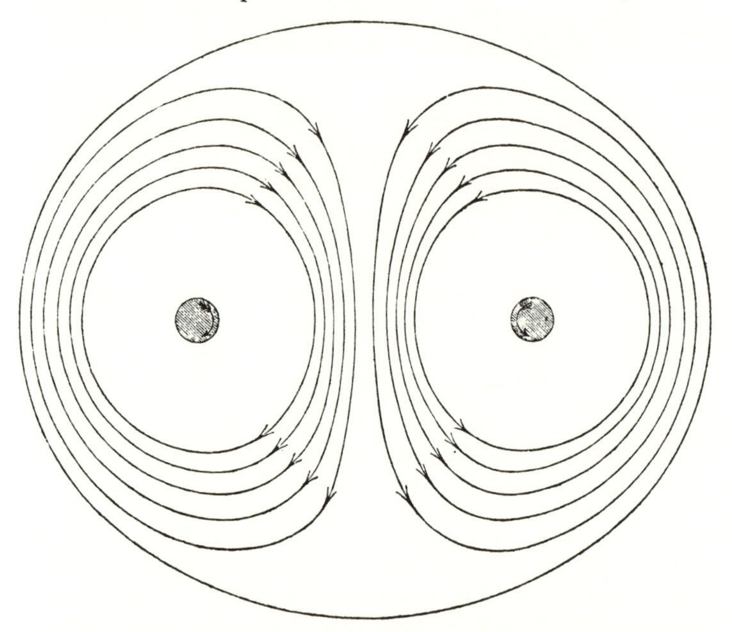

convex outline indicates the bounding surface between the quantity of smoke which is carried forward with the ring in its motion, and the surrounding air which yields to let it pass. It is not so easy to distinguish the corresponding convex outline behind the ring, because a confused trail of smoke is generally left in the rear. In a perfect fluid the bounding surface of the portion carried forward would necessarily be quite symmetrical on the anterior and posterior sides of the middle plane of the ring. The motion of the surrounding fluid must be precisely the same as it would be if the space within this surface were occupied by a smooth solid; but in reality the air within it is in a state of

* The diagram represents precisely the convex outline referred to, and the lines of motion of the interior fluid carried along by the vortex, for the case of a double vortex consisting of two infinitely long, parallel, straight vortices of equal rotations in opposite directions. The curves have been drawn by Mr. D. M'Farlane, from calculations which he has performed by means of the equation of the system of curves, which is

$$\frac{y^2}{a} = \frac{2x}{a} \cdot \frac{N+1}{N-1} - \left(1 + \frac{x^2}{a^2}\right), \text{ where } \log_e N = \frac{x+b}{a}.$$

The proof will be given in the mathematical paper which the author intends to communicate in a short time to the Royal Society of Edinburgh.

rapid motion, circulating round the circular axis of the ring with increasing velocity on the circuits nearer and nearer to the ring itself. The circumstances of the actual motion may be imagined thus:—Let a solid column of india-rubber, of circular section, with a diameter small in proportion to its length, be bent into a circle, and its two ends properly spliced together so that it may keep the circular shape when left to itself; let the aperture of the ring be closed by an infinitely thin film; let an impulsive pressure be applied all over this film, of intensity so distributed as to produce the definite motion of the fluid, specified as follows, and instantly thereafter let the film be all liquefied. This motion is, in accordance with one of Helmholtz's laws, to be along those curves which would be the lines of force, if, in place of the india-rubber circle, were substituted a ring electromagnet*; and the velocities at different points are to be in proportion to the intensities of the magnetic forces in the corresponding points of the magnetic field. The motion, as has long been known, will fulfil this definition, and will continue fulfilling it if the initiating velocities at every point of the film perpendicular to its own plane be in proportion to the intensities of the magnetic force in the corresponding points of the magnetic field. Let now the ring be moved perpendicular to its own plane in the direction *with* the motion of the fluid through the middle of the ring, with a velocity very small in comparison with that of the fluid at the centre of the ring. A large approximately globular portion of the fluid will be carried forward with the ring. Let the velocity of the ring be increased; the volume of fluid carried forward will be diminished in every diameter, but most in the axial or fore-and-aft diameter, and its shape will thus become sensibly oblate. By increasing the velocity of the ring forward more and more, this oblateness will increase, until, instead of being wholly convex, it will be concave before and behind, round the two ends of the axis. If the forward velocity of the ring be increased until it is just equal to the velocity of the fluid through the centre of the ring, the axial section of the outline of the portion of fluid carried forward will become a lemniscate. If the ring be carried still faster forward, the portion of it carried with the india-rubber ring will be itself annular; and, relatively to the ring, the motion of the fluid will be backwards through the centre. In all cases the figure of the portion of fluid carried forward and the lines of motion will be symmetrical, both relatively to the axis and relatively to the two sides of the equatorial plane. Any one of the states of motion thus described might of course be produced either in the order described, or by first giving a velocity to the

* That is to say, a circular conductor with a current of electricity maintained circulating through it.

ring and then setting the fluid in motion by aid of an instantaneous film, or by applying the two initiative actions simultaneously. The whole amount of the impulse required, or, as we may call it, the effective momentum of the motion, or simply the momentum of the motion, is the sum of the integral values of the impulses on the ring and on the film required to produce one or other of the two components of the whole motion. Now it is obvious that as the diameter of the ring is very small in comparison with the diameter of the circular axis, the impulse on the ring must be very small in comparison with the impulse on the film, unless the velocity given to the ring is much greater than that given to the central parts of the film. Hence, unless the velocity given to the ring is so very great as to reduce the volume of the fluid carried forward with it to something not incomparably greater than the volume of the solid ring itself, the momenta of the several configurations of motions we have been considering will exceed by but insensible quantities the momentum when the ring is fixed. The value of this momentum is easily found by a proper application of Green's formulæ. Thus the actual momentum of the portion of fluid carried forward (being the same as that of a solid of the same density moving with the same velocity), together with an equivalent for the inertia of the fluid yielding to let it pass, is approximately the same in all these cases, and is equal to a Green's integral expressing the whole initial impulse on the film. The equality of the effective momentum for different velocities of the ring is easily verified without analysis for velocities not so great as to cause sensible deviations from spherical figure in the portion of fluid carried forward. Thus in every case the length of the axis of the portion of the fluid carried forward is determined by finding the point in the axis of the ring at which the velocity is equal to the velocity of the ring. At great distances from the plane of the ring that velocity varies, as does the magnetic force of an infinitesimal magnet on a point in its axis, inversely as the cube of the distance from the centre. Hence the cube of the radius of the approximately globular portion carried forward is in simple inverse proportion to the velocity of the ring, and therefore its momentum is constant for different velocities of the ring. To this must be added, as was proved by Poisson, a quantity equal to half its own amount, as an equivalent for the inertia of the external fluid; and the sum is the whole effective momentum of the motion. Hence we see not only that the whole effective momentum is independent of the velocity of the ring, but that its amount is the same as the magnetic moment in the corresponding ring electromagnet. The same result is of course obtained by the Green's integral referred to above.

The synthetical method just explained is not confined to the case of a single circular ring specially referred to, but is equally applicable to a number of rings of any form, detached from one another, or linked through one another in any way, or to a single line knotted to any degree and quality of "multiple continuity," and joined continuously so as to have no end. In every possible such case the motion of the fluid at every point, whether of the vortex core or of the fluid filling all space round it, is perfectly determined by Helmholtz's formulæ when the shape of the core is given. And the synthetic investigation now explained proves that the effective momentum of the whole fluid motion agrees in magnitude and direction with the magnetic moment of the corresponding electromagnet. Hence, still considering for simplicity only an infinitely thin line of core, let this line be projected on each of three planes at right angles to one another. The areas of the plane circuit thus obtained (to be reckoned according to De Morgan's rule when autotomic, as they will generally be) are the components of momentum perpendicular to these three planes. The verification of this result will be a good exercise on "multiple continuity." The author is not yet sufficiently acquainted with Riemann's remarkable researches on this branch of analytical geometry to know whether or not all the kinds of "multiple continuity" now suggested are included in his classification and nomenclature.

That part of the synthetical investigation in which a thin solid wire ring is supposed to be moving in any direction through a fluid with the free vortex motion previously excited in it, requires the diameter of the wire at every point to be infinitely small in comparison with the radius of curvature of its axis and with the distance of the nearest of any other part of the circuit from that point of the wire. But when the effective moment of the whole fluid motion has been found for a vortex with infinitely thin core, we may suppose any number of such vortices, however near one another, to be excited simultaneously; and the whole effective momentum in magnitude and direction will be the resultant of the momenta of the different component vortices each estimated separately. Hence we have the remarkable proposition that the effective momentum of any possible motion in an infinite incompressible fluid agrees in direction and magnitude with the magnetic moment of the corresponding electromagnet in Helmholtz's theory. The author hopes to give the mathematical formulæ expressing and proving this statement in the more detailed paper, which he expects soon to be able to lay before the Royal Society.

The question early occurs to any one either observing the phenomena of smoke-rings or investigating the theory,—What conditions determine the size of the ring in any case? Helmholtz's investigation proves that the angular vortex velocity of the core varies directly as its length, or inversely as its sectional area. Hence the strength of the electric current in the electro-magnet, corresponding to an infinitely thin vortex core, remains constant, however much its length may be altered in the course of the transformations which it experiences by the motion of the fluid. Hence it is obvious that the larger the diameter of the ring for the same volume and strength of vortex motions in an ordinary Helmholtz ring, the greater is the whole kinetic energy of the fluid, and the greater is the momentum; and we therefore see that the dimensions of a Helmholtz ring are determinate when the volume and strength of the vortex motion are given, and, besides, either the kinetic energy or the momentum of the whole fluid motion due to it. Hence if, after any number of collisions or influences, a Helmholtz ring escapes to a great distance from others and is then free, or nearly free, from vibrations, its diameter will have been increased or diminished according as it has taken energy from, or given energy to, the others. A full theory of the swelling of vortex atoms by elevation of temperature is to be worked out from this principle.

Professor Tait's plan of exhibiting smoke-rings is as follows:— A large rectangular box, open at one side, has a circular hole of 6 or 8 inches diameter cut in the opposite side. A common rough packing-box of 2 feet cube, or thereabout, will answer the purpose very well. The open side of the box is closed by a stout towel or piece of cloth, or by a sheet of india-rubber stretched across it. A blow on this flexible side causes a circular vortex ring to shoot out from the hole on the other side. The vortex rings thus generated are visible if the box is filled with smoke. One of the most convenient ways of doing this is to use two retorts with their necks thrust into holes made for the purpose in one of the sides of the box. A small quantity of muriatic acid is put into one of these retorts, and of strong liquid ammonia into the other. By a spirit-lamp applied from time to time to one or other of these retorts, a thick cloud of sal-ammoniac is readily maintained in the inside of the box. A curious and interesting experiment may be made with two boxes thus arranged, and placed either side by side close to one another or facing one another so as to project smoke-rings meeting from opposite directions—or in various relative positions, so as to give smoke-rings proceeding in paths inclined to one another at any angle, and passing one another at various distances. An interesting variation of the experiment may be made by using clear air without smoke in one of the boxes. The invisible vortex rings projected from it render their existence startlingly sensible when they come near any of the smoke-rings proceeding from the other box.

WILLIAMSON, ALEXANDER WILLIAM. (b. Wandworth, London, England, 1 May 1824; d. Hindhead, Surrey, England, 6 May 1904)

Williamson studied chemistry with Leopold Gmelin after undertaking a medical education at the University of Heidelberg in 1840. He worked with Liebig at Giessen from 1844–1846 and established a private laboratory in Paris from 1846–1849. There he took private lessons in mathematics from Auguste Comte and he formed acquaintances with Laurent, Gerhardt, Wurtz, and Dumas. Williamson held the chair of practical chemistry at

University College, London from 1849–1887; after 1855, he filled the chair of general chemistry as well. After the mid-1850s Williamson interested himself principally in institutional and administrative matters, including the creation, at University College, of a chair of applied chemistry (chemical engineering).

His work on etherification is a classical exemplar of chemical research which combines careful experimentation with the elucidation of important theoretical problems. In 1850 he announced that ether is not formed by the loss of water from alcohol, but by the substitution of an alkyl radical into alcohol (the "Williamson synthesis"). His familiarity with the ideas of Laurent and Gerhardt led Williamson to suggest, in 1851, that water can be viewed as a general type of compound from which other inorganic and organic bodies are formed: one or more hydrogen atoms may be replaced by the equivalent of simple or compound radicals. His work on ether supported the two-volume formula for water (H_2O, against HO or H_4O_2). Further, his explanation of continuous etherification by the action of sulfuric acid on alcohol supported the hypothesis that atoms and molecules are kinetic rather than static. He preferred the view that sulfuric acid acts as a catalyst by forming an active intermediary compound, rather than by exerting a catalytic, contact force.

Williamson was an active member of the Royal Society, the British Association for the Advancement of Science, and the London Chemical Society. His debates with Benjamin Brodie in meetings of the Chemical Society provided a forum in England for discussion of competing ideas about atoms, molecules, and equivalents and also about the relative advantages of different systems of chemical notation. Epistemological issues regarding the aims of scientific theories were crucial in these debates.

XXXI.—*On the Atomic Theory.*

By Prof. A. W. WILLIAMSON, Pres. Chem. Soc., F.R.S., &c.

THERE are considerable differences, not to say discrepancies, between the statements made by different chemists on the subject of the atomic theory. In some text books of the science the replacing values of so-called equivalent weights of elements are described as being the atomic weights of those elements, while in the same and in other books, statements are made respecting the principles of the determination of atomic weights which lead to different numbers from those representing the replacing values.

It sometimes happens that chemists of high authority refer publicly to the atomic theory as something which they would be glad to dispense with, and which they are rather ashamed of using. They seem to look upon it as something distinct from the general facts of chemistry, and something which the science would gain by throwing off entirely.

Thus, in one book we find the statement:—"It appears from this that 2 × 8, or 16 parts of oxygen is the smallest quantity of oxygen that can be supposed to enter into the reaction just considered, if we would avoid speaking of fractions of equivalents; and we shall find hereafter that the same is true with regard to all other well defined reactions in which oxygen takes part. Hence, this quantity of oxygen, 16 parts by weight (hydrogen being the unit), is called an indivisible weight, or atomic weight, or one atom of oxygen."

And again:—

"The question whether matter is infinitely divisible, or whether its divisibility is limited, remains, at the present day, in the same state as when it first engaged the attention of the Greek philosophers, or perhaps that of the sages of Egypt and Hindostan long before them."

Another author says:—

"The law of multiple proportions being founded on experimental facts, stands as a fixed bulwark of the science, which must ever remain true; whereas the atomic theory by which we now explain this great law may possibly, in time, give place to one more perfectly suited to the explanation of new facts."

When we refer to the author's enunciation of this great law, in a paragraph which is referred to as containing a statement of the law, we find it stated that in the compounds of oxygen and nitrogen the oxygen is in the proportion of 1, 2, 3, 4, 5 to one and the same quantity of nitrogen, and that no compounds exist containing any intermediate quantity of oxygen.

Another distinguished chemist also lays down that the atomic weight of each element is made to agree as far as possible with the three following conditions :—

1st. The smallest proportion by weight in which the element enters into, or is expelled from a chemical compound, the weight of hydrogen so entering or leaving a chemical compound being taken as unity.

2nd. The weight of the element in the solid condition, which, at any given temperature, contains the same amount of heat as seven parts by weight of lithium at the same temperature.

3rd. The weight of the element which, in the form of gas or vapour, occupies, under like conditions of temperature and pressure, the same volume is one part by weight of hydrogen.

Now, if we endeavour to determine according to these rules the atomic weight of phosphorus, we may compare the quantity of hydric chloride containing one part by weight of hydrogen with that weight of phosphoric chloride PCl_5, which contains the same weight of chlorine, viz., $\frac{P}{5}Cl$. In this latter compound, $\frac{P}{5}$ or 6·2 parts of phosphorus, has entered into combination with 35·5 parts by weight of chlorine, instead of one part by weight of hydrogen, which unites with 35·5 of chlorine in forming hydric chloride. 6·2 is, therefore, according to this rule, taken by itself, the atomic weight of phosphorus. We should not be justified in adopting the weight of phosphorus derived from an analysis of the lower chloride, PCl_3, because the rule directs us to take the smallest proportion, and the proportion of phosphorus in PCl_5 is smaller than in PCl_3.

If instead of the chloride we take the acid P_2O_5, or a phosphate such as PO_4Na_3, and compare them respectively with $5H_2O$ and $H_5Na_3O_4$, we obtain the same atomic weight for phosphorus, 6·2. By the second rule (if we interpret it as meaning the capacity for heat between two given temperatures), we find that seven parts of lithium, if heated from 0° C. to 1° C., absorb 6·59 degrees of heat, while one part by weight of solid phosphorus absorbs ·1885. We have, consequently, the atomic weight 34·9, which, considering the nature of the determinations, may be accepted as a reasonable approach to the atomic weight 31, now adopted by chemists.

According to the third rule, we obtain for phosphorus the atomic weight 62, inasmuch as P^6 occupy in the state of vapour under like conditions the same volume as H^2.

In the case of all elements, the first rule gives the smallest weight. Thus, calcium, in the calcic nitrate, $Ca(NO_3)_2$, replaces hydrogen in hydric nitrate in the proportion of 20 to 1. Oxygen in carbonic oxide replaces hydrogen by marsh-gas in the proportion of 4 to 1; in most of its compounds in the proportion of 8 to 1. Nitrogen in sal ammoniac replaces 5 atoms of hydrogen in the molecules 4HH and HCl, in the proportion of $2\frac{4}{5}$ to 1.

Another distinguished author describes—

"The law of definite proportions;

"The law of multiple proportions; and

"The law of equivalent proportions."

He subsequently describes the "hypothesis of the atomic constitution of matter," the word hypothesis being no doubt intended to indicate an opinion on his part that the atomic constitution of matter is open to more doubt than the so-called law of multiple proportions, &c. Again, in another able book we find it stated that to each element is assigned a "particular number, termed its proportionate number, which expresses the least indivisible proportion of the element that is found to enter into a combination," hydrogen being taken as the unit,—a very intelligible description of the atomic weight of the element. In illustration of these so-called proportional numbers, the author gives Gerhardt's table of atomic weights, which were in use at the time his work was written. He avoids the word atom in describing his indivisible proportions, although later on he occasionally falls into the use of the common words "atomic weight," and atom. When he comes to explain molecules and equivalents, there is no more talk about proportional numbers.

It certainly does seem strange that men accustomed to consult nature by experiment so constantly as chemists do, should make use of a system of ideas of which such things can be said. I think I am not overstating the fact, when I say, that, on the one hand, all chemists use the atomic theory, and that, on the other hand, a considerable number of them view it with mistrust, some with positive dislike. If the theory really is as uncertain and unnecessary as they imagine it to be, let its defects be laid bare and examined. Let them be remedied if possible, or let the theory be rejected, and some other theory used in its stead, if its defects are really as irremediable and as grave as is implied by the sneers of its detractors.

But if the theory be a general expression of the best ascertained relations of matter in its chemical changes, the only general expression which those relations have as yet found, and be hypothetical only in so far as it presupposes among unknown substances relations analogous to those discovered among those which are known, then it must be classed among the best and most precious trophies which the human mind has earned, and its development must be fostered as one of the highest aims and objects of our science.

It cannot be desirable to leave the question on its present footing; and if we have any opportunity of bringing the existing doubts and misgivings to a distinct issue, it cannot be right to delay such a consummation.

Such are the feelings which induced me to accept the invitation with which the Council has honoured me, and to bring before you an analysis of the evidence of the present atomic theory, as I conceive it to exist dispersed in the minds and among the hands of chemists at the present time.

I propose to consider the subject under three successive headings.

1st. The proportions by weight in which elements replace one another, or their so-called equivalent weights, and the multiples of those proportions.

2nd. The units of chemical action, or the so-called molecules.

3rd. Atomic values.

Equivalent Weights.

Quantitative analyses have shown that the weight of oxygen which combines with one part by weight of hydrogen to form water, is eight times as great as that of the hydrogen ; that the weight of chlorine which combines with one part by weight of hydrogen to form hydric chloride is 35·5 times as great as that of the hydrogen. Whenever we displace oxygen from water by chlorine, forming hydric chloride, we find that for every eight parts by weight of oxygen so displaced, thirty-five and a half parts by weight of chlorine are taken up; and whenever we perform the opposite substitution, viz., displace chlorine from hydric chloride by oxygen so as to form water, we find the inverse ratio, viz., eight parts by weight of oxygen, replacing 35·5 parts by weight of chlorine. These numbers, 8 and 35·5, represent the relative weights in which oxygen and chlorine combine with the unit weight of hydrogen, and they are called the equivalent weights. The decomposition of ammonia by heat yields nitrogen gas and hydrogen in the proportion of $4\frac{2}{3}$ parts by weight of nitrogen to one of hydrogen, and in like manner the decomposition of marsh-gas yields three parts by weight of carbon to every one of hydrogen. Moreover, when we displace nitrogen from ammonia by chlorine, forming hydric chloride, we find that chlorine takes the place of the nitrogen in the proportion of 35·5 parts to $4\frac{2}{3}$; and in like manner when chlorine displaces carbon from marsh-gas forming hydric chloride, it does so in the proportion of 35·5 parts to 3 of carbon. So also when oxygen replaces carbon in marsh-gas forming water, we find that eight parts of oxygen take the place of three of carbon.

We have accordingly, from the comparison of these hydrogen compounds,—

35·5	as the equivalent	weight of	chlorine,
8	,,	,,	oxygen,
$4\frac{2}{3}$	,,	,,	nitrogen,
3	,,	,,	carbon.

But chlorine can be made to replace the hydrogen in marsh-gas by successive steps, and we find that it does so in the proportion of 35·5 parts by weight to one part of hydrogen, and in the product so formed hydrogen can be made to replace chlorine, reproducing marsh-gas always in the proportion of one part by weight to 35·5 of chlorine.

Upon this and similar evidence we attribute to hydrogen, in relation to chlorine, the equivalent weight 35·5, and by so doing we extend our meaning of the word equivalent. 35·5 parts of chlorine are equivalent to 8 of oxygen, because they combine with the unit weight of hydrogen; and 35·5 of chlorine are equivalent to one of hydrogen, because the elements replace one another in that proportion by weight in their compounds. In like manner oxygen can be made to replace hydrogen in ammonia and in marsh-gas, forming nitrous acid and carbonic acid respectively. It there reacts in the proportion of the same equivalent weight which it has in relation to chlorine. Nitrogen and carbon can, partly by such-like direct reactions, and partly by indirect reactions, be proved to replace hydrogen in the proportion of the same weights in which they combine with it. So that the equivalent weights of chlorine, oxygen, nitrogen, and carbon are the weights of those elements capable of combining (directly or indirectly) with the unit weight of hydrogen, or of replacing the unit weight of hydrogen.

Several other metals have been found to be analogous to hydrogen, viz., lithium, sodium, potassium, silver, &c., and to replace it in combination with chlorine or oxygen in the proportions corresponding to the following equivalent weights:—Lithium 7, sodium 23, potassium 39, silver 108, &c.; and in like manner bromine and iodine have been found analogous to chlorine, and to replace it in equivalent weights corresponding to 80 of bromine and 127 of iodine.

In the compounds ClH, ClK, BrH, BrK, $ClAg$, ICl, $BrLi$, OH_2, OHK, ONa_2, $OClH$, NH_3, NH_2K, NK_3, CH_4, CH_3Cl, CH_2Cl_2, $CHCl_3$, CO_2, the elements are contained in proportions corresponding to the equivalent weights above mentioned. But there are other compounds in which they occur in different proportions; thus iodine combines with chlorine in the proportion of 127 to 35·5, and also in the proportion of $42\frac{1}{3}$ to 35·5; nitrogen also combines with hydrogen and chlorine in the common compound, sal-ammoniac, and in the other compounds like it, in the proportion of $2\frac{4}{5}$ to every one of hydrogen or the equivalent weight of another element. Carbon combines with oxygen in a second proportion in carbonic oxide, viz., that of 3 to 4, and with hydrogen in many other proportions besides that which occurs in marsh-gas: viz., in methyl, four parts of carbon to one of hydrogen, in olefiant gas, six parts of carbon to one of hydrogen, and in acetylene, twelve parts of carbon to one of hydrogen, besides many others intermediate between these.

Now, Dalton knew the two oxygen-compounds of carbon, besides marsh-gas and olefiant gas. He represented the oxide as containing one equivalent of oxygen weighing 8, with one of carbon weighing 6; the acid as containing twice 8 parts by weight of oxygen, combined with 6 of carbon.

In like manner, he considered olefiant gas as a compound of one equivalent of carbon, weighing 6, with the unit of hydrogen, marsh-gas as 6 of carbon with two units of hydrogen.

Some other compounds he found, partly by his own analyses, partly by analyses made by other chemists, to contain their elements in proportions approaching to simple multiples of certain weights peculiar to those elements respectively. He explained this coincidence by representing compounds as built up of atoms of their elements in various proportions to one another.

Many other elements are now known to combine in various proportions, which may be so represented, and it is common to refer to the fact which Dalton explained by atoms, as the law of multiple proportions.

It is desirable that we analyse the evidence which is now before chemists of these multiple proportions, and further their connection with the atomic theory.

The analyses known to Dalton have by the progress of chemistry been, for the most part, replaced by others more accurate and more trustworthy, and some of the numbers representing the combining proportions of elements, have undergone considerable modifications in consequence of our improved methods of observation; but both Dalton and we agree in one respect, viz., that taking the best numbers we can get by analyses, we calculate the composition of every compound, on the assumption that it must be built up of integral multiples of certain numbers belonging to the elements respectively; and it must be admitted as a *primâ facie* confirmation of the justice of this assumption, that the more perfect our methods of observation become, and the more carefully they are applied, the more nearly do our experimental results agree with multiple proportions of such numbers.

In the case of compounds of very simple constitution, the agreement between our experimental results and the theory of multiple proportions is so close, that it would be unreasonable to attribute the coincidences to chance. But, on the other hand, the progress of research leads frequently to the discovery of compounds of which the constituents are present in less simple proportions, and others again in which the proportion can be represented only by high numbers. And in many cases the difference between the numbers corresponding to the formula which is admitted, and the numbers corresponding to some other formula, is not greater than the unavoidable errors of experiment. So that analysis does not decide which of the formulæ is the true one, nor does it furnish an argument in favour of any formula. Thus a high term of the marsh-gas series, such as $C_{27}H_{56}$, cannot be analysed with certainty. The numbers found by combustion were—

C 85·5
H 14·9.

The formula	$C_{27}H_{56}$	the formula	$C_{26}H_{54}$	and	$C_{27}H_{54}$
requires	C 85·27	requires	C 85·25	requires	C 85·71
	H 14·73,		H 14·75,		H 14·29

The actual result of observation agrees with none of these formulæ, yet, in examining it, chemists assume that it ought to agree with some such formula. They select the most probable, and attribute to errors of observation the divergence from it.

The most important evidence brought to bear on the composition of this body was of two kinds: first, an examination of a silver-salt, formed from the product of its cautious oxidation, and containing the following percentages, viz.:—

C 62·36
H 10·31
Ag 20·92,

which approach more nearly to the numbers—

C 62·66
H 10·25
Ag 20·9,

required by the formula $C_{27}H_{53}AgO_2$, than to any other probable formula; and, secondly, a study of the action of bromine upon it proving that it was a marsh-gas and not an olefiant gas, like that last mentioned.

The hydrocarbon itself approached nearer in composition to the formula $C_{27}H_{56}$ than to any other, and the study of two of its reactions thus agrees in indicating that formula for the original hydrocarbon as more probable than any other atomic formula.

It would be quite correct to say that *some* elements are found capable of uniting with one another in the proportion of simple multiples of certain weights, and not in intermediate proportions. Thus carbon unites with oxygen in the proportion of 12 parts of carbon to 16 of oxygen, and also in the proportion of 12 of carbon to 2 × 16 of oxygen. If we take less oxygen than 16 parts to every 12 of carbon, some of the carbon remains as such uncombined, and no compound of the two elements has ever been got with less oxygen than 16 parts to every 12 of carbon. But if we take more and more oxygen in proportion, we find that carbon can unite with it till it has got 2 × 16 parts. Yet we do not admit that it unites in all proportions intermediate between 16 and 2 × 16, for by examining these products, formed by adding more and more oxygen to the compounds of 16 grammes with 12 carbon, we find that they are mixtures of carbonic acid and carbonic oxide; the former being rapidly soluble in aqueous potash, whereas the latter remains undissolved.

But the theory of multiple proportions is not limited in so modest a manner. It is applied to all elements and to all compounds of them, in spite of the fact that the usual results of observation require straining to agree with any multiple formula, and the immense majority of substances have not as yet been reduced by the process to any definite formulæ whatever. The very great majority of mineral and of organic substances which we meet with on the surface of the earth, have not been proved to have a composition agreeing with any formula.

When multiple proportions are spoken of, it is not usually explained whether multiples of equivalent weights are meant, or multiples of the atomic weights of the elements. It is sometimes asserted that we have a law of multiple proportions, which is a direct representation of experimental facts, whereas the atomic theory is a hypothesis, and independent of such law. Those who make this statement must be understood to refer to multiples of equivalent weights, and not to multiples of atomic weights; for if, as they assert, atomic weights are hypothetical, then any relation between the multiples of atomic weights are only multiples of hypothetical numbers, and cannot have greater certainty than the atomic weights themselves.

If we knew the empirical proportions of elements, and did

not believe in atoms, we should describe carbonic oxide as the compound of three parts of carbon with four of oxygen, and carbonic acid as the compound of three parts, by weight, of carbon with eight of oxygen. The hydrocarbons mentioned would be described as follows, viz.: marsh-gas, as a compound of three parts of carbon with one of hydrogen; olefiant gas, a compound of six parts of carbon with one of hydrogen; methyl, a compound of four parts of carbon with one of hydrogen; ethyl, a compound of twenty-four parts of carbon with five of hydrogen; and so on. We should probably use the simplest entire numbers corresponding to the actual proportion between the elements; but we should not use the higher numbers standing to one another in the same proportions which are introduced by the atomic theory.

It is quite true that the elements of some simple, well-known compounds are shown by analysis to be present in proportions corresponding to simple multiples of a particular weight belonging to each of them respectively; but in compounds of less simple constitution, no such relation is obvious. The analysis of most organic bodies does not suggest any simple proportion between the numbers of those weights of their constituents, and it sometimes suggests a relation which is not the true one. Thus, the only simple representation of the results of analyses of the high term of the marsh-gas series above mentioned, would be to describe it as containing six parts, by weight, of carbon to one of hydrogen. The actual process by which we establish the composition of such complex bodies is by *assuming* that the composition of each of them must correspond to entire multiples of the atomic weights of their elements, and by treating as errors of observation any divergence between the proportions discovered by analysis and such atomic proportions.

It is not easy to judge of the evidence of one part of a system by itself. The doctrines of equivalence, of molecular weights, and of atoms mutually support one another, and are habitually used by chemists in connection with one another. Our present task is to examine the doctrine of multiple proportions of equivalents; first, in respect of the evidence upon which it rests; secondly, in respect of the proof which it affords, if any, of the existence of atoms.

The second question may be best investigated in the form of a comparison between the atomic method of multiples, and what I may be permitted to call the tomic method of submultiples; carbonic acid is represented in the atomic plan as containing twice as much oxygen as is contained in carbonic oxide; whereas, on the other plan, it may be represented as differing from carbonic oxide by containing half as much carbon as is contained in the oxide. It is true that by combining oxygen with carbonic oxide we get carbonic acid; but it is equally true that by combining carbon with carbonic acid we obtain carbonic oxide. As far as the proportion of the elements is concerned, we have no better right to suppose that the carbon is indivisible, and that the acid contains twice as much oxygen as the oxide, than we have to suppose that the carbon is divisible, and that half of the carbon is taken out of carbonic oxide in forming carbonic acid.

We cannot be too careful in considering this fact, for many important consequences follow from it. Our proposition relates to an undefined quantity of carbonic oxide. It is known that by combining four parts, by weight, of oxygen with seven parts by weight of carbonic oxide, we get exactly eleven parts of carbonic acid; or, what is the same thing from another point of view, that if we take away four parts of oxygen from eleven of carbonic acid, we get seven of carbonic oxide. It is also known that if we take away 1·5 parts of carbon from 7 of carbonic oxide, we get 5·5 parts of carbonic acid; and, reciprocally, that if we add 1·5 parts of carbon to 5·5 of carbonic acid, we get 7 of carbonic oxide. These parts by weight may be tons, or pounds, or ounces, or grains, or any other actual experimental weight. A philosopher disbelieving in the existence of atoms, would point out that whereas the divisibility of matter is infinite, we find that by the chemical processes available for the removal of carbon from carbonic oxide, the carbon divides into equal portions, one remaining with the oxygen, the other leaving it; and his statement that this half of the carbon is added on again when the process is reversed, is as consistent with the evidence as that of the believer in atoms, who asserts that in forming carbonic acid, oxygen is added in quantity equal to the oxygen in the carbonic oxide, while the formation of carbonic oxide from the acid consists in taking away half the oxygen. What has been here said of the proportions between the elements in carbonic oxide and carbonic acid may be said with equal force of the proportions between the elements contained in other compounds, such as the following:—

$FeCl_2$, $FeCl_3$, CH_4, CH_3, CH_2, C_2H_3, CH, NH_3, NH_4Cl, SH_2, SCl, OH_2, OH, PI_2, PCl_3, PCl_5, ICl, ICl_3, $SiCl_4$, SiF_6K_2, $PtCl_2$, $PtCl_4$, $PtCl_6K_2$, BF_3, BF_4K, AuCl, $AuCl_3$, $AuCl_4Na$, $CrCl_2$, $CrCl_3$, CrF_6 $SnCl_2$, $SnCl_4$, HgCl, $HgCl_2$, CuI, $CuCl_2$.

If we examine consistently with the principles above adopted, the equivalent weight of each element contained in these compounds, we find that we must describe it as the weight which in any compound is combined with the unit of hydrogen, or might be replaced by it. This is easily obtained by dividing the number represented by the atomic symbol by the number of equivalents of the hydrogen or chlorine families with which it is combined.

We thus obtain the values :

$\frac{Fe}{2}$, $\frac{Fe}{3}$, $\frac{C}{4}$, $\frac{C}{3}$, $\frac{C}{2}$, $\frac{2C}{3}$, C, $\frac{N}{3}$, $\frac{N}{5}$, $\frac{S}{2}$, S, $\frac{O}{2}$, O, $\frac{P}{2}$, $\frac{P}{3}$, $\frac{P}{5}$, I, $\frac{I}{3}$, $\frac{Si}{4}$, $\frac{Si}{8}$, $\frac{Pt}{2}$, $\frac{Pt}{4}$, $\frac{Pt}{8}$, $\frac{B}{3}$, $\frac{B}{5}$, Au, $\frac{Au}{3}$, $\frac{Au}{5}$, $\frac{Cr}{2}$, $\frac{Cr}{3}$, $\frac{Cr}{6}$, $\frac{Sn}{2}$, $\frac{Sn}{4}$, Hg, $\frac{Hg}{2}$, Cu, $\frac{Cu}{2}$.

I have taken only a few of the well known compounds, yet among them there are ten elements, each of which has two distinct equivalent weights, four of the elements have got three equivalent weights each, and another (carbon) is shown to have five different equivalent weights. It might easily be shown to have many more.

Now it is interesting to observe that whenever examples are given of the supposed law of multiple proportions, Chemists take not these fractional expressions representing equivalents, but usually the atomic symbols themselves. They describe, in fact, *atoms* as occurring thus combined with one another in the proportions of entire multiples of their weight; in fact, the so-called law of multiple proportions has no existence apart from the atomic theory; those who adopt it seem not to be aware that they are using the notion of atoms, or else they are shy of mentioning it.

The fact that two elements, such as carbon and oxygen, are

capable of combining in more than one proportion is in reality (when we consider it by itself) quite as much an argument against the atomic theory as an argument in favour of it. That the carbon in carbonic oxide can be divided in the proportion of $\frac{1}{2}$, or the oxygen can be multiplied in the proportion of 2 to 1, are equally true statements of the proportion. If oxygen could neither take more nor less carbon than in the proportion of three parts by weight to every four of oxygen, we could say with certainty that that proportional weight of carbon cannot be divided, or in other words that it is an atomic proportion; but the existence of the two compounds, although not inconsistent with the atomic theory, points as much to proportional divisibility of carbon as to proportional multiplicability of oxygen.

When one of those who profess to disbelieve in the atomic theory has ascertained by analysis the percentage composition of a compound, and wants to find its formula, he divides the percentage weight of each element by its atomic weight. He seeks for the smallest integral numbers which represent the proportion of atoms, and he attributes to impurity of his sample or to errors of analysis any deviation from the atomic formula thus obtained. He looks to the reactions of the body for aid in constructing his atomic formula, and controls his analyses by considerations derived from well established reactions, but whenever he is led by any of these considerations to a formula which contains a fraction of any atomic weight, he takes a multiple of the formula sufficiently high to be entirely free from such fractions. In no case does he reason on a basis independent of the atomic theory.

Existence of Molecules.

The atomic theory led to the discovery of molecules. Chemists saw in the phenomena of combination in multiple proportions, processes which were in accordance with their pre-conceived belief in atoms, and they studied combining proportions from the point of view of atoms. They acted on the assumption that each element consists of small indivisible particles of like nature; that the atoms of each element are different in many chemical properties, and usually in weight, from those of every other element; that the unlike atoms combine together in various relative numbers to form clusters, or so-called molecules, every pure compound consisting of such molecules, each one like the rest, but differing from the molecules of every other compound. Thus a given quantity of carbonic oxide was supposed to consist of an aggregate of molecules, each one composed of an atom of carbon united with an atom of oxygen; carbonic acid, to consist of molecules, each one built up of three atoms, viz., one of carbon, and two of oxygen. In calculating the molecular weight of any compound from the ascertained proportion of its elements, they assumed that each molecule must contain at least such a number of atoms of each of its elements as would, when multiplied by the atomic weight of the respective elements, represent the actual proportion by weight in which the elements had combined. Thus 44 parts by weight of carbonic acid have been found to contain 32 of oxygen, and twelve of carbon. Two atoms of oxygen, each weighing 16, with one atom of carbon weighing 12, must be contained in the smallest existing particle of carbonic acid.

Ammonia contains hydrogen and nitrogen in the proportion of 3 parts of the former to 14 of the latter. 3 atoms of hydrogen, each weighing 1, with one atom of nitrogen weighing 14, represent the simplest molecular constitution corresponding to that proportion of the elements. The molecule of olefiant gas, in like manner, must contain at least one atom of carbon with two atoms of hydrogen, while that of marsh gas contains at least one of carbon with four of hydrogen.

An acid hydric salt, the tartrate, was found by analysis to contain carbon, hydrogen, and oxygen, in proportions corresponding to the minimum formula $C_2H_3O_3$. When neutralised by potash, its solution yielded a potash salt corresponding to the minimum formula $KC_2H_2O_3$, and it is known that the weight of hydric sulphate indicated by the formula $\frac{SO_4H_2}{2}$ or $\frac{S}{2}O_2H$ would react on the same weight of potash forming a neutral salt. This relation between the tartrate and sulphate may be expressed by saying that 75 parts by weight of the tartrate are equivalent to 49 parts by weight of the sulphate.

In like manner a compound of basic properties, called quinia, was found to contain carbon, hydrogen, nitrogen, and oxygen, in proportions which can be represented by the minimum formula $C_{10}H_{12}NO$. It was found that the base unites with hydric chloride, and forms a platinum-salt, analogous to the salt formed by ammonia, and known to have the composition $PtCl_6H_2(NH_3)_2$. The weights of these bases denoted by the formulæ NH_3 and $C_{10}H_{12}NO$, are accordingly equivalent to one another in the sense in which we have hitherto used that word. It is well known that, in the actual development of our knowledge of these relations, the notion of equivalence sprang up among acids and bases, and was extended from them to elementary bodies.

It is interesting and instructive to observe that, for a considerable time, the proportions in which compounds react on one another were represented by formulæ according to the atomic theory as it then prevailed, which recorded the empirical proportions of the constituents, but were not correct or consistent representations of the molecular weights. These labours served to collect an immense amount of evidence for the establishment of molecular weights, but the end was not distinctly in view, and the only rational guide was the imperfect atomic theory then existing.

Thus the great Berzelius, the master spirit of that period, used formulæ which in our system of atomic weights may be thus rendered:

Water	H_2O
Potash	K_2O
Potassic hydrate	H_2OK_2O
Nitric acid....................	N_2O_5
Potassic nitrate	$K_2ON_2O_5$
Hydro-potassic sulphate	$H_2OSO_3K_2OSO_3$
Ether....................	$C_4H_{10}O$
Alcohol	$C_4H_{10}OH_2O$
Carbonic acid	CO_2
Carbonic chloride	CCl_4
Phosgene	CO_2CCl_4
Chloro-sulphuric acid	$2SO_3 + SCl_6$
Chloro-chromic acid............	$2CrO_3 + CrCl_6$
Bismuthic oxychloride..........	$Br_2O_3BrCl_3$
Hydric chlor-acetate ($C_2Cl_3HO_2$)..	$C_2O_3C_2Cl_6H_2O$
Hydric oxamate ($C_2NH_3O_3$)	$C_2O_3H_2OC_2O_2N_2H_4$
Bichlorinated ether $(C_2H_3Cl_2)_2O$..	$2C_4H_6Cl_6C_4H_6O_3$.

Some of his atomic weights were inconsistent with our molecular formulæ. Thus hydrogen, chlorine, nitrogen, and some other elements, he really treated as having atomic weights twice as great in relation to oxygen as we now consider them to have. For his couple of volumes of hydrogen, denoted by the symbol Ħ, were not allowed to separate from one another, so that his formula ĦO for water was in its uses similar to the formula HO, and analogous to his formula for potash, which was written KO.

The theory of radicals did much towards classifying compounds. The names ammonium and cyanogen were given to groups of elements which exhibited analogies with single elements. Ethyl was another radical which did admirable service; and benzoyl afforded an instance of a radical of another kind.

The theory of organic radicals is in reality an extension of the atomic theory, and it was needed as a step towards the proof of that theory.

The equivalent proportions of elements, and the multiples of those proportions in which they combine, afforded no proof of the existence of atoms, but they led to the discovery of compound atoms analogous to undecomposed atoms. At the same time, many of these organic radicals, although exhibiting the closest analogy with elements, are known to be only capable of existing under certain conditions, and are easily burnt or decomposed.

It was a great step to extend the use of the word atom to groups of elements known to hold together only under certain limited conditions. For by including in one common term, atom, the smallest particles of the elements, and the smallest particles of these compounds which behave like elements, we deprive the word atom of the only objectionable peculiarity of which it might have been accused. It is no longer an absolute term; and in its application to the elements it denotes the fact that they do not undergo decomposition under any conditions known to us. If anybody use the word in its absolute sense in its present applications, he is guilty of a manifest inconsistency.

It must not, however, be supposed that the existence of groups of elements, which perform functions analogous to those of single elements, is any proof of the existence of atoms, for compounds like C_2H_5Cl, $C_{16}H_{33}Cl$, NH_4Cl, C_6H_7Cl, &c., may be represented as containing submultiples of the less numerous atoms, just as NH_3 can be represented as $\frac{N}{3}H$, or CO_2 as $\frac{C}{2}O$.

One of the most fruitful ideas in the work of establishing the molecular constitution of compounds, was the idea of types—and the correlative idea of substitution. If we maintain unity of type, the knowledge of a few molecular formulæ serves as a guide to many others. Thus, if we take water as H_2O, alcohol can be proved by normal substitutions to be C_2H_6O. We find that phosgene reacts upon alcohol according to the equation $C_2H_5OH + COCl_2 = C_2H_5OCOCl + HCl$, whereas, if the molecular formula of Berzelius for phosgene were true, the first action would be $C_2H_6O + C_2O_2Cl_4 = C_2H_5OC_2O_2Cl_3 + HCl$; the second action would be the formation of the compound $C_2O_2Cl_2O_2(C_2H_5)_2$.

The process is a mere replacement of chlorine by C_2H_5O, forming a compound of the molecular formula $C_3H_5ClO_2$, instead of one of double that formula. Now, having found that as small a weight as $COCl_2$ can take part in one distinct reaction, we know that the molecular weight is not greater than that. Again, the molecular formula of water being known, we have the reaction $SO_2Cl_2 + H_2O = SO_3HCl + HCl$, and $SO_3HCl + H_2O = SO_4H_2 + HCl$, which establish for chlorosulphuric acid the maximum formula SO_2Cl_2, and for hydric sulphate the formula SO_4H_2. So also the hydric chloracetate reacts on a molecule of potassic hydrate thus, $C_2HCl_3O_2 + HOK = C_2KCl_3O_2 + H_2O$. In these reactions there are alternative explanations: thus, if phosgene is $C_2O_2Cl_4$, then the ether $C_3H_5ClO_2$ must have double that molecular weight: for the reaction is a regular double decomposition, in which no indication of increase or diminution in the number of molecules takes place.

With the chloro-sulphuric acid there is a similar alternative. $S_3O_{12}H_6$ must be the molecular formula of hydric sulphate, if $S_3O_6Cl_6$ be that of the chloro-acid: for the above reaction is as normal as possible, and, moreover, we can recover the chloro-acid from the hydric salt by a couple of steps as normal, as far as the sulphate is concerned:—

$$SO_2{}^{OH}_{OH} + PCl_5 = SO_2{}^{OH}_{Cl} + POCl_3 + HCl.$$

$$SO_2{}^{OH}_{Cl} + PCl_5 = SO_2Cl_2 + POCl_3 + HCl.$$

In order to judge of the evidence of molecular weights obtained by a study of chemical reactions, we ought to consider, on the one hand, what properties must belong to compounds built up of atoms, and, on the other hand, what properties might belong to them if there were no limit to the divisibility of each kind of elementary matter.

Now the first and most essential characteristic of every molecule is that it possesses a weight equal at least to the sum of the atomic weights of its constituents, in such number as represents the simplest proportion consistent with a correct analysis. Thus a compound containing iron and oxygen in such proportions by weight as correspond to three atoms of oxygen to every two atoms of iron, must have at least as great a molecular weight as $2 \times 56 + 3 \times 16$. On the other hand, if iron and oxygen, when combined in this proportion, were infinitely divisible, one would be able to get smaller weights of the oxide to take part in reactions, than such as are 160 times as heavy as the unit weight of hydrogen. We can replace the oxygen in this oxide by chlorine; we can combine it more or less fully with sulphuric acid, or with other acids; we can decompose these salts by other compounds; we can combine them with other salts; and we can compare each of these products with the original oxide, quantifying the proportions of the materials which take part in each transformation: and the result of it all is that we get a chain of evidence proving that less than 160 parts of ferric oxide never take part in any reaction.

So again with other bodies.

Thus a base composed of carbon, hydrogen, and nitrogen was obtained, which could not be represented by a simpler formula than $C_6H_{15}N$. The atomic theory tells us that less than $72 + 15 + 14 = 101$ parts by weight of this base cannot possibly take part in any reaction. We bring it in contact with hydric chloride in varying proportions, beginning with very little of the chloride. By examining the product, we find it is a mixture of the unchanged base with a compound containing carbon, hydrogen, nitrogen, and chlorine, in proportions corresponding to the formula $C_6H_{16}NCl$, and containing exactly 101 parts of the base with one part by weight of additional hydrogen and 35·5 parts of chlorine. If we add our hydric chloride to the base, exactly in this proportion, we have nothing left of either material uncombined with the other; and if we add more hydric chloride, such excess is left uncombined and unchanged. The base does react on hydric chloride in the proportions indicated by its atomic formula, and has not been found to react in any other proportion.

Take an acid salt like hydric citrate, $C_6H_8O_7$ being the simplest atomic formula representing its composition. By bringing it in various proportions in contact with a solution of potassic hydrate, we find that it is capable of reacting in three distinct proportions on that basic hydrate—

$$C_6H_8O_7 + HOK = C_6H_7KO_7 + H_2O$$
$$C_6H_7KO_7 + HOK = C_6H_6K_2O_7 + H_2O$$
$$C_6H_6K_2O_7 + HOK = C_6H_5K_3O_7 + H_2O.$$

If we look through the reactions of well-known bodies with one another, we find that there are many which are not made out quantitatively; but not one of the enormous number that we know accurately and with certainty, has shown a combining proportion of molecules at variance with the atomic theory, and corresponding to the idea that matter is infinitely divisible.

The density of gases and vapours has afforded valuable independent confirmation of the truth of the atomic constitution of molecules, for every compound which evaporates without decomposition has a vapour-density proportional to its molecular weight.

The great benefits conferred upon our science by the doctrine of types would have been impossible without the doctrine of radicals. The two theories were for a considerable time held by different chemists, and were supposed to be inconsistent with one another. They have proved to be essential to one another; so that it may confidently be asserted that those who showed that NH_4 may replace an atom of hydrogen, of potassium, &c., and that C_2H_5 may also perform such functions, prepared the way for the comparisons of bodies of one type.

Another class of considerations which have done admirable work in establishing molecular weights, are those relating to the number of products of substitution obtainable by altering the proportion of materials.

Thus hydric sulphate not only forms the neutral potassic sulphate which we considered in relation to the neutral tartrate, but it also forms a double salt containing hydrogen and potassium, according to the formula $HKSO_4$, when it is brought in contact with half or less than half as much potassic hydrate as is needed to form the neutral salt; and the fact that the hydrogen in the sulphate can be replaced by potassium, in the proportion of half or all, gives an independent proof that the minimum formula of the hydric salt must be H_2SO_4. So also the hydric tartrate forms, not only the neutral salt above mentioned, but also an acid double salt of hydrogen and potassium, for which the simplest formula is $C_4H_5KO_6$, and a double salt with sodium and potassium $C_4H_4NaKO_6$.

The common phosphates were proved by such evidence as this to be tribasic, and the oxalates, in like manner, to be bibasic. Other reactions confirm the conclusions thus arrived at. For instance, the bibasic salts yield two classes of amides—normal amides, such as oxamide $C_2O_2N_2H_4$, and compounds of amide and salt, like potassic oxamate $C_2O_2{}^{NH_2}_{OK}$; and monobasic salts are distinguished from these not only by their inability to react in more than one proportion on potash or soda, but also by their inability to form compounds of amide and salt, like the oxamate. Again, bibasic hydrogen salts lose their water by heat, and monobasic hydrogen salts do not.

The classification of compounds by their analogies has been one of the most important operations for enabling us to see their distinctive resemblances and differences, and the establishment of series of homologous organic compounds has been, perhaps, the most perfect and useful case of such classification, as well as one of the most fruitful of benefit to the progress of science.

So well did the analogies and reactions among organic compounds serve to connect them with one another, by simple and natural principles, that the simpler part of the science has gained chiefly of late years by adopting methods and conclusions established in the more varied field of organic compounds. Thus, the chief arguments adduced by Gerhardt for the molecular weight of carbonic acid which we now adopt, and for the corresponding molecular weight of water, were to the effect that in no reaction between well known molecules is a smaller proportion of carbonic acid given off or taken up than that corresponding to the molecular weight 44; and, in like manner, less than 18 parts, by weight, of water are never taken up nor given off in any well known reaction between molecules of known weight. The examples of molecules in each case were mainly taken from organic chemistry. The molecular weights of the gaseous elements, hydrogen, chlorine, nitrogen, oxygen, &c., were studied in great part with the aid of the light obtained from the comparison of organic reactions, in which they take part, especially the double decompositions which take place when chlorine normally replaces hydrogen.

A study of the phenomena of electrolysis, and some other inorganic processes, had led to metals being considered as possessing, in the free state, a molecular constitution analogous to that of compounds; and Brodie's reduction of oxides by oxygen gave evidence of combinations between atoms of oxygen in forming the free gas.

Amongst the general processes which were instrumental in judging of the correctness of molecular formulæ was Gerhardt's so-called law of even numbers. Among molecular formulæ which he considered well established, he observed that the sum of the atoms of a certain list which he supplied was always an even number; whereas, the sum of the atoms of elements not belonging to his list was sometimes an even, sometimes an uneven number.

The circumstance is now known to be one among several consequences of a difference in replacing value, which constitutes one of the most important characteristics of atoms, and we also know that many elements which Gerhardt classed with hydrogen, nitrogen, &c., do not belong to that class; but even in its original imperfect form the rule rendered very important services. Thus, such formulæ as $C_2H_3O_3$, $C_{10}H_{12}NO$, Gerhardt knew could not be molecular formulæ, and must be at least doubled. A study of their reactions amply confirms his verdict. We also know that such formulæ as $FeCl_3$, $HgCl$, $AlCl_3$, cannot represent molecules, and must be corrected so as to contain at least even numbers of chlorine atoms without including the metallic atoms in the account.

To do justice to the completeness of the evidence which is obtained respecting many molecular weights, by these purely chemical methods of research, it would be necessary to go into the full particulars of the operations and observations in the respective cases, and that would task, still more unduly than I am now tasking it, the indulgence of the Society. But I feel convinced that the brief indications which I am able to give of the kind of evidence obtained under these various headings, will recall to the minds of chemists the particulars which it would be needful to describe specially to persons not intimately acquainted with the science.

There are, however, other properties of molecules which have been noticed since chemists discovered the existence of the molecules themselves, and these are of kinds to be more easily understood by all—properties which involve no change of composition in the molecules, but which belong to them as long as they subsist.

Of these physical properties, one of the most important is the volume of molecules in the state of gas or vapour. Whenever we examine perfect vapours composed of like molecules, we find that whatever the molecules may be, they occupy the same volume under like conditions. In other words, these little units of matter, found by the guiding aid of the atomic theory, are found also to be units of force when examined under those conditions. There have, in several cases, been important discrepancies between the chemical evidence of molecular constitution and the evidence afforded by vapour-volume. Thus, hydric sulphate and ammonic chloride are among the compounds of whose molecular weights we have the best and most varied chemical evidence, yet the vapour formed from each of these occupies about double the volume of a gas molecule. A careful examination of the vapour formed from each of them showed, however, that it is a mixture of molecules of two sorts formed by the decomposition of the original compound. The fact that errors in the indications supplied by vapour-density could be thus pointed out by the chemical evidence of molecular weight is surely no slight confirmation of the perfection of the evidence of molecular weights obtainable by chemical methods; and, on the other hand, the immense number of coincidences between the indications obtained from the two independent sources is a strong evidence of the truth and parallelism of the two.

There are, also, crystalline compounds, which chemistry has found to be closely analogous, which occupy, in the crystalline state, equal, or nearly equal, volumes.

Among other physical properties which confirm the truth of

the atomic views of molecular constitution are boiling points. Whenever truly analogous elements are comparable in respect to boiling points, it is found that those composed of heavier atoms boil at higher temperatures than those composed of lighter atoms. A comparison of the successive terms of the series—chlorine, bromine, and iodine, illustrates this difference very strikingly. So also a comparison of the several terms of the series—oxygen, sulphur, selenium, and tellurium, and not less so the elements nitrogen, phosphorus, arsenic, antimony, and bismuth.

Now, whenever volatile organic compounds belonging to a homologous series are compared, it is found that of two such compounds, the one having the higher molecular weight has also the higher boiling point. The glycols present, however, an exception to this general observation.

The melting points of homologous organic compounds also show differences running parallel in like manner to their differences of molecular weight.

The relative velocities of motion of particles, as shown by the processes of diffusion, afford another confirmation of the general truth of the molecular weights; for, on the one hand, it is known that heavy particles diffuse more slowly than light particles; on the other hand, a comparison of the relative velocities of movement of molecules of relative weights, previously determined upon chemical evidence, shows that the heavier molecule of chemistry is also the heavier molecule in diffusion.

At this point of my study of this question I am in a serious difficulty, for molecules have no *locus standi* in the absence of the atomic theory. They are, to use the words of Dumas, physical atoms; their existence is a necessary consequence of the atomic theory, and all chemical reactions agree in proving their existence. They are also discovered by an examination of the mechanical properties of gases.

The opponents of the atomic theory are bound to explain, in some other way, the facts which point so distinctly to the existence of molecules, if they wish to advance from the position of mere contradictors to that of chemists.

Hitherto they have not done so, and the case stands thus: on the one hand, we have a simple theory which explains in a consistent manner the most general results of accurate observation in chemistry, and is daily being extended and consolidated by the discovery of new facts which range themselves naturally under it. On the other hand, we have a mere negation: for the statements of those who say that our evidence of the existence of atoms is not conclusive, and yet omit to show any alternative theory, are nothing more. In discussing the relative composition of carbonic acid and carbonic oxide, I have endeavoured to put the negation in a tangible form, as an affirmation that there are no limits to the divisibility of elements, and to represent compounds containing multiples of equivalent weights as containing sub-multiples corresponding to the same empirical proportions.

If elements, in combining with one another, merely undergo more minute subdivision, so that a compound, like amylic alcohol, $CH_{\frac{12}{5}}O_{\frac{1}{5}}$, is formed from methylic alcohol by removing hydrogen and oxygen from the compound CH_4O, then we could expect a weight of it corresponding to a formula containing less hydrogen and oxygen to perform the functions of the molecule CH_4O.

Of two things, one: either the existence of molecules is denied, or it is admitted.

In the former case, the vast and consistent body of chemical evidence of the existence of molecules must be set aside and disproved, and the physical confirmations of their existence must also be proved to be erroneous.

In the latter case it must be shown by quantitative analysis that those complex molecules to which we attribute a great weight, are really lighter than their simpler analogues.

Now, the difficulty is to choose arguments for an opponent. It would probably matter little to us which they might select; but when they content themselves with saying that we are wrong, without either showing in what respect our evidence or our reasonings are at fault, and without showing any other evidence or any other reasonings which they consider preferable, it is difficult to know what else to do with them than to state our case and leave them to their reflections.

Atomic Values.

The most important and weighty chemical property of atoms as yet discovered, is one which can only be perceived by the aid of a knowledge of the equivalent weights of the elements and of the molecular weight of a considerable number of compounds.

Thus if we know the molecular formulæ of the following compounds, we can compare them with one another and perceive their differences of composition :

Hydric acetate has the molecular formula	$C_2H_4O_2$
Hydric monochloracetate	$C_2H_3ClO_2$
„ dichloracetate	$C_2H_2Cl_2O_2$
„ trichloracetate	$C_2HCl_3O_2$.

Again,

Marsh-gas has the molecular formula....	CH_4
Methylic chloride	CH_3Cl
Methylenic chloride	CH_2Cl_2
Chloroform	$CHCl_3$
Carbonic chloride	CCl_4.

It is apparent from these series of formulæ that one equivalent weight of chlorine can take the place of one of hydrogen in a molecule, or two of chlorine can replace two of hydrogen, or three replace three, and so on. Hydrogen-compounds containing in each molecule a much greater number of hydrogen equivalents, have been found to allow, in like manner, the hydrogen to be replaced, one equivalent at a time, by chlorine, and to the extent of a great number of equivalents in each molecule.

In like manner, potassium replaces hydrogen in molecules, sometimes in the proportion of an even number of equivalents, sometimes in the proportion of an uneven number of equivalents, as will be seen by the following molecular formulæ:—

H_2O	H_2SO_4	H_3PO_4	H_3N	$H_2C_4H_4O_6$
HKO	$HKSO_4$	H_2KPO_4	KH_2N	$HKC_4H_4O_6$
K_2O	K_2SO_4	HK_2PO_4		$K_2C_4H_4O_6$
		K_3PO_4		

When, however, oxygen takes the place of hydrogen, without change of equivalent weight, it is only in the proportion of two equivalents of oxygen replacing two of hydrogen, four of oxygen replacing four of hydrogen, six replacing six, and so on; always an even number of hydrogen equivalents replaced in each reaction of the kind. Thus carbonic acid is formed from marsh-gas by the normal reaction $CH_4 + 2O_2 = CO_2 + 2H_2O$, in which four equivalent weights of oxygen replace four of hydrogen. Hydric nitrate has a constitution similar to ammonic hydrate, the difference being that it contains four equivalent weights of oxygen instead of four equivalent weights of hydrogen in the ammonia, as will be easily seen by a comparison of the formulæ $HONH_4$ and $HONO_2$.

Again, phosgene is methylenic chloride in which two equivalents of hydrogen are replaced by two of oxygen—

$$CH_2Cl_2, \ COCl_2.$$

The compounds ..	ammonia..............	$2H_3N$
	nitrous acid............	O_3N_2
	glycoll................	$C_2H_6O_2$
	hydric glycollate	$C_2H_4O_3$
	hydric oxalate..........	$C_2H_2O_4$,

show a replacement of hydrogen by oxygen in steps, each one being two hydrogen replaced by two of oxygen. Now, it is not the existence of these compounds which proves any different value in oxygen atoms from chlorine atoms; it is the fact that no case is known of a molecule containing a single equivalent of oxygen in place of one of hydrogen, nor of three equivalents of oxygen in place of three of hydrogen, nor, in short, of any uneven number of hydrogen equivalents replaced by an uneven number of oxygen equivalents.

The same difference between oxygen and chlorine may be seen by comparing all well known molecular formulæ containing either one of those elements, apart from the particular process by which they may have been formed. Thus we have, by way of illustration, some bodies taken at random from various classes of compounds—

C_2H_5Cl	H_2O
$C_2H_4Cl_2$	CO
$C_2H_3Cl_3$	CO_2
C_2H_6ClN	$C_2H_2O_4$
$C_2H_5Cl_2N$	C_2H_6O
PCl_3	$C_5H_{12}O$
NH_4Cl	$C_4H_6O_4$
CrO_2Cl_2	$C_3H_8O_3$
SO_2Cl_2	C_2H_3OCl
SO_4H_2	$C_2H_4OCl_2$
PO_4HNa_2	
PbO_2H_2	
PbO	
PbO_2.	

There are, however, some cases which at first sight do not agree with this general conclusion. Thus, on comparing potassic chloride with potassic oxide, it might be said that the equivalent weight of potassium in the former is combined with an equivalent weight of chlorine, and in the latter with an equivalent weight of oxygen; thus KCl, $KO_{\frac{1}{2}}$. But the answer is that this comparison is made in defiance of the known molecular formula of potash (K_2O), established by reactions and analogies which leave no doubt of its correctness. Whenever oxygen replaces chlorine in such a chloride, it binds together two basic equivalents, one from one molecule of the chloride one from another; and in like manner when chlorine replaces oxygen in any molecule, such as water H_2O or potassic hydrate, it loosens into two molecules what the oxygen held bound up in one. Thus, by passing oxygen and hydric chloride through a red-hot tube, we have a replacement of chlorine by oxygen, taking place, as far as it goes, according to the molecular equation $O_2 + 4HCl = 2H_2O + 2Cl_2$, half as many molecules of water being formed as molecules of hydric chloride are decomposed; and on the other hand if we send chlorine and steam through a hot tube, we break up a part of the steam, forming two molecules of chloride from every one of oxide.

When we apply the word atom to two equivalent weights of oxygen, and say that an atom of oxygen has a replacing value equal to that of two atoms of chlorine or of hydrogen, we give to a couple of equivalent weights of oxygen the name atom, as a record of the fact that we have never known them to divide, and that we must in all our considerations respecting oxygen bear in mind this peculiarity of it. In carbonic oxide it may be argued that an atom of oxygen takes the place of four equivalents of hydrogen of marsh-gas, and that it has there the equivalent weight 4, instead of its usual equivalent weight 8. But on the other hand the gas may be represented as containing carbon with an equivalent weight of double its usual one, viz., 6 instead of 3. In any case, the gas affords an instance of a change of atomic value, and if we describe the value of the atom of oxygen in disregard of this case, it is because we have to record its commonest value.

What has been here said of oxygen applies to sulphur equally. No molecule has ever been found containing an uneven number of equivalent weights of sulphur, and the atom of sulphur has certainly the value of two chlorine or hydrogen. There are, however, strong reasons for believing that in some of its compounds its atom has a replacing value of four or even six. Selenium and tellurium are so closely allied to oxygen and sulphur in their reactions that their atoms are always classed together. In like manner bromine and iodine go with chlorine, and lithium, sodium, &c., go with hydrogen and potassium. Silver also has an atom of uneven value.

Nitrogen is found upon similar evidence to divide in no smaller proportion than that of three times $4\frac{2}{3}$, or 14 parts by weight, for whenever it has an equivalent weight of $2\frac{4}{5}$, as in NH_4Cl, its atom has five values.

It seems to me of paramount importance to the due understanding of the functions of this element, to state in the most direct and matter of fact way possible, the actual proportion in which its atom replaces hydrogen. If the replacing value of an atom be measured by the number of hydrogen atoms needed to replace it, then nitrogen has in ammonia the atomic value three, for three atoms of hydrogen are needed in order to replace one of nitrogen in NH_3, forming three molecules of hydrogen H_3H_3; and in sal ammoniac it has just as distinctly the atomic value five, for five atoms of hydrogen are needed in order to take the place of one of nitrogen, forming one molecule of hydric chloride and four molecules of hydrogen: $ClNH_4$ becomes $\overbrace{ClH} + \overbrace{H_4H_4}$. Sal-ammoniac is as well known a molecule as ammonia itself; each of them undergoes decomposition at high temperatures, the complex molecule not requiring so high a temperature for its decomposition as the simpler one. Those chemists who represent sal-ammoniac as containing one atom of nitrogen of the same value as in ammonia, seem to me to incur the disadvantage of stating rather what, according to their own theoretical notions *ought* to be, than what actually *is*, and thereby to lose sight of one of the most important of all properties of atoms, viz., the property of changing value under particular chemical and physical conditions which it is our business to study and describe.

I shall return to this important point somewhat later, and endeavour to trace an outline of the conditions of these changes.

Together with nitrogen are always classed the elements phosphorus, arsenic, antimony, and bismuth. Their atoms have frequently the value of three, but frequently also that of five. Boron has in its chloride and fluoride the value three, but in its double fluoride the value five—BCl_3, BF_3, KBF_4. Gold is not proved to have any smaller atomic value than three, for we have no evidence of the molecular weight of the compound AuCl. Compounds corresponding to $AuCl_3$ and $NaAuCl_4$, are common and tolerably well known.

Carbon is at present perhaps that element of which the compounds are best investigated. A comparison of the composition of an immense number of carbon-compounds has not led to the discovery of a single molecule containing a number of its equivalent weights, other than such as can be divided by four. Its atom is accordingly considered to weigh 12, and to have the value 4. In carbonic oxide, the atom of carbon has a somewhat uncertain value, according as we consider oxygen to vary in value or not. Silicium, titanium, and tin are in many respects so closely allied to carbon that it is to be expected that their atomic weight must be such as to give the atoms the value 4 in their ordinary compounds. The atom tin in stannous chloride appears at first sight to be worth 2, but the molecular formula of the compound is not known, and may possibly be Sn_2Cl_4 like olefiant gas. Silicium in its double fluoride, SiF_6K_2, is evidently worth 8, and titanium and tin have also doubtless a higher value than 4 in their double salts.

Platinum seems to vary from four values to eight, as in the compounds $PtCl_4$ and $PtCl_6K_2$. Its lower chloride $PtCl_2$ ought probably to have a higher molecular formula.

Not only does the empirical comparison of the number of equivalent weights of each element contained in well-known molecules lead to a discovery of the relative weights of atoms, but there are other chemical evidences which concur with these. Thus the univalent alkali-metals have long been known to be unable to form basic salts, whereas many multivalent metals form such salts with peculiar facility; witness antimony and bismuth, mercury and lead, in the compounds SbOCl, $Bi_2O_2SO_4$, $Pb(HO)(NO_3)$. In such compounds as these, the metal performs the function of binding together several radicals or simple atoms, just as in water or potassic hydrate the oxygen binds the two atoms together; and when such a metal is removed, and an equivalent quantity of a single-value metal supplied in its place, the atoms which had been bound together pass off, each in combination with its own atom of metal. Thus, in the reaction $Pb{}^{HO}_{NO_3} + Ag_2SO_4 = PbSO_4 + AgOH + AgNO_3$ we have a severance of AgOH from $AgNO_3$, which would not take place if silver were of double its atomic weight, and, like lead, of the atomic value 2.

Double amides, such as $Hg{}^{Cl}_{NH_2}$, and double salts, such as $Ba{}^{NO_2}_{C_2H_3O_2}$, $Cu{}^{AsO_2}_{C_2H_3O_2}$ are also evidences of the biacid character of the base, just as $KNaC_4H_4O_6$ is an evidence of the bibasic character of the acid.

Again, compounds of metals with organic radicals have furnished admirable evidences of the atomic weights of various metals whose equivalent weights are known. Thus the compounds $Zn(C_2H_5)_2$, $Zn{}^{C_2H_5}_{I}$, $Hg(C_2H_4)_2$, $Hg{}^{C_2H_5}_{I}$, $Sn(C_2H_5)_4$, $Sn{}^{(C_2H_5)_3}_{Cl}$, $Sn{}^{(C_2H_5)_2}_{Cl_2}$, $Pb(C_2H_5)_4$, $Pb{}^{(C_2H_5)_3}_{Cl}$, &c., have been among the best chemical evidences of the atomic values of the respective metals.

Similar proofs of the value of the atom of silicium have been given by the formation of molecules containing atoms of ethyl and chlorine together numbering *four*, just as in the molecule of marsh-gas the hydrogen can be replaced by chlorine in four distinct proportions.

Atomic Heat.

Perhaps the most direct confirmation of the atomic weights established upon evidence of the various kinds which we have been considering is afforded by the capacity for heat of elementary bodies. A comparison of the relative capacity for heat of solid elements showed that all atoms have nearly the same capacity for heat. Thus 23 grammes of sodium absorb, when heated from 0° C. to 1° C., almost exactly the same quantity of heat as 39 grammes of potassium when heated from 0° C. to 1° C.; and 210 grammes of bismuth absorb that same quantity when heated from 0° C. to 1° C. Silicium, carbon, and boron form exceptions to this law, but in all other cases the approximation to equality between the numbers found for the several solid elements is admitted to be as close as can be expected from determinations involving so many difficulties, and including so many sources of error.

Like every other confirmation of the atomic theory, the law of atomic heat has contributed to the development and systematisation of that theory. It is well known that a considerable number of metals, such as lead, mercury, iron, tin, barium, &c., appeared, according to the atomic weights originally assigned to them, to have about half the atomic heat of the metals before mentioned, and that this circumstance was urged by an illustrious Italian chemist as an argument for doubling their atomic weights, so as to bring them within the law. It is also too well known to need illustration to you, that the chemical properties of those metals amply confirm the truth of Cannizzaro's proposition, that their atomic weights are twice as great as was formerly supposed. Indeed so much chemical evidence has now been accumulated, that we should certainly retain the higher atomic weights for lead, tin, iron, barium, &c., even if we knew nothing of atomic heat.

Atomic values are not so variable as equivalent weights. A most important light has been thrown on the phenomena of equivalence by our knowledge of atomic values and the general principle of their changes. The foundation of this consideration was laid by those chemists who proved that the molecule of chlorine consists of two atoms of chlorine combined with one another, and that the molecule of oxygen consists of two atoms of oxygen combined with one another.

It is well known that free carbon at high temperatures unites with hydrogen, forming acetylene, which has the molecular formula C^2H_2. I will write the number belonging to each atomic symbol above, as an exponent, when I wish to represent the atoms so numbered as combined with one another, O^2, &c., meaning a compound of an atom of oxygen with an atom of oxygen; C^2 a compound of an atom of carbon with another atom of carbon. On the other hand, I will write the number belonging to any atomic symbol below, when I wish to represent the atoms as not combined with one another. Thus H_2 means two atoms of hydrogen, and OH_2 means a compound of one atom of oxygen with two atoms of hydrogen, each of which atoms of hydrogen is merely united with the atom of oxygen. It is merely an abbreviated form of writing HOH or $\begin{matrix} HO \\ H \end{matrix}$, or $\begin{matrix} H \\ O \\ H \end{matrix}$ &c. By adding two atoms of hydrogen to acetylene we get olefiant gas, C^2H_4, and this again, in its turn, can be transformed into methyl C^2H_6. These bodies, C^2H_2, C^2H_4, C^2H_6, are well known to have molecular weights corresponding to two atoms of carbon in each, and the hydrogen is held in each molecule by being combined with the carbon. With more hydrogen the double atom of carbon cannot combine, for the next hydrocarbon to methyl is marsh-gas CH_4 containing a single atom of carbon.

Again, if we remove hydrogen from marsh-gas in the smallest possible proportion, we find that the atom of carbon in one molecule combines with one atom of carbon in another molecule, while they are losing their hydrogen. Methylic iodide reacts on zinc-iodomethide by a normal interchange of iodine for methyl $H_3CI + IZn\,CH_3 = H_3CCH_3 + ZnI_2$. It is admitted that the explanation of the molecular weight of methyl lies in the fact that it contains two atoms of carbon combined with one another to the extent of one-quarter of their value, and that this combination of carbon with carbon is the essential condition for the formation of such a complex molecule. But, if this be true, the other terms of the series are explained by a mere extension of the same principle. Ethylene contains two atoms of carbon H_2CCH_2, each combined with two atoms of hydrogen, and each one combined with the other to the extent of two of its values. Acetylene, HCCH, is in like manner a compound of two atoms of carbon, each one combined with the other to the extent of three values.

It was known that an atom of carbon may combine partly with hydrogen, partly with some other element or elements, and it was also known that like atoms combine with one another; so that it was quite consistent with known facts to suppose that an atom of carbon may combine partly with another atom of carbon, partly with some other element. The simplification introduced into the chemistry of the compounds of carbon by this theory is, however, truly admirable, and, by itself, entitles the distinguished author of the idea to the high rank among theorists which he holds.

Carbon in all its compounds may have the same atomic value consistently with our present knowledge of those compounds. In carbonic oxide alone we are compelled to represent it as having only the value two, unless we attribute to oxygen the value four.

Among the instances of increase of equivalent weight in elements, there are a great number in which the element exhibiting that increase, by giving up some of the unlike with which it was united, unites with the like while doing so.

Thus, SCl_2, by the mere action of heat, decomposes with evolution of chlorine, and by an atom of sulphur combining with another atom of sulphur, $2SCl_2 = S_2Cl_2 + Cl_2$. Water and hydric peroxide are doubtless related to one another, in the same manner, each of the two atoms of oxygen in the molecule of the peroxide being half combined with oxygen, half with hydrogen. Mercuric chloride loses half its chlorine, forming calomel, while the mercury unites to the extent of half its value with mercury, and cupric chloride is reduced to cuprous chloride by a similar process. On the other hand, phosphoric chloride loses chlorine, and forms phosphorous chloride, without any combination of phosphorus with phosphorus taking place. The equation $PCl_5 = PCl_3 + Cl^2$ represents the process; and until it can be shown that the molecule of phosphorous chloride is a multiple of PCl_3, the reaction must be accepted as direct evidence of a change in the atomic value of phosphorus from 5 to 3. In like manner, if ammonia be admitted to have the molecular formula NH_3, the decomposition of sal-ammoniac by heat, according to the equation $ClNH_4 = NH_3 + HCl$, must be admitted as evidence of a change of atomic value in nitrogen from 5 to 3. When auric chloride is decomposed, with evolution of chlorine and formation of aurous chloride, the interpretation of the reaction requires evidence which we do not possess respecting the molecular weight of the lower chloride. If aurous chloride has the molecular formula AuCl, its formation is accompanied by a change of atomic value in the gold; but if its molecule contains more than one atom of gold, the metal may exist in both compounds with the same atomic value.

Among complex organic molecules there is, of course, considerable room for variety in the arrangement of atoms, which we may consider consistently with the empirical equivalent weight of the carbon.

The hydrocarbon benzole is of so much interest from its derivatives, that it has attracted a good deal of attention, and to explain its molecular constitution, the six atoms of carbon have been represented as arranged in a ring, each atom being combined with the atom of carbon at one side of it to the extent of two values, and with the atom of carbon on the other side of it to the extent of one value.

Another arrangement, which was mentioned a good many years ago at this Society, is to consider each atom of carbon combined with three others, and with one atom of hydrogen. Such an arrangement is easily represented in space. Let the accompanying formula represent as unfolded in one plane, the group which would be formed by the carbon atom at one end of each line moving back through the paper until equidistant from the two others in the line.

H | H | H
C C C
C C C
H | H | H

I put bars between symbols of elements, which, for convenience, are juxtaposed, but which I do not wish to represent as combined with one another.

Naphthalin may, in like manner, be represented by the formula C^5H_4, as containing four atoms of carbon, each united with two of the others and with one atom of hydrogen, while the fifth atom of carbon is combined with each of the first four.

HC CH
C
HC CH

With respect to the conditions under which changes of atomic value take place, some valuable information may be gained by comparing molecules known to exist in different states. Let any chemist write for himself a list of molecules known to exist in the liquid state alone, and not volatile without decomposition. Let him compare with that, a list of molecules known to evaporate unchanged; and thirdly, let him look at the molecular formulæ of the hitherto uncondensed gases. He will notice that in passing from the non-volatile to the uncondensed, he goes from complex molecules to simpler and simpler ones. Let him compare the formulæ of products formed by destructive distillation with those of the materials from which they are formed. The examples may be taken from mineral and from organic chemistry. He will find some cases of the formation of a more complex molecule by distillation; but for one such he will find an enormous number of cases of the breaking up of complex molecules into more and more simple ones. Take, for instance, the series CO_3H_2, SO_3H_2, NH_5O, $SbCl_5$, PCl_5, SO_4H_2, PO_4H_3, $C_2O_4H_2$, NH_3, CH_4, H_2O, CO_2, beginning with compounds so exceedingly unstable, that, working at ordinary temperatures, we hardly know them; and finishing by compounds of great stability, which are only broken up into simpler molecules at the most intensely high temperatures. The fact stares us in the face, that simple molecules correspond to high temperatures, complex molecules correspond to low temperatures. Several of these changes of composition are certainly accompanied by changes of atomic value. Whenever there is a change of atomic value effected by change of temperature, the lower atomic value corresponds to the higher temperature. These phenomena have attracted some attention of late, in consequence of Deville's researches on dissociation. I will give an example of the application of this principle to the study of reactions in illustration of the importance of seeing and recording differences of atomic value in elements.

I will again use the notation which I referred to above. The following formulæ give samples of nitrogen-compounds in which nitrogen has the atomic value 5: $ClNH_4$, $HONH_4$, $HON(C^2H_5)_4$, $(H_4NO)_2CO$, $(H_4NO)_2C^2O_2$, $H_4NO\underset{O}{C}CH_3$. The action of heat upon them gives rise to NH_3 in two cases, $N(C^2H_5)_3$, $(H_2N)_2CO$, $(H_2N)_2C^2O_2$, $H_2N\underset{O}{C}CH_3$. In each case the nitrogen-atom is reduced to the value of 3. In fact it will be seen that in ammonium-salts nitrogen is worth 5, in amides only worth 3. Ammonia is an amide in this sense.

Among chemical evidences of atoms, the discovery of the distinction between direct and indirect combination is worth consideration, more especially as it is independent of the quantitative comparisons which have hitherto guided us. Many elements which have never been obtained directly combined, are bound together by a third; thus we have no direct compound of hydrogen and potassium, but in potassic hydrate an atom of hydrogen is united with oxygen, and an atom of potassium is united with this same oxygen HOK. Hydrogen in this hydrate is indirectly united with potassium, an atom of oxygen being the connecting link between them.

Some elements can be either directly or indirectly united with one another, and exhibit different properties in each condition. In organic compounds we know an immense number in which hydrogen is directly united with carbon. Thus, methylic oxide H_3COCH_3 has all its six atoms of hydrogen united with carbon directly, whilst vinic alcohol $H_3\underset{H_2}{C}COH$ has five of its atoms of hydrogen united with carbon directly, and one atom indirectly, through oxygen.

In consequence of this difference, we find that the hydrogen, which is indirectly united with carbon through the intervention of oxygen, separates from it as soon as by any process the oxygen is replaced by chlorine or bromine; whereas the hydrogen which is directly united with carbon undergoes no change by such replacement. H_3COCH_3 by the action of PCl_5 becomes $H_3CCl + ClCH_3$, whereas $H_3C\underset{H_2}{C}OH$ becomes $H_3C\underset{H_2}{C}Cl + ClH$. When the hydrocarbons are liberated from these compounds, it is well known that a direct combination of carbon in one atom with carbon in another takes place, as in the formation of C^2H_6 from a methyl-compound, or $H_3C\underset{H_2}{C}|\underset{H_2}{C}CH_3$ from an ethyl-compound.

In like manner to oxygen, an atom of zinc holds two atoms of ethyl in combination in the molecule $H_3C\underset{H_2}{C}Zn\underset{H_2}{C}CH_3$, and by the action of water, the atom of zinc is replaced by two atoms of hydrogen, each atom of ethyl getting an atom of hydrogen in exchange for the half-atom of zinc which it held at first.

These and many similar reactions are now so familiar to chemists that I only allude to them to recall their bearing on the order of combination of atoms.

Not only are they known and admitted to be due to differences of arrangement, such as I have alluded to, but the idea which they have established is found a sure and faithful guide in the investigation of the intricate varieties presented by isomeric organic bodies in their varied reactions.

Thus it is well known from Hofmann's admirable researches on the isomeric cyanides and sulphocyanates, that, by the action of ethylia on chloroform, an ethylic cyanide is formed, isomeric with the compound formerly known. It breaks up by the action of hydric chloride in presence of water, into hydric formate, and the compound of ethylia and hydric chloride, instead of breaking up, as the other cyanide does, under the influence of alkalis, into ammonia and a propionate.

The following equations represent the two reactions:—

$$\underset{H_2}{CN^vCCH_3} + 2H_2O + HCl = \underset{H_3|H_2}{ClN^vCCH_3} + \underset{H}{HOCO}$$

$$\underset{H_2}{N'''CCCH_3} + H_2O + HOK = N'''H_3 + \underset{O|H_2}{KOCCCH_3}$$

And in like manner his new sulphocyanates break up under the influence of nascent hydrogen, according to the equation—

$$\underset{H_2}{H_3CCN^vCS} + 2H_2 = \underset{H_2}{H_3CCNH_2} + H_2CS$$

whilst their isomerics on the SH_2 type break up thus:—

$$\underset{H_2}{H_3CCSCN'''} + H^2 = \underset{H_2}{H_3CCSH} + HCN$$

In the new cyanides, the carbon of the cyanogen is indirectly united with the carbon of the ethyl through an atom of nitrogen of the value 5; whereas, in the old cyanide of ethyl, the carbon of the cyanogen is directly combined with the carbon of the ethyl, and in the sulphocyanate carbon is indirectly combined with carbon through the intervention of sulphur.

What concerns us in these reactions is to see the evidence which they afford, that the binding element is an atom. The very fact itself amounts to that. In potassic hydrate, oxygen is combined with hydrogen; it is also combined with potassium, but the hydrogen cannot pass off, even at a red heat, in combination with its half of the oxygen. The two halves are inseparable, and when I say that in the molecule of potassic hydrate there is a single atom of oxygen, I merely state that fact.

I would gladly proceed now to other and certainly not less interesting considerations based upon the atomic theory, which have received verification from experiment, considerations for which I think it would be at least rather difficult to find any equivalent if matter were not composed of atoms; but I must content myself with indicating them at present, and reserving for a future occasion the task of giving more development to them. I allude to atomic motion—the dynamics of chemistry. It has been shown that heterogeneous molecules, when mixed with one another in the fluid state, interchange their analogous constituents, and that the products thus formed, if they remain mixed in the fluid state, also exchange their analogous constituents, reproducing the original materials. In a solution of two salts there are formed two new salts by the exchange of bases; and it depends on the relative velocities with which the decomposing change and the reproducing change go on in the liquid, what proportion the quantity of the original materials bears in the final mixture to that of the new products.

In using the atomic language and atomic ideas, it seems to me of great importance that we should limit our words as much as possible to statements of fact, and put aside into the realm of imagination all that is not in evidence. Thus the question whether our elementary atoms are in their nature indivisible, or whether they are built up of smaller particles, is one upon which I, as a chemist, have no hold whatever, and I may say that in chemistry the question is not raised by any evidence whatever.

They may be vortices, such as Thomson has spoken of; they may be little hard indivisible particles of regular or irregular form. I know nothing about it; and I am sure that we can best extend and consolidate our knowledge of atoms by examining their reactions, and studying the physical properties of their various products, looking back frequently at the facts acquired, arranging them according to their analogies, and striving to express in language as concise as possible the general relations which are observed among them.

In conclusion I must say that the vast body of evidence of the most various kinds, and from the most various sources, all pointing to the one central idea of atoms, does seem to me a truly admirable result of human industry and thought. Our atomic theory is the consistent general expression of all the best known and best arranged facts of the science, and certainly it is the very life of chemistry.

Discussion on Dr. Williamson's Lecture on The Atomic Theory, November 4th, 1869.

Sir BENJAMIN C. BRODIE, Bart., F.R.S., in the Chair.

THE Chairman said the business before the Society was to consider a paper "On the Atomic Theory," which had already been read before the Society by Professor Williamson. This theory was not to be regarded as a fixed and definite system of ideas, but one that had undergone great variation and change. There was the atomic theory of Dalton; then that of Berzelius, and also, that of Laurent and Gerhardt; and now they had come to a further form of this theory, in which new ideas had arisen in regard to the nature of atoms and their properties; and he (the Chairman) thought it would greatly facilitate the discussion if Dr. Williamson would explain what was the precise form of the atomic theory which he was prepared to maintain and defend, and how they might discriminate that from other forms of the theory, some of which Dr. Williamson himself would admit to be neither reasonable nor rational. First, there was the question whether there were atoms at all; then, being atoms, what those atoms were, and whether they were endowed with the properties assigned to them in regard to their replacing power, and to what was termed their value. These were the questions raised by the lecture of Dr. Williamson, and which they were now to discuss.

Professor Williamson said that he did not admit that, among chemists, there had been several atomic theories. If there had been, he had yet to learn of them. There had prevailed, he believed, from time immemorial, a belief that matter was built up of small particles; and he might also say that no contrary opinion had ever been embodied in any definite form. Dalton doubtless looked for evidences of what he believed to be true, and saw, in the simple multiple proportions which he specially alluded to, facts which were in accordance with that preconceived conviction; but his was the mere germ of an atomic theory, and the changes which have taken place in it since are consistent with the original notions. The very rapid changes had, in the main, been additions to the starting-point given by Dalton. The line of work so vigorously pursued of late years by many chemists, regarding the order in which atoms are arranged in their compounds, was mentioned by the speaker. His endeavour had been to put together more fully than had been done of late, what is known of the chemical evidence: firstly, respecting the limits to the divisibility of matter, and, secondly, with regard to the properties of those finite particles which constitute the limits of divisibility, solely from the point of view habitually employed by chemists in their ordinary working. He had also endeavoured to separate those conclusions which appeared to him warranted by facts from everything else. Whether the smallest particles of matter have a spherical form or not, whether they are in their nature indivisible, whether they are in reality the ultimate atoms of matter, or like the planets of this system, he knew not, nor did such questions exist for him as a chemist. He therefore thought it wise to exclude them, important as they were, from the actually existing atomic theory. The work of Berzelius was perhaps the most significant transition from the original notion of Dalton. Silently these atoms had been accepted by chemists, and the whole course of investigation had been one grand confirmation of the assumption, that compounds must have molecular weights corresponding to at least the smallest atomic proportion which would represent the actual numbers of analysis. The perfectly independent observations which had been made with reference to the boiling points of homologous liquids, the phenomena of diffusion, and the equality of volumes of masses containing an equal number of these molecules under similar conditions in the gaseous state: all concurred to corroborate the conclusions necessitated by the atomic theory. Thus, independent workmen hewed stones which, when hewn, were found to fit in exactly with the others, forming a perfectly homogeneous whole. Dr. Williamson concluded his remarks by referring to the atomic values, a term which, he thought, represented in a more precise way than some others what is called atomicity; and he also referred to the difference between direct and indirect combination, a fact which he conceived to be essentially atomic in its nature.

Dr. Frankland said that he was not present when Professor Williamson delivered his lecture, and he had only cursorily looked over the printed report; but, so far as he could gather, the object of the author had been to establish the atomic theory as an absolute truth. There had been, he thought, so little, if any opposition to the application of the atomic theory to the phenomena of chemistry, that a discussion could scarcely be raised upon any other feature of the question than that of its absoluteness. He considered it impossible to get at the truth, as to whether matter was composed of small and indivisible particles, or whether it was continuous—the question belonged to what metaphysicians termed "the unknowable;" but he acknowledged the importance of the fullest use of the theory as a kind of ladder to assist the chemist in progressing from one position to another in his science. He was, however, averse to accepting the theory as an absolute truth. Any attempt to realise by its help the action of attractive or repulsive forces upon matter was excessively difficult; indeed to realise such an action through a perfectly void space was to him quite impossible. The same difficulty presented itself to Faraday, who was often obliged to throw this atomic theory on one side, as an obstacle to his progress. He said that if matter be assumed to be thus composed of solid particles separated by a void space, then, in considering the phenomena of electricity in connection with that view of matter, this space existing between the atoms must be either a conductor or a non-conductor of electricity. If a conductor, such a thing as an insulator was obviously an impossibility; if it was a non-conductor, such a thing as a conductor was equally inconceivable; so that, apart from chemistry, there were considerations which made it very undesirable that this theory should be represented as an absolute truth. The speaker then referred to the combination which, in certain cases, is effected between gases by the application of heat, and said that, according to the atomic hypothesis, they must assume that the gases, when heated, have their particles driven further asunder; but how, then, could they more readily enter into combination when heated, as in the case of oxygen and hydrogen, and in many other similar instances? No such difficulty presented itself itself if the gases were regarded as continuous matter. He admired the atomic theory as much as Professor Williamson, and he thought no one could blame him for not making a sufficient use of it; but he did not wish to be considered a blind believer in the theory, or as unwilling to renounce it if anything better presented itself to assist him in his work.

Dr. Odling said that, as a chemist, he was not particularly interested in the question whether matter is infinitely divisible or not, but he did not think that Dr. Williamson's argument had established that conclusion. Dalton was always spoken of as the discoverer of the law, or doctrine, or theory of combination in definite and multiple proportions, to account for which his invention or adoption of the atomic theory was founded; but he conceived that Dalton's great discovery might be better expressed as the law of combination in *reciprocal* and multiple, instead of in *definite* and multiple, proportions. Dalton, in fact, discovered that, if we have three different kinds of matter, A, B, and C, and we find that x times A unites with y times B, and with z times C, then if B and C can combine with one another, it is in the proportion of y and z, or some simple multiples thereof. These were the great facts which Dalton discovered; he showed that they were applicable to all the cases known in his day, and they have proved applicable to all the cases that have been discovered since. Dr. Williamson had said of chemists, that all their modes of thought, and all the government of their actions, are based upon the atomic theory; be he (Dr. Odling) maintained that they are based upon the observed fact that certain bodies combine in certain proportions. Dr. Williamson also argued that the atomic theory is based upon the existence of molecules, and that molecules have no *locus standi* in the absence of the atomic theory. He (Dr. Odling) disputed that position; for the fact that the hydrogen existing in marsh-gas is divisible into four parts, and the hydrogen existing in ammonia only into three parts, is, he conceived, a fact quite independent of the atomic theory. The laws of combination, as far as they go, were compared by Sir Humphry Davy to Kepler's laws of planetary motion, which were simply a general expression of observed facts; and, in the same way, these laws of chemical combination are general expressions of observed facts upon which the atomic hypothesis is superinduced.

Professor Miller thought that Dr. Odling's arguments had not disproved the atomic theory. If we have not the atomic hypothesis, it may fairly be asked what explanation we possess of the laws of combination? Certain facts are admitted on all hands, as determined by experiment, and when an important and extensive series of facts has been ascertained, the endeavour to explain them by some supposition which embraces the facts, and enables us to anticipate new ones if possible, is a necessary result of our mental constitution. Unless some hypothesis is accepted, it is absolutely impossible to reason upon the facts; and, even if we cannot adopt this hypothesis as absolutely true, such a view of the constitution of matter at least explains all the chemical facts that have hitherto been presented. It appeared to him that those who deny the atomic hypothesis are bound to supply something to enable them to interpret phenomena with the same regularity and order as that hypothesis, the accuracy of which they deny and which they desire to displace. No one would suppose that the absence of an undulatory theory for light would conduce to the explanation of optical science; and he held that what the undulatory theory is to the phenomena of light, the atomic theory is to the phenomena of chemistry.

Professor Foster thought the question before the Society was not so much the utility of the atomic theory as its truth. To say that the atomic theory has been very useful, and that, if we refuse to admit it, we have nothing to put in its place, is not exactly the point. A false theory may for the time be an exceedingly useful one, but the present question is whether the atomic theory is true or false, not whether it is a useful or useless theory. With the notions which we are in the habit of entertaining as to the nature of combination, we can indeed scarcely avoid the atomic theory. Naumann has stated, in his "Relations of Heat to Chemistry," that, if we assume that in any compound body the components really occupy distinct portions of space, we must assume that body to be incapable of infinite divisibility. To take a concrete case, he speaks of vermilion. Assuming that in this body the sulphur and mercury occupy separate portions of space, and do not interpenetrate, then we know that so far as any actual division goes, we can separate the portion of vermilion into smaller and smaller bits, but each bit is in itself still a perfect bit of vermilion, containing sulphur and mercury. If, however, we could carry the division further, we must ultimately come to the point at which any further division would give us dissimilar portions of matter, at which, indeed, we should actually have broken off the sulphur from the mercury, and thus have arrived at an atom of this compound body vermilion. He (Professor Foster) thought, however, that the ideas generally entertained of the existence of the constituents of bodies in compounds are not absolutely necessary ideas.

The fundamental phenomenon of every chemical change may perhaps be stated in some such terms as these. We begin with some kind or kinds of matter, say oxygen and hydrogen; we perform a certain operation; these kinds of matter disappear, and another kind of matter, viz., water, begins to exist. But it is, perhaps, after all, a conventional mode of expressing the fact, to say that the oxygen and hydrogen are contained in the water. We know that between the bodies which disappear and the body which appears, there are certain relations, not only qualitative but quantitative, the total mass of the disappearing substances being equal to that of the appearing substances; but we may perhaps return, sometimes, at any rate, with great benefit, to the notion that one portion of matter is actually transmuted into another; that it ceases to exist as such, but something else comes in place of it. From such ideas the existence of atoms would not follow as a necessity, but with our present mode of stating and reasoning about chemical changes, an atomic hypothesis or basis appears to be inevitable.

Professor Tyndall sympathized with the ideas of Professor Williamson; his thoughts had, to some extent, been expressed by Dr. Miller, who had made a strong point in referring to the undulatory theory. It would be quite possible, to detach the phenomena of light from the physical theory of an ether; to say, indeed, that the undulatory theory is something superinduced upon the facts, and a mere imagination. "Still," said Dr. Tyndall, "I do not think we should suffer ourselves to be fettered in that way, and I did certainly expect that Dr. Odling would have gone one step further when he referred to Davy's allusion to the laws of Kepler. Those laws stated the facts of planetary motion as clearly as we know them at the present time; but facts are not sufficient to satisfy the human mind. Another man followed Kepler, and superinduced something upon the facts—a something which we now know as the theory of gravitation. I say it is the nature of the human mind, whenever it attains to any degree of depth—it is the nature of a profound thinker when he ponders upon phenomena of this nature, to seek for some underlying principle—something altogether outside the region of facts—from which he deduces the facts as consequences. Professor Foster has very justly remarked that the usefulness of a theory is not a justification for holding what Professor Frankland has called an absolute truth, and that a bad theory is very often useful. Undoubtedly it is. The emission theory of light was useful in the hands of Newton, Malus, Laplace, and especially of Brewster; and still the emission theory fell. Why did it fall? Because in the progress of investigation phenomena were observed which were perfectly inexplicable upon the emission theory, but which have been completely explained by the undulatory theory. The undulatory theory will stand in front of all opposition as long as it is competent to explain the facts of optics; and I apprehend that the atomic theory will stand as long as it is competent to explain the facts of chemistry."

Dr. Mills observed that to prove the truth of the atomic theory, one of the most important points to establish would be that the general analogy of nature is in favour of indivisibles, whereas the very contrary is the fact. The great kingdoms of nature, animal, vegetable, and mineral, have no positive lines, but pass into one another by imperceptible gradations, as likewise do the forms of nature, geographical, physiological, and crystallographical. In comparing the atomic theory with the undulatory theory of light, it should be remembered that the latter has this advantage, that we have often seen the phenomena of waves, and therefore if an ether exists, it may have a wave-like motion. The existence of waves is a fact; but the existence of atoms, or, in short, of a limit of any sort in nature, is not a fact. No one has ever seen an atom, and in this respect the atomic theory has an exact parallel in the theory of phlogiston. Dr. Mills further remarked that it seems to be forgotten by many writers that matter might be infinitely divisible, and yet that definite proportions might exist; for between two infinites there may be a finite ratio, so that the atomic theory is not perfectly necessary to chemistry. He believed that motion is the highest generalization of modern science, and it therefore affords the only criterion by which all theories, including those of chemistry, must be judged, and the notion of a limit is practically inconsistent with the idea of motion.

Dr. Williamson, in his reply, remarked that the benefit accruing from looking at the question from another point of view would be incalculable, and if another view were developed to supersede the present one, he would be among the first to rejoice at it, but he had not to do with the future of this matter. He had endeavoured to put together the actual evidence of the present view, and none of its opponents had been able to find a fallacy in that evidence. He thought it most important to keep the theory, at each period of its growth, in the modest position which it at present occupies, that of generalizing the best ascertained relations of the elements in their chemical changes.

The Chairman said that he liked to have clear and definite ideas, as far he was able to realize them, on scientific subjects, and he was unable to consider even an atomic theory in the abstract. The atomic theory, as he understood it, was—that any given portion of matter is made up of a number of finite particles, the aggregate of which could be divided and subdivided until at last only one indivisible particle would remain, and this view might be derived from the perusal and common assent of every work on chemistry that defined the atomic theory. He thought that such a view as the physical indivisibility of matter must be separated from the facts and basis of chemistry. Dr. Williamson seemed to think the theory and facts were one, but he (the Chairman) asserted that they are two distinct things. He must also express his dissent from the views expressed by Drs. Frankland and Miller, with reference to the use of a doctrine or theory in which they do not believe. He could not understand using a theory and denying it at the same time. This theory had, he thought, been often the means of deluding chemists into the belief that they understood things about which they knew nothing. He found the works of Kekulé and Naquet scribbled over with pictures of molecules and atoms, arranged in all imaginable ways, for which no adequate reason was given, and if there was no reason for this, it was a mischievous thing to do, for it led to a confusion of ideas, and to mixing up fictions with facts. Many students, he believed, thought the atomic pictures in Naquet's work were the fundamental things to be studied in chemistry; they did not draw the distinction between the facts and theory of the science. His view was that the ultimate constitution of this material universe was one of those questions upon which no light had yet been thrown, and he agreed with Dr. Odling when he said that the science of chemistry did not require or prove the atomic theory. Chemists were bound to have some real and adequate means of working at the science of chemistry, and discovering its laws, and they could effect this, either through the investigation of the laws of gaseous combination, or the study of the capacity of bodies for heat. When they approached the science through the study of gaseous combinations, they discovered that the combining proportions of bodies are capable of being represented by integral numbers, and they also found an analogous fact in studying the capacity of bodies for heat. Either of these methods might be made the basis of the science, and lead to the same general system of ideas, in which we retain what is valuable in the atomic theory, without committing ourselves to a number of assertions incapable of proof.

MACH, ERNST. (b. Chirlitz-Turas near Brno, Moravia [now Chrlice-Turany, Czechoslovakia], 18 February 1838; d. Vaterstetten, near Haar, Germany, 19 February 1916)

After taking his doctoral degree in physics in 1860 at the University of Vienna, Privatdozent Mach gave popular scientific lectures in Vienna on optics, musical acoustics, and psychophysics, and presented university lectures on the principles of mechanics and physics for medical students. He was professor of mathematics at the University of Graz from 1864–1867, professor of experimental physics at Charles-Ferdinand University in Prague from 1867–1895, and rector there from 1882–1884 when the University separated into German and Czech Faculties. Mach returned to the University of Vienna in 1895 with the title of Professor of the History and Theory of the Inductive Sciences. Some of his best-known books appeared after this time: *Principien der Wärmelehre* (1896), dedicated to J. B. Stallo; *Erkenntnis und Irrtum* (1905); and *Die Principien der physikalischen Optik* (1921, posthumously).

Mach's name is perhaps most familiar today among physics and engineering students for his study of sound waves, wave propulsion, and the gas dynamics of projectiles. (The "Mach number" is the ratio of the speed of an object to the speed of sound in the undisturbed medium in which the object is travelling.) His work in physics included conventional studies in optics, electricity, and mechanics; in physiology and psychophysics he studied aural accommodation, the sense of time, kinesthetic sensation and equilibrium, and spatial vision. The discovery of what later were called "Mach bands" (a phenomenon that relates the physiological effect of spatially distributed light stimuli to visual perception) was an important result of this phase of his work.

As a historian and philosopher, Mach was a phenomenalist whose ideas strongly influenced a younger generation of physicists, including Max Planck and Albert Einstein, and the development of the logical positivism of the Vienna Circle of the 1920s was influenced by his empirical positivism. Mach laid the physical and physiological basis for his epistemological views in the *Analyse der Empfindungen* (1886), but his views on sensation, economy of thought, and the inadequacy of Newtonian mechanics already are well-developed in the *Gęschichte und Die Wurzel des Satzes der Erhaltung der Arbeit* published in 1872. The historian Erwin Hiebert has noted that while Mach routinely employed the Newtonian mechanics of point particles and the atomism of kinetic theory in his early lectures, he, nevertheless, slowly developed a more and more incisive critique of this approach in physics. The 1872 book on the conservation of energy was among the most influential of his writings in the nineteenth century.

HISTORY AND ROOT

OF THE PRINCIPLE OF THE

CONSERVATION OF ENERGY

BY

ERNST MACH

TRANSLATED FROM THE GERMAN AND ANNOTATED BY

PHILIP E. B. JOURDAIN, M.A. (Cantab.)

CHICAGO
THE OPEN COURT PUBLISHING CO
LONDON
KEGAN PAUL, TRENCH, TRÜBNER & CO., LTD.
1911

MECHANICAL PHYSICS

THE attempt to extend the mechanical theorem of the conservation of work to the theorem of excluded perpetual motion is connected with the rise of the mechanical conceptions of nature, which again was especially stimulated by the progress of the mechanical theory of heat. Let us, then, glance at the theory of heat.

The modern mechanical theory of heat and its view that heat is motion principally rest on the fact that the quantity of heat present decreases in the measure that work is performed and increases in the measure that work is used, provided that this work does not appear in another form. I say the modern theory of heat, for it is well known that the explanation of heat by means of motion had already more than once been given and lost sight of.

If, now, people say, heat vanishes in the measure that it performs work, it cannot be material, and consequently must be motion.

S. Carnot found that whenever heat performs work, a certain quantity of heat goes from a higher temperature-level to a lower one. He supposed in this that the quantity of heat remains constant. A simple analogy is this: If water (say, by means of a water-mill) is to perform work, a certain quantity of it must flow from a higher to a lower level; the quantity of water remains constant during the process.

When wood swells with dampness, it can perform work, burst open rocks, for example; and some people, as the ancient Egyptians, have used it for that purpose. Now, it would have been easy for an Egyptian wiseacre to have set up a mechanical theory of humidity. If wetness is to do work, it must go from a wetter body to one less wet. Evidently the wiseacre could have added that the quantity of wetness remains constant.

Electricity can perform work when it flows from a body of higher potential to one of lower potential; the quantity of the electricity remains constant.

A body in motion can perform work if it transfers some of its *vis viva* to a body moving more slowly. *Vis viva* can perform work by passing from a higher velocity-level to a lower one; the *vis viva* then decreases.

It would not be difficult to produce such an analogy from every branch of physics. I have intentionally chosen the last, because complete analogy breaks down.

When Clausius brought Carnot's theorem into connexion with the reflexions and experiments of Mayer, Joule, and others, he found that the addition "the quantity of heat remains constant" must be given up. One must, on the other hand, say that a quantity of heat proportional to the work performed vanishes.

"The quantity of water remains constant while work is performed, because it is a substance. The quantity of heat varies because it is not a substance."

These two statements will appear satisfactory to most scientific investigators; and yet both are quite worthless and signify nothing.

We will make this clear by the following question which bright students have sometimes put to me. Is there a mechanical equivalent of electricity as there is a mechanical equivalent of heat? Yes, and no. There is no mechanical equivalent of *quantity* of electricity as there is an equivalent of *quantity* of heat, because the same quantity of electricity has a very different capacity for work, according to the circumstances in which it is placed; but there *is* a mechanical equivalent of electrical energy.

Let us ask another question. Is there a mechanical equivalent of water? No, there is no mechanical equivalent of quantity of water, but there is a mechanical equivalent of weight of water multiplied by its distance of descent.

When a Leyden jar is discharged and work thereby performed, we do not picture to ourselves that the quantity of electricity disappears as work is done, but we simply assume that the electricities come into different positions, equal quantities of positive and negative electricity being united with one another.

What, now, is the reason of this difference of view in our treatment of heat and of electricity? The reason is purely historical, wholly conventional, and, what is still more important, is wholly indifferent. I may be allowed to establish this assertion.

In 1785 Coulomb constructed his torsion balance, by which he was enabled to measure the repulsion of electrified bodies. Suppose we have two small balls, A, B, which over their whole extent are similarly electrified. These two balls will exert on one another, at a certain distance r of their centres from one another, a certain repulsion p. We bring into contact with B, now, a ball C, suffer both to be equally electrified, and then measure the repulsion of B from A and of C from A at the same distance r. The sum of these repulsions is again p. Accordingly something has remained constant. If we ascribe this effect to a substance, then we infer naturally its constancy. But the essential point of the exposition is the divisibility of the electric force p and not the simile of substance.

In 1838 Riess constructed his electrical air-thermometer (the thermoelectrometer). This gives a measure of the quantity of heat produced by the discharge of jars. This quantity of heat is not proportional to the quantity of electricity contained in the jar by Coulomb's measure, but if q be this quantity and s be the capacity, is proportional to q^2/s, or, more simply still, to the energy of the charged jar. If, now, we discharge the jar completely through the thermometer, we obtain a certain quantity of heat, W. But if we make the discharge through the thermometer into a second jar, we obtain a quantity less than W. But we may obtain the remainder by completely discharging both jars through the air-thermometer, when it will again be proportional to the energy of the two jars. On the first, incomplete discharge, accordingly, a part of the electricity's capacity for work was lost.

When the charge of a jar produces heat, its energy

is changed and its value by Riess's thermometer is decreased. But by Coulomb's measure the quantity remains unaltered.

Now let us imagine that Riess's thermometer had been invented before Coulomb's torsion balance, which is not a difficult feat of imagination, since both inventions are independent of each other; what would be more natural than that the "quantity" of electricity contained in a jar should be measured by the heat produced in the thermometer? But then, this so-called quantity of electricity would decrease on the production of heat or on the performance of work, whereas it now remains unchanged; in the first case, therefore, electricity would not be a *substance* but a *motion*, whereas now it is still a substance. The reason, therefore, why we have other notions of electricity than we have of heat, is purely historical, accidental, and conventional.

This is also the case with other physical things. Water does not disappear when work is done. Why? Because we measure quantity of water with scales, just as we do electricity. But suppose the capacity of water for work were called quantity, and had to be measured, therefore, by a mill instead of by scales; then this quantity also would disappear as it performed the work. It may, now, be easily conceived that many substances are not so easily got at as water. In that case we should be unable to carry out the one kind of measurement with the scales while many other modes of measurement would still be left us.

In the case of heat, now, the historically established

measure of "quantity" is accidentally the work-value of the heat. Accordingly, its quantity disappears when work is done. But that heat is not a substance follows from this as little as does the opposite conclusion that it is a substance. In Black's case the quantity of heat remains constant because the heat passes into no *other* form of energy.

If anyone to-day should still wish to think of heat as a substance, we might allow that person this liberty with little ado. He would only have to assume that that which we call quantity of heat was the energy of a substance whose quantity remained unaltered, but whose energy changed. In point of fact we might much better say, in analogy with the other terms of physics, energy of heat, instead of quantity of heat.

By means of this reflection, the peculiar character of the second principal theorem of the mechanical theory of heat quite vanishes, and I have shown in another place that we can at once apply it to electrical and other phenomena if we put "potential" instead of "quantity of heat" and "potential function" instead of "absolute temperature." (See note 3, p. 85.)

If, then, we are astonished at the discovery that heat is motion, we are astonished at something which has never been discovered. It is quite irrelevant for scientific purposes whether we think of heat as a substance or not.

If a physicist wished to deceive himself by means of the notation that he himself has chosen—a state of things which cannot be supposed to be—he would behave similarly to many musicians who, after they have long forgotten how musical notation and softened pitch arose, are actually of the opinion that a piece marked in the key of six flats (G♭) must sound differently from one marked in the key of six sharps (F♯).

If it were not too much for the patience of scientific people, one could easily make good the following statement. Heat is a substance just as much as oxygen is, and it is not a substance just as little as oxygen. Substance is possible phenomenon, a convenient word for a gap in our thoughts.

To us investigators, the concept "soul" is irrelevant and a matter for laughter. But matter is an abstraction of exactly the same kind, just as good and just as bad as it is. We know as much about the soul as we do of matter.

If we explode a mixture of oxygen and hydrogen in an eudiometer-tube, the phenomena of oxygen and hydrogen vanish and are replaced by those of water. We, say, now, that water *consists* of oxygen and hydrogen; but this oxygen and this hydrogen are merely two thoughts or names which, at the sight of water, we keep ready, to describe phenomena which are not present, but which will appear again whenever, as we say, we decompose water.

It is just the same case with oxygen as with latent heat. Both can appear when, at the moment, they cannot yet be remarked. If latent heat is not a substance, oxygen need not be one.

The indestructibility and conservation of matter cannot be urged against me. Let us rather say conservation of *weight;* then we have a pure fact, and we see at once that it has nothing to do with any theory. This cannot here be carried out any farther.

One thing we maintain, and that is, that in the investigation of nature, we have to deal only with knowledge of the connexion of appearances with one another. What we represent to ourselves behind the appearances exists *only* in our understanding, and has for us only the value of a *memoria technica* or formula, whose form, because it is arbitrary and irrelevant, varies very easily with the standpoint of our culture.

If, now, we merely keep our hold on the new laws as to the connexion between heat and work, it does not matter how we think of heat itself; and similarly in all physics. This way of presentation does not alter the facts in the least. But if this way of presentation is so limited and inflexible that it no longer allows us to follow the many-sidedness of phenomena, it should not be used any more as a formula and will begin to be a hindrance to us in the knowledge of phenomena.

This happens, I think, in the mechanical conception of physics. Let us glance at this conception that all physical phenomena reduce to the equilibrium and movement of molecules and atoms.

According to Wundt, all changes of nature are mere changes of place. All causes are motional causes.[17] Any discussion of the philosophical grounds on which Wundt supports his theory would lead us deep into the speculations of the Eleatics and the Herbartians. Change of place, Wundt holds, is the *only* change of a thing in which a thing remains identical with

[17] *Op. cit.*, p. 26.

itself. If a thing changed *qualitatively*, we should be obliged to imagine that something was annihilated and something else created in its place, which is not to be reconciled with our idea of the identity of the object observed and of the indestructibility of matter. But we have only to remember that the Eleatics encountered difficulties of exactly the same sort in motion. Can we not also imagine that a thing is destroyed in *one* place and in *another* an exactly similar thing created?

It is a bad sign for the mechanical view of the world that it wishes to support itself on such preposterous things, which are thousands of years old. If the ideas of matter, which were made at a lower stage of culture, are not suitable for dealing with the phenomena accessible to those on a higher plane of knowledge, it follows for the true investigator of nature that these ideas must be given up; not that only those phenomena exist, for which ideas that are out of order and have been outlived are suited.

But let us suppose for a moment that all physical events can be reduced to spatial motions of material particles (molecules). What can we do with that supposition? Thereby we suppose that things which can never be seen or touched and only exist in our imagination and understanding, can have the properties and relations only of things which can be touched. We impose on the creations of thought the limitations of the visible and tangible.

Now, there are also other forms of perception of other senses, and these forms are perfectly analogous to space—for example, the tone-series for hearing, which corresponds to a space of one dimension—and we do not allow ourselves a like liberty with them. We do not think of all things as sounding and do not figure to ourselves molecular events musically, in relations of heights of tones, although we are as justified in doing this as in thinking of them spatially.

This, therefore, teaches us what an unnecessary restriction we here impose upon ourselves. There is no more necessity to think of what is merely a product of thought spatially, that is to say, with the relations of the visible and tangible, than there is to think of these things in a definite position in the scale of tones.

And I will immediately show the sort of drawback that this limitation has. A system of n points is in form and magnitude determined in a space of r dimensions, if e distances between pairs of points are given, where e is given by the following table:

	e_1	e_2
1	$n-1$	$2n-3$
2	$2n-3$	$3n-6$
3	$3n-6$	$4n-10$
4	$4n-10$	$5n-15$
5	$5n-15$	$6n-21$
r	$rn-\frac{r(r+1)}{2}$	$(r+1)n-\frac{(r+1)(r+2)}{2}$

In this table, the column marked by e_1 is to be used for e if we have made conditions about the sense of the given distances, for example, that in the straight line all points are reckoned according to one direction; in the plane all towards one side of the straight line through the first two points; in space all towards one side of the plane through the first three points; and so on. The column marked by e_2 is to be used if merely the absolute magnitude of the distance is given.

Between n points, combining them in pairs, $\frac{n(n-1)}{1.2}$ distances are thinkable, and therefore in general more than a space of a given number of dimensions can satisfy. If, for example, we suppose the e_1-column to be the one to be used, we find in a space of r dimensions the difference between the number of thinkable distances and those possible in this space to be

$$\frac{n(n-1)}{1.2}-rn+\frac{r(r+1)}{2}=k,$$

or

$$n(n-1)-2rn+r(r+1)=2k,$$

which can be brought to the form

$$(r-n)^2+(r-n)=2k.$$

This difference is, now, zero, if

$$(r-n)^2+(r-n)=0, \text{ or } (r-n)+1=0, \text{ or } n=r+1.$$

For a space of three dimensions, the number of distances thinkable is greater than the number of distances possible in this space when the number of points is greater than four. Let us imagine, for example, a molecule consisting of five atoms, A, B, C, D, and E, then between them, ten distances are thinkable, but, in a space of three dimensions, only nine are possible, that is to say, if we choose nine such distances, the tenth thinkable one is determined by means of the nature of this space, and it is no longer arbitrary. If AB, BC, CA, AD, DB, DC, are given me, I get a tetrahedron of fixed form. If, now,

I add E with the distances *E A*, *E B*, and *E C* determined, then *D E* is determined by them. Thus it would be impossible gradually to alter the distance *D E* without the other distances being thereby altered. Thus, there might be serious difficulties in the way of imagining many pent-atomic isomeric molecules which merely differ from one another by the relation of *D* and *E*. This difficulty vanishes in our example, when we think the pent-atomic molecule in a space of four dimensions; then ten independent distances are thinkable and also ten distances can be set up.

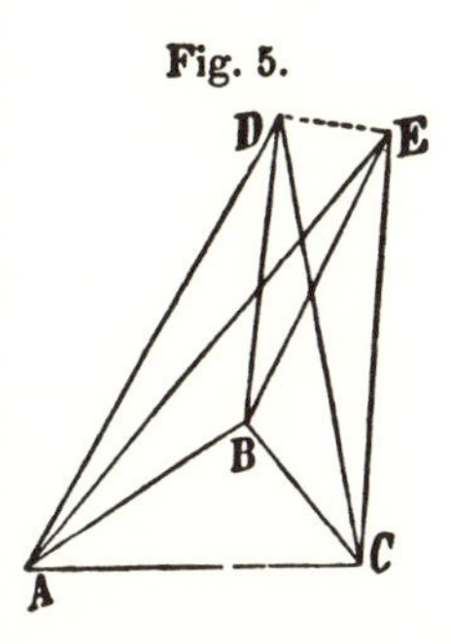

Fig. 5.

Now, the greater the number of atoms in a molecule, the higher the number of the dimensions of space do we need to make actual all the thinkable possibilities of such combinations. This is only an example to show under what limitations we proceed when we imagine the chemical elements lying side by side in a space of three dimensions, and how a crowd of the relations of the elements can escape us thereby if we wish to represent them in a formula which cannot comprise them. (See note 4, p. 86.)

It is clear how we can study the nature of chemical combinations without giving ourselves up to the conception mentioned, and how, indeed, people have now begun to study them. The heat of combustion generated by a combination gives us a clearer idea of the stability and manner of combination than any pictorial representation. If, then, it were possible, in any molecule composed of n parts, to determine the $\frac{n(n-1)}{1.2}$ heats of combination of every two parts, the nature of the combination would be characterized thereby. According to this view, we would have to determine $\frac{n(n-1)}{1.2}$ heats of combination, whereas, if the molecules were thought spatially, $3n-6$ heats of combination suffice. Perhaps, too, a more rational manner of writing chemical combinations can be founded on this. We would write the components in a circle, draw a line from each to each, and write on the latter the respective heat of combination.

Perhaps the reason why, hitherto, people have not succeeded in establishing a satisfactory theory of electricity is because they wished to explain electrical phenomena by means of molecular events in a space of three dimensions.

Herewith I believe that I have shown that one can hold, treasure, and also turn to good account the results of modern natural science without being a supporter of the mechanical conception of nature, that this conception is not necessary for the knowledge of the phenomena and can be replaced just as well by another theory, and that the mechanical conceptions can even be a hindrance to the knowledge of phenomena.

Let me add a view on scientific theories in general: If all the individual facts—all the individual phenomena, knowledge of which we desire—were immediately accessible to us, a science would never have arisen.

Because the mental power, the memory, of the individual is limited, the material must be arranged. If, for example, to every time of falling, we knew the corresponding space fallen through, we could be satisfied with that. Only, what a gigantic memory would be needed to contain the table of the correspondences of s and t. Instead of this we remember the formula $s=\frac{gt^2}{2}$, that is to say, the rule of derivation by means of which we find, from a given t, the corresponding s, and this replaces the table just mentioned in a very complete, convenient, and compendious manner.

This rule of derivation, this formula, this "law," has, now, not in the least more real value than the aggregate of the individual facts. Its value for us lies merely in the convenience of its use: it has an economical value. (See note 5, p. 88.)

Besides this collection of as many facts as possible in a synoptical form, natural science has yet another problem which is also economical in nature. It has to resolve the more complicated facts into as few and as simple ones as possible. This we call explaining. These simplest facts, to which we reduce the more complicated ones, are always unintelligible in themselves, that is to say, they are not further resolvable. An example of this is the fact that one mass imparts an acceleration to another.

Now, it is only, on the one hand, an economical question, and, on the other, a question of taste, at what unintelligibilities we stop. People usually deceive themselves in thinking that they have reduced the unintelligible to the intelligible. Understanding consists in analysis alone; and people usually reduce uncommon unintelligibilities to common ones. They always get, finally, to propositions of the form: if A is, B is, therefore to propositions which must follow from intuition, and, therefore, are not further intelligible.

What facts one will allow to rank as fundamental facts, at which one rests, depends on custom and on history. For the lowest stage of knowledge there is no more sufficient explanation than pressure and impact.

The Newtonian theory of gravitation, on its appearance, disturbed almost all investigators of nature because it was founded on an uncommon unintelligibility. People tried to reduce gravitation to pressure and impact. At the present day gravitation no longer disturbs anybody: it has become a *common* unintelligibility.

It is well known that action at a distance has caused difficulties to very eminent thinkers. "A body can only act where it ɪs"; therefore there is only pressure and impact, and no action at a distance. But where is a body? Is it only where we touch it? Let us invert the matter: a body is where it acts. A little space is taken for touching, a greater for hearing, and a still greater for seeing. How did it come about that the sense of touch alone dictates to us where a body is? Moreover, contact-action can be regarded as a special case of action at a distance.

It is the result of a misconception, to believe, as people do at the present time, that mechanical facts are more intelligible than others, and that they can provide the foundation for other physical facts. This belief arises from the fact that the history of mechanics is older and richer than that of physics, so that we have been on terms of intimacy with mechanical facts for a longer time. Who can say that, at some future time, electrical and thermal phenomena will not appear to us like that, when we have come to know and to be familiar with their simplest rules?

In the investigation of nature, we always and alone have to do with the finding of the best and simplest rules for the derivation of phenomena from one another. One fundamental fact is not at all more intelligible than another: the choice of fundamental facts is a matter of convenience, history, and custom.

The ultimate unintelligibilities on which science is founded must be facts, or, if they are hypotheses, must be capable of becoming facts. If the hypotheses are so chosen that their subject (*Gegenstand*) can never appeal to the senses and therefore also can never be tested, as is the case with the mechanical molecular theory, the investigator has done more than science, whose aim is facts, requires of him—and this work of supererogation is an evil.

Perhaps one might think that rules for phenomena, which cannot be perceived in the phenomena themselves, can be discovered by means of the molecular theory. Only that is not so. In a complete theory, to all details of the phenomenon details of the hypothesis must correspond, and all rules for these hypothetical things must also be directly transferable to the phenomenon. But then molecules are merely a valueless image.

Accordingly, we must say with J. R. Mayer: "If a fact is known on all its sides, it is, by that knowledge, explained, and the problem of science is ended."[18]

[18] *Mechanik der Wärme*, Stuttgart, 1867, p. 239.

(*NOTES*)

3. (See p. 47.) The note in question appeared in the number for February, 1871, of the Prague journal, *Lotos*, but was, however, drawn up a year earlier. This is a complete reproduction of it:

The second law of thermodynamics can, as is well known, be expressed for a simple case by the equation

$$-\frac{Q}{T}+Q'\left(\frac{1}{T'}-\frac{1}{T}\right)=0,$$

where Q denotes the quantity of heat transformed into work, at the absolute temperature T, and Q' the quantity of heat which simultaneously sunk from the higher temperature T to the temperature T'.

Now, we have not far to seek for the observation that this theorem is not limited to the phenomena of heat, but can be transferred to other natural phenomena, if, instead of the quantity of heat, we put the potential of whatever is active in the phenomenon, and, instead of the absolute temperature, the potential function. Then the theorem may be expressed thus:

If a certain potential-value P of an agent at the potential-level V passes over into another form—for example, if the potential of an electrical discharge is transferred into heat—then another potential-value, P', of the same agent sinks simultaneously from the higher potential-level V to the lower one V'. And the said values are connected with one another by the equation

$$-\frac{P}{V}+P'\left(\frac{1}{V'}-\frac{1}{V}\right)=0.$$

In the application of the theorem, the only questions are, what is to be conceived as potential (as equivalent of mechanical work), and what is the potential-function. In many cases this is self-evident and long established, in others it can easily be found. If, for example, we wish to apply the theorem to the impact of inert masses, obviously the *vis viva* of these masses is to be conceived as the potential, and their velocity as the potential-function. Masses of equal velocity can communicate no *vis viva* to one another—they are at the same potential-level.

I must reserve for another occasion the development of these theorems.

PRAGUE
February 16, 1870

4. (See p. 53.) The manner in which I was led to the view that we need not necessarily represent to ourselves molecular-processes spatially, at least not in a space of three dimensions, was as follows:

In the year 1862, I drew up a compendium of physics for medical men, in which, because I strove after a certain philosophical satisfaction, I carried out rigorously the mechanical atomic theory. This work first made me conscious of the insufficiency of this theory, and this was clearly expressed in the preface and at the end of the book, where I spoke of a total reformation of our views on the foundations of physics.

I was busied, at the same time, with psychophysics and with Herbart's works, and so I became convinced that the intuition of space is bound up with the organization of the senses, and, consequently, that we are not justified in ascribing spatial properties to things which are not perceived by the senses. In my lectures on psychophysics,[3] I already stated clearly that we are not justified in thinking of atoms spatially. Also, in my theory of the organ of hearing,[4] I brought before my readers the series of tones as an analogue of space of one dimension. At the same time the quite arbitrary and, on this account, faulty limitation of the number of dimensions in Herbart's derivation of "intelligible" space struck me. By that, now, it became clear to me that, for the understanding, relations like those of space, and of any number of dimensions, are thinkable.

My attempts to explain mechanically the spectra of the chemical elements and the divergence of the theory with experience strengthened my view that we must not represent to ourselves the chemical elements in a space of three dimensions. I did not venture, however, to speak of this candidly before orthodox physicists. My notices in Schlömilch's *Zeitschrift* of 1863 and 1864 contained only an indication of it.

All the views on space and time developed in this pamphlet were first communicated in my course of lectures on mechanics in the summer of 1864 and in my course on psychophysics delivered in the winter of 1864–1865, which latter course was attended by large audiences, and also by many professors of the University of Graz. The most important and most general results of these considerations were published by me in the form of short notes in Fichte's *Zeitschrift für Philosophie* of

3 *Oesterr. Zeitschr. für praktische Heilkunde*, 1863.

4 *Sitzber. der Wiener Akademie*, 1863.

1865 and 1866. In this, external stimuli were entirely lacking, for Riemann's paper, which first appeared in 1867,[5] was quite unknown to me.

5. (See p. 55.) The view that in science we are chiefly concerned with the convenience and saving of thought, I have maintained since the beginning of my work as a teacher. Physics, with its formulae and potential-function, is especially suited to put this clearly before me. The moment of inertia, the central ellipsoid, and so on, are simply examples of substitutes by means of which we conveniently save ourselves the consideration of the single mass-points. I also found this view developed with especial clearness in the case of my friend the political economist E. Herrmann. From him I have taken what seems to me a very suitable expression: "Science has a problem of economy or thrift."

[5] [Riemann's work *Ueber die Hypothesen, welche der Geometrie zu Grunde liegen* was written and read to a small circle in 1854, first published posthumously in 1867, and reprinted in his *Ges. Werke*, pp. 255–268.]

LOCKYER, JOSEPH NORMAN. (b. Rugby, England, 17 May 1836; d. Salcombe Regis, England, 16 August 1920)

Lockyer was educated in the English Midlands and travelled in Switzerland and France before entering the civil service in the War Office in 1857. His success as an amateur astronomer led, from 1870, to governmental scientific posts culminating in his appointment to the Solar Physics Observatory in South Kensington, London. There he remained until it was transferred to Cambridge in 1911. He then established an observatory near his home in Salcombe Regis, Devonshire, and this observatory is now attached to the University of Exeter. Lockyer founded the scientific journal *Nature* in 1869 and edited it for the next fifty years, during which time it became the world's leading general scientific periodical.

Lockyer's spectroscopic observations of the sun gave him a distinguished reputation. In 1868 he successfully observed the solar prominences at a time other than during a total solar eclipse, obtaining the bright lines of the prominence spectrum without interference from the spectrum of diffused sunlight in the atmosphere. This was done independently and almost simultaneously by Jules Janssen, and the work of the two men was honored as a codiscovery by the Paris Academy of Sciences. Lockyer detected a yellow line in the prominence spectrum and in the outer atmosphere of the sun (which he named the chromosphere); he attributed the line to an unknown element, to which he gave the name helium. It was not until 1895 that William Ramsay verified the existence of helium as a terrestrial element.

Lockyer was a man given to speculation, and was interested in cosmic and stellar evolution and in the possible astronomical significance of ancient eastern temples and prehistoric stone circles. One of his controversial hypotheses was the notion of the "dissociation" (into subatoms) of the chemical elements under the influence of strong energy sources, like those found in the sun. This view was based on his observation that the spectra of an element varies with energy conditions and his interpretation that spectra of different elements contain lines in common. More accurate measuring devices revealed the apparent coincidences of lines to be illusory. Nevertheless, Lockyer's ideas and evidence were at the focus of many discussions, in the late nineteenth century, about the possibility of a substructure for elemental atoms.

CHAPTER IV.

ATOMS AND MOLECULES SPECTROSCOPICALLY CONSIDERED.

WHAT are atoms and what are molecules? A chemist discourses on the atomic weight of certain elements, and defines and talks about molecular volumes, and the like. Here is a definition given by Dr. Frankland in his book on Chemistry ("Lecture Notes," p. 2): "An atom is the smallest proportion by weight in which the element (that is to say the element to which the atom under discussion belongs) enters into or is expelled from a chemical compound." He then points out that when atoms are isolated—that is, when they are separated from other kinds of matter—they do not necessarily exist as atoms in the old sense; they go about in company, generally being associated in pairs. He then defines such a combination of atoms as an elementary molecule. Here, then, is put before us authoritatively a chemist's view of the difference between an atom and a molecule.

Let us now go to the physicist and see if we can gather from him his idea of atoms or molecules. It is remarkable that, in Prof. Clerk-Maxwell's "Theory of Heat," in which we find much that is known by physicists about molecular theories, the word "atom" is not used at all. We are at once introduced to the word "molecule," which is defined to be "a small mass of matter the parts of which do not part company during the excursions which the molecule makes when the body to which it belongs is hot."

Prof. Clerk-Maxwell goes on to give us ideas about these "molecules," which have resulted from the investigations of himself and others. Here are some of them (p. 286): "All bodies consist of a number of small parts called molecules. Every molecule consists of a definite quantity of matter, which is exactly the same for all the molecules of the same substance. The mode in which the molecule is bound together is the same for all molecules of the same substance. A molecule may consist of several distinct portions of matter held together by chemical bonds, and may be set in vibration, rotation, or any other kind of relative motion, but so long as the different portions do not part company but travel together in the excursions made by the molecule, our theory calls the whole connected mass a single molecule." Here, then, we have our definition of a molecule enlarged.

The next point insisted upon is that the molecules of all bodies are in a state of continual agitation.

That this agitation or motion exists in the smallest parts of bodies is partly made clear by the fact that we cannot see the bodies themselves move.

What, then, on this theory, is the difference between the solid, liquid, and gaseous states of matter?

In a solid body the molecule never gets beyond a certain distance from its initial position. The path it describes is often within a very small region of space. Prof. Clifford, in a lecture upon atoms, has illustrated this very clearly. He supposes a body in the middle of the room held by elastic bands to the ceiling and the floor, and in the same manner to each side of the room. Now pull the body from its place; it will vibrate, but always about a mean position; it will not travel bodily out of its place. It will always go back again.

We next come to fluids: concerning these we read—"In fluids, on the other hand, there is no such restriction to the excursions of a molecule. It is true that the molecule generally can travel but a very small distance before its path is disturbed by an encounter with some other molecule; but after this encounter there is nothing which determines the molecule rather to return towards the place from whence it came than to push its way into new regions. Hence in fluids the path of a molecule is not confined within a limited region, as in the case of solids, but may penetrate to any part of the space occupied by the fluid."

Now we have the motion of the molecule in the solid and the fluid. How about the movement in a gas? "A gaseous body is supposed to consist of a large number of molecules moving very rapidly." For instance, the molecules of air travel about twenty miles in a minute. "During the greater part of their course these molecules are not acted upon by any sensible force, and therefore move in straight lines with uniform velocity. When two molecules come within a certain distance of each other, a mutual action takes place between them which may be compared to the collision of two billiard balls. Each molecule has its course changed and starts in a new path."

The collision between two molecules is defined as an "Encounter;" the course of a molecule between encounters a "Free path." It is then pointed out that "in ordinary gases the free motion of a molecule takes up much more time than is occupied by an encounter. As the density of the gas increases the free path diminishes, and in liquids no part of the course of a molecule can be spoken of as its free path."

The Kinetic Theory of Gases, on which theory these statements are made, has this great advantage about it, that it explains certain facts which had been got at experimentally, facts which had been established over and over again, but which lacked explanation altogether till this molecular theory, which takes for granted the existence of certain small things which are moving rapidly in gases, less rapidly in fluids, and still less in solids, was launched. The theory, in fact, explains in a most ample manner, many phenomena so well known that they are termed Laws. It explains Boyle's law, and others.

This theory, which takes for its basis the existence of molecules and their motions, explains pressure by likening it to the bombardment of the sides of the containing vessel by the molecules in motion; or it tells us that the temperature of a gas depends upon the velocity of the agitation of the molecules, and that this velocity of the molecules in the same gas is the same for the same temperature, whatever be the density. When the density varies, the pressure varies in the same proportion. This is Boyle's law. Further, the densities of two gases at the same temperature and pressure are proportional to the masses of their individual molecules, or, when two gases are at the same pressure and temperature, the number of molecules in unit of volume is the same. This is the law of Gay Lussac.

We are now, then, fairly introduced to the "atom" of the chemist and the "molecule" of the physicist; we see at once that the methods of study employed by chemical and physical investigators are widely different. The chemist never thinks about encounters, and the physicist is careless as to atomic weight; in his mind's eye he sees a perpetual clashing and rushing of particles of matter, and he deals rather with the quality of the various motions than with the material.

In Prof. Clerk-Maxwell's book (p. 306) it is assumed that while the molecule is traversing its free path after an encounter, it vibrates according to its own law, the law being determined by the construction of the molecule, or let us say its chemical nature, so that the vibration of one particle of sodium would be like that of another particle of sodium, but unlike that of a particle of another chemical substance, let us say iron. If the interval between encounters is long, the molecule may have used up its vibrations before the second encounter, and may not vibrate at all for a certain time previous to it. The extent of the vibration will depend upon the kind of encounter, and will in a certain sense be independent of the number of encounters.

We can imagine a small number of feeble encounters, a large number of feeble encounters, a small number of strong encounters, and a large number of strong encounters.

In the case of feeble encounters, we pass from a small number to a large one by increasing the density.

In the case of strong encounters we pass from low temperature with small density to high temperature with great density.

Increase of density will reduce "free path."

Increase of temperature will increase the intensity of the vibrations.

The shorter the free path the more complex the vibrations.

The more decided the encounter the more will the vibration of the molecule be brought out, not merely the *fundamental vibrations*, as we may term them, which we get when the free path is longest, but all those which are possible to each molecule.

Now why have these detailed statements concerning the vibrations of molecules been necessary? Because we believe that each molecular vibration disturbs the ether; that spectra are thus begotten; each wave-length of light resulting from a molecular tremor of corresponding wave-length. The molecule is, in fact, the sender, the ether the wire, and the eye the receiving instrument, in this new telegraphy.

As before, let us endeavour to see if our ideas may be rendered more clear by any of the more familiar phenomena in sound.

Let us call the motions, of whatever nature they may be, set up in a molecule of matter, vibrations. Then, to get the most concrete notion of such a light source, let us compare it with the most simple sound source, a tuning-fork.

The same tuning-fork will always give us the same sound, but the sound is more complex than might at first sight be imagined. In addition to the prevailing note, which depends upon the number of vibrations per second, as we have seen, there are other tones which have a definite relationship to the prevailing, or, as it is termed, the fundamental note. The loudness of all these, as we have also seen, depends upon the amplitude of the vibration of the tuning-fork.

But there are more complicated cases of vibration than these.

In the violin we have a convenient example which shows us that the quality of the sound produced, that is the quality of the vibration of the string set up, depends upon the manner in which the bow is drawn over the string.

Similarly, the same vibrating plate, when damped at different points, vibrates in quite a different manner.

Now does the spectroscope throw any light upon molecular questions? is there any hope that the spectroscope, as researches with it are extended, may aid the study of a subject which lies at the root of chemical and physical investigation?

In order to endeavour to answer these questions, I will now proceed to lay down some propositions embracing the knowledge which has been acquired up to this time from this point of view. These propositions I shall take one by one. I shall state the experimental basis on which the statements rest, and shall refer to the methods by which the results have been obtained.

I shall for a time use the word "particle" to represent a small mass of matter, because it does not tie me to the "atom," or the "molecule" of the chemist, or to the "molecule" of the physicist. "Particle" is a neutral term, which I hope none will quarrel with.

§ 1. *Radiation.*

1. *When bodies retain a solid or liquid form when incandescent, their constituent molecules give out rays of light such that the spectrum is continuous as far as it goes.*

This was Kirchoff's first generalization.

It surely is an important fact from the point of view of the molecular theory, that all solids and liquids, with their particles moving in the manner already stated, do give us a perfectly distinct spectrum from that which we get when we deal with any rare gas or vapour whatever. A poker put into the fire becomes first of a dull red heat, after a time a white heat is arrived at. So far as the vibrations exist they are continuous, there are no breaks in the series of wave-lengths. We may also drive a platinum wire to incandescence in the same way by means of electricity. Analyse the light by means of the spectroscope, the spectrum is the same as that of the poker. Further, we can go to the sun, and divest it in imagination of the atmosphere which absorbs so much of its light, and we know that, with a small exception, we shall get a perfectly continuous spectrum similar to that in the case of the poker or platinum wire.

In this continuous spectrum we have a spectroscopic fact connected with that kind of molecular motion which physicists attribute to particles so long as they are closely packed together in the solid state, and so long as they have but a small free path, as in the fluid state.

2. *When particles are in a state of gas or vapour, and are rendered incandescent by high tension electricity, line-spectra are produced in the case of all the chemical elements.*

These line-spectra are only to be obtained from gases and vapours, and, with few exceptions, only when we employ high-tension electricity.

We get a spectroscopic result perfectly distinct from the one we had before, precisely in the case where according to the physicists we have an enormous motion and agitation of particles.

3. *The characteristic vibration of a particle is independent of length of free path.*

In the Kinetic theory, as generally enunciated, there is nothing to show that the same particle may not be in question in the solid, liquid, and gaseous states, the only change of condition being in the amount of free path.

Now as the spectra of solids and gases present a complete difference in kind (see 1 and 2), if the particle were always the same, it would be necessary to assume that, under different conditions of free path, the same particle can be thrown into different states of vibration and give us different spectra.

The question is, have we at the present time any facts at our disposal? I think we have, although some may not consider them sufficient in number or cogency; but it must always be remembered that it is precisely in such questions as these that experiments on an extended scale become almost impossible.

All the facts we have, however, tend to show that a known change of molecular condition is always accompanied by a change of spectrum, *e.g.*, sulphur vapour above and below 1,000° C.—at which point its vapour density changes—has distinct spectra.

Salts have spectra of their own, in which no lines, either of the constituent metal or metalloid, are to be found.

Another line of argument. In some cases we can mix vapours with liquids and the spectrum of the vapour remains unchanged in character; that is, the circumambient particles of the liquid behave in one case in exactly the same manner as the circumambient particles of the air do in another.

Iodine in bisulphide of carbon, $N_2 O_4$ in water, and didymium salts in water are illustrations.

Nay, we may even enclose, or appear to enclose, some substances in glass (salts of didymium, erbium, nickel, cobalt), and we get a spectrum so special in each case that we know that the particles are still going through their motions, are still vibrating, in spite of the absence of free path, and in spite of the "solid" state of their surroundings.

I shall, elsewhere, use other lines of argument to show that the reason that we so rarely see these characteristic spectra in connection with the solid state lies in the fact that the solid state is one reached not only by reduction of free path, which enables the molecules to lie nearer together, but by a reduction of molecular agitation, which in all probability enables them to combine *inter se.*

4. *In some cases particles in a state of gas or vapour can be set swinging by heat waves.*

Salts of sodium and strontium, subjected to the heat of a Bunsen burner, are at once dissociated, and the particles of the metals are set swinging by the heat waves, and we get the lines in the spectra of their vapours. Now that is not only true for salts of strontium and sodium, but for some of the elements themselves. But if salts of iron, or of the other heavy metals are placed in the flame, we do not get bright lines. Or again, in some other vapours, such as sulphur, we only get a spectrum, not of lines, but continuous over a limited part of the spectrum. In fact it may be said, with the exception of these elements which easily reverse themselves, this heat is absolutely imcompetent to give anything like a bright line.

5. *The spectra of both elementary and compound bodies vary with varying degrees of heat.*

It has already been stated that a Bunsen burner is enough to set an atom of sodium free from its combination with chlorine and make its vapour give us a bright line, while we cannot do this in the case of iron and other substances. We may say then that we have there a first stage of temperature. Many monad metals give us their line spectra at a low degree of heat. Take some dyad metals, such as zinc and cadmium; this first stage of temperature will only make them red or white hot, a much higher temperature is required to drive them into vapour. We get the line spectrum from sodium; do we get that from cadmium when we have melted cadmium? We do not. This is an excessively important point. The first stage of temperature, which gives us a line spectrum in the case of sodium, is powerless to give us such a spectrum in the case of cadmium.

A second stage of heat at least is therefore required to get a line spectrum. If I take sulphur, dealing with it by means of absorption, and heat it, I get a continuous spectrum at the first stage. I increase the heat to the second stage, what do I get then? A line spectrum, as I do in the case of sodium? No! A spectrum like that of carbon, not a line spectrum at all. I apply still a higher, a third, stage of temperature and then I get a line spectrum. In the case of the metalloids we have thus three stages of heat with three spectra. If there is such a thing as a particle at all, are we not justified in asking whether there is not some difference between the "particular" arrangements of the metalloids, and those of the metals? and some connection between temperature and the "atomic weights" of the chemist?

Before I go further I will throw these results into a tabular form, which will show that through these various heat stages, in the case of metals like sodium there is a great preponderance of line spectrum, and in the case of metalloids like sulphur, there is a great preponderance of fluted spectrum.

	Na.	Cd.	S.
Fifth stage—spark	line spectrum	line	line
Fourth stage—arc	line	line	fluted
Third stage—white heat	line	(?)	"
Second stage—bright red heat	line	continuous absorption in the blue	"
First stage of heat—dull red heat	line	continuous spectrum	continuous absorption in the blue

6. *From the fact that we have lines in the spectra of compound gases, it would be hazardous to affirm that the aggregate, which, with the highest dissociating power*

we can employ, gives us a line spectrum of a so-called element, could not be broken up if a still higher dissociating power could be employed.

This proposition has a bearing not only on the celestial but also on the terrestrial side of the inquiry, and is referred to at length in the chapter on dissociation.

7. *There is spectroscopic evidence which seems to show that, starting with a mass of solid elemental matter, such mass of matter is continually broken up as the temperature (including in this term the action of electricity) is raised.*

The evidence upon which I rely is furnished by the spectroscope in the region of the visible spectrum.

To begin by the extreme cases, all solids give us continuous spectra; all vapours produced by the high-tension spark give us line-spectra.

Now the continuous spectrum may be, and as a matter of fact is, observed in the case of chemical compounds, whereas all compounds known as such are resolved by the high-tension spark into their constituent elements. We have a right, therefore, to assume that an element in the solid state is a more complex mass than the element in a state of vapour, as its spectrum is the same as that of a mass which is known to be more complex.

The spectroscope supplies us with intermediate stages between these extremes.

(*a*) The spectra vary as we pass from the induced

current with the jar to the spark without the jar, to the voltaic arc, or to the highest temperature produced by combustion. The change is always in the same direction; and here, again, the spectrum we obtain from elements in a state of vapour (a spectrum characterized by spaces and bands) is similar to that we obtain from vapours of which the compound nature is unquestioned.

(β) At high temperatures, produced by combustion, the vapours of some elements (which give us neither line- nor channelled space-spectra at those temperatures, although we undoubtedly get line-spectra when electricity is employed, as before stated) give us a continuous spectrum at the more refrangible end, the less refrangible end being unaffected.

(γ) At ordinary temperatures, in some cases, as in selenium, the more refrangible end is absorbed; in others the continuous spectrum in the blue is accompanied by a continuous spectrum in the red. On the application of heat, the spectrum in the red disappears, that in the blue remains; and further, as Faraday has shown in his researches on gold-leaf, the masses which absorb in the blue may be isolated from those which absorb in the red. It is well known that many substances known to be compounds in solution give us absorption in the blue or blue and red; and, also, that the addition of a substance known to be compound (such as water) to substances known to be compound which absorb the blue, superadds an absorption in the red.

In those cases which do not conform to what has

been stated, the limited range of the visible spectrum must be borne in mind. Thus I have little doubt that the simple gases, at the ordinary conditions of temperature and pressure, have an absorption in the ultra-violet, and that highly compound vapours are often colourless because their absorption is beyond the red, with or without an absorption in the ultra-violet. Glass is a good case in point; others will certainly suggest themselves as opposed to the opacity of the metals.

If we assume, in accordance with what has been stated, that the various spectra to which I have referred are really due to different molecular aggregations, we shall have the following series, going from the more simple to the more complex.

First stage of complexity of molecule . . .	Line-spectrum.
Second Stage	Fluted-spectrum.
Third stage	Continuous absorption at the blue end not reaching to the less refrangible end. (This absorption may break up into a fluted-spectrum.)
Fourth stage.	Continuous absorption at the red end not reaching to the more refrangible end. (This absorption may break up into a fluted-spectrum.)
Fifth stage	Unique continuous absorption.

One or two instances of the passage of spectra from one stage to another, beginning at the fifth stage, may be given.

From the fifth stage to the fourth—The absorption

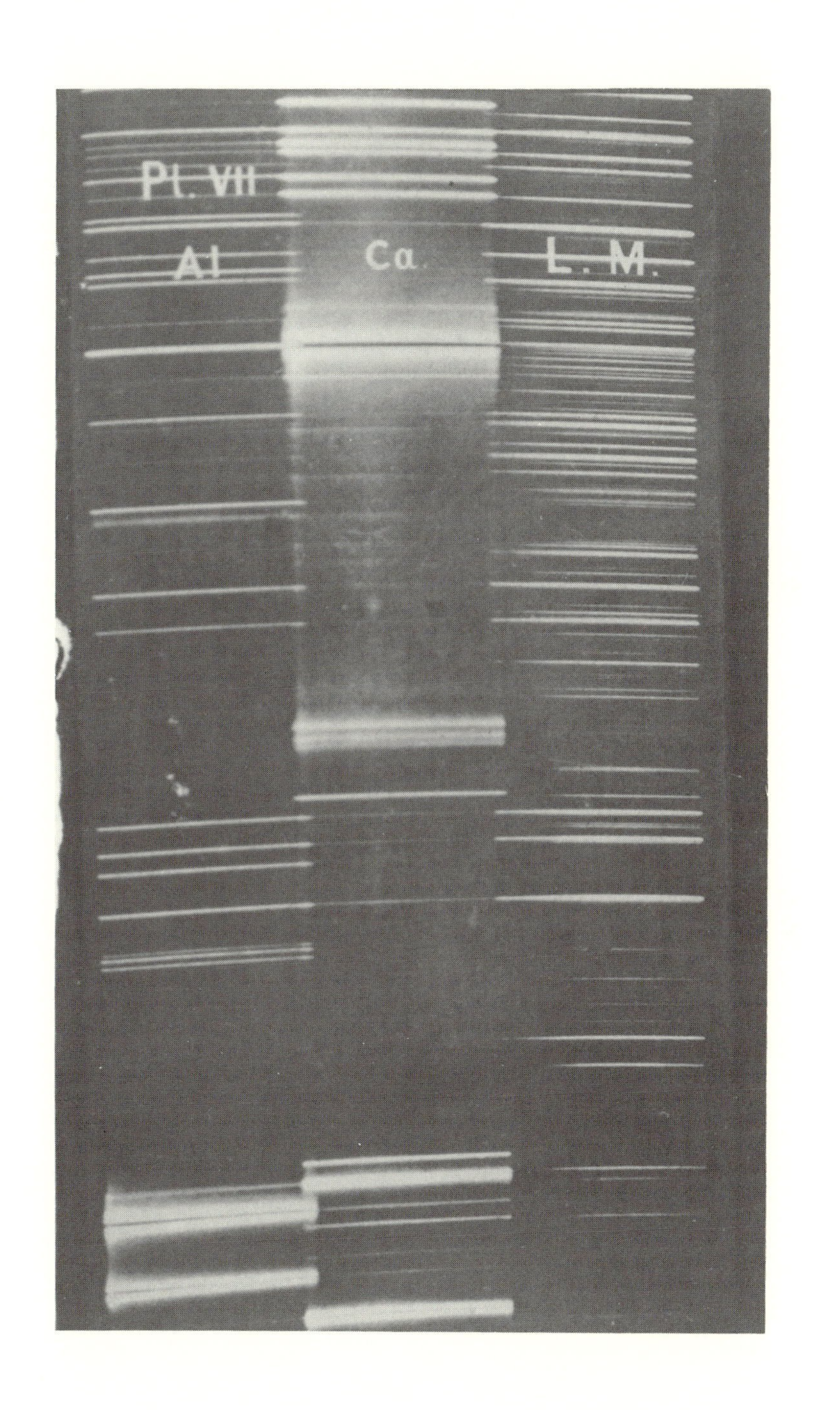
Pl. VII
Al
Ca.
L. M.

of the vapours of potassium in a red-hot tube is at first continuous. As the action of the heat is continued, this continuous spectrum breaks in the middle; one part of it retreats to the blue, the other to the red.

From the fourth stage to the third—Faraday's researches on gold-leaf best illustrate this; but I hold that my explanation of them by masses of two degrees of complexity only is sufficient without his conclusion ("Researches in Chemistry," p. 417), that they exist "of intermediate sizes or proportions."

Gold is generally yellow, as you know, but gold is also blue and sometimes red. It must be perfectly clear to all, that if particles vibrate the colours of substances must have something to do with the vibrations. If the colours have anything to do with the particles it must be with their vibrations. Now as the spectrum in the main consists of red, yellow, and blue, the red and the blue rays are doing something in a substance which only transmits or reflects the yellow light; if we put gold leaf in front of the lime light, we can see whether the yellow light does or does not suffer any change. The yellow disappears; we have a green colour; the red and blue are absent. The gold leaf is of excessive thickness. What would happen could I make it thinner? Its colour would become more violet. This I have proved by using aqua regia. But we can obtain a solution of fine gold, which lets the red light through. Its particles are doing something with the blue vibrations. We can obtain another solution which only transmits the blue. Now what is the difference—

the "particular" differences—between the gold in these solutions, and that which is yellow by reflected, and green or violet by transmitted light? It is a question worthy of much study. Here are some more experiments. Take some chloride of cobalt, which is blue, put it into a test-tube, to which add water. It turns red. I content myself by asking why it turns red? We take some chloride of nickel, which is yellow, and put it into another test-tube: we add water, and it turns green. First question—Why this change? Second question—Has the green colour of this solution anything to do with the red colour of the solution of gold?

From the third stage to the second—Sulphur-vapour first gives a continuous absorption at the blue end; on heating, this breaks up into a fluted-spectrum.

The fluted spectra of potassium and sodium make their appearance after the continuous absorption in the blue and red vanishes.

From the second stage to the first—In many metalloids the spectra, without the jar, are fluted, on throwing the jar into the circuit the line-spectrum is produced, while the cooler exterior vapour gives a fluted absorption-spectrum.

The fluted spectra of potassium and sodium change into the line-spectrum (with thick lines which thin subsequently) as the heat is continued.

These various molecular combinations may go far to explain the law of multiple proportions of the chemist.

8. *Line spectra become more complicated with increased density or temperature, provided the state of gas or vapour be retained.*

The importance of this observed fact in connection with the molecular theory cannot be overrated. In the solid the particles can only oscillate round their mean position ; in the gas they can go through with enormous rapidity a tremendous number of various movements of rotation and vibration, and along their free path ; and spectroscopically we can follow these movements by differences in the phenomena observed. We get a solid or liquid condition, and a continuous spectrum ; we get the most tenuous gaseous condition, and then the phenomenon is changed, and the spectrum consists of a single line. The present point is this, that, so far as the visible spectrum goes, it is possible by working with a gas at low pressure, and not too high temperature, to get a spectrum from any gas or vapour of only a single line. As we increase the density, and thus increase encounters ; or increase temperature, and thus increase the energy of each encounter, so does the spectrum get more and more complicated.

9. *In the case of metals there are two different ways in which this complexity comes about.*

We may picture to ourselves the particles cooling and losing their energy, as we get further from the source of supply ; we see that the nearer the particles are to the centre the more they bang about and the more lines we get in the spectrum. It is important to notice that vibration once begun always goes on; it never gives *place* to others, although it may give *rise* to others; so that we get the largest number of lines in the centre, where the particles are closest together.

Here we have specially to refer to the fact that the way in which the complicated spectrum is built up varies in different substances. Plate VII. reproduces a photograph of the spectrum of aluminium and calcium compared with that of the Lenarto meteorite. The spectra of calcium and aluminium differ generically from that of the meteorite. I want to draw attention to the thick or winged lines we get in the case of aluminium and calcium. These spectra are good specimens of those which give a more brilliant spectrum by thickening the lines, while the elements in the meteorite afford good examples of those which produce a brilliant spectrum by increasing the number of their lines.

10. *When low temperatures are employed, important difference in kind is generally observed between the spectra of metals and those of metalloids, taken as a whole.*

Spectroscopically it is more easy to define the difference between these two great classes of elements than the chemists would imagine. A portion of the spectrum of the carbon vapour always present in the electric arc affords as good a representation of the spectrum of a metalloid as anything I can give. It is rhythmic. It is a "fluted" spectrum.

11. *Many phenomena observed when the molecular vibration is brought about by electricity seem to indicate that in some cases all, and in others only some, of the molecules are affected.*

When the jar spark is used with gas in a Geissler tube at low pressure, it is pretty certain that all the molecules of the included gas are affected, although, if the tube be of irregular figure or contain capillary parts,

FIG. 40.—Copy of Professor Tait's photograph of the spark in air.

in all probability all the molecules will not be secondarily affected. But in almost all cases in which the spark is used without the jar, except when the pressure is very low, all the molecules do not seem to be affected. Why this should be so is seen clearly by an inspection of the beautiful photographs which Professor Tait has obtained of the spark in air.

Salet, in his beautiful researches on hydrogen, has shown that if such a spark can be got to pass through hydrogen at high pressure, the lines are thin.. If we take the general view that ordinary oxygen is diatomic, we cannot understand the formation of ozone, unless we assume that only a small number of the molecules come under the influence of the spark.

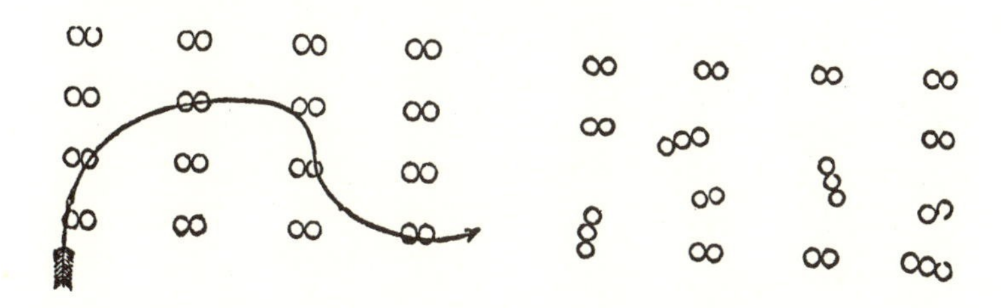

FIG. 41. FIG. 42.

The accompanying rough illustrations will explain my meaning. On Fig. 41 we see a spark having sixteen diatomic molecules of oxygen; in Fig. 42 we see the production of these triatomic molecules of ozone as a result.

In the above cases we have assumed the molecules traversed by the spark to be all alike, to begin with, but there are phenomena observed when mixtures of gases and acid vapours are dealt with which merit much more attention than they have received up to the present time. To take an instance. In a mixture of mercury, vapour, and hydrogen, by varying the tension of the current and the pressure of the gas, it is possible to get the mercury spectrum alone, or the hydrogen spectrum alone, or both combined.

12. *Particles, the amplitudes of vibrations of which may either be so slight that no visible light proceeds from them, or so great that they give out light of their own, absorb light of the same wave-length and of greater amplitude passing through them.*

Here we are in presence of one of the grand generalizations of this century, with which the names of Stokes, Ångström, Balfour Stewart, and Kirchhoff will for ever be associated.

Consider how beautiful this statement is when we regard it in the light of its teaching with regard to molecular motions. We throw sodium into a flame and get a yellow light; we place it on the poles of our electric lamp and render it incandescent, and its light is rich yellow.

We have similarly incandescent sodium outside the sun, through which incandescent sodium the rays of sunlight pass outwards towards the earth, and we may have non-luminous sodium vapour in a test-tube; through which we may throw the white light of an electric lamp. The vibration of the sodium vapour in the case of the sun would be competent to give us the complete bright line spectrum of sodium were the interior sun abolished and only the vaporous envelope left, while the vapour of the tube is invisible in the dark. The passage of the light in both cases, however, is accompanied by absorption, and, instead of bright lines, we obtain dark ones.

Our knowledge of the elements existing in the sun and stars depends entirely, as we have already seen, upon the principle first suggested by Stokes, that particles are set swinging when waves pass through them with the particular rate of vibration which they affect.

13. *Those elements which increase the complexity of the spectrum by widening their lines, most easily produce the phenomena of absorption.*

This is another remarkable fact connected with the foregoing. A thin dark line is observed in the centre of the thick bright lines; this is due to the absorption by the rarer cooler vapour lying outside the interior hotter vapour. This is almost invariably observed in the substances giving us the lines thickening, while iron and the allied metals does not give us any such reversal. It is well to see if one can group facts together. That is the first business of a man of science. It is extraordinary that in all the substances I have yet examined, the question of specific gravity decides not only whether the substance should have its spectrum complicated by thickening or increasing its lines, but whether such reversal shall be easily obtained. The specific gravity of iron is high. In the case of aluminium, magnesium,

Pl. VI
Mn. & ☉

sodium, and others where this is low, we have the widening of the lines and the easy reversal.

14. *In those cases in which the light of winged lines is absorbed by the cooler exterior vapour, the shorter lines are not reversed.*

Some examples may be given. In the spectrum of coal gas, dissociated by an inducted current, only the F line of hydrogen is reversed. In the spectrum of sodium-vapour produced by the passage of the voltaic arc only D is reversed, all the other lines are bright.

In the spectrum of manganese-vapour produced in like manner, of the four adjacent lines in the violet only three are absorbed, as shown in Plate VI.

A corollary of this proposition, so far as the manganese experiments is concerned, is that the molecular motion of the vapour of manganese in the sun, is much greater than that in the cooler portions of the arc, as may be seen by reference to the solar spectrum shown in the same plate for purposes of comparison.

We find all the lines in question reversed in the latter spectrum, hence the molecular motion in the sun is greater than that in the core of the arc.

15. *A compound particle—that is a particle known to consist of two distinct elements—has a vibration which is as peculiar to itself as the vibration of a particle of an element is peculiar to itself.*

Thus to begin, with a definite example, the salts of strontium have each a distinct spectrum. Take the particle of

N_2O_4. The absorption spectrum of this gas, or rather mixture of gases, is very complex ; its particles have a vibration quite of their own. Now it is a gas which it is perfectly easy to dissociate. It is easy to turn it from N_2O_4 to NO_2. We introduce a new spectrum. These facts—and they might easily be multiplied—show then that a compound particle is a perfectly distinct physical thing, with vibrations, rotations, and free paths of its own. There is no apparent connection between the vibrations of a compound particle and those of any of the substances which make up that compound particle.

16. *In the case of metalloids, and compound gases containing them, the spectrum to a large extent depends upon the thickness of the vapour through which the light passes, and often, if not invariably, the absorption increases towards the red end as the thickness is increased.*

Here is one of the points of the most extreme theoretical importance, and one about which least is known. There is a statement in Prof. Maxwell's book, that if we take a metallic vapour and employ a great thickness of it, we shall get from it the same spectrum as from a small thickness of great density. This is Prof. Maxwell's statement. I venture to think that it requires much qualification, for in questions of thickness the spectroscope can offer the physicist a million of miles or a millimetre to work with, and one would think that such a difference should be enough.

If I take a tube with a bore of the size of the lead

in a pencil, enclose some hydrogen and render it incandescent, we see a line of a certain thickness, with a certain pressure. Looking through the sun's coronal atmosphere in an eclipse, we pierce seven or eight hundred thousand miles of hydrogen gas. The thickness of the line is the same. Various thicknesses of sodium vapour do not alter the thickness of the lines, provided that thickness is the only variable condition. But if we pass from metals to the metalloids, then the statement seems more justified. There is considerable interest attached to the question whether there is or is not any chlorine in the sun's outer atmosphere. I have endeavoured to settle this question by contrasting the absorption chlorine spectrum with the solar spectrum; different thicknesses of chlorine have been employed, and the spectrum becomes much more decided with each change of thickness. It seems that, if we take the metalloids, the absorption of a small thickness often takes place in the violet portion of the spectrum.

Now can these results be harmonised ? Here I acknowledge we tread on very difficult ground, and with our present knowledge it would be perhaps best to say nothing ; but I am not sure that this would not be scientific cowardice, so I will ask, under all reserve, whether the following explanation may not be a probable one ? With metallic vapours the lines, though not widened by thickness as they are widened by great density, are certainly darkened, but all the lines are not visible—only the longest, generally. Now if we assume that the fluted spectrum of the metalloids is really, even where it appears continuous, built up of lines,* then the darkening of these lines by greater thickness will not only make those darker that we see with a small thickness but bring others into visibility; and if this goes on till we have a very great thickness we may have an immense difference in the appearance of the spectrum. In short it would seem that there is a much closer connection between the stronger and feebler lines in the fluted spectra of the metalloids than there is between long and short lines in the case of the metals.

17. *In encounters of dissimilar molecules the vibrations of each are damped.*

We saw in 8, that the complexity of the spectrum of any substance increased with the number of encounters among similar molecules. The sympathetic encounter seems to fill the molecule, as it were, with fresh energy of the spectrum producing kind, and hence it is that the brilliancy increases either with increased density or temperature. But if we operate upon a mixture of dissimilar molecules, then experiment shows that unsympathetic encounters deprive both molecules of a part of this energy. If we confine our attention to any one of the constituents, then each increase in the quantity of another is followed by a dimming of the spectrum of the first.

* Thalén's beautiful researches on the spectrum of iodine quite bear out this view.

18. *If we are to hold that the lines, both "fundamental" and "short," which we get in a metallic spectrum, are due to encounters, then, as neither the quantity of the encounters nor their quality is necessarily altered by increasing the thickness of the stratum, the assumption that a great* thickness *of a gas or vapour causes its radiation, and therefore its absorption, to assume more and more the character of a continuous spectrum as the thickness is increased, seems devoid of true theoretical foundation.*

To test this point I made the following experiments:—

1. An iron tube about 5 feet long was filled with dry hydrogen; pieces of sodium were carefully placed at intervals along the whole length of the tube, except close to the ends. The ends were closed with glass plates. The tube was placed in two gas-furnaces in line and heated. An electric lamp was placed at one end of the tube and a spectroscope at the other.

When the tube was red-hot and filled with sodium-vapour throughout, as nearly as possible, its whole length, a stream of hydrogen slowly passing through the tube, the line D was seen to be absorbed; it was no thicker than when seen under similar conditions in a test-tube, and far thinner than the line absorbed by sodium-vapour in a test-tube, if the density be only slightly increased.

Only the longest "fundamental" line was absorbed.

The line was thicker than the D line in the solar spectrum, in which spectrum all the short lines are reversed.

2. As it was difficult largely to increase either the

temperature or the density of the sodium-vapour, I made another series of experiments with iodine-vapour.

I have already pointed out the differences indicated by the spectroscope between the quality of the vibrations of the "atom" of a metal and of the "subatom" of a metalloid (by which term I define that mass of matter which gives us a spectrum of fluted spaces, and builds up the continuous spectrum in its own way). Thus, in iodine, the short lines, brought about by increase of density in an atomic spectrum, are represented by the addition of a system of well-defined "beats" and broad bands of continuous absorption to the simplest spectrum, which is one exquisitely rhythmical, the intervals increasing from the blue to the red, and in which the beats are scarcely noticeable.

On increasing the density of a very small thickness by a gentle heating, the beats and bands are introduced, and, as the density was still further increased, the absorption became continuous throughout the whole of the visible spectrum.

The absorption of a thickness of 5 feet 6 inches of iodine-vapour at a temperature of 59° F. gave me no indication of bands, while the beats were so faint that they were scarcely visible.

19. *On the whole, certain kinds of particles affect certain parts of the spectrum.*

Take the bright lines of the metals; if we were to mix together all the known metals in the sun, make a compound which should consist of all of them, put it into the lower pole of an electric lamp and photograph the spectrum, then we should find the majority of the lines would be in the violet end of the spectrum, scarcely any in the red end. That is the reason why the spectrum of the sun, which contains so many of the metals, is so complicated in the violet. If we combine a metal and a metalloid, we find, in many cases at all events, that the vibrations will lie in the red end of the spectrum ; we shall also find that there is a connection between the atomic weight of the metalloids and the region of the spectrum in which their lines appear under similar conditions.

We have, in fact, simple particles and short waves, compound particles and long waves. Nor is this all. In many cases we find both ends of the spectrum, and in many cases the more refrangible end only, blocked by continuous absorption. This occurs so often in absorption spectra that one is led to suspect that it is due to some arrangement of particles.

20. *Some of the vibrations are very closely connected with others, as evidenced by repetitions of similar groups of lines in different parts of the spectrum.*

Here we are brought face to face with a revelation of the vibrations of particles, which, if I am not mistaken, will be made much of by the mathematical physicist in the future.

I will content myself by giving two or three striking instances, first noticed by Mascart. We shall see that the longest line is at work in all of them.

In sodium we may say that the longest line is double; I refer to D′ and D″. All the lines are double.

In magnesium the longest line is a triple combination. This is repeated exactly in the violet.

In manganese we may almost say that the same thing happens, but the phenomenon is much more absolute in the case of those particles such as sodium and magnesium, which, on other grounds, I suspect to be of the simplest structure.

21. *Our knowledge of the vibrations of particles will be incomplete until the vibration is known from the extreme violet (invisible) to the extreme red (invisible).*

In the meantime great help may be got from inferences, and, in the case of metalloids at low temperatures, from the position of their continuous absorption: and it is a question whether light may not be thus thrown upon the opacity of some solid substances and the transparency of others.

I think it not too much to say that already, in the case of some gases and vapours which are apparently transparent, it is as certain in some cases that their absorption is in the ultra red, as it is certain that in the case of others the absorption is in the ultra violet. And further, it can scarcely be that this absorption is not of the continuous or fluted kind—in other words, that no gas is "atomic" in the chemist's sense, except when subjected to the action of electricity, or, in the case of hydrogen, to a high temperature.

MAXWELL, JAMES CLERK. (b. Edinburgh, Scotland, 13 June 1831; d. Cambridge, England, 5 November 1879)

When Maxwell was ten years old, he entered Edinburgh Academy; his education continued at Edinburgh University where he studied from 1847–1850, and then at Cambridge University from 1850–1854. He held professorships at Marischal College in Aberdeen and at King's College in London from 1856 to 1865, at which time he retired to his family estate in Galloway, Scotland to write the *Treatise on Electricity and Magnetism* (1873). He was made a fellow of Trinity College at Cambridge University in 1855, and, in 1871, he was appointed the first professor of experimental physics at Cambridge University, where he planned and developed the Cavendish Laboratory, soon to become one of the leading physics laboratories in the world.

Maxwell's first scientific paper appeared when he was fifteen years old; later, at the age of thirty-one, he received the Adams Prize at Cambridge (1859) for his essay "On the Stability of Saturn's Rings." From 1855 to 1872 he published a series of papers on the perception of color and color-blindness. His first important memoir on electricity ("On Faraday's Lines of Force") appeared in 1855–1856, originating from a long correspondence with William Thomson, who, in 1846, had treated the electric force, at any point, as analogous to the flux of heat from sources distributed in the same manner as hypothetical electrical particles. Maxwell's goal was to develop further Faraday's "field" ideas, taking off from Thomson's treatment of electrical laws and the laws of action-at-a-distance. This Maxwell did by using the generalized coordinate system of Lagrange, reducing all electric and magnetic phenomena to stresses and motions of a material medium, the ether. The power of the theory was established by successful predictions: first, that the velocity of light in vacuo is identical to the ratio of the electromagnetic and electrostatic units; and secondly, that the speed of transmission of electromagnetic radiations is equal to that of light.

While undermining the action-at-a-distance tradition on the one hand, Maxwell also developed powerful tools for testing and confirming the kinetic theory of gases and the atomic view of matter. His elementary treatise *Matter and Motion* (1876) became a classic for physical atomism and the dynamical theory of heat. Influenced by his reading of Rudolf Clausius's 1857–1858 papers dealing with the velocities and mean-free paths of molecules in a gas, Maxwell drew upon statistics to develop a treatment for the distribution of particle velocities in a gas. The application of statistics to physics was of extraordinary influence in the future development of the discipline and provided a key to experimental tests for the molecular-kinetic theory. These tests confirmed Avogadro's equal volumes-equal numbers hypothesis from new and unexpected directions, via Maxwell's work.

PAPERS READ BEFORE THE CHEMICAL SOCIETY.

XXII.—*On the Dynamical Evidence of the Molecular Constitution of Bodies.*

By J. Clerk-Maxwell.

Of all hypotheses as to the constitution of bodies, that is surely the most warrantable which assumes no more than that they are material systems, and proposes to deduce from the observed phenomena just as much information about the conditions and connections of the material system as these phenomena can legitimately furnish.

In studying the constitution of bodies we are forced from the very beginning to deal with particles which we cannot observe. For whatever may be our ultimate conclusions as to molecules and atoms, we have experimental proof that bodies may be divided into parts so small that we cannot perceive them.

Hence, if we are careful to remember that the word particle means a small part of a body, and that it does not involve any hypothesis as to the ultimate divisibility of matter, we may consider a body as made up of particles, and we may also assert that in bodies or parts of bodies of measurable dimensions, the number of particles is very great indeed.

The next thing required is a dynamical method of studying a material system consisting of an immense number of particles, by forming an idea of their configuration and motion, and of the forces acting on the particles, and deducing from the dynamical theory those phenomena which, though depending on the configuration and motion of the invisible particles, are capable of being observed in visible portions of the system.

The dynamical principles necessary for this study were developed by the fathers of dynamics, from Galileo and Newton to Lagrange and Laplace; but the special adaptation of these principles to molecular studies has been to a great extent the work of Prof. Clausius of Bonn, who has recently laid us under still deeper obligations by giving us, in addition to the results of his elaborate calculations, a new dynamical idea, by the aid of which I hope we shall be able to establish several important conclusions without much symbolical calculation.

The equation of Clausius, to which I must now call your attention, is of the following form:—

$$pV = \frac{2}{3}T - \frac{2}{3}\Sigma\Sigma\left(\frac{1}{2}Rr\right).$$

Here p denotes the pressure of a fluid, and V the volume of the vessel which contains it. The product $p\,V$, in the case of gases at constant temperature, remains, as Boyle's Law tells us, nearly constant for different volumes and pressures. This member of the equation, therefore, is the product of two quantities, each of which can be directly measured.

The other member of the equation consists of two terms, the first depending on the motion of the particles, and the second on the forces with which they act on each other.

The quantity T is the kinetic energy of the system, or, in other words, that part of the energy which is due to the motion of the parts of the system.

The kinetic energy of a particle is half the product of its mass into the square of its velocity, and the kinetic energy of the system is the sum of the kinetic energy of its parts.

In the second term, r is the distance between any two particles, and R is the attraction between them. (If the force is a repulsion or a pressure, R is to be reckoned negative.)

The quantity $\frac{1}{2}\,R\,r$, or half the product of the attraction into the distance across which the attraction is exerted, is defined by Clausius as the *virial* of the attraction. (In the case of pressure or repulsion, the virial is negative.)

The importance of this quantity was first pointed out by Clausius, who, by giving it a name, has greatly facilitated the application of his method to physical exposition.

The virial of the system is the sum of the virials belonging to every pair of particles which exist in the system. This is expressed by the double sum $\Sigma\,\Sigma\,(\frac{1}{2}\,R\,r)$, which indicates that the value of $\frac{1}{2}\,R\,r$ is to be found for every pair of particles, and the results added together.

Clausius has established this equation by a very simple mathematical process, with which I need not trouble you. We may see, however, that it indicates two causes which may affect the pressure of the fluid on the vessel which contains it: the motion of its particles, which tends to increase the pressure, and the attraction of its particles, which tends to diminish the pressure.

We may therefore attribute the pressure of a fluid either to the motion of its particles or to a repulsion between them.

Let us test by means of this result of Clausius the theory that the pressure of a gas arises entirely from the repulsion which one particle exerts on another, these particles, in the case of gas in a fixed vessel, being really at rest.

In this case the virial must be negative, and since by Boyle's Law the product of pressure and volume is constant, the virial also must be constant, whatever the volume, in the same quantity of gas at constant temperature. It follows from this that Rr, the product of the repulsion of two particles into the distance between them, must be constant, or in other words that the repulsion must be inversely as the distance, a law which Newton has shown to be inadmissible in the case of molecular forces, as it would make the action between distant parts of bodies greater than that between contiguous parts. In fact, we have only to observe that if Rr is constant, the virial of every pair of particles must be the same, so that the virial of the system must be proportional to the number of pairs of particles in the system—that is, to the square of the number of particles, or in other words to the square of the quantity of gas in the vessel. The pressure, according to this law, would not be the same in different vessels of gas at the same density, but would be greater in a large vessel than in a small one, and greater in the open air than in any ordinary vessel.

The pressure of a gas cannot therefore be explained by assuming repulsive forces between the particles. It must, therefore, depend, in whole or in part, on the motion of the particles.

If we suppose the particles not to act on each other at all, there will be no virial, and the equation will be reduced to the form

$$Vp = \frac{2}{3} T.$$

If M is the mass of the whole quantity of gas, and c is the mean square of the velocity of a particle, we may write the equation—

$$Vp = \frac{1}{3} Mc^2,$$

or in words, the product of the volume and the pressure is one-third of the mass multiplied by the mean square of the velocity. If we now assume, what we shall afterwards prove by an independent process, that the mean square of the velocity depends only on the temperature, this equation exactly represents Boyle's Law.

But we know that most ordinary gases deviate from Boyle's Law, especially at low temperatures and great densities. Let us see whether the hypothesis of forces between the particles, which we rejected when brought forward as the sole cause of gaseous pressure, may not be consistent with experiment when considered as the cause of this deviation from Boyle's Law.

When a gas is in an extremely rarefied condition, the number of particles within a given distance of any one particle will be proportional to the density of gas. Hence the virial arising from the action of one particle on the rest will vary as the density, and the whole virial in unit of volume will vary as the square of the density.

Calling the density ρ, and dividing the equation by V, we get—

$$p = \frac{1}{3}\rho c^2 - \frac{2}{3} A \rho^2$$

where A is a quantity which is nearly constant for small densities.

2 M 2

Now, the experiments of Regnault show that in most gases, as the density increases the pressure falls below the value calculated by Boyle's Law. Hence the virial must be positive; that is to say, the mutual action of the particles must be in the main attractive, and the effect of this action in diminishing the pressure must be at first very nearly as the square of the density.

On the other hand, when the pressure is made still greater the substance at length reaches a state in which an enormous increase of pressure produces but a very small increase of density. This indicates that the virial is now negative, or, in other words, the action between the particles is now, in the main, repulsive. We may therefore conclude that the action between two particles at any sensible distance is quite insensible. As the particles approach each other the action first shows itself as an attraction, which reaches a maximum, then diminishes, and at length becomes a repulsion so great that no attainable force can reduce the distance of the particles to zero.

The relation between pressure and density arising from such an action between the particles is of this kind.

As the density increases from zero, the pressure at first depends almost entirely on the motion of the particles, and therefore varies almost exactly as the pressure, according to Boyle's Law. As the density continues to increase, the effect of the mutual attraction of the particles becomes sensible, and this causes the rise of pressure to be less than that given by Boyle's Law. If the temperature is low, the effect of attraction may become so large in proportion to the effect of motion that the pressure, instead of always rising as the density increases, may reach a maximum, and then begin to diminish.

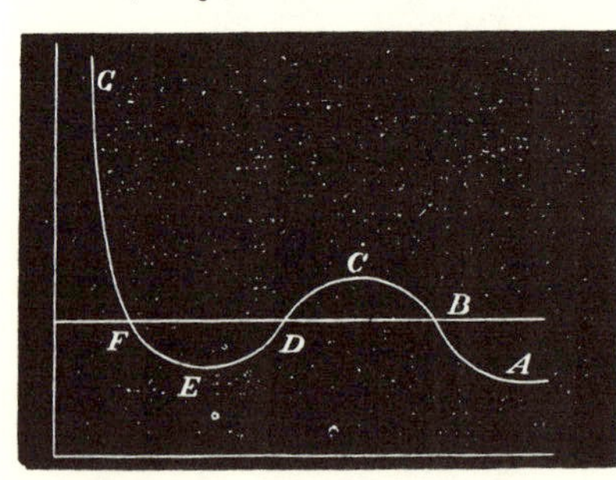

At length, however, as the average distance of the particles is still further diminished, the effect of repulsion will prevail over that of attraction, and the pressure will increase so as not only to be greater than that given by Boyle's Law, but so that an exceedingly small increase of density will produce an enormous increase of pressure.

Hence the relation between pressure and volume may be represented by the curve *A B C D E F G*, where the horizontal ordinate represents the volume, and the vertical ordinate represents the pressure.

As the volume diminishes, the pressure increases up to the point *C*, then diminishes to the point *E*, and finally increases without limit as the volume diminishes.

We have hitherto supposed the experiment to be conducted in such a way that the density is the same in every part of the medium. This, however, is impossible in practice, as the only condition we can impose on the medium from without is that the whole of the medium shall be contained within a certain vessel. Hence, if it is possible for the medium to arrange itself so that part has one density and part another, we cannot prevent it from doing so.

Now the points B and F represent two states of the medium in which the pressure is the same but the density very different. The whole of the medium may pass from the state B to the state F, not through the intermediate states $C\ D\ E$, but by small successive portions passing directly from the state B to the state F. In this way the successive states of the medium as a whole will be represented by points on the straight line $B\ F$, the point B representing it when entirely in the rarefied state, and F representing it when entirely condensed. This is what takes place when a gas or vapour is liquefied.

Under ordinary circumstances, therefore, the relation between pressure and volume at constant temperature is represented by the broken line $A\ B\ F\ G$. If, however, the medium when liquefied is carefully kept from contact with vapour, it may be preserved in the liquid condition and brought into states represented by the portion of the curve between F and E. It is also possible that methods may be devised whereby vapour may be prevented from condensing, and brought into states represented by the points in $B\ C$.

The portion of the hypothetical curve from C to E represents states which are essentially unstable, and which cannot therefore be realised.

Now let us suppose the medium to pass from B to F along the hypothetical curve $B\ C\ D\ E\ F$ in a state always homogeneous, and to return along the straight line $F\ B$ in the form of a mixture of liquid and vapour. Since the temperature has been constant throughout, no heat can have been transformed into work. Now the heat transformed into work is represented by the excess of the area $F\ D\ E$ over $B\ C\ D$. Hence the condition which determines the maximum pressure of the vapour at given temperature is that the line $B\ F$ cuts off equal areas from the curve above and below.

The higher the temperature, the greater the part of the pressure which depends on motion, as compared with that which depends on forces between the particles. Hence, as the temperature rises, the dip in the curve becomes less marked, and at a certain temperature the curve, instead of dipping, merely becomes horizontal at a certain point, and then slopes upward as before. This point is called the critical point. It has been determined for carbonic acid by the masterly researches of Andrews. It corresponds to a definite temperature, pressure, and density.

At higher temperatures the curve slopes upwards throughout, and there is nothing corresponding to liquefaction in passing from the rarest to the densest state.

The molecular theory of the continuity of the liquid and gaseous states forms the subject of an exceedingly ingenious thesis by Mr. Johannes Diderik van der Waals,* a graduate of Leyden. There are certain points in which I think he has fallen into mathematical errors, and his final result is certainly not a complete expression for the interaction of real molecules, but his attack on this difficult question is so able and so brave, that it cannot fail to give a notable impulse to molecular science.

The purely thermodynamical relations of the different states of matter do not belong to our subject, as they are independent of particular theories about molecules. I must not, however, omit to mention a most important American contribution to this part of thermodynamics by Prof. Willard Gibbs,† of Yale College, U.S., who has given us a remarkably simple and thoroughly satisfactory method of representing the relations of the different states of matter by means of a model. By means of this model, problems which had long resisted the efforts of myself and others may be solved at once.

Let us now return to the case of a highly rarefied gas in which the pressure is due entirely to the motion of its particles. It is easy to calculate the mean square of the velocity of the particles from the equation of Clausius, since the volume, the pressure, and the mass are all measureable quantities. Supposing the velocity of every particle the same, the velocity of a molecule of oxygen would be 461 metres per second, of nitrogen 492, and of hydrogen 1844, at the temperature of 0° C.

The explanation of the pressure of a gas on the vessel which contains it by the impact of its particles on the surface of the vessel has been suggested at various times by various writers. The fact, however, that gases are not observed to disseminate themselves through the atmosphere with velocities at all approaching those just mentioned, remained unexplained, till Clausius, by a thorough study of the motions of an immense number of particles, developed the methods and ideas of modern molecular science.

To him we are indebted for the conception of the mean length of the path of a molecule of a gas between its successive encounters with other

* Over de continuiteit van den gas en vloeistof toestand. Leiden: A. W. Sijthoff, 1873.

† "A Method of Geometrical Representation of the Thermodynamic Properties of Substances by means of Surfaces."—*Transactions of the Connecticut Academy of Arts and Sciences*, vol. ii, part 2.

molecules. As soon as it was seen how each molecule, after describing an exceedingly short path, encounters another, and then describes a new path in a quite different direction, it became evident that the rate of diffusion of gases depends not merely on the velocity of the molecules, but on the distance they travel between each encounter.

I shall have more to say about the special contributions of Clausius to molecular science. The main fact, however, is, that he opened up a new field of mathematical physics by showing how to deal mathematically with moving systems of innumerable molecules.

Clausius, in his earlier investigations at least, did not attempt to determine whether the velocities of all the molecules of the same gas are equal, or whether, if unequal, there is any law according to which they are distributed. He therefore, as a first hypothesis, seems to have assumed that the velocities are equal. But it is easy to see that if encounters take place among a great number of molecules, their velocities, even if originally equal, will become unequal, for, except under conditions which can be only rarely satisfied, two molecules having equal velocities before their encounter will acquire unequal velocities after the encounter. By distributing the molecules into groups according to their velocities, we may substitute for the impossible task of following every individual molecule through all its encounters, that of registering the increase or decrease of the number of molecules in the different groups.

By following this method, which is the only one available either experimentally or mathematically, we pass from the methods of strict dynamics to those of statistics and probability.

When an encounter takes place between two molecules, they are transferred from one pair of groups to another, but by the time that a great many encounters have taken place, the number which enter each group is, on an average, neither more nor less than the number which leave it during the same time. When the system has reached this state, the numbers in each group must be distributed according to some definite law.

As soon as I became acquainted with the investigations of Clausius, I endeavoured to ascertain this law.

The result which I published in 1860 has since been subjected to a more strict investigation by Dr. Ludwig Boltzmann, who has also applied his method to the study of the motion of compound molecules. The mathematical investigation, though, like all parts of the science of probabilities and statistics, it is somewhat difficult, does not appear faulty. On the physical side, however, it leads to consequences, some of which, being manifestly true, seem to indicate that the hypotheses are well chosen, while others seem to be so irreconcilable with known experimental results, that we are compelled to admit that something essential to the complete statement of the physical theory of molecular encounters must have hitherto escaped us.

I must now attempt to give you some account of the present state of these investigations, without, however, entering into their mathematical demonstration.

I may begin by stating the general law of the distribution of velocity among molecules of the same kind.

If we construct a diagram of velocity by taking a fixed point, drawing from it a line representing in direction and magnitude the velocity of a molecule, and making a dot at the end of the line, the position of the dot will indicate the state of motion of the molecule.

If we do the same for all the other molecules, the diagram will be dotted all over, the dots being more numerous in certain places than in others.

The law of distribution of the dots may be shown to be the same as that which prevails among errors of observation or of adjustment.

DIAGRAM OF VELOCITIES.

The dots in this diagram may be taken to represent the velocities of molecules, the different observations of the position of the same star, or the bullet-holes round the bull's-eye of a target, all of which are distributed in the same manner.

The velocities of the molecules have values ranging from zero to infinity, so that in speaking of the average velocity of the molecules we must define what we mean.

The most useful quantity for purposes of comparison and calculation is called the "velocity of mean square." It is that velocity whose square is the average of the squares of the velocities of all the molecules.

This is the velocity given above as calculated from the properties of different gases. A molecule moving with the velocity of mean square has a kinetic energy equal to the average kinetic energy of all the molecules in the medium, and if a single mass equal to that of the whole quantity of gas were moving with this velocity, it would have the same kinetic energy as the gas actually has, only it would be in a visible form and directly available for doing work.

If in the same vessel there are different kinds of molecules, some of greater mass than others, it appears from this investigation that their velocities will be so distributed that the average kinetic energy of a molecule will be the same, whether its mass be great or small.

Here we have perhaps the most important application which has yet been made of dynamical methods to chemical science. For, suppose that we have two gases in the same vessel. The ultimate distribution of agitation among the molecules is such that the average kinetic energy of an individual molecule is the same in either gas. This ultimate state is also, as we know, a state of equal temperature. Hence the condition that two gases shall have the same temperature is that the average kinetic energy of a single molecule shall be the same in the two gases.

Now, we have already shown that the pressure of a gas is two-thirds of the kinetic energy in unit of volume. Hence, if the pressure as well as the temperature be the same in the two gases, the kinetic energy per unit of volume is the same, as well as the kinetic energy per molecule. There must, therefore, be the same number of molecules in unit of volume in the two gases.

This result coincides with the law of equivalent volumes established by Gay-Lussac. This law, however, has hitherto rested on purely chemical evidence, the relative masses of the molecules of different substances having been deduced from the proportions in which the substances enter into chemical combination. It is now demonstrated on dynamical principles. The molecule is defined as that small portion of the substance which moves as one lump during the motion of agitation. This is a purely dynamical definition, independent of any experiments on combination. The density of a gaseous medium, at standard temperature and pressure, is proportional to the mass of one of its molecules as thus defined.

We have thus a safe method of estimating the relative masses of molecules of different substances when in the gaseous state. This method is more to be depended on than those founded on electrolysis or on specific heat, because our knowledge of the conditions of the motion of agitation is more complete than our knowledge of electrolysis, or of the internal motions of the constituents of a molecule.

I must now say something about these internal motions, because the greatest difficulty which the kinetic theory of gases has yet encountered belongs to this part of the subject.

We have hitherto considered only the motion of the centre of mass of the molecule. We have now to consider the motion of the constituents of the molecule relative to the centre of mass.

If we suppose that the constituents of a molecule are atoms, and that each atom is what is called a material point, then each atom may move in three different and independent ways, corresponding to the three dimensions of space, so that the number of variables required to determine the position and configuration of all the atoms of the molecule is three times the number of atoms.

It is not essential, however, to the mathematical investigation to assume that the molecule is made up of atoms. All that is assumed is that the position and configuration of the molecule can be completely expressed by a certain number of variables.

Let us call this number n.

Of these variables, three are required to determine the position of the centre of mass of the molecule, and the remaining $n - 3$ to determine its configuration relative to its centre of mass.

To each of the n variables corresponds a different kind of motion.

The motion of translation of the centre of mass has three components.

The motions of the parts relative to the centre of mass have $n - 3$ components.

The kinetic energy of the molecule may be regarded as made up of two parts—that of the mass of the molecule supposed to be concentrated at its centre of mass, and that of the motions of the parts relative to the centre of mass. The first part is called the energy of translation, the second that of rotation and vibration. The sum of these is the whole energy of motion of the molecule.

The pressure of the gas depends, as we have seen, on the energy of translation alone. The specific heat depends on the rate at which the whole energy, kinetic and potential, increases as the temperature rises.

Clausius had long ago pointed out that the ratio of the increment of the whole energy to that of the energy of translation may be determined if we know by experiment the ratio of the specific heat at constant pressure to that at constant volume.

He did not, however, attempt to determine *à priori* the ratio of the two parts of the energy, though he suggested, as an extremely probable hypothesis, that the average values of the two parts of the energy in a given substance always adjust themselves to the same ratio. He left. the numerical value of this ratio to be determined by experiment.

In 1860 I investigated the ratio of the two parts of the energy on the hypothesis that the molecules are elastic bodies of invariable form. I found, to my great surprise, that whatever be the shape of the molecules, provided they are not perfectly smooth and spherical, the ratio of the two parts of the energy must be always the same, the two parts being in fact equal.

This result is confirmed by the researches of Boltzmann, who has worked out the general case of a molecule having n variables.

He finds that while the average energy of translation is the same for molecules of all kinds at the same temperature, the whole energy of motion is to the energy of translation as n to 3.

For a rigid body $n = 6$, which makes the whole energy of motion twice the energy of translation.

But if the molecule is capable of changing its form under the action of impressed forces, it must be capable of storing up potential energy, and if the forces are such as to ensure the stability of the molecule, the average potential energy will increase when the average energy of internal motion increases.

Hence. as the temperature rises, the increments of the energy of translation, the energy of internal motion, and the potential energy are as 3, $(n - 3)$, and e respectively, where e is a positive quantity of unknown value depending on the law of the force which binds together the constituents of the molecule.

When the volume of the substance is maintained constant, the effect of the application of heat is to increase the whole energy. We thus find for the specific heat of a gas at constant volume—

$$\frac{1}{2J}\frac{p_0 V_0}{273^\circ}(n + e)$$

where p_0 and V_0 are the pressure and volume of unit of mass at zero centigrade, or 273° absolute temperature, and J is the dynamical equivalent of heat. The specific heat at constant pressure is—

$$\frac{1}{2J}\frac{p_0 V_0}{273^\circ}(n + 2 + e).$$

In gases whose molecules have the same degree of complexity the value of n is the same, and that of e *may* be the same.

If this is the case, the specific heat is inversely as the specific gravity, according to the law of Dulong and Petit, which is, to a certain degree of approximation, verified by experiment.

But if we take the actual values of the specific heat as found by Regnault, and compare them with this formula, we find that $n + e$ for air and several other gases cannot be more than 4·9. For carbonic acid and steam it is greater. We obtain the same result if we compare the ratio of the calculated specific heats

$$\frac{2 + n + e}{n + e}$$

with the ratio as determined by experiment for various gases, namely, 1·408.

And here we are brought face to face with the greatest difficulty which the molecular theory has yet encountered, namely, the interpretation of the equation $n + e = 4{\cdot}9$.

If we suppose that the molecules are atoms—mere material points, incapable of rotatory energy or internal motion—then n is 3 and e is zero, and the ratio of the specific heats is 1·66, which is too great for any real gas.

But we learn from the spectroscope that a molecule can execute vibrations of constant period. It cannot therefore be a mere material point, but a system capable of changing its form. Such a system cannot have less than six variables. This would make the greatest value of the ratio of the specific heats 1·33, which is too small for hydrogen, oxygen, nitrogen, carbonic oxide, nitrous oxide, and hydrochloric acid.

But the spectroscope tells us that some molecules can execute a great many different kinds of vibrations. They must, therefore, be systems of a very considerable degree of complexity, having far more than six variables. Now, every additional variable introduces an additional amount of capacity for internal motion without affecting the external pressure. Every additional variable, therefore, increases the specific heat, whether reckoned at constant pressure or at constant volume. So does any capacity which the molecule may have for storing up energy in the potential form. But the calculated specific heat is already too great when we suppose the molecule to consist of two atoms only. Hence every additional degree of complexity which we attribute to the molecule can only increase the difficulty of reconciling the observed with the calculated value of the specific heat.

I have now put before you what I consider to be the greatest difficulty yet encountered by the molecular theory. Boltzmann has suggested that we are to look for the explanation in the mutual action between the molecules and the æthereal medium which surrounds them. I am afraid, however, that if we call in the help of this medium, we shall only increase the calculated specific heat, which is already too great.

The theorem of Boltzmann may be applied not only to determine the distribution of velocity among the molecules, but to determine the distribution of the molecules themselves in a region in which they are acted on by external forces. It tells us that the density of distribution of the molecules at a point where the potential energy of a molecule is ψ, is proportional to $e^{-\frac{\psi}{\kappa\theta}}$ where θ is the absolute temperature, and κ is a constant for all gases. It follows from this, that if several gases in the same vessel are subject to an external force like that of gravity, the distribution of each gas is the same as if no other gas were present. This result agrees with the law assumed by Dalton, according to which the atmosphere may be regarded as consisting of two independent atmospheres, one of oxygen, and the other of nitrogen; the density of the oxygen diminishing faster than that of the nitrogen, as we ascend.

This would be the case if the atmosphere were never disturbed, but the effect of winds is to mix up the atmosphere and to render its composition more uniform than it would be if left at rest.

Another consequence of Boltzmann's theorem is, that the temperature tends to become equal throughout a vertical column of gas at rest.

In the case of the atmosphere, the effect of wind is to cause the temperature to vary as that of a mass of air would do if it were carried vertically upwards, expanding and cooling as it ascends.

I have confined my remarks to a very small part of the field of molecular investigation. I have said nothing about the molecular theory of the diffusion of matter, motion, and energy, for though the results, especially in the diffusion of matter and the transpiration of fluids, are of great interest to many chemists, and though from them we deduce important molecular data, they belong to a part of our study the data of which, depending on the conditions of the encounter of two molecules, are necessarily very hypothetical. I have thought it better to exhibit the evidence that the parts of fluids are in motion, and to describe the manner in which that motion is distributed among molecules of different masses.

To show that all the molecules of the same substance are equal in mass, we may refer to the methods of dialysis introduced by Graham, by which two gases of different densities may be separated by percolation through a porous plug.

If in a single gas there were molecules of different masses, the same process of dialysis, repeated a sufficient number of times, would furnish us with two portions of the gas, in one of which the average mass of the molecules would be greater than in the other. The density and the combining weight of these two portions would be different. Now, it may be said that no one has carried out this experiment in a sufficiently elaborate manner for every chemical substance. But the processes of nature are continually carrying out experiments of the same kind; and if there were molecules of the same substance nearly alike, but differing slightly in mass, the greater molecules would be selected in preference to form one compound, and the smaller to form another. But hydrogen is of the same density, whether we obtain it from water or from a hydrocarbon, so that it seems that neither oxygen nor carbon can select from a quantity of hydrogen molecules greater or smaller than the average.

The estimates which have been made of the actual size of molecules are founded on a comparison of the volumes of bodies in the liquid or solid state, with their volumes in the gaseous state. In the study of molecular volumes we meet with many difficulties, but at the same time there are a sufficient number of consistent results to make the study a hopeful one.

The theory of the possible vibrations of a molecule has not yet been studied as it ought, with the help of a continual comparison between the dynamical theory and the evidence of the spectroscope. An intelligent student, armed with the calculus and the spectroscope, can hardly fail to discover some important fact about the internal constitution of a molecule.

The observed transparency of gases may seem hardly consistent with the results of molecular investigations: for a model of the molecules of a gas consisting of marbles scattered at distances bearing the proper proportion to their diameters, would allow very little light to penetrate through a hundred feet.

But if we remember the small size of the molecules compared with the length of a wave of light, we may apply certain theoretical investigations of Lord Rayleigh's about the mutual action between waves and small spheres, which show that the transparency of the atmosphere, if affected only by the presence of molecules, would be far greater than we have any reason to believe it to be.

A much more difficult investigation, which has hardly yet been attempted, relates to the electric properties of gases. No one has yet explained why dense gases are such good insulators, and why, when rarefied or heated, they permit the discharge of electricity, whereas a perfect vacuum is the best of all insulators.

It is true that the diffusion of molecules goes on faster in a rarefied gas, because the mean path of a molecule is inversely as the density.

But the electrical difference between dense and rare gas appears so be too great to be accounted for in this way.

But while I think it right to point out the hitherto unconquered difficulties of this molecular theory, I must not forget to remind you of the numerous facts which it satisfactorily explains. We have already mentioned the gaseous laws, as they are called, which express the relations between volume, pressure, and temperature, and Gay-Lussac's very important law of equivalent volumes. The explanation of these may be regarded as complete. The law of molecular specific heats is less accurately verified by experiment, and its full explanation depends on a more perfect knowledge of the internal structure of a molecule than we as yet possess.

But the most important result of these inquiries is a more distinct conception of thermal phenomena. We learn how to distinguish that kind of motion which we call heat from other kinds of motion. The peculiarity of the motion called heat is that it is perfectly irregular; that is to say, that the direction and magnitude of the velocity of a molecule at a given time cannot be expressed as depending on the present position of the molecule and the time, but depends also on the particular molecule considered.

In the visible motion of a body, on the other hand, the velocity of the centre of mass of all the molecules in any visible portion of the body is the observed velocity of that portion, though the molecules may have also an irregular agitation on account of the body being hot.

In the transmission of sound, too, the different portions of the body have a motion which is generally too minute and too rapidly alternating to be directly observed. But in the motion which constitutes the physical phenomenon of sound, the velocity of each portion of the medium at any time can be expressed as depending on the position and the time elapsed; so that the motion of a medium during the passage of a sound-wave is regular, and must be distinguished from that which we call heat.

If, however, the sound-wave, instead of travelling onwards in an orderly manner, and leaving the medium behind it at rest, meets with resistances which fritter away its motion into irregular agitations, this irregular molecular motion becomes no longer capable of being propagated swiftly in one direction as sound, but lingers in the medium in the form of heat till it is communicated to colder parts of the medium by the slow process of conduction.

The motion which we call light, though still more minute and rapidly alternating than that of sound, is, like that of sound, perfectly regular, and therefore is not heat. What was formerly called radiant heat is a phenomenon physically identical with light.

When the radiation arrives at a certain portion of the medium, it enters it and passes through it, emerging at the other side. As long as the medium is engaged in transmitting the radiation it is in a certain state of motion, but as soon as the radiation has passed through it, the medium returns to its former state, the motion being entirely transferred to a new portion of the medium.

Now, the motion which we call heat can never of itself pass from one body to another unless the first body is, during the whole process, hotter than the second. The motion of radiation, therefore, which passes entirely out of one portion of the medium and enters another, cannot be properly called heat.

MARIGNAC, JEAN CHARLES G. DE. (b. Geneva, Switzerland, 24 April 1817; d. Geneva, 15 April 1894)

In 1835 Marignac entered the Ecole Polytechnique in Paris, and from 1837–1839 studied engineering and minerology at the Ecole des Mines. While in Paris, he attended the lecture courses of J. B. Dumas. He also gained practical laboratory experience at Liebig's laboratory in Giessen and at the porcelain factory at Sèvres. From 1841–1875 Marignac held the chair of chemistry at the Académie de Genève and after 1845, the chair of mineralogy as well. From 1846 to 1857 he was a joint editor of the Swiss journal *Archives des Sciences*.

Marignac's reputation rests on his very accurate work in analytic chemistry, including the determination of the atomic weights of nearly thirty elements and his separation of elements in the rare-earth series. When his own determination of atomic weights did not support Prout's hypothesis (that atomic weights are whole-number multiples of that of hydrogen), Marignac suggested, in 1843, that the real subunit might be only half the weight of hydrogen. He further speculated that some special effect might be responsible for the fact that the weights of hypothetical particles composing an element do not add up to an element's "atomic weight." This hypothesis calls to mind F. W. Aston's studies in the 1920s of the difference between the exact mass of an element and the closest whole number.

Marignac began using the two-volume H_2O formula in 1858 and later urged chemists to adopt the 0 = 16 scale. In 1877, when several meetings of the Paris Academy of Sciences were scenes for debate about the atomic and equivalent systems of chemistry, Marignac supported the arguments of Wurtz and published his reasons in the French journal *Moniteur Scientifique.*

ART. XIII.—*Chemical Equivalents and Atomic Weights considered as bases of a system of Notation;* by C. MARIGNAC.

(Translated from Moniteur Scientifique of Quesneville for September, 1877, by Mr. P. Casamajor.)

THE Academy of Sciences of Paris has witnessed lately, at several of its sittings, an interesting discussion, in which several eminent chemists, among its members, have taken part.* This discussion related to two questions which have often been brought before it, and which will probably be brought before it again many times.

One subject of discussion was a principle, stated in 1811 by

* Messrs. Sainte Claire Deville, Würtz, Berthelot, Fizeau.

an Italian philosopher, Avogadro, on the equality of the molecules of all bodies in a gaseous state. This principle is often placed in opposition to the law of Gay Lussac, on the simple relations which exist between volumes of gases, capable of combining with one another, which law was established a few years before, and which, to tell the truth, is not in contradiction with the hypothesis of Avogadro. From this question arose another, on the relative merits of the chemical notations, expressed in equivalents or in atoms.

For the present, I will not discuss the first of these questions. The truth of the principle of Avogadro can only be admitted on the condition of supposing that the atoms of simple gases cannot exist in a free state, but are welded together in pairs, forming molecules occupying two volumes, like the molecules of compound bodies. Exceptions should be made for mercury and cadmium, whose molecules are formed of only one atom, and for phosphorus and arsenic, whose molecules must contain four atoms.

This hypothesis is not absurd in itself. It may account for certain chemical facts, such, for instance, as the greater energy of action that bodies possess in a nascent state, or before their atoms have combined two by two to form molecules; also for the ease with which certain reactions take place, as pointed out by M. Würtz. It also explains several physical facts, such as the equality of specific heat for the same volume of simple or compound gases, whose molecule is formed of two atoms, as carbonic oxide, hydrochloric acid. It found lately an important confirmation in the researches of Messrs. Kundt and Warburg* on the specific heat of vapor of mercury, which show that this heat agrees with the mechanical theory of heat for monatomic gases, and that this agreement does not exist for other simple gases. We must acknowledge, however, that these considerations do not constitute sufficient proofs.

On the other hand, there are some compound bodies whose vapor densities are in contradiction with the principle of Avogadro. We should be forced to admit that all these compounds suffer decomposition when they seem to be reduced to vapor, so that, instead of measuring their volume, we measure that of their elements, or of the products of their decomposition. Although this decomposition has been ascertained in some cases, it has not in all.

As may be seen, the principle of Avogadro gives rise to serious objections, and, without being convinced of its worthlessness, like my eminent friend, M. Deville, I acknowledge that it is as yet but an hypothesis, in contradiction with facts, which have not been satisfactorily explained. But, I repeat

* Berichte der deutschen Chemischen Gesellschaft, 1875, p. 945.

it, I have no wish to enter into this discussion at present, as it would require to be extensively developed, and it can only be definitely settled by long and difficult experiments. I have recalled this discussion because its solution must exert a certain influence on chemical notations, although the connection between the two questions is not necessarily very close.

As to the best system of notations, it may be necessary to explain why such a question is raised and can only be raised in France. In every other country the question has solved itself gradually, as chemists have, one after another, accepted the atomic notations, and given up the formulas by equivalents in their writings and in their teachings* In this gradual manner, in almost every country, by the successive assent of the great majority of chemists, atomic formulas have been substituted for the others without any formal struggle. In France, however, it has been very different. I am not aware whether the regulations of the University† forbid a professor from adopting the method of instruction which he thinks best, or whether teachers adopt a uniform system from the belief that otherwise their pupils would be placed in a relatively inferior position, if they did not adopt the system most in favor with examiners; at any rate, such an important change, as the introduction of a system of chemical notation, can only be generally introduced when it has been judged necessary, not only by the majority of teachers, but also by the Superior Councils, which govern the University, and, in these, chemists are not the only persons who have influence. It is easily understood that, under these conditions, the partisans of both systems wish to have them discussed in the presence of the scientific body which has the greatest authority, with the object of maintaining the established system, or of introducing the other.

To enter into this discussion, it is doubtless advisable, in the first instance, to define what is understood by these equivalents and these atomic weights, which are placed in opposition to one another.

As to equivalents, I see that M. Berthelot tells us that "their definition is a clear conception." Unfortunately he did not give this definition, and I confess that I do not know of any, at least of any that is precise and general. Doubtless when we compare, with one another, elements, such as chlorine, bromine and iodine, the definition of their relative equivalents is perfectly clear, as the term equivalent is its own definition. But when we deal with bodies which have not a similar

* Fresenius still considers the formulas in equivalents as the best.—*Translator.*

† The entire educational system of France is consolidated under one organization, called the University, comprising faculties of letters, medicine, law and theology, *lycées* for secondary instruction and schools of primary instruction.—*Translator.*

analogy, and particularly if they do not perform the same functions, the idea of equivalence has no meaning. I defy anybody to give a general definition of equivalents which justifies the weight 14, adopted for nitrogen. In volume, it corresponds to the equivalents of hydrogen and of chlorine, but it has not the same chemical value. It has the same chemical value as the equivalents of phosphorus and arsenic, but it does not occupy the same volume. Moreover, it corresponds neither in volume nor in chemical value to the equivalents of oxygen, of sulphur and of most metals. Why then should this number exist?

If, instead of starting from a general definition, which does not exist, we try to find the meaning of equivalents in the methods employed in determining them, we are led to the following conclusion:

It is proved by experience that we may assign to a body, be it simple or compound, various weights, multiples of the same number, and that these weights express the proportions according to which all bodies combine with one another. We may choose one of these weights to express the equivalent of the body. All combinations may then be represented as the union of a certain number of equivalents of the elements, and, if the equivalent is represented by a symbol (in general the first letter of the name of each element), combinations may be represented by formulas which are not generally complicated. This is, after all, the only condition required of equivalents, and hence the only general definition, although not very precise, which can be given is that the equivalent represents for every element or compound body one of the weights which may combine with other equivalents. Theoretically it matters little which of the weights is chosen. Practically, however, one of the weights is preferred, taking as a guide one of the following rules, which cannot be considered as very rigid, as they do not all lead, in all cases, to the same result:

1. When bodies are analogous, and have the same chemical character, their equivalents are represented by the weights which replace each other in analogous combinations. Let us note, however, that this rule is not followed for compound bodies, such as bases and acids, whose so-called equivalents are weights which often have very different values of combination, and we are thereby led to very singular anomalies of statement, such as these: two equivalents of alumina corresponds to three equivalents of magnesia; one equivalent of phosphoric acid to three of nitric acid, &c. In reality the fundamental principle of equivalents has been entirely abandoned for compound bodies, and, in its stead, a method has been adopted, which has been borrowed from the atomic

theory, by taking for their weights the sum of the equivalents of the elements which they contain.

I believe I am not in error when I affirm this, as M. Berthelot* says: "One equivalent of phosphoric acid corresponds to three equivalents of nitric acid, when it forms a tribasic phosphate.

2. Equivalents are chosen in such a way that compounds, which offer the greatest analogies, are represented by similar formulas. This principle served as a guide in determining the equivalents of aluminum and of copper. It is often in contradiction with the preceding. For instance, aluminum and magnesium, which are both powerful deoxidizing agents, do not replace each other in the proportions indicated by the equivalents adopted for these two metals.

3d. When neither of these rules is applicable, or when they lead to complicated formulas, the equivalent of a body is chosen in such a way as to give the simplest possible formulas for its most important combinations. This rule justifies the adoption of the equivalents of nitrogen, phosphorous, arsenic and of some other elements.

We may see by the above that the equivalents constitute a purely conventional and arbitrary system, without any scientific value.

The explanation I have given of equivalents is somewhat different from that which my illustrious teacher, M. Dumas, gave in his lessons of chemical philosophy. This eminent chemist took as his starting point the equivalents of bases, as determined by their true chemical equivalence, founded on the same quantity of oxygen contained in the base. The equivalents of acids are rigorously deducted from the weights necessary to neutralize an equivalent of base. Afterwards, he seeks to establish the equivalents of the elements by considerations which, he acknowledges, are often arbitrary. This method of determining equivalents however, has been either never adopted, or entirely abandoned, doubtless because it led to formulas which are inadmissible. I have given the meaning of equivalents, such as they have been adopted, and not such as they might have been.

We may now consider the definition of atomic weights. If the precise definition of equivalents is impossible, while their determination is comparatively easy, for they are adopted by arbitrary rules, it is the reverse with atomic weights.

* Meeting of the Académie des Sciences of June 4th, 1877. I cannot in any manner accept what he says, in the same place, that this equivalent of phosphoric acid corresponds to one equivalent of nitric acid in monobasic phosphates, or two equivalents in bibasic phosphates. To admit such expressions, we must deny to water the part that all chemists attribute to it in these salts, since the publication of Graham's researches.

If we refer to the fundamental hypothesis of the atomic theory, which supposes that the divisibility of bodies is not indefinite, but that they are formed by the agglomeration of excessively small but indivisible particles, or atoms, the theoretical definition of atomic weights is of the simplest, as they are the relative weights of these ultimate particles. But, however simple the definition may be, the determination of the weights is surrounded with great difficulties.

The hypothesis of the existence of atoms accounts in such a simple manner for that of chemically equivalent proportions for elements which play the same part, that we are naturally led, at first sight, to consider these proportions as representing their relative atomic weights, although this consequence is not rigorously necessary. It is evident, however, that as neither this consideration of chemical equivalence, nor any other consideration drawn from chemistry alone, has led to a complete and logical system of chemical equivalents, we cannot by such considerations be guided in the choice of all the atomic weights, and as these, on account of the hypothesis that is made on their nature, cannot be arbitrary, like equivalents, it has become necessary to study the physical properties of the elements and of compound bodies to find motives for this determination of the atomic weights. Among the properties which can be appealed to, the most important are the densities of gases and vapors, the specific heats and isomorphism.

I acknowledge that in some very rare cases these three orders of physical properties do not lead to the same result, and I agree with Mr. Berthelot that between these three data we must make a choice. I am, however, in complete disagreement from him in the conclusion that I draw from this. If he does not say so expressly, his whole argument proves that, in his opinion, no account is to be taken of these physical properties, when they disturb the usage established for weights that have been adopted for a long time in chemical notations. On the contrary, I think that great account should be taken of these physical properties, and that when they all agree we must have no fear of modifying a few formulas which have only long usage in their favor, particularly if the necessary modification is unimportant. If, moreover, the physical properties do not agree, it is necessary to study the facts with the greatest care, and see if, in some cases, a disagreement can be explained and then choose the weight which agrees the best with the general properties of the elements and its combinations.

Is it impossible to do this? The best proof that it is not, and that there is even no serious difficulty in determining the atomic weight which agrees the best with the physical properties, is to be found in this circumstance that there is no disagreement among chemists, who accept this system of notation, as to the atomic weights, except for a few bodies that are not, as yet, sufficiently known; whose physical properties have not been sufficiently studied, and for which, besides, the idea of equivalents is quite as uncertain as that of atomic weights.

I am perfectly aware that the majority of chemists, who have adopted atomic formulas, believe that they are now able to give a rigorous definition of atomic weights. Starting from molecules, which they define as the smallest quantity of a body, simple or compound, which can exist in the state of liberty; admitting as an axiom the principle of Avogadro, which states the equality of volume of all molecules in a gaseous state, from which may be deducted their relative weights, they define the atom as the smallest quantity of a body which may enter into the composition of a molecule. This definition allows them to determine the atomic weights with certainty, at least for those bodies that enter into volatile combinations. I have not given great weight to this consideration because, not more than Messrs. Deville and Berthelot, do I regard the principle of Avogadro as absolutely demonstrated. But I wish it to be specially noticed that there is not, so far as I know, a single case in which the application of the above definition of atomic weight has been used to change an atomic weight previously obtained by considerations based on the physical properties. Perhaps I should except boron and silicon; but the atomic weights of these elements had never been considered as firmly established, nor indeed had their equivalents. This observation might be appealed to as the strongest proof of the accuracy of the definition of atomic weights, but I have no wish to admit it, as constituting a sufficiently sure base for the determination of atomic weights.

I have here to answer an objection, which I acknowledge to be serious, and which I believe is at the bottom of the opposition of M. Berthelot. The atomic weights rest on an hypothesis which has never been, and, in fact, can never be demonstrated, which many scientific men do not consider as verisimilar, that of the existence of atoms.

I am nearly ready to agree with M. Berthelot in his opposition, and I have certainly no idea of defending the atomic theory, but merely the chemical notations founded on the atomic weights. My answer to the objection stated above is that the existence of atoms is only useful in justifying the name of *atomic weights*, which, for my part, I would very willingly have replaced by any other. I know of no case in which an atomic weight has been determined by a method founded on the indivisibility of atoms; consequently we may consider atomic weights as entirely independent of this indivisibility.

In reality, I consider atomic weights, and I believe that many chemists agree in this, as being only equivalents, in the determination of which arbitrary conventions have been replaced by scientific considerations, based on the study of physical properties.

Let us now sum up the advantages that atomic notations present from this point of view.

For the elements, in the first place, the atomic weights represent equal volumes of all simple gases, so that their ratios of combination in volumes are directly expressed by atomic formulas, while the formulas in equivalents do not offer this advantage. This law presents some exceptions for vapors, in the cases of phosphorus, arsenic, mercury and cadmium, but the same divergence exists for equivalents.

Atomic weights are exactly proportional to the specific heats of simple gases, that are not liquifiable, which agreement does not exist for equivalents. According to the law of Dulong and Petit, the specific heats of the atoms of all simple bodies, either solid or liquid, are nearly the same, except for three bodies, carbon, boron and silicon, whose physical properties offer numerous irregularities, and in which the specific heat varies with the temperature in a manner unknown in other bodies. Equivalents do not offer this concordance. I will not insist on the objection raised by M. Berthelot, and founded on this, that the equality of specific-heats of atoms is far from being absolute, as he was sufficiently answered by MM. Würtz and Fizeau. I will merely add that if we only admitted physical laws that are absolute, we should have to reject them all. Even the law of volumes of Gay Lussac would have to be dropped, as it has been ascertained that all gases have not the same coëfficient of expansion, so that the existence of simple ratios in combinations by volume are not strictly accurate.

As to compound bodies, the molecular formulas, based on the use of atomic weights, present the same advantages, perhaps to a higher degree, when we compare them to the formulas in equivalents.

The use of atomic weights allows us to simplify the formulas of a great number of compounds by dividing them by two. Particularly is this the case with organic compounds. Not only does the formula become simpler, but there is an important advantage gained, that the formulas of almost all compounds correspond to the same volume, which is double the volume of the simple atom. The only exceptions are for a very limited number of bodies, generally belonging to types of complex composition, such as salts of ammonia and of bases derived from ammonia; even for these it has not been proved that they are not regulated by any law, even if M. Deville is right in thinking that the irregularities they present are not due to the decomposition of their vapors. On the other hand, the formulas by equivalents teach us nothing on the vapor densities of compound bodies, as their equivalents may correspond to two, four or eight volumes of vapor, perhaps even of six, if the old equivalent of silicon is kept, as is done by many of those who prefer the notations by equivalents. Molecular formulas also agree with the specific heats of compound bodies in the solid state. According to the law of Woestyn, molecular heats are proportional to the number of atoms contained in the molecule, which law has the same degree of approximation as that of Dulong and Petit. Formulas by equivalents do not show these properties.

Finally, the system of notation, based on atomic weights, gives the explanation of several cases of isomorphism which are incomprehensible with the notation based on equivalents. For instance, in the case of perchlorates and permanganates, and in the case of chloride and sulphide of silver when compared to protochloride and protosulphide of copper. I may also recall that it was by considerations of the same kind that I was led to discover oxygen in fluorine compounds of niobium, where its presence had not been suspected, and that the formulas of these compounds, expressed in equivalents, would never have suggested this idea.

In presence of these advantages, we may ask: what are those that are offered by the system of equivalents and its resulting notation? I believe I can indicate two.

In the first place, as the system is conventional, it does not of itself contain any necessary reason for changes, and it may remain invariable. As there was no serious motive for choosing the number fourteen as the equivalent of nitrogen, rather than seven, which would have given it the same volume as oxygen, or $\frac{14}{3}$ which would have accounted for its value of combination toward hydrogen and the metals, we may readily believe that there will never be a sufficient motive to replace it by one of these numbers. The determination of equivalents not being governed by any fixed rule, they will not be necessarily modified when we come to have a more accurate knowledge of the properties of bodies.

In the second place, as, in their determination, no account is taken of the physical properties of bodies, greater attention can be given to their chemical equivalence, when it exists. This presents some advantages in practical chemistry.

These considerations are doubtless of some value; but if we examine things a little closer, we may easily see that, in this respect, there is really very little difference between the two systems.

It is true that there was a time when atomic weights had to be changed, and it is doubtless, on this account, that atomic weights were dropped and equivalents adopted. Nevertheless, the history of chemistry shows that for more than thirty years no changes have been judged necessary for well known bodies, and that those which have been admitted for elements, whose properties or whose combinations had previously been imperfectly known, were so thoroughly justified by their chemical properties, that even the equivalents of these bodies have had to be modified. Such was the case for bismuth, uranium, vanadium, tantalum and niobium. In reality, the only important change that atomic weights have had to suffer, since their introduction in chemical science, has been the reduction to half of the weights of silver and of the alkaline metals, a reduction based on their specific heat in the solid state; on the specific heat of their combinations, or on isomorphism as was done in the first instance by M. Regnault.* We may see by this that, on the score of invariability, the two systems are on a par.

As to the advantage which results from the fact that equivalents express ratios of real chemical equivalence, in cases where they are not indicated by atomic weights, it would be an important one if chemical equivalence were indicated in all cases; but we know that this is not so. It is really not more difficult to conceive and to remember that an atom of oxygen is worth two of chlorine, and an atom of lead two of silver than to know that an equivalent of nitrogen is worth three of oxygen, and that two equivalents of aluminium are worth three of magnesium. So there is really no advantage, on these two heads, which can counterbalance those which I have shown for atomic notations.

It may be said that the preceding is a contradiction of what I said before. I said that the system of equivalents presents conditions of invariability that are not presented by atomic weights. Further on I have shown that every change of atomic weight had necessitated a corresponding change in equivalents.

If we look for the cause of this apparent contradiction, it seems to me that we shall be led to make an observation which gives the key to the discussion actually going on. It is that, in reality, if we keep out of sight every question as to the *origin* of the terms *equivalents* and *atomic weights*, there is no difference between the two systems, and the partisans of equivalents are willing enough to accept the principles which serve to determine atomic weights, except when the necessity arises of changing the formulas of bodies that are of great importance and occur with great frequency.

* Annales de Chimie et de Physique, 1841, III, vol. i, p. 191.

BERTHELOT, PIERRE EUGÈNE MARCELLIN. (b. Paris, France 25 October 1827; d. Paris, 18 March 1907)

Berthelot studied at the Collège Henri IV in Paris and then attended courses in the Paris Faculties of Medicine and Sciences. He studied chemistry in the private laboratory of T. J. Pelouse, became a demonstrator for Antoine Balard at the Collège de France, and completed his doctoral thesis on glycerine chemistry in 1854. He took a pharmacy degree in 1858, and, in 1859, stepped into a new chair of organic chemistry at the Paris Ecole de Pharmacie. From 1865 until his death, he held a chair of organic chemistry at the Collège de France, where his laboratory became one of the most important in France. Active in political affairs, Berthelot was a member of governmental education committees and, from 1871 on, a member of the Senate. He did work of military importance on explosives and also wrote on the history of alchemy and chemistry, including the book *La Révolution chimique. Lavoisier* (1889).

Berthelot was a staunch opponent of the idea that the formation of organic substances requires the intervention of a special vital force, and he opposed Pasteur's vitalistic interpretation of yeast fermentation. Berthelot's *La Chimie organique fondée sur la synthèse* (1860) was one of his many influential texts in organic chemistry which summed up his work on hydrocarbon and alcohol synthesis, glycerin derivatives and sugars, and the implications of organic synthesis for physiological chemistry. Berthelot rediscovered acetylene in 1860 and prepared from their elements acetylene, benzene, and other aromatic compounds usually derived from coal tar. In the late 1860s his interests turned increasingly to thermochemistry. With his pupils Berthelot established a body of reliable data relating to heats of combustion, solution, and neutralization; and he introduced the terms *exothermic* and *endothermic*. His "law of maximum work" (that chemical change accomplished without the intervention of external energy results in a system of bodies which produces the most heat) was shown later to be true only at a temperature of absolute zero.

Berthelot opposed the notational reforms suggested at the Karlsruhe Congress and remained firmly committed through most of his career to the one-volume (HO) formulas and to a chemical notation based on equivalent combining weights. He defended this point of view against Adolphe Wurtz in meetings at the Paris Academy of Sciences in 1877 and in his reply to Marignac in the *Moniteur Scientifique*. It was not until 1897 that Berthelot began using modern structural formulas.

ART. XXIII.—*On Systems of Chemical Notation.* Letter of M. BERTHELOT to M. Marignac* (from the Moniteur Scientifique of December, 1877).

ALLOW me, in the first place, to correct the opinions concerning the teaching of chemistry in France, which I find expressed in your article, and which have been propagated, perhaps intentionally, in foreign countries. There is no regulation which makes it obligatory on professors of the Faculties to adopt any particular notation, and, as a matter of fact, both notations, by equivalents and by atoms, are about equally represented in our lectures. At the Sorbonne, MM. Würtz and Friedel have adopted the system of atoms; the College of France presented last year a professor who uses the atomic notation; I was commissioned to make the report to the Minister, who gave his approval and made the nomination. In examinations both notations are equally accepted, and, if any pressure exists on the candidates, it is exerted by the partisans of atoms rather than by the others. Officially, therefore, perfect freedom exists on this question. If the atomic notation has not been generally accepted in France, it is because it has not succeeded, so far, in obtaining the good opinion of the majority of scientists; but, notwithstanding, imputations have not been spared that the partisans of equivalents are animated with a retrograde spirit.

* An answer to the paper in Moniteur Scientifique, September, 1877. (See this Journal, February, 1878, p. 89.

Allow me, in the next place, to point out some observations on the fundamental part of the question under discussion. It presented itself before the Paris Academy of Sciences under two heads: the system of atoms, and the language or notation of atomic weights. You were right in separating these two things. I had tried to do the same thing, but with less distinctness, in my last work, *On Chemical Synthesis*, in which I explained the system very fully, but without adopting it, and I said that the notation by atoms possesses certain advantages, but also some disadvantages. The discussion recently raised could not, in the nature of things, assume this methodical form; but I believe that I kept about the same ground, as I always said that the two languages expressed the same ideas in the same way, in most cases, except that special advantages belonged to each system of notation. Your conclusions seem to be about the same as mine.

The definition of equivalents, which you accuse me of not giving, was nevertheless presented during the discussion, and I will take the liberty of reproducing it: "Equivalents express, in my opinion, the ratios of weight according to which bodies combine or substitute themselves for one another." These ratios may be determined by the balance with infinitely greater precision than can be ascribed to most physical laws. As, however, experience proves that bodies combine according to several proportions, which are multiples of one another, it follows that equivalents themselves are only determined within an approximation of a multiple of a certain unity, precisely as axes are determined in crystallography. The choice of the unity belonging to each body is therefore somewhat arbitrary. It may be determined from purely chemical considerations, which are never wanting, by taking the weight which agrees the best with the general reactions of the body, which affords the simplest form, and that which conforms the best with analogies. These analogies are generally expressed by precise rules which are founded on the reciprocal substitution of metals and metalloids, the formation of oxides and acids, and their reciprocal combinations, and the multiple proportions according to which elements combine. In only one case, that of alumina, have we had to appeal to more delicate analogies, drawn from the existence of a remarkable class of double salts, which have been corroborated by the general resemblance of the salts of this base with those of the sesquioxides. It is only in a subordinate way, and for the purpose of giving greater precision to chemical analogies, which are often somewhat vague, that physical properties have been introduced, such as the gaseous density, the specific heat, the crystalline form, the molecular volume in the solid state, etc.

The part which physical properties are to play in the determination of equivalents, and the relative importance of these properties seem to me to constitute the only difference between your views and mine. If it was possible to make equivalents agree exactly with gaseous densities, as the old atomic school had hoped to accomplish, the numbers obtained would probably be adopted by all chemists. This has happened in organic chemistry, in which the same equivalent weights are accepted by every body. Unfortunately, however, this concordance does not exist in mineral chemistry, whence the attempt of the new school to establish atomic weights by means of specific heats. But I persist in the opinion, although I am sorry to find that it is opposed to yours, that this base has not sufficient theoretical solidity when it is in contradiction with the gaseous densities. The specific heats of simple gases, which obey the laws of Mariotte and of Gay Lussac (an obedience which is expressed by the constancy of their gaseous densities), are necessarily the same under the same volume, because the specific heat measures the work accomplished in fulfilling these laws. If the specific heats of the elements in the solid state do not observe the same ratios as in the gaseous state, it is for one of the two following reasons, either the specific heats of the solid elements change unequally with the temperature, as I believe is the case, or two gaseous molecules are united in one solid molecule, as the atomists suppose. In either case, it seems to me that the specific heats of solids must be put aside in the determination of absolute equivalents.*

I insist the more on this point that the new equivalents, if we attribute to this word the extensive meaning that you rightly give to it, introduce an undeniable complication in chemical reactions. In your classical researches on the specific heats of saline solution you found yourself obliged to double the atomic weights of hydrochloric and of nitric acid and of their salts, with the object of expressing with greater clearness the analogies and parallelism of their properties. You wrote:

$$H^2Cl^2;\ Na^2Cl^2;\ N^2O^5,\ H^2O;\ N^2O^5,\ K^2O,$$

and in the same manner you were led to double acetic acid and the acetates:

$$C^4H^6O^3,\ H^2O;\ C^4H^6O^3,\ K^2O.$$

The same necessity has been felt by all those who have had to express the equivalent ratios of acids, of water and of bases,

*** I cannot accept your opinion on the absolute value of the law of Wœstyn in the calculation of the specific heats of solid compounds. You know very well that M. Kopp, who went to the bottom of this question in 1864, found himself obliged, in verifying this relation, to attribute to the solid elements in their combination specific heats varying from 6·4 (silver, chlorine, nitrogen), down to 4 (oxygen), 2·3 (hydrogen), and 1·8 (carbon).**

as may be seen in the remarkable papers of Mr. Thompson on Thermo-chemistry, and even in the new edition of Gmelin, now publishing in Germany (see, among other things, iodic acid I^2O^5, H^2O).

The agreement of the numbers adopted by the partisans of atomic weights is then more apparent than real.

But I have no wish to prolong this controversy, particularly, as, between us, the only question is as to ascribing one value or another to the unity, of which the various equivalents of the same body are multiples. If we confine the question within these limits, it certainly does not present the excessive importance which has been ascribed to it for the last twenty years. The new atomic school has not, it appears to me, justified its pretension of changing the very base of chemical doctrines, and of founding a *new chemistry*, essentially different from the old. The only thing it has done has been to intermix the meshes of its hypotheses with our demonstrated laws, and this to the great detriment of the teaching of positive science. I believe that it would be advisable, in the future, to set aside all these systems, and to turn the minds of young scientists towards the really new views offered by molecular mechanics, which promise such rich harvests of discoveries.

Benzeval-sur-Dives (Calvados), August 10th, 1877.

Answer of M. Marignac.

I cannot but esteem myself very happy that the remarks which I recently presented in the *Moniteur Scientifique* on systems of chemical notations have given rise to the interesting article which precedes. I would like, however, to add a few remarks.

I found this fault with chemical equivalents, that they are not susceptible of a precise and general definition. M. Berthelot, while opposing this doctrine, seems to me to confirm it, as he was obliged to give a double definition. Sometimes these equivalents represent the ratios according to which bodies substitute themselves to one another; this is certainly a precise conception, which I have called *true chemical equivalence*, but it can only be applied to restricted cases. At other times, they are the ratios according to which bodies combine, or rather one of the ratios, which are multiples of one another, according to which combinations take place; in this case, the question of equivalence remains undetermined and more or less arbitrary. Besides, we may add that it is not always easy to see why one or the other of these definitions can be applied to the equivalent of a body. For instance, aluminium, compared to the metals which are most nearly related to it, those of the earths and alkaline earths, has a perfectly determinate value of substitution, and still this is not the value that has been chosen for its equivalent.

As M. Berthelot himself observes, we only differ in the opinion that each of us has formed on the part and relative importance which are to be ascribed to physical properties in the determination of equivalents. I may possibly have an exaggerated idea of the importance of these properties. But does not M. Berthelot, on his side, labor under a delusion when he thinks that chemical considerations alone are a sufficient guide to a chemist in this determination; is he not under the influence, which he has often discovered in his opponents, of a state of mind in which things seem very natural because we are accustomed to accept them? If the equivalent adopted for aluminium was not confirmed by the specific heat of this metal, by the vapor density of its chloride, and by numerous considerations of isomorphism, is he very sure that we would not hesitate between the formulas Al^2O^3 and AlO for alumina? What would make us suppose that hesitation would be permissible is the great number of bodies in which the determination of the equivalent has remained doubtful as long as physical properties have not served as guides, such, for instance, as in silicon, zirconium, glucinum, and the numerous group of the metals of cerite and gadolinite. The only thing I ask is to allow the same importance to physical properties in the case of bodies which occur with frequency.

M. Berthelot finds fault with me for being in apparent contradiction with myself, because in my researches on the specific heats of saline solutions, I doubled the molecular weights of some compounds, to better express the parallelism of the properties of some compound groups. It is true that, in comparing with one another the salts of the same acid, it has seemed to me more natural to refer their properties to equivalent quantities, or to quantities containing always the same proportion of acid. But if this has led me to group together two molecules of an alkaline chloride or nitrate, I was obliged, for the same reason, when taking the specific heats of sulphate of alumina or of alkaline phosphates to take, as unities of the weights of these salts, quantities which are really equivalent of other bodies, but which only represent fractions of the admitted equivalents of these bodies. Notwithstanding, M. Berthelot does not conclude that this is a proof that alumina should be written AlO and phosphoric acid $PhO\frac{5}{3}$, and that the equivalents of aluminium and phosphorus should be modified accordingly.

It may sometimes be interesting to compare certain properties of bodies by referring them to chemically equivalent weights, in cases where these correspond neither to molecular weights nor to the equivalents usually adopted, but we cannot conclude from this that those weights ought to be adopted as symbols of notations.

Apart from all these things, I agree with M. Berthelot that it is not advisable to exaggerate the importance of these questions, the solution of which cannot affect the important laws and theories of chemistry, and about which we can only reach a conclusion when we have arrived at more complete knowledge of the molecular constitution of compound bodies. This constitution itself will be doubtless revealed to us by researches on molecular mechanics, such as those on Thermo-chemistry, through which this eminent scientist aids so powerfully the advancement of science.

HELMHOLTZ, HERMANN VON. (b. Potsdam, Germany, 31 August 1821; d. Berlin, Germany, 8 September 1894)

Helmholtz studied medicine at the Friedrich Wilhelm Institute in Berlin, and also took courses at the University of Berlin. He worked on a dissertation in physiology under Johannes Müller, and, in 1842, he received an M.D. degree. During the period in which he was a surgeon to the military regiment in Potsdam, he maintained ties with Ernst Brücke, Emile duBois-Reymond, and Karl Ludwig, who were in Berlin and who were all committed to a physiological program of explaining life processes without appeal to "vital forces." In 1847 Helmholtz completed the classic memoir "Ueber die Erhaltung der Kraft," setting out the mathematical principles of the law of conservation of energy. In

1848 he was appointed associate professor of physiology at Königsberg. He moved to the chair of physiology at Bonn from 1855–1858, then to the chair of physiology at Heidelberg from 1858–1871, and from 1871 until his death, to the chair of physics at Berlin. At both Heidelberg and Berlin he established two new research institutes, one in physiology and the other in physics. In 1887 he accepted the directorship of the new Physikalisch-technische Reichsanstalt at Charlottenberg, on the outskirts of Berlin.

Working in physiology, Helmholtz invented the ophthalmometer and investigated the radii of curvature of the crystalline lens of the eye, explaining the mechanism of visual accommodation, and giving new strength to the theory of color vision showing the three primary colors to be red, green, and violet. He studied physiological acoustics from a mechanical standpoint and accounted for our perception of quality of tone. His *Physiological Optics* (1856–1855) and *Sensations of Tone* (1862) are classics.

His most important work in thermodynamics, in addition to the 1847 paper, includes an 1882 memoir distinguishing "bound" and "free" energy in chemical reactions, where the free energy, not the total energy change measured by the evolution of heat, is given by an equation now known as the "Gibbs-Helmholtz equation." His work in hydrodynamics in the 1850s resulted in an 1858 paper which described mathematically the properties of fluid motion in vortex tubes. This paper became the basis for William Thomson's theory of the vortex atom.

Electrodynamics became of increasing interest to Helmholtz after 1870, when he published a critique of Weber's law of electrostatic forces. Experiments carried out in Helmholtz's laboratory inclined him to regard the dielectric ether as a necessity for an adequate electrical theory. In his Faraday Lecture of 1881 he expressed support for Maxwell's theory and predicted the decline of action-at-a-distance theories. On the basis of work in electrolysis, he further expressed the view that electricity consists of discrete charges of "atoms of electricity" and that chemical forces are ultimately electrical in nature. This view was fully in consonance with the unifying philosophical principles which introduce his first (1847) memoir on the conservation of energy.

XLII.—*On the Modern Development of Faraday's Conception of Electricity.*

By Professor HELMHOLTZ.

[The Faraday Lecture, delivered before the Fellows of the Chemical Society, in the Theatre of the Royal Institution, on Tuesday, April 5, 1881.]

LADIES AND GENTLEMEN,

As I have the honour of speaking to you in memory of the great man, who from the very place where I stand has so often revealed to his admiring auditors the most unexpected secrets of nature, I hope at the outset to gain your assent if I limit my exposition to that side of his activity, which I know the best from my own experience and studies: I mean the theory of Electricity. The majority, indeed, of Faraday's own researches were connected directly or indirectly with questions regarding the nature of electricity, and his most important and most renowned discoveries lay in this field. The facts which he discovered are universally known. Every physicist, at present, is acquainted with the rotation of the plane of polarisation of light by magnetism, with dielectric tension and diamagnetism, and with the measurement of the intensity of galvanic currents by the voltameter, whilst induced currents act on the telephone, are applied to paralysed muscles, and nourish the electric light. Nevertheless, the fundamental conceptions by which Faraday was led to these much admired discoveries, have not received an equal amount of consideration. They were very divergent from the trodden path of scientific theory, and appeared rather startling to his contemporaries. His principal aim was to express in his new conceptions only facts, with the least possible use of hypothetical substances and forces. This was really an advance in general scientific method, destined to purify science from the last remnants of metaphysics. Faraday was not the first, and not the only man, who has worked in this direction, but perhaps nobody else at his time did it so radically. But every reform of fundamental and leading principles introduces new kinds of abstract notions, the sense of which the reader does not catch in the first instance.

Under such circumstances it is often less difficult for a man of original thought to discover new truth than to discover why other people do not understand and do not follow him. This difficulty must increase in Faraday's case because he had not gone through the same common course of scientific education as the majority of his readers. Now that the mathematical interpretation of Faraday's conceptions regarding the nature of electric and magnetic forces has been given by Clerk Maxwell, we see how great a degree of exactness and precision was really hidden behind the words, which to Faraday's contemporaries appeared either vague or obscure; and it is in the highest degree astonishing to see what a large number of general theorems, the methodical deduction of which requires the highest powers of mathematical analysis, he found by a kind of intuition, with the security of instinct, without the help of a single mathematical formula. I have no intention of blaming his contemporaries, for I confess that many times I have myself sat hopelessly looking upon some paragraph of Faraday's descriptions of lines of force, or of the galvanic current being an axis of power, &c. A single remarkable discovery may, of course, be the result of a happy accident, and may not indicate the possession of any special gift on the part of the discoverer; but it is against all rules of probability, that the train of thought which has led to such a series of surprising and unexpected discoveries, as were those of Faraday, should be without a firm, although perhaps hidden, foundation of truth. We must also in his case acquiesce in the fact that the greatest benefactors of mankind usually do not obtain a full reward during their lifetime, and that new ideas need the more time for gaining general assent the more really original they are, and the more power they have to change the broad path of human knowledge.

Faraday's electrical researches, although embracing a great number of apparently minute and disconnected questions, all of which he has treated with the same careful attention and conscientiousness, really always aim at two fundamental problems of natural philosophy, the one, more regarding the nature of the forces termed physical, or of forces working at a distance; the other, in the same way, regarding chemical forces, or those which act from molecule to molecule, and the relation between these and the first.

I shall give you only a short exposition on the degree of development which has been reached in the present state of science with regard to the first of these problems. The discussion of this question amongst scientific men is not yet finished, although, I think, it approaches its end. It is entangled with many geometrical and mechanical difficulties. How these are to be solved, and what are the arguments *pro* and *contra*, I cannot undertake to explain in a short public lecture, with any hope of gaining your scientific conviction. I can therefore give only a short statement of this side of the question, representing my own opinions; but I must not conceal the fact, that several men of great scientific merit, principally among my own countrymen, do not yet agree with me.

The great fundamental problem which Faraday called up anew for discussion was the existence of forces working directly at a distance without any intervening medium. During the last and the beginning of the present century, the model, after the likeness of which nearly all physical theories had been formed, was the force of gravitation acting between the sun, the planets, and their satellites. It is known with how much caution and even reluctance, Sir Isaac Newton himself proposed his grand hypothesis, which was destined to become the first great and imposing example of the power of true scientific method. We need not wonder that Newton's successors attempted at first to gain the same success by introducing analogous assumptions into all the various branches of natural philosophy. Electrostatic and magnetic phenomena especially appeared as near relations to gravitation, because electric and magnetic attractions and repulsions, according to Coulomb's measurements, diminish in the same proportion as gravity with increasing distance.

But then came Oerstedt's discovery of the motions of magnets under the influence of electric currents. The force acting in these phenomena had a new and very singular character. It seemed as if this force would drive a single isolated pole of a magnet in a circle around the wire conducting the current, on and on, without end, never coming to rest. And although it is not possible really to separate one pole of a magnet from the other, Ampère succeeded in producing such continuous circular motions by making a part of the current itself moveable with the magnet.

This was the starting point for Faraday's researches on electricity. He saw that a motion of this kind could not be produced by any force of attraction or repulsion, working from point to point. The first motive which guided him seems to have been an instinctive foreboding of the law of conservation of energy, which many attentive observers of nature had entertained before it was brought by Joule to a precise scientific definition. If the current is able to increase the velocity of the magnet, the magnet must react on the current. So he made the experiment, and discovered induced currents. He traced them out through all the various conditions under which they ought to appear. He found that an electromotive force striving to produce these currents arises wherever and whenever magnetic force is generated or destroyed. He concluded that in a part of space traversed by magnetic force there ought to exist a peculiar state of tension, and that every change of this tension produces electromotive force. This unknown hypothetical state he called provisionally the electrotonic state, and he was occupied for years and years in finding out what this electrotonic state was. He first discovered in 1838 the dielectric polarisation of electric insulators subject to electric forces. Such bodies, under the influence of electric forces, exhibit phenomena perfectly analogous to those observed in soft iron under the influence of the magnetic force. Eleven years later, in 1849, he was able to demonstrate that all ponderable matter is magnetised under the influence of sufficiently intense magnetic force, and at the same time he discovered the phenomena of diamagnetism, which indicated that even space, devoid of all ponderable matter, is magnetisable. The most simple explanation of these phenomena, indeed, is, that diamagnetic bodies are less magnetisable than a vacuous space, or than the luminiferous ether filling that space. In this way real changes corresponding to that hypothetical electrotonic state were demonstrated, and now, with quite a wonderful sagacity and intellectual precision, Faraday performed in his brain the work of a great mathematician without using a single mathematical formula. He saw with his mind's eye that magnetised and dielectric bodies ought to have a tendency to contract in the direction of the lines of force, and to dilate in all directions perpendicular to the former, and that by these systems of tensions and pressures in the space which surrounds electrified bodies, magnets or wires conducting electric currents, all the phenomena of electro-static, magnetic, electro-magnetic attraction, repulsion, and induction could be explained, without recurring at all to forces acting directly at a distance. This was the part of his path where so few could follow him; perhaps a Clerk Maxwell, a second man of the same power and independence of intellect, was needed to reconstruct in the normal methods of science the great building, the plan of which Faraday had conceived in his mind, and attempted to make visible to his contemporaries.

Nobody can deny that this new theory of electricity and magnetism, originated by Faraday and developed by Maxwell, is in itself well consistent, in perfect and exact harmony with all the known facts of experience, and does not contradict any one of the general axioms of dynamics, which have been hitherto considered as the fundamental truths of all natural science, because they have been found valid, without any exception, in all known processes of nature. A confirmation of great importance was given to this theory by the circumstance demonstrated by Clerk Maxwell, that the qualities which it must attribute to the imponderable medium filling space are able to produce and sustain magnetic and electric oscillations, propagating like waves, and with a velocity exactly equal to that of light. Several parts even of the theory of light are deduced with less difficulty from this new theory than from the well-known undulatory theory of Huyghens, which ascribes to the luminiferous ether the qualities of a rigid elastic body.

Nevertheless, the adherents of direct action at a distance have not yet ceased to search for solutions of the electromagnetic problem. The moving forces exerted upon each other by two wires conducting galvanic currents had long ago been reduced in a very ingenious way by Ampère, to attracting or repelling forces belonging to the linear elements of every current. The intensity of these forces is considered to depend not only on the distance of both parts of the current, but also in a rather complicated manner on the angles which the directions of the two currents make with each other and with the straight line joining them both. Ampère was not acquainted with induced currents, but the phenomena of these could be derived from the law of Ampère, connecting it with the general law, deduced by Faraday from his experiments, that the current induced by the motion of a magnet or of another current always resists this motion. The general mathematical expression of this law was established by Professor Neumann, of Königsberg. It gave directly, not the value of the forces, but the value of their mechanical work, the value of what mathematicians call an electrodynamic potential, and it reduced electromagnetic phenomena to forces acting, not from point to point, but from one linear element of a current to another. Linear elements of a wire conducting a galvanic current are, of course, complicated structures compared with atoms. I have myself elaborated several mathematical papers to prove that this formula of Professor Neumann was in harmony with all the known phenomena exhibited by closed galvanic circuits, and that it did not come into contradiction with the general axioms of mechanics in any case of electric motion. I succeeded in finding an experimental method of observing electrostatic effects of electromagnetic induction under conditions in which closed circuits could not be generated. This experiment decided against the supposition that Neumann's theory was complete so long as only the electric motions in metallic or fluid conductors were considered as active currents, but it was in accordance with the theory of Faraday and Maxwell, who supposed that from the extremities of conducting bodies, where an electric charge collects, electric motion is continued through the insulating media separating them.

Other eminent men have tried to reduce electromagnetic phenomena to forces acting directly between distant quantities of the hypothetical electric fluids, with an intensity which depends not only on the distance, but also on the velocities and accelerations of those electric quantities. Such theories have been proposed by Professor W. Weber, of Göttingen, by Riemann, the too early-deceased mathematician, and by Professor Clausius, of Bonn. All these theories explain very satisfactorily the phenomena of closed galvanic currents. But applied to other electric motions, they all come into contradiction with the general axioms of dynamics.

The hypothesis of Professor Weber makes the equilibrium of electricity unstable in any conductor of moderate dimensions, and renders possible the development of infinite quantities of work from finite bodies. I do not find that the objections, brought forward at first by Sir W. Thomson and Professor Tait in their Treatise on Natural Philosophy, and discussed and specialised afterwards by myself, have been invalidated by the discussions going on about this question. The hypothesis of Riemann, which he did not himself publish during his lifetime, labours under the same objection, and is at the same time in contradiction to Newton's axiom, which established the equality of action and reaction for all natural forces.

The hypothesis of Professor Clausius avoids the first objection, but not the second, and the author himself has conceded that this objection could be removed only by the assumption of a medium filling all space, between which and the electric fluids the forces acted.

The present development of science shows then, I think, a state of things very favourable to the hope that Faraday's fundamental conceptions may in the immediate future receive general assent. His theory, indeed, is the only one existing which is at the same time in perfect harmony with the facts as far as they are observed, and does not beyond the reach of facts lead into any contradiction of the general axioms of dynamics.

Clerk Maxwell himself has developed his theory only for closed conducting circuits. I have endeavoured during the last few years to investigate the results of this theory also for conductors not forming closed circuits. I can already say that the theory is in harmony with all the observations we have on the phenomena of open circuits: I mean (1) the oscillatory discharge of a condenser through a coil of wire, (2) my own experiments on electromagnetically induced charges on a rotating condenser, and (3) Mr. Rowland's observations on the electromagnetic effect of a rotatory disc charged with one kind of electricity.

The deciding assumption which removes the theoretical difficulties is that introduced by Faraday, who assumed that any electric motion in a conducting body which charges its surface with electricity is continued in the surrounding insulating medium as beginning or ending dielectric polarisation with an intensity equivalent to that of the current. A second inference from this supposition is, that the forces working at a distance do not exist, or are at least unimportant, when compared with the tensions and pressures of the dielectric medium.

It is not at all necessary to accept any definite opinion about the ultimate nature of the agent which we call electricity. Faraday himself avoided as much as possible giving any affirmative assertion regarding this problem, although he did not conceal his disinclination to believe in the existence of two opposite electric fluids.

For our own discussion of the electrochemical phenomena, to which we shall now turn, I beg permission to use the language of the old dualistic theory, which considers positive and negative electricity as two imponderable substances, because we shall have to speak principally on relations of quantity.

We shall try to imitate Faraday as well as we can by keeping carefully within the domain of phenomena, and, therefore, need not speculate about the real nature of that which we call a quantity of positive or negative electricity. Calling them substances of opposite sign, we imply with this name nothing else than the fact that a positive quantity never appears or vanishes without an equal negative quantity appearing or vanishing at the same time in the immediate neighbourhood. In this respect they behave really as if they were two substances, which cannot be either generated or destroyed, but which can be neutralised and become imperceptible by their union.

I see very well that this assumption of two imponderable fluids of opposite qualities is a rather complicated and artificial machinery, and that the mathematical language of Clerk Maxwell's theory expresses the laws of the phenomena very simply and very truly, with a much smaller amount of hypothetical implications; but I confess I should really be at a loss to explain without the use of mathematical formulæ, what he considers as a quantity of electricity, and why such a quantity is constant, like that of a substance. The original old notion of substance is not at all identical with that of matter. It signifies, indeed, that which behind the changing phenomena lasts as invariable, which can be neither generated nor destroyed, and in this oldest sense of the word we may really call the two electricities substances.

I prefer the dualistic theory, because it expresses clearly the perfect symmetry between the positive and negative side of electric phenomena, and I keep the well known supposition that as much negative electricity enters where positive goes away, because we are not acquainted with any phenomena which could be interpreted as corresponding with an increase or a diminution of the total electricity contained in any body. The unitary theory, which assumes the existence of only one imponderable electrical substance, and ascribes the effects of opposite kind to ponderable matter itself, affords a far less convenient basis for an electrochemical theory.

I now turn to the second fundamental problem aimed at by Faraday, the connection between electric and chemical force.

Already before Faraday went to work, an elaborate electrochemical theory had been established by the renowned Swedish chemist, Berzelius, which formed the connecting link of the great work of his life, the systematisation of the chemical knowledge of his time. His starting point was the series in which Volta had arranged the metals according to the electric tension which they exhibit after contact with each other. Metals easily oxidised occupied the positive end of this series, those with small affinity for oxygen the negative end. Metals widely distant in the series develope stronger electric charges than those near to each other. A strong positive charge of one metal, and a strong negative of the other, must cause them to attract each other and to cling to each other. The same faculty of exciting each other electrically was ascribed by Berzelius to all the other elements; he arranged them all into a series, at the positive end of which he placed potassium, sodium, barium, calcium, &c.; at the negative end oxygen, chlorine, bromine, &c. Two atoms of different elements coming into contact are supposed to excite each other electrically, like the metals in Volta's experiment. Berzelius' conceptions about the distribution of opposite electricities in the molecules, and his deductions regarding the intensity of these forces, were not very clear, and not in harmony with the laws of electric forces which had already been developed by Green and Gauss. A fundamental point which Faraday's experiment contradicted was the supposition that the quantity of electricity collected in each atom was dependent on their mutual electrochemical differences, which Berzelius considered as the cause of their apparently greater chemical affinity.

His theory of the binary character of all chemical compounds was also connected with this electrochemical theory. Two elements, as he supposed, one positive, the other negative, could unite to a compound of the first degree, a basic oxide or an acid; two such compounds into a compound of the second degree, a salt. But there was nothing to prevent one atom of every positive element from uniting as directly with two, three, or even seven of another negative element as with one. The same was assumed by Berzelius for negative elements. The modern experience of chemistry directly contradicts these statements. But although the fundamental conceptions of Berzelius' theory have been forsaken, chemists have not ceased to speak of positive and negative constituents of a compound body. Nobody can overlook that such a contrast of qualities as was expressed in Berzelius' theory really exists, well developed at the extremities, less evident in the middle terms of the series, and playing an important part in all chemical actions, although often subordinated to other influences.

When Faraday began to study the phenomena of decomposition by the galvanic current, which of course were considered by Berzelius as amongst the firmest supports of his theory, he put a very simple question, the first question, indeed, which every chemist speculating about electrolysis ought to have thought of. He asked, what is the quantity of electrolytic decomposition if the same quantity of electricity is sent through several electrolytic cells. By this investigation he discovered that most important law, generally known under his name, but called by him the law of definite electrolytic action.

When he began his experiments, neither Daniell's nor Grove's battery was known, and there were no means of producing currents of constant intensity; the methods of measuring this intensity were also in their infancy. This may excuse his predecessors. Faraday overcame this difficulty by sending the same current of electricity for the same time through a series of two or more electrolytic cells. He proved at first that the dimensions of the cell, and the size of the metallic plates through which the current entered and left the cell, had no visible influence upon the quantity of the products of decomposition. Cells containing the same electrolytic fluid between plates of the same metals gave always the same quantity, after being traversed by the same current. Then he compared the amount of decomposition in cells containing different electrolytes, and he found it exactly proportional to the chemical equivalents of the elements, which were either separated or converted into new compounds.

Faraday concluded from his experiments that a definite quantity of electricity cannot pass a voltametric cell containing acidulated water between electrodes of platinum, without setting free at the negative electrode a corresponding definite amount of hydrogen, and at the positive electrode the equivalent quantity of oxygen, one atom of oxygen for every pair of atoms of hydrogen. If instead of hydrogen any other element capable of replacing hydrogen is separated from the electrolyte, this is done also in a quantity exactly equivalent to the quantity of hydrogen which would have been evolved by the same electric current. According to the modern chemical theory of quantivalence, therefore, the same quantity of electricity passing through an electrolyte either sets free, or transfers to other combinations, always the same number of units of affinity at both electrodes; for instance,

instead of $\left.\begin{matrix}H\\H\end{matrix}\right\}$, either $\left.\begin{matrix}K\\K\end{matrix}\right\}$, or $\left.\begin{matrix}Na\\Na\end{matrix}\right\}$, or Ba}, or Ca}, or Zn},

or Cu} from cupric salts,

or $\left.\begin{matrix}Cu\\Cu\end{matrix}\right\}$ from cuprous-salts, &c.

The simple or compound halogens separating at the other electrodes, are equivalent of course to the quantity of the metallic element with which they were formerly combined.

According to Berzelius' theoretical views, the quantity of electricity collected at the point of union of two atoms ought to increase with the strength of their affinity. Faraday demonstrated by experiment, that so far as this electricity came forth in electrolytic decomposition, its quantity did not at all depend on the degree of affinity. This was really a fatal blow to Berzelius' theory.

Since that time our experimental methods and our knowledge of the laws of electrical phenomena have made enormous progress, and a great many obstacles have now been removed which entangled every one of Faraday's steps, and obliged him to fight with the confused ideas and ill-applied theoretical conceptions of some of his contemporaries. The original voltameter of Faraday, an instrument which measured the quantity of gases evolved by the decomposition of water, in order to determine with it the intensity of the galvanic current, has been replaced by the silver voltameter of Poggendorff, which permits of much more exact determinations by the quantity of silver deposited from a solution of silver nitrate on a strip of platinum. We have galvanometers which not only indicate that there is a galvanic current, but likewise measure its electromagnetic intensity very exactly and in a very short time, and do this as well for the highest as for the lowest degrees of intensity. We have electrometers, like the quadrant electrometer of Sir W. Thomson, able to measure differences of electric potential corresponding to less than one-hundredth of a Daniell's cell. As for the frequently-used term of electric potential, a term introduced by Green, you may translate it as signifying the electric pressure to which the positive unit of electricity is subject at a certain place. We need not hesitate to say that the more experimental methods were refined, the more completely were the exactness and generality of Faraday's law confirmed.

In the beginning Berzelius and the adherents of Volta's original theory of galvanism, based on the effects of metallic contact, raised many objections against Faraday's law.

By the combination of Nobili's astatic pairs of magnetic needles with Schweigger's multiplier, a coil of copper wire with numerous circumvolutions, galvanometers became so delicate that they were able to indicate the electrochemical equivalent of currents so feeble as to be quite imperceptible by all chemical methods. With the newest galvanometers you can very well observe currents which would require to last a century before decomposing one milligram of water, the smallest quantity that is usually weighed on chemical balances. You see that if such a current lasts only some seconds or some minutes, there is not the slightest hope of discovering its products of decomposition by chemical analysis. And even if it should last a long time, the minute quantities of hydrogen collected at the negative electrode may vanish, because they combine with the traces of atmospheric oxygen absorbed by the liquid. Under such conditions a feeble current may continue as long as you like without producing any visible trace of electrolysis, not even of galvanic polarisation, the appearance of which can be used as an indication of previous electrolysis. Galvanic polarisation, as you know, is an altered state of the metallic plates which have been used as electrodes during the decomposition of an electrolyte. Polarised electrodes, when connected by a galvanometer, give a current which they did not give before being polarised. By this current the plates are discharged again and returned to their original state of equality. The most probable explanation of this polarisation is that molecules of the electrolyte, charged with electricity, are carried by the current to the surface of the metal, itself charged with opposite electricity, and are retained there by electric attraction. That really constituent atoms of the electrolyte partake in the production of galvanic polarisation cannot well be doubted, because this state can be produced and also destroyed purely by chemical means. If hydrogen has been carried to an electrode by the current, contact with the atmospheric oxygen removes the state of polarisation.

The depolarising current is indeed a most delicate means of discovering previous decomposition. But even this may fail if the nascent polarisation is destroyed by an intervening chemical action, like that of the oxygen of the air. To avoid this, delicate experiments on this subject cannot be performed except in vessels carefully purified of all gases.

I have lately succeeded in doing this in a far more perfect way than before, by using the hermetically sealed cell (Fig. 1), which contains

FIG. 1.

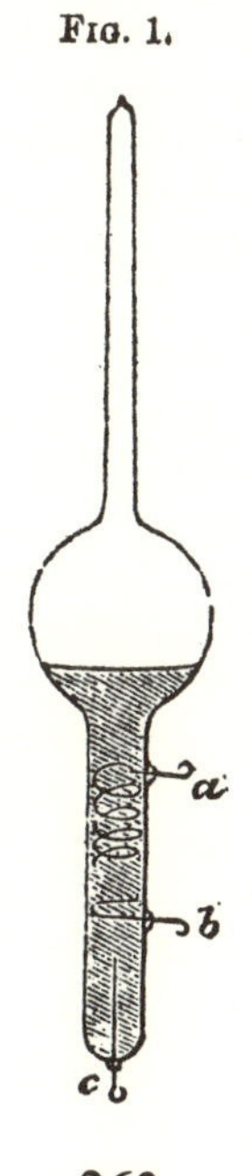

water acidulated with sulphuric acid. Two platinum wires, *b* and *c*, and a third platinum wire, *a*, which in the interior is connected with a spiral of palladium, can be used as electrodes. The tube, before it had been closed, was connected with a water-pump, and at the same time oxygen was evolved from *b* and *c* by two Grove's elements; the hydrogen carried to the palladium wire, *a*, was occluded in the metal. In this way the liquid in the tube is washed out with oxygen under low pressure and freed from all other gases. After the closing of the tube, the remaining small traces of electrolytic oxygen combine slowly with the hydrogen of the palladium. Traces of hydrogen occluded in the platinum wires *b* and *c* can be transferred by a feeble electromotive force into the palladium; and even new quantities of electrolytic gases, evolved after closing the tube, can be removed again by a Daniell's cell, which carries hydrogen to the palladium, where it is occluded, and oxygen to *b* and *c*, where it combines with hydrogen, as long as traces of this gas are dissolved in the liquid. The rest of the oxygen absorbed by the liquid combines with the occluded hydrogen.

I have ascertained with this apparatus that under favourable conditions one can observe the polarisation produced during a few seconds by a current which would decompose only one milligram of water in a century.

But even if the appearance of galvanic polarisation should not be acknowledged by opponents as a sufficient indication of previous decomposition, it is not difficult at present to reduce the indications of a good galvanometer to absolute measure, and to calculate the amount of decomposition which ought to be expected according to Faraday's law, and to verify that in all the cases in which no products of electrolysis can be discovered, their amount is too small for chemical analysis.

Products of decomposition cannot appear at the electrodes without the occurrence of motions of the constituent molecules of the electrolyte throughout the whole length of the liquid. On this point the majority of Faraday's predecessors were already agreed, but they differed from each other as soon as they came to the question what those motions were. Faraday saw very clearly the importance of this problem, and again appealed to experiment. He filled two cells with an electrolytic fluid, connecting them by a thread of asbestos wetted with the same fluid, in order to determine separately the quantity of all the chemical constituents transferred to the one and the other extremity of the electrolytic conductor. You know that he proposed for these atoms or groups of atoms transported by the current through the fluid the Greek word "Ions," the "Travellers;" and comparing the current of positive electricity with a stream of water, he called "cations" those atoms which went down the stream in the same direction with the positive electricity to the cathode, the metallic plate through which this electricity left the fluid. The "anions," on the contrary, go up the stream to the anode, the metal plate which is the source of the current of + E. Cations generally are atoms which are substitutes of hydrogen; anions are halogens.

This subject has been studied very carefully and for a great number of liquids, by Professor Hittorff, of Münster, and Professor G. Wiedemann, of Leipsic. They found that generally the velocity of the cation and the anion is different. Professor F. Kohlrausch, of Würzburg, has brought to light the very important fact that in diluted solutions of salts, including hydrates of acids and hydrates of caustic alkalis, every ion under the influence of currents of the same density moves on with its own peculiar velocity, independently of other ions moving at the same time in the same or in opposite directions.

Among the cations hydrogen has the greatest velocity, then follow potassium, ammonium, silver, sodium, afterwards the bivalent atoms of barium, copper, strontium, calcium, magnesium, zinc; near to the latter appears univalent lithium. Among the anions hydroxyl (OH) is the first, then follow the other univalent atoms iodine, bromine, cyanogen, chlorine, the compounds NO_3, ClO_3, the bivalent halogens of sulphuric and carbonic acid; after these fluorine and the halogen of acetic acid ($C_2H_3O_2$). The only exception to this rule is the difference observed between the decomposition of univalent and bivalent compounds. Generally the velocity of any ion when separated from a bivalent mate is less than when separated from one or two univalent mates.

It seems possible that the majority of molecules SO_4H_2 may be divided electrolytically into SO_4 and H_2, some of them on the other hand into SO_4H and H. By the latter some hydrogen would be carried backwards, and therefore the velocity of the total amount might appear diminished.

If both ions are moving, we shall find liberated at each electrode (1) that part of the corresponding ion which has been newly carried to that side: (2) another part which has been left by the opposite ion, with which it had been formerly combined. The total amount of chemical motion in every section of the fluid is, therefore, represented by the sum of the equivalents of the cation gone forwards, and of the anion gone backwards, in the same way as in the dualistic theory of electricity the total amount of electricity flowing through a section of the conductor corresponds to the sum of positive electricity going forwards and of negative electricity going backwards.

This established, Faraday's law tells us that through each section of an electrolytic conductor we have always equivalent electrical and chemical motion. The same definite quantity of either positive or negative electricity moves always with each univalent ion, or with every unit of affinity of a multivalent ion, and accompanies it during all its motions through the interior of the electrolytic fluid. This quantity we may call the electric charge of the atom.

I beg to remark that hitherto we have spoken only of phenomena. The motion of electricity can be observed and measured. Independently of this, the motion of the chemical constituents can also be measured. Equivalents of chemical elements and equivalent quantities of electricity are numbers which express real relations of natural objects and actions. That the equivalent relation of chemical elements depends on the pre-existence of atoms may be hypothetical; but we have not yet any theory sufficiently developed which can explain all the facts of chemistry as simply and as consistently as the atomic theory developed in modern chemistry.

Now the most startling result of Faraday's law is perhaps this. If we accept the hypothesis that the elementary substances are composed of atoms, we cannot avoid concluding that electricity also, positive as well as negative, is divided into definite elementary portions, which behave like atoms of electricity. As long as it moves about on the electrolytic liquid, each ion remains united with its electric equivalent or equivalents. At the surface of the electrodes decomposition can take place if there is sufficient electromotive force, and then the ions give off their electric charges and become electrically neutral.

The same atom can be charged in different compounds with equivalents of positive or of negative electricity. Faraday pointed out sulphur as being an element which can act either as anion or as cation. It is anion in sulphide of silver, a cation perhaps in strong sulphuric acid. Afterwards he suspected that the deposition of sulphur from sulphuric acid might be a secondary result. The cation may be hydrogen, which combines with the oxygen of the acid, and drives out the sulphur. But if this is the case, hydrogen recombined with oxygen to form water must retain its positive charge, and it is the sulphur, which in our case must give off positive equivalents to the cathode. Therefore this sulphur of sulphuric acid must be charged with positive equivalents of electricity. The same may be applied to a great many other instances. Any atom or group of atoms which can be substituted by secondary decomposition for an ion must be capable of giving off the corresponding equivalent of electricity.

When the positively charged atoms of hydrogen or any other cation are liberated from their combination and evolved as gas, the gas becomes electrically neutral; that is, according to the language of the dualistic theory, it contains equal quantities of positive and negative electricity; either every single atom is electrically neutralised, or one atom, remaining positive, combines with another charged negatively.

This latter assumption agrees with the inference from Avogadro's law, that the molecule of free hydrogen is really composed of two atoms.

Now arises the question: Are all these relations between electricity and chemical combination limited to that class of bodies which we know as electrolytes. In order to produce a current of sufficient strength to collect enough of the products of decomposition without producing too much heat in the electrolyte, the substance which we try to decompose ought not to offer too much resistance to the current. But this resistance may be very great, and the motion of the ions may be very slow, so slow indeed that we should need to allow it to go on for hundreds of years before we should be able to collect even traces of the products of decomposition; nevertheless all the essential attributes of the process of electrolysis could subsist. In fact we find the most various degrees of conducting power in various liquids. For a great number of them, down to distilled water and pure alcohol, we can observe the passage of the current with a sensitive galvanometer. But if we turn to oil of turpentine, benzene, and similar substances, the galvanometer becomes silent. Nevertheless these fluids also are not without a certain degree of conducting power. If you connect an electrified conductor with one of the electrodes of a cell filled with oil of turpentine, the other with the earth, you will find that the electricity of the conductor is discharged unmistakably more rapidly through the oil of turpentine than if you take it away and fill the cell only with air.

We may in this case also observe polarisation of the electrodes as a symptom of previous electrolysis. Connect the two pieces of platinum in oil of turpentine with a battery of eight Daniells, let it stay 24 hours, then take away the battery, and connect the electrodes with a quadrant electrometer; it will indicate that the two surfaces of platinum, which were homogeneous before, produce an electromotive force which deflects the needle of the electrometer. The electromotive force of this polarisation has been determined in some instances by Mr. Picker in the Laboratory of the University of Berlin; he has found that the polarisation of alcohol decreases with the proportion of water which it contains, and that that of the purest alcohol, ether, and oil of turpentine, is about 0·3, that of benzene 0·8 of a Daniell's element.

Another sign of electrolytic conduction is, that liquids placed between two different metals produce an electromotive force. This is never done by metals of equal temperature, or by other conductors which, like metals, let electricity pass without being decomposed. The production of an electromotive force is observed even with a great many rigid bodies, although very few of them allow us to observe electrolytic conduction with the galvanometer, and even these only at temperatures near their melting points. I remind you of the galvanic pile of Zamboni, in which pieces of dry paper are intercalated between thin leaves of metal. If the connection lasts long enough, even glass, resin, shellac, paraffin, sulphur—the best insulators we know—do the same. It is nearly impossible to prevent the quadrants of a delicate electrometer from being charged by the insulating bodies by which they are supported.

In all the cases which I have quoted, one might suspect that traces of humidity absorbed by the substance or adhering to their surface were the electrolytes. I show you, therefore, this little Daniell's cell, Fig. 2, constructed by my former assistant, Dr. Giese, in which a solu-

Fig. 2.

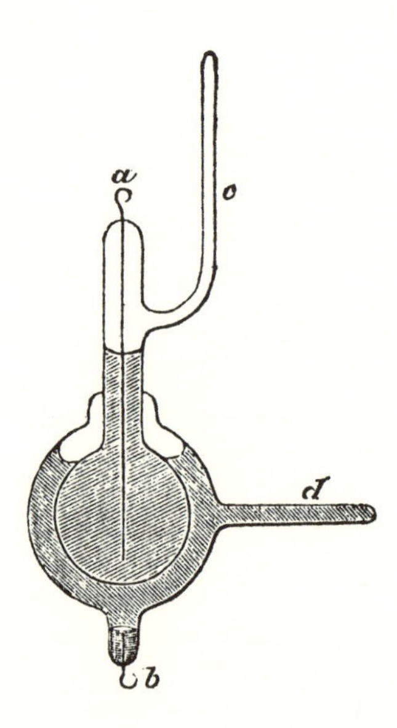

tion of sulphate of copper with a platinum wire, *a*, as an electrode, is enclosed in a bulb of glass hermetically sealed. This is surrounded by a second cavity, sealed in the same way, which contains a solution of zinc sulphate and some amalgam of zinc, to which a second platinum wire, *b*, enters through the glass. The tubes *c* and *d* have served to introduce the liquids, and have been sealed afterwards. It is, therefore, like a Daniell's cell, in which the porous septum has been replaced by a thin stratum of glass. Externally all is symmetrical at the two poles; there is nothing in contact with the air but a closed surface of glass, through which two wires of platinum penetrate. The whole charges the electrometer exactly like a Daniell's cell of very great resistance, and this it would not do if the septum of glass did not behave like an electrolyte; for a metallic conductor would completely destroy the action of the cell by its polarisation.

All these facts show that electrolytic conduction is not at all limited to solutions of acids or salts. It will, however, be rather a difficult problem to find out how far the electrolytic conduction is extended, and I am not yet prepared to give a positive answer. What I intended to remind you of was only that the faculty to be decomposed by electric motion is not necessarily connected with a small resistance to the current. It is easier for us to study the cases of small resistance, but the illustration which they give us about the connection of electric and chemical force is not at all limited to the acid and saline solutions usually employed.

Hitherto we have studied the motions of ponderable matter as well as of electricity, going on in an electrolyte. Let us now study the forces which are able to produce these motions. It has always appeared somewhat startling to everybody who knows the mighty power of chemical forces, and the enormous quantity of heat and mechanical work which they are able to produce, how exceedingly small is the electric attraction at the poles of a battery of two Daniell's cells, which nevertheless is able to decompose water. 1 gram of water, produced by burning hydrogen with oxygen, developes so much heat, that this heat transformed by a steam engine into mechanical work would raise the same weight to a height of 1,600,000 mètres. And on the contrary we require to use the most delicate contrivances to show that a gold leaf or a little piece of aluminium hanging on a silk fibre can be at all moved by the electric attraction of the battery. The solution of this riddle is found if we look at the quantities of electricity with which the atoms appear to be charged.

The quantity of electricity which can be conveyed by a very small quantity of hydrogen, when measured by its electrostatic forces, is exceedingly great. Faraday saw this, and endeavoured in various ways to give at least an approximate determination. He ascertained that even the most powerful batteries of Leyden jars, discharged through a voltameter, give scarcely any visible traces of gases. At present we can give definite numbers. The electrochemical equivalent of the electromagnetic unit of the galvanic current has been determined by Bunsen, and more recently by other physicists. This determination was followed by the very difficult comparison of the electromagnetic and electrostatic effects of electricity, accomplished at first by Professor W. Weber, and afterwards under the auspices of the British Association by Professor Clerk Maxwell.* The result is,

* According to the latest and most careful measurements of the electrochemical

that the electricity of 1 mgrm. of water, separated and communicated to two balls, 1 kilomt. distant, would produce an attraction between them equal to the weight of 26,800 kilogr.

As I have already remarked, the law that the intensity of the force is inversely proportional to the square of the distance, and directly proportional to the quantities of attracting and of attracted mass, holds good as well in the case of gravitation as in that of electric attraction and repulsion. We may, therefore, compare the gravitation acting between two quantities of hydrogen and oxygen with the attraction of their electric charges. The result will be independent of the size and the distance of these quantities. We find that the electric force is as great as the gravitation of ponderable masses, being 71,000 billion times greater than that of the oxygen and hydrogen* containing these electric charges.

effect, performed by Professor Kohlrausch, junr., the electromagnetic unit of the galvanic current, as it was defined by W. Weber (= 0·1 of the British Association unit), decomposes 0·009476 mgrm. of water per second. This same unit-current of Weber transfers about $300 \cdot 10^9$ electrostatic units of electricity (Weber himself gave $311 \cdot 10^9$, Maxwell $288 \cdot 10^9$) through each section of the circuit per second, half of them being positive, half of them negative, the two halves going in opposite directions. The electrostatic unit introduced by Gauss and W. Weber is that quantity of electricity which repels the same quantity at the distance of one millimeter with unit force, that is, with a force which, acting during one second upon one milligram, transmits to it a velocity of one millimeter per second. Gravity acting upon one milligram produces an acceleration 9,809 times as great. Weber's unit force, therefore, is equal to $\frac{1}{9809}$ of the weight of one milligram.

The force, F, measured by weight, with which the electric quantity, + E, measured in electrostatic units, attracts the opposite quantity − E, at a distance r is equal to—

$$F = \frac{E^2}{r^2} \cdot \frac{1}{9809} \text{ mgrm.}$$

If E is the quantity carried to each electrode during the decomposition of one milligram of water, it is according to the determinations quoted before $= 1618 \cdot 10^{10}$ units, and if we put r = 1 kilometer = 1,000,000 millimeters, we get the result quoted above.

* The gravity of a weight m is the force of attraction between it and the mass of the earth, which may be considered as concentrated in the central point of the earth. If h is the mean density of the earth, and r its radius, the mass of the earth will be—

$$\frac{4\pi}{3} r^3 . h.$$

If G is the force of attraction between two units of mass at unit of distance, the attraction of the mass m by the earth will be—

$$gm = \frac{4\pi}{3} rh . Gm.$$

Take—

$$\frac{\pi}{2} r = 10^7 \text{ meters} = 10^{10} \text{ mm.}$$

$$h = 5{\cdot}62 \text{ mgrms. per cub. mm.,}$$

The total force exerted by the attraction of an electrified body upon another charged with opposite electricity is always proportional to the quantity of electricity contained on the attracting as on the attracted body.

Although, therefore, the attracting forces exerted by the poles of a little battery able to decompose water on such electric charges as we can produce with our electric machines, are very moderate, the forces exerted by the same little apparatus on the enormous charges of the atoms in one milligram of water may very well compete with the mightiest chemical affinity.

If we now turn to investigate how the motions of the ponderable molecules are dependent upon the action of these forces, we must distinguish two different cases. At first we may ask what forces are wanted to call forth motions of the ions with their charge through the interior of the fluid; secondly, what are wanted to separate the ion from the fluid and its previous combinations?

Let us begin with the case in which the conducting liquid is surrounded everywhere by insulating bodies. Then no electricity can enter, none can go out through its surface, but positive electricity can be driven to one side, negative to the other, by the attracting and repelling forces of external electrified bodies. This process going on as well in every metallic conductor is called "electrostatic induction." Liquid conductors behave quite like metals under these conditions. Great quantities of electricities are collected, if large parts of the surfaces of the two bodies are very near to each other. Such an arrangement is called an electric condenser. We can arrange electric condensers in which one of the surfaces is that of a liquid, as Messrs. Ayrton and Perry have done lately. The water-dropping collector of electricity, invented by Sir W. Thomson, is a peculiar form of such a condenser, which can be charged with perfect regularity by the slightest electromotive force perceptible only to the most sensitive electrometer.

the attraction between $\frac{8}{9}$ mgrm. of oxygen and $\frac{1}{9}$ mgrm. of hydrogen contained in 1 mgrm. of water, and separated by a distance of 1 mm. will be—

$$\frac{1}{9} \cdot \frac{8}{9} \cdot G = \frac{g}{27{\cdot}5, 62{\cdot}10^{10}},$$

or equal to the weight of—

$$\frac{65904}{10^{17}} \text{ mgrm.}$$

The attraction of the electric charges, which has been calculated before for a distance of 1 kilometer, reduced to that of 1 mm., will be equal to the weight of—

$$20800{\cdot}10^{18} \text{ mgrms.}$$

From this it follows that the attraction by gravitation of those two masses would become equal to the attraction of their electric charges, if each of the masses could be increased 71,300 billions of times.

that the electricity of 1 mgrm. of water, separated and communicated to two balls, 1 kilomt. distant, would produce an attraction between them equal to the weight of 26,800 kilogr.

As I have already remarked, the law that the intensity of the force is inversely proportional to the square of the distance, and directly proportional to the quantities of attracting and of attracted mass, holds good as well in the case of gravitation as in that of electric attraction and repulsion. We may, therefore, compare the gravitation acting between two quantities of hydrogen and oxygen with the attraction of their electric charges. The result will be independent of the size and the distance of these quantities. We find that the electric force is as great as the gravitation of ponderable masses, being 71,000 billion times greater than that of the oxygen and hydrogen* containing these electric charges.

effect, performed by Professor Kohlrausch, junr., the electromagnetic unit of the galvanic current, as it was defined by W. Weber (= 0·1 of the British Association unit), decomposes 0·009476 mgrm. of water per second. This same unit-current of Weber transfers about $300 \cdot 10^9$ electrostatic units of electricity (Weber himself gave $311 \cdot 10^9$, Maxwell $288 \cdot 10^9$) through each section of the circuit per second, half of them being positive, half of them negative, the two halves going in opposite directions. The electrostatic unit introduced by Gauss and W. Weber is that quantity of electricity which repels the same quantity at the distance of one millimeter with unit force, that is, with a force which, acting during one second upon one milligram, transmits to it a velocity of one millimeter per second. Gravity acting upon one milligram produces an acceleration 9,809 times as great. Weber's unit force, therefore, is equal to $\frac{1}{9809}$ of the weight of one milligram.

The force, F, measured by weight, with which the electric quantity, + E, measured in electrostatic units, attracts the opposite quantity − E, at a distance r is equal to—

$$F = \frac{E^2}{r^2} \cdot \frac{1}{9809} \text{ mgrm.}$$

If E is the quantity carried to each electrode during the decomposition of one milligram of water, it is according to the determinations quoted before $= 1618 \cdot 10^{10}$ units, and if we put r = 1 kilometer = 1,000,000 millimeters, we get the result quoted above.

* The gravity of a weight m is the force of attraction between it and the mass of the earth, which may be considered as concentrated in the central point of the earth. If h is the mean density of the earth, and r its radius, the mass of the earth will be—

$$\frac{4\pi}{3} r^3 . h.$$

If G is the force of attraction between two units of mass at unit of distance, the attraction of the mass m by the earth will be—

$$gm = \frac{4\pi}{3} rh . Gm.$$

Take—

$$\frac{\pi}{2} r = 10^7 \text{ meters} = 10^{10} \text{ mm.}$$

$$h = 5 \cdot 62 \text{ mgrms. per cub. mm.,}$$

The total force exerted by the attraction of an electrified body upon another charged with opposite electricity is always proportional to the quantity of electricity contained on the attracting as on the attracted body.

Although, therefore, the attracting forces exerted by the poles of a little battery able to decompose water on such electric charges as we can produce with our electric machines, are very moderate, the forces exerted by the same little apparatus on the enormous charges of the atoms in one milligram of water may very well compete with the mightiest chemical affinity.

If we now turn to investigate how the motions of the ponderable molecules are dependent upon the action of these forces, we must distinguish two different cases. At first we may ask what forces are wanted to call forth motions of the ions with their charge through the interior of the fluid; secondly, what are wanted to separate the ion from the fluid and its previous combinations?

Let us begin with the case in which the conducting liquid is surrounded everywhere by insulating bodies. Then no electricity can enter, none can go out through its surface, but positive electricity can be driven to one side, negative to the other, by the attracting and repelling forces of external electrified bodies. This process going on as well in every metallic conductor is called "electrostatic induction." Liquid conductors behave quite like metals under these conditions. Great quantities of electricities are collected, if large parts of the surfaces of the two bodies are very near to each other. Such an arrangement is called an electric condenser. We can arrange electric condensers in which one of the surfaces is that of a liquid, as Messrs. Ayrton and Perry have done lately. The water-dropping collector of electricity, invented by Sir W. Thomson, is a peculiar form of such a condenser, which can be charged with perfect regularity by the slightest electromotive force perceptible only to the most sensitive electrometer.

the attraction between $\frac{8}{9}$ mgrm. of oxygen and $\frac{1}{9}$ mgrm. of hydrogen contained in 1 mgrm. of water, and separated by a distance of 1 mm. will be—

$$\frac{1}{9} \cdot \frac{8}{9} \cdot G = \frac{g}{27 \cdot 5, 62 \cdot 10^{10}},$$

or equal to the weight of—

$$\frac{65904}{10^{17}} \text{ mgrm.}$$

The attraction of the electric charges, which has been calculated before for a distance of 1 kilometer, reduced to that of 1 mm., will be equal to the weight of—

$$20800 \cdot 10^{18} \text{ mgrms.}$$

From this it follows that the attraction by gravitation of those two masses would become equal to the attraction of their electric charges, if each of the masses could be increased 71,300 billions of times.

Professor Wüllner has proved that even our best insulators, exposed to electric forces for a long time, are ultimately charged quite in the same way as metals would be charged in an instant. There can be no doubt that even electromotive forces less than $\frac{1}{100}$ Daniell produce perfect electrical equilibrium in the interior of an electrolytic liquid.

Another somewhat modified instance of the same effect is afforded by a voltametric cell containing two electrodes of platinum, which are connected with a Daniell's cell the electromotive force of which is insufficient to decompose the electrolyte. Under this condition the ions carried to the electrodes cannot give off their electric charges. The whole apparatus behaves, as was first accentuated by Sir W. Thomson, like a condenser of enormous capacity. The quantity of electricity, indeed, collected in a condenser under the same electromotive force is inversely proportional to the distance of the plates. If this is diminished to $\frac{1}{100}$, the condenser takes in 100 times as much electricity as before. Now, bringing the two surfaces of platinum and of the liquid into immediate contact, we reduce their interval to molecular distances. The capacity of such a condenser has been measured by Messrs. Varley, Kohlrausch, and Colley. I have myself made some determinations which show that oxygen absorbed in the fluid is of great influence on the apparent value. By removing all traces of gas I have got a value a little smaller than that of Kohlrausch, which shows that if we divide equally the total value of the polarisation between two platinum plates of equal size, the distance between the two strata of positive and negative electricity, the one lying on the last molecules of the metal, the other on those of the fluid, ought to be a ten millionth part (Kohlrausch $\frac{1}{15000000}$) of a millimeter. We always come nearly to the same limit, when we calculate the distances through which molecular forces are able to act, as already shown in several other instances by Sir W. Thomson.

Owing to the enormous capacity of such an electrolytic condenser, the quantity of electricity which enters into it, if it is charged even by a feeble electromotive force, is sufficiently great to be indicated easily by a galvanometer. What I now call charging the condenser, I have before called polarising the metallic plate. Both, indeed, are the same process, because electric motion is always accompanied in the electrolytes by chemical decomposition.

Observing the polarising and depolarising currents in a cell like that represented in Fig. 1, we can observe these phenomena with the most feeble electromotive forces of $\frac{1}{1000}$ Daniell, and I found that down to this limit the quantity of electricity entering into the condenser was proportional to the electromotive force by which it was collected. By taking larger surfaces of platinum, I suppose it will be possible to reach a limit much lower than that. If any chemical force existed besides that of the electrical charges which could bind all the pairs of opposite ions together, and required any amount of work to be vanquished, an inferior limit ought to exist to such electromotive forces as are able to attract the ions to the electrodes, and to charge these as condensers. No phenomenon indicating such a limit has as yet been discovered, and we must, therefore, conclude that no other force resists the motions of the ions through the interior of the liquid than the mutual attractions of their electric charges. These are able to prevent the atoms of the same kind which repel each other from collecting at one place, and atoms of the other kind attracted by the former from collecting at any other part of the fluid, as long as no external electric force favours such distribution. The electric attraction, therefore, is able to produce an equal distribution of the opposite constituent atoms throughout the liquid, so that all parts of it are neutralised electrically as well as chemically.

On the contrary, as soon as an ion is to be separated from its electrical charge, we find that the electrical forces of the battery meet with a powerful resistance, the overpowering of which requires a good deal of work to be done. Usually the ions, losing their electric charges, are at the same time separated from the liquid; some of them are evolved as gases, others are deposited as rigid strata on the surface of the electrodes, like galvanoplastic copper. But the union of two constituents having powerful affinity to form a chemical compound, always produces, as you know very well, a great amount of heat, and heat is equivalent to work. On the contrary, decomposition of the compound substance requires work, because it restores the energy of the chemical forces which has been spent by the act of combination. Oxygen and hydrogen separated from each other contain a store of energy, for on burning the hydrogen in the oxygen they unite, form water, and develope a great amount of heat. In the water the two elements are contained, and their chemical attraction continues to work as before, keeps them firmly united, but can no more produce any change, any positive action. We must reduce the combined elements into their first state, we must separate them, applying a force which is capable of vanquishing their affinity before they are ready to renew their first activity. The amount of heat produced by the chemical combination is the equivalent of the work done by the chemical forces brought into action. It requires the same amount of work to separate the compound and to restore hydrogen and oxygen uncombined. I have already given the value of this amount calculated as a weight raised against the force of gravity.

Metals uniting with oxygen or halogens produce heat in the same way, some of them, like potassium, sodium, and zinc, even more than an equivalent quantity of hydrogen; less oxidisable metals, like copper, silver and platinum, less. We find, therefore, that heat is generated when zinc drives copper out of its combination with the compound halogen of sulphuric acid, as is the case in a Daniell's cell.

If a galvanic current passes through any conductor, a metallic wire, or an electrolytic fluid, it evolves heat. Dr. Joule was the first who proved experimentally that if no other work is done by the current, the total amount of heat evolved in a galvanic circuit during a certain time is exactly equal to that which ought to have been generated by the chemical actions which have been performed during that time. But this heat is not evolved at the surface of the electrodes where these chemical actions take place, but is evolved in all the parts of the circuit, proportionally to the galvanic resistance of every part. From this it is evident that the heat evolved is an immediate effect, not of the chemical action, but of the galvanic current, and that the chemical work of the battery has been spent to produce only the electric action.

To keep up an electric current through an electric conductor, indeed, requires work to be done. New stores of positive electricity must be continually introduced at the positive end of the conductor, the repulsive force acting upon them having to be overcome; negative electricity, in the same way, into the negative end. This can be done by mere mechanical force, with an electric machine working by friction, or by electrostatic or by electromagnetic induction. In a galvanic current it is done by chemical force, but the work required remains the same.

If we apply Faraday's law, a definite amount of electricity passing through the circuit corresponds with a definite amount of chemical decomposition going on in every electrolytic cell of the same circuit. According to the theory of electricity, the work done by such a definite quantity of electricity which passes, producing a current, is proportionate to the electromotive force acting between both ends of the conductor. You see, therefore, that the electromotive force of a galvanic circuit must be, and is indeed, proportional to the heat generated by the sum of all the chemical actions going on in all the electrolytic cells during the passage of the same quantity of electricity. In the cells of the galvanic battery chemical forces are brought into action able to produce work; in cells in which decomposition is occurring, work must be done against opposing chemical forces; the rest of the work done appears as heat evolved by the current, as far as it is not used up to produce motions of magnets or other equivalents of work.

You see the law of the conservation of energy requires that the electromotive force of every cell must correspond exactly with the total amount of chemical forces brought into play, not only the mutual affinities of the ions, but also those minor molecular attractions produced by the water and other constituents of the fluid. These minor attractions have lately formed the subject of most valuable and extended calorimetric researches by Messrs. Andrews, Thomsen, and Berthelot. But even influences too minute to be measured by calorimetric methods may be discovered by measuring the electromotive force. I have myself deduced from the mechanical theory of heat the influence which the quantity of water contained in a solution of metallic salts has on the electromotive force. The chemical attraction between salt and water can be measured in this instance by the diminution of the tension of the aqueous vapours over the liquid, and the results of the theoretical deduction have been confirmed in a very satisfactory manner by the observations of Dr. James Moser.

Hitherto we have supposed that the ion with its electric charge is separated from the fluid. But the ponderable atoms can give off their electricity to the electrode, and remain in the liquid, being now electrically neutral. This makes scarcely any difference in the value of the electromotive force. For instance, if chlorine is separated at the anode, it will at first remain absorbed by the liquid; if the solution becomes saturated, or if we make a vacuum over the liquid, the gas will rise in bubbles. The electromotive force remains unaltered. The same may be observed with all the other gases. You see in this case that the change of electrically negative chlorine into neutral chlorine is the process which requires so great an amount of work, even if the ponderable matter of the atoms remains where it was.

On the contrary, if the electric attraction does not suffice to deprive the ions collecting at the surface of the electrodes of their electric charge, you will find the cation attracted and retained by the cathode, the anion by the anode, with a force far too great to be overpowered by the expansive force of gases. You may make a vacuum as perfect as you like over a cathode polarised with hydrogen, or an anode polarised with oxygen; you will not obtain the smallest bubble of gas. Increase the electric potential of the electrodes, so that the electric force becomes powerful enough to draw the electric charge of the ions over to the electrode: the ions will be liberated and free to leave the electrode, passing into the gaseous state or spreading in the liquid by diffusion. One cannot assume, therefore, that their ponderable matter is attracted by the electrode; if this were the case, this attraction ought to last after discharge as before. We must conclude, therefore, that the ions are drawn to the electrode only because they are charged electrically.

The more the surface of the positive electrode is covered with negative atoms of the anion, and the negative with the positive ones of the cation, the more is the attracting force of the electrodes exerted upon the ions of the liquid diminished by this second stratum of opposite electricity covering them. On the contrary, the force with which the positive electricity of an atom of hydrogen situated at the surface of the electrode itself is attracted towards the negatively charged metal increases in proportion as more negative electricity collects before it on the metal, and behind it in the fluid.

The electrical force acting on equal quantities of electricity situated at the inside of one of the electric strata of a condenser is proportional to the electromotive force which has charged the condenser, and inversely proportional to the distance of the charged surfaces. If these are $\frac{1}{100}$ mm. apart, it is 100 times as great as if they are one millimeter apart. If we come, therefore, to molecular distances, like those calculated from the measurement of the capacity of polarised electrodes, the force is ten million times as great, and becomes able, even with a moderate electromotive force, to compete with the powerful chemical forces which combine every atom with its electric charge, and hold the atoms bound to the liquid.

Such is the mechanism by which electric force is concentrated at the surface of the electrodes and increased in its intensity to such a degree that it becomes able to overpower the mightiest chemical affinities we know of. If this can be done by a polarised surface, acting like a condenser charged by a very moderate electromotive force, can the attractions between the enormous electric charges of anions and cations be an unimportant and indifferent part of chemical affinity?

In a decomposing cell the ions resist external forces striving to separate them from their electric charges. Let the current go in the opposite direction, and you will have an opposite effect. In a Daniell's cell neutral zinc enters as cation into the electrolyte, taking with it only positive electricity, and leaving its negative electricity to the metallic plate. At the copper electrode positive copper separates from the electrolyte and is neutralised, giving off its charge to the electrode. But the Daniell's cell in which this goes on does work, as we have seen. We must conclude, therefore, that an equivalent of positive electricity, on charging an atom of zinc, does more work than the same equivalent does on charging an atom of copper.

You see, therefore, if we use the language of the dualistic theory and treat positive and negative electricities as two substances, the phenomena are the same as if equivalents of positive and negative electricity were attracted by different atoms, and perhaps also by the different values of affinity belonging to the same atom with different force. Potassium, sodium, zinc, must have strong attraction to a positive charge; oxygen, chlorine, bromine to a negative charge.

Do we perceive effects of such an attraction in other cases? Here we come to the much discussed question of Volta's assumption that electricity is produced by contact of two metals. About the fact there can be no doubt. If we produce metallic contact between a piece of copper and a piece of zinc, opposed to each other like the two plates of a condenser, and carried by isolating rods of shell-lac we find that after contact the zinc is charged positively, the copper negatively. This is just the effect we ought to expect if zinc has a higher attracting force to positive electricity, this force working only through molecular distances. I have proposed this explanation of Volta's experiments in my little pamphlet on the "Conservation of Energy," published in 1847. All the facts observed with different combinations of metallic conductors are perfectly in harmony with it. Volta's law of the series of tension comprising all metallic conductors is easily deduced from it. If only metals come into play, their galvanic attractions produce instantaneously a state of electric equilibrium, so that no lasting current can occur. Electrolytic conductors, on the contrary, are decomposed chemically by every motion of electricity through their surface. Electric equilibrium, therefore, will not be possible before this decomposition has been finished, and till that stage is reached, the electric motion must continue. This point has been accentuated already by Faraday, as the essential difference between the two classes of conductors.

The original theory of Volta was incomplete in an essential point, because he was not acquainted with the fact of electrolytic decomposition. His original conception of the force of contact is, therefore, in contradiction to the law of Conservation of Energy; and even before this law was established and enunciated with scientific precision, there were many chemists and physicists, amongst them Faraday, who had the right instinct that this could not be the true explanation. The opponents of Volta's opinions tried to give chemical explanations also of those experiments of his, which were carried out exclusively with metal conductors. They might be oxidised by the oxygen of the air, and the amount of oxidation required for a very slight electric charge was so infinitesimal, that no chemical analysis could ever discover it; so small, that even in the highest vacuum, and in the purest specimens of hydrogen or nitrogen with which we might surround the plates, there was oxygen enough to continue the effect for years. From this point of view the chemical theory cannot be refuted. On the contrary, the so-called chemical theory of Volta's fundamental experiments was rather indefinite; it scarcely did more than tell us: here is the possibility of a chemical process, here electricity can be produced. But which kind, how much, to which potential, remained indefinite. I have not found in all the papers which have been written for the defence of the chemical theory, a clear explanation why zinc opposed to copper in liquids, where zinc really is oxidized and dissolved, becomes negative, and why in air and other gases it becomes positive, if the same cause, namely oxidation, is at work. The hypothesis, on the contrary, of a different degree of affinity between the metals and the two electricities, gives a perfectly definite answer. I do not see why an actual chemical process should be wanted to charge the zinc and copper on contact. But you see that the forces, which according to their hypothesis produce the electric effect, are the same as those which must be considered as the cause of a main part of all chemical actions.

Again, the electric charges produced by contact of zinc and copper are very feeble. They have become measurable only with the help of the latest improvements introduced into the construction of electrometers by Sir W. Thomson; but the cause of their feeble intensity is evident. If you bring into narrow contact two plain and well-polished plates of zinc and copper, the quantity of electricity collected at both sides of their common surface is probably very great; but you cannot observe it before having separated the plates. Now it is impossible to separate them at the same instant over the whole extent of their surface. The charge which they retain will correspond with the inclined position which they have at the moment when the last point of contact is broken; then all the other parts of the surfaces are already at a distance from each other infinitely greater than molecular distance, and conduction in metals always establishes nearly instantaneously the electric equilibrium corresponding to the actual situation. If you wish to avoid this discharge during the separation of the plates, one of them must be isolated; then indeed we get a far more striking series of phenomena, those belonging to electricity of friction. Friction, probably, is only the means of producing a very close contact between the two bodies. If the surfaces are very clean and free from air, as for instance in a Geissler's tube, the slightest rolling contact is sufficient to develop the electric charge. I can show you two such tubes exhausted so far that very high electric tension is necessary to make the vacuum luminous, one containing a small quantity of mercury, the other the fluid compound of potassium and sodium. In the first the negative metal is intensely negative relatively to glass, in the second the metal is on the positive extremity of Volta's series; the glass proves to be more positive also in this case, but the difference is much smaller than with mercury, and the charge is feeble.

Faraday very often recurs to this to express his conviction that the forces termed chemical affinity and electricity are one and the same. I have endeavoured to give you a survey of the facts connected with the question, and to avoid as far as possible the introduction of hypotheses, except the atomic theory of modern chemistry. I think the facts leave no doubt that the very mightiest among the chemical forces are of electric origin. The atoms cling to their electric charges, and opposite electric charges cling to each other; but I do not suppose that other molecular forces are excluded, working directly from atom to atom. Several of our leading chemists have lately begun to distinguish two classes of compounds, viz., molecular aggregates and typical compounds, the latter being united by atomic affinities, the former not.

Electrolytes belong to the latter class. If we conclude from the facts that every unit of affinity is charged with one equivalent either of positive or of negative electricity, they can form compounds, being electrically neutral only if every unit charged positively unites under the influence of a mighty electric attraction with another unit charged negatively. You see that this ought to produce compounds in which every unit of affinity of every atom is connected with one and only one other unit of another atom. This, as you will see immediately, is the modern chemical theory of quantivalence, comprising all the saturated compounds. The fact that even elementary substances, with few exceptions, have molecules composed of two atoms, makes it probable that even in these cases electric neutralisation is produced by the combination of two atoms, each charged with its full electric equivalent, not by neutralisation of every single unit of affinity.

Unsaturated compounds with an even number of unconnected units of affinity offer no objection to such an hypothesis; they may be charged with equal equivalents of opposite electricity. Unsaturated compounds with one unconnected unit, existing only at high temperatures, may be explained as dissociated by intense molecular motion of heat in spite of their electric attractions. But there remains one single instance of a compound which, according to the law of Avogadro, must be considered as unsaturated even at the lowest temperature, namely, nitric oxide (NO), a substance offering several very uncommon peculiarities, the behaviour of which will be perhaps explained by future researches.

But I abstain from entering into further specialities; perhaps I have already gone too far. I would not have dared to do it, had I not felt myself sheltered by the authority of that great man who was guided by a never-erring instinct of truth. I thought that the best I could do for his memory was to recall to the minds of the men by whose energy and intelligence chemistry has undergone its modern astonishing development, what important treasures of knowledge lie still hidden in the works of that wonderful genius. I am not sufficiently acquainted with chemistry to be confident that I have given the right interpretation, the interpretation which Faraday himself would have given, if he had been acquainted with the law of chemical quantivalence. Without the knowledge of this law I do not see how a consistent and comprehensive electrochemical theory could be established. Faraday did not try to develope a complete theory of this kind. It is a characteristic of a man of high intellect to see where to avoid going further in his theoretical speculations for want of facts, as to see how to proceed when he finds the way open. We ought therefore to admire Faraday also in his cautious reticence, although now, standing on his shoulders, and assisted by the wonderful development of organic chemistry, we are able, perhaps, to see farther than he did. I shall consider my work of to-day well rewarded if I have succeeded in kindling anew the interest of chemists in the electrochemical part of their science.

ARRHENIUS, SVANTE AUGUST (b. Vik, Sweden, 19 February 1859; d. Stockholm, Sweden, 2 October 1927)

Arrhenius studied at the University of Uppsala from 1876 until 1878 and completed a doctoral dissertation in 1884 after working with the physicist Erik Edlund at the Swedish Academy of Sciences in Stockholm. His dissertation on the electrolytic theory of dissociation was not well received at Uppsala, but attracted the attention of Wilhelm Ostwald in Riga and Jacobus Henricus Van't Hoff in Amsterdam. In 1884 Arrhenius was appointed lecturer in physical chemistry at Uppsala, and during 1886–1890 he studied throughout Europe in the laboratories of Ostwald, Kohlrausch, Boltzmann, and Van't Hoff. From 1891 to 1905 he taught at the Technical High School in Stockholm, first as lecturer (1891–1895), then as professor of physics (after 1895), and rector (1896–1905). In 1905 he became Director of the Physical Chemistry Department of the newly founded Nobel Institute in Stockholm, and he received the Nobel Prize in Chemistry in 1903. He carried great influence with the Swedish Academy in the selection of Nobel laureates in chemistry and physics. In later years Arrhenius became interested in immunology and in meterological and cosmic phenomena. He published the first textbook on cosmic physics in 1903, the *Lehrbuch der kosmischen Physik*.

Along with Ostwald and Van't Hoff, Arrhenius is regarded as a founder of the "ionist" school of physical chemistry. In his 1884 doctoral thesis he presented an experimental section on conductivity of dilute solutions and a theoretical section on the dissociation of salt molecules dissolved in water. In this latter section, Arrhenius drew upon two seemingly unconnected ideas: first, Williamson's theory of etherification, that a group of chemical molecules is a system in dynamic equilibrium (molecules continually exchange radicals or atoms with each other); and second, Rudolf Clausius's hypothesis that a small fraction of a dissolved salt is dissociated into ions even when no current is passing through the solution. Arrhenius distinguished between the "active" (electrolytic) and "inactive" (nonelectrolytic) parts into which the complex molecules of a salt solution dissociate, suggesting that the percentage of "active" ions increases with increased dilution. After learning of the experimental work of François Raoult on freezing-point depressions in salt solutions and reading Van't Hoff's 1885 paper "On the Laws of Chemical Equilibrium in Dilute Gases or Solutions," Arrhenius suggested that deviations in dilute salt solutions from the gas law might be explained by the fact that a salt is partially dissociated into its ions. In his 1887 paper published in the *Zeitschrift für physikalische Chemie*, Arrhenius set kinetic and atomic theory on a a stronger footing with his new conception of "ionization": that at infinite dilution, all molecules of electrolytes in water break up into charged "ions" and exert equal action as conductors.

ON THE DISSOCIATION OF SUBSTANCES IN AQUEOUS SOLUTION.[1]

By Svante Arrhenius.

In a memoir submitted to the Swedish Academy of Sciences on 14th October 1885 van't Hoff established both experimentally and theoretically the following very important generalisation of Avogadro's Law.[2]

"La pression exercée par les gaz à une température déterminée si un même nombre de molécules en occupe un volume donné, est égale à la pression osmotique qu'exerce dans les mêmes circonstances la grande majorité des corps, dissous dans les liquides quelconques."

This law was proved by van't Hoff so conclusively

[1] [From *Zeitschrift für physikalische Chemie*, vol. i., pp. 631-648. This memoir embodies two communications read before the Swedish Academy of Sciences and published in *Öfversigt af Kongl. Svenska Vetenskaps-Akademiens Forhandlingar*, 1887 :—

1. An attempt to calculate the dissociation (activity coefficients) of substances in aqueous solution. (Read 8th June 1887), p. 405.
2. On additive properties of dilute salt solutions. (Read 9th November 1887), p. 561.

The first of these papers is in Swedish ; the second in German.]

[2] Van't Hoff, Une propriété générale de la matière diluée, p. 43. *Sv. Vet.-Ak. Handlingar*, **21**, No. 17, 1886. (Also in *Archives néerlandaises* for 1885.)

that little doubt can exist as to its entire accuracy. There remained, however, a certain difficulty to be overcome, namely, that the law was only valid for the "great majority of substances," a very considerable number of the aqueous solutions examined forming exceptions, since they exert a much higher osmotic pressure than is required by the law.

When a gas shows a similar deviation from Avogadro's Law, the abnormality is explained by assuming that the gas is in a state of dissociation. A well-known example is afforded by the behaviour of chlorine, bromine, and iodine at high temperatures, these substances being regarded under the given conditions as split up into atoms.

The same expedient might be adopted to explain the exceptions to van't Hoff's Law, but so far it has not been used, probably owing to the novelty of the subject, the large number of known exceptions, and the grave objections on the chemical side which might be raised to such an explanation. The object of the present communication is to show that the hypothesis of dissociation of certain substances in aqueous solution is strongly supported by conclusions drawn from their electrical properties, and also that on closer consideration the chemical objections are appreciably lessened.

For the explanation of electrolytic phenomena we must postulate with Clausius[1] that a portion of the molecules of an electrolyte is dissociated into ions which move independently of one another. Since now the "osmotic pressure" which a substance dissolved in a liquid exerts on the walls of the containing vessel must be attributed, in accordance with modern kinetic conceptions, to the impacts of the ultimate particles of this substance on

[1] Clausius, *Pogg. Ann.*, **101**, 347 (1857). Wiedemann, *Elektricität*, **2**, 941.

the boundary walls in the course of their motion, we must in harmony with this assume that the pressure exerted on the walls of the vessel by a molecule dissociated in the above sense will be the same as that exerted by its ions in the free state. If then we could calculate what proportion of the molecules of an electrolyte had undergone dissociation into ions, we could also calculate by van't Hoff's Law the value of the osmotic pressure.

In a previous communication "Sur la conductibilité galvanique des électrolytes" I have called those molecules, whose ions have independent motion, *active molecules*, and those whose ions are firmly bound together, *inactive molecules*. I have likewise emphasised the probability that at the most extreme dilution all the inactive molecules of an electrolyte are converted into active molecules.[1] On this assumption I will base the calculations made in the sequel. The ratio of the number of active molecules to the total number of molecules, active and inactive, I have called the activity coefficient.[2] The activity coefficient of an electrolyte at infinite dilution is thus assumed to be unity. At smaller dilutions it is less than unity, and, on the principles laid down in the work above cited, is equal to the quotient of the actual molecular conductivity of the solution by the limiting value to which the molecular conductivity of the same solution approaches with increasing dilution. This method of calculation is only applicable if the solutions

[1] *Bihang till Kongl. Vet.-Ak. Handlingar*, 8, No. 13, p. 61; No. 14, p. 5 and p. 13 (1884).

[2] [Le coëfficient d'activité d'un électrolyte est le nombre exprimant le rapport du nombre d'iones qu'il y a réelement dans l'électrolyte, au nombre d'iones qui y seraient renfermées, si l'électrolyte était totalement transformé en molécules électrolytiques simples (*loc. cit.*, No. 14, p. 5).]

considered are not too concentrated, *i.e.*, to solutions in which disturbing factors, such as internal friction, etc., can be neglected.

If the activity coefficient α is known, we can derive from it, as follows, the value of van't Hoff's coefficient i, which is the ratio of the osmotic pressure actually exerted by a substance to the osmotic pressure which it would exert if it consisted entirely of inactive (undissociated) molecules. The coefficient i is obviously equal to the sum of the number of inactive molecules and the number of ions, divided by the total number of molecules, active and inactive. Thus if m is the number of inactive and n the number of active molecules, and if k is the number of ions into which each active molecule is dissociated (*e.g.* for KCl, $k = 2$, viz. K and Cl; for $BaCl_2$ or K_2SO_4, $k = 3$, viz. Ba, Cl and Cl, or K, K and SO_4), then we have

$$i = \frac{m + kn}{m + n}$$

The activity coefficient α is equal to $\frac{n}{m+n}$,

and consequently

$$i = 1 + (k - 1)\alpha.$$

according to which formula the numbers in the last column of the following table have been calculated.

On the other hand the value of i can be calculated by van't Hoff's method from the results of Raoult's experiments on the freezing points of solutions as follows. The depression t of the freezing point of water in degrees centigrade caused by the solution of one gram molecule of the given substance in a litre of water is divided by 18.5, and the values of $i = t/18.5$ thus obtained are tabulated in the second-last column. All numbers given below are calculated on the assumption that 10 grams of the substance were dissolved in 1 litre of water as was actually the case in Raoult's experiments.

In the following table the first two columns give the name and chemical formula of the substance, the third the value of the activity coefficient (Lodge's dissociation-ratio)[1] and the last two the values of i calculated by the two different methods ($i = t/18.5$ and $i = 1+(k-1)\alpha$). The substances studied are divided into four main groups: (1) non-conductors, (2) bases, (3) acids, (4) salts.

1.—NON-CONDUCTORS.

Substance.	Formula.	α.	$i = t/18{\cdot}5$.	$i = 1+(k-1)\alpha$.
Methyl alcohol . .	CH_3OH	0	0·94	1
Ethyl alcohol . .	C_2H_5OH	0	0·94	1
Butyl alcohol . .	C_4H_9OH	0	0·93	1
Glycerol . . .	$C_3H_5(OH)_3$	0	0·92	1
Mannitol . . .	$C_6H_{14}O_6$	0	0·97	1
Invert sugar . . .	$C_6H_{12}O_6$	0	1·04	1
Cane sugar . . .	$C_{12}H_{22}O_{11}$	0	1·00	1
Phenol	C_6H_5OH	0	0·84	1
Acetone	C_3H_6O	0	0·92	1
Ethyl ether . . .	$(C_2H_5)_2O$	0	0·90	1
Ethyl acetate . . .	$C_4H_8O_2$	0	0·96	1
Acetamide . . .	$C_2H_3ONH_2$	0	0 96	1

[1] Lodge, On Electrolysis, *Rep. of Brit. Assoc.*, Aberdeen Meeting, 1885, p. 756.

2.—BASES.

Substance.	Formula.	a.	$i = t/18 \cdot 5$.	$i = 1 + (k-1)a$.
Baryta	$Ba(OH)_2$	0·84	2·69	2·67
Strontia	$Sr(OH)_2$	0·86	2·61	2·72
Lime	$Ca(OH)_2$	0·80	2·59	2·59
Lithia	$LiOH$	0·83	2·02	1·83
Soda	$NaOH$	0·88	1·96	1·88
Potash	KOH	0·93	1·91	1·93
Thallium hydroxide	$TlOH$	0·90	1·79	1·90
Tetramethyl-ammonium hydroxide	$N(CH_3)_4OH$	...	1·99	...
Tetraethyl - ammonium hydroxide	$N(C_2H_5)_4OH$	0·92	...	1·92
Ammonia	NH_3	0·01	1·03	1·01
Methylamine	CH_3NH_2	0·03	1·00	1·03
Trimethylamine	$(CH_3)_3N$	0·03	1·09	1·03
Ethylamine	$C_2H_5NH_2$	0·04	1·00	1·04
Propylamine	$C_3H_7NH_2$	0·04	1·00	1·04
Aniline	$C_6H_5NH_2$	0·00	0·83	1·00

3.—ACIDS.

Substance.	Formula.	a.	$i = t/18 \cdot 5$.	$i = 1 + (k-1)a$.
Hydrochloric	HCl	0·90	1·98	1·90
Hydrobromic	HBr	0·94	2·03	1·94
Hydriodic	HI	0·96	2·03	1·96
Hydrofluosilicic	H_2SiF_6	0·75	2·46	1·75
Nitric	HNO_3	0·92	1·94	1·92
Chloric	$HClO_3$	0·91	1·97	1·91
Perchloric	$HClO_4$	0·94	2·09	1·94
Sulphuric	H_2SO_4	0·60	2·06	2·19
Selenic	H_2SeO_4	0·66	2·10	2·31
Phosphoric	H_3PO_4	0·08	2·32	1·24
Sulphurous	H_2SO_3	0·14	1·03	1·28
Hydrogen sulphide	H_2S	0·00	1·04	1·00
Iodic	HIO_3	0·73	1·30	1·73
Phosphorous	H_3PO_3	0·46	1·29	1·46
Boric	H_3BO_3	0·00	1·11	1·00
Hydrocyanic	HCN	0·00	1·05	1·00
Formic	$HCOOH$	0·03	1·04	1·03
Acetic	CH_3COOH	0·01	1·03	1·01
Butyric	C_3H_7COOH	0·01	1·01	1·01
Oxalic	$(COOH)_2$	0·25	1·25	1·49
Tartaric	$C_4H_6O_6$	0·06	1·05	1·11
Malic	$C_4H_6O_5$	0·04	1·08	1·07
Lactic	$C_3H_6O_3$	0·03	1·01	1·03

4.—SALTS.

Substance.	Formula.	α.	$i = t/18{\cdot}5$.	$i = 1 + (k-1)\alpha$.
Potassium chloride	KCl	0·86	1·82	1·86
Sodium chloride	$NaCl$	0·82	1·90	1·82
Lithium chloride	$LiCl$	0·75	1·99	1·75
Ammonium chloride	NH_4Cl	0·84	1·88	1·84
Potassium iodide	KI	0·92	1·90	1·92
Potassium bromide	KBr	0·92	1·90	1·92
Potassium cyanide	KCN	0·88	1·74	1·88
Potassium nitrate	KNO_3	0·81	1·67	1·81
Sodium nitrate	$NaNO_3$	0·82	1·82	1·82
Ammonium nitrate	NH_4NO_3	0·81	1·73	1·81
Potassium acetate	$KC_2H_3O_2$	0·83	1·86	1·83
Sodium acetate	$NaC_2H_3O_2$	0·79	1·73	1·79
Potassium formate	$KCHO_2$	0·83	1·90	1·83
Silver nitrate	$AgNO_3$	0·86	1·60	1·86
Potassium chlorate	$KClO_3$	0·83	1·78	1·83
Potassium carbonate	K_2CO_3	0·69	2·26	2·38
Sodium carbonate	Na_2CO_3	0·61	2·18	2·22
Potassium sulphate	K_2SO_4	0·67	2·11	2·33
Sodium sulphate	Na_2SO_4	0·62	1·91	2·24
Ammonium sulphate	$(NH_4)_2SO_4$	0·59	2·00	2·17
Potassium oxalate	$K_2C_2O_4$	0·66	2·43	2·32
Barium chloride	$BaCl_2$	0·77	2·63	2·54
Strontium chloride	$SrCl_2$	0·75	2·76	2·50
Calcium chloride	$CaCl_2$	0·75	2·70	2·50
Cupric chloride }	$CuCl_2$	...	2·58	...
Zinc chloride }	$ZnCl_2$	0·70	...	2·40
Barium nitrate	$Ba(NO_3)_2$	0·57	2·19	2·13
Strontium nitrate	$Sr(NO_3)_2$	0·62	2·23	2·23
Calcium nitrate	$Ca(NO_3)_2$	0·67	2·02	2·33
Lead nitrate	$Pb(NO_3)_2$	0·54	2·02	2·08
Magnesium sulphate	$MgSO_4$	0·40	1·04	1·40
Ferrous sulphate	$FeSO_4$	0·35	1·00	1·35
Cupric sulphate	$CuSO_4$	0·35	0·97	1·35
Zinc sulphate	$ZnSO_4$	0·38	0·98	1·38
Cupric acetate	$Cu(C_2H_3O_2)_2$	0·33	1·68	1·66
Magnesium chloride	$MgCl_2$	0·70	2·64	2·40
Mercuric chloride	$HgCl_2$	0·03	1·11	1·05
Cadmium iodide	CdI_2	0·28	0·94	1·56
Cadmium nitrate	$Cd(NO_3)_2$	0·73	2·32	2·46
Cadmium sulphate	$CdSO_4$	0·35	0·75	1·35

The last three entries in the second-last column are not, as are all the others, taken from Raoult's[1] work, but from older data of Rüdorff,[2] who used very large quantities of dissolved material, so that those three values cannot lay claim to great accuracy. The values of α are calculated from the data of Kohlrausch,[3] Ostwald[4] (acids and bases), and a few due to Grotrian[5] and Klein.[6] The values calculated from Ostwald's data are by far the most exact, because both the magnitudes involved in α are here easily determined with great accuracy. The error in i calculated from these values of α can scarcely be more than 5 per cent. The values of α and i calculated from Kohlrausch's data are somewhat less certain, chiefly on account of the difficulty of estimating exactly the maximum value of the molecular conductivity. This is still more difficult for α and i calculated from the experimental data of Grotrian and of Klein, which may in unfavourable cases show an error of 10 or 15 per cent. The degree of accuracy of Raoult's numbers is difficult to estimate: judging from the data themselves for very nearly related substances errors of 5 per cent. or even somewhat more are not improbable.

It should be noted that in the above tables there have been included for the sake of completeness *all* substances for which a moderately accurate calculation of i by the two methods has been found possible. When

[1] Raoult, *Ann. de Ch. et de Phys.* [5] **28**, 133 (1883); [6] **2**, 66, 99, 115 (1884); [6] **4**, 401 (1885).

[2] Rüdorff, cited in Ostwald's *Lehrbuch der allg. Chemie* [1st Edition], **1**, 414.

[3] Kohlrausch, *Wied. Ann.*, **6**, 1 and 145 (1879); **26**, 161 (1885).

[4] Ostwald, *Journ. f. pr. Ch.* [2] **32**, 300 (1885); [2] **33**, 352 (1886); *Zeit. physikal. Chem.*, **1**, 74 and 97 (1887).

[5] Grotrian, *Wied. Ann.*, **18**, 177 (1883)

[6] Klein, *Wied. Ann.*, **27**, 151 (1886).

conductivity data for a substance were wanting, *e.g.*, cupric chloride and tetramethylammonium hydroxide, for purposes of comparison the calculation was made from the data for a nearly related substance (zinc chloride and tetra-ethylammonium hydroxide) whose electrical properties could not be notably different from those of the substance in question.

From the values of i which exhibit a very great difference between the two methods of calculation, that for H_2SiF_6 should probably at the outset be excluded, for Ostwald has shown that in all probability this acid in aqueous solution partially decomposes into $6HF$ and SiO_2, which would explain the high value of i found by Raoult's method.[1]

There is one circumstance which to a small extent tends to invalidate the comparison between the last two columns, namely, that the values in strictness hold good for different temperatures. The values in the second-last column are all for temperatures differing little from 0° C. as they have been derived from experiments on slight lowerings of the freezing point of water. On the other hand the values in the last column for acids and bases (Ostwald's experiments) were obtained at 25° C., the others at 18° C. The figures in the last column for non-conductors are of course also valid for 0° C., since these substances at this temperature also contain no appreciable quantity of dissociated (active) molecules.

There appears nevertheless on a comparison of the numbers in the last two columns a very marked parallelism between them.[2] This shows a posteriori that in all probability the assumptions on which I

[1] [Compare, however, E. Baur, *Zeit. physikal. Chem.*, 48, 483 (1904).]

[2] For some salts forming definite exceptions, see below, p. 53.

have based the calculation are in the main correct. These assumptions were:—

(1) That van't Hoff's law holds good not merely for the *majority* but for *all* substances, including those formerly regarded as exceptions (electrolytes in aqueous solution).

(2) That every electrolyte in aqueous solution consists in part of molecules electrolytically and chemically active and in part of inactive molecules which, however, on dilution change into active molecules, so that at infinite dilution only active molecules are present.

The objections which may probably be urged from the chemical side, are in the main the same as those raised against Clausius' hypothesis, and these I have already endeavoured to show are completely untenable.[1] A repetition of these objections may therefore be regarded as superfluous. I will emphasise only one point. Although the dissolved substance exerts on the walls of the vessel an osmotic pressure precisely as if it were partly dissociated into ions, yet the dissociation here in question is not quite the same as that, for instance, which is shown by the decomposition of an ammonium salt at a high temperature. In the first case the products of dissociation (the ions) have very large electrical charges of opposite sign, and thus are subject to certain conditions (the incompressibility of electricity), from which it follows that the ions cannot without a great expenditure of energy be separated from each other in any marked degree.[2] In ordinary dissociation, on the other hand, where such conditions do not occur, the products of dissociation can generally be separated from each other.

These two assumptions are of the most far-reaching

[1] *Loc. cit.*, No. 14, pp. 6 and 31.

[2] *Ibid.*, No. 14, p. 8.

significance, not only from the theoretical point of view, which will receive discussion below, but also in the highest degree from the practical standpoint. If it can be shown—and I have endeavoured to make it highly probable—that the law of van't Hoff is universally applicable, then the chemist has at hand an extraordinarily convenient means of determining the molecular weight of any substance soluble in a liquid.[1]

I may also draw attention to the fact that the above equation (1) gives the connection between the two magnitudes i and α which play the chief parts in the two theories recently developed by van't Hoff and by myself.

In the above calculation of i I have tacitly assumed that the inactive molecules exist in the solution as simple molecules and not united to form larger molecular complexes. The results of this calculation (*i.e.*, the figures in the last column) compared with the results of direct observation (the figures in the second-last column) show that in general the supposition is completely justified. Were it not so, the numbers in the second-last column would be smaller than those of the last. Exceptions in which this is without doubt the case are found in the sulphates of the magnesium series ($MgSO_4$, $FeSO_4$, $CuSO_4$, $ZnSO_4$, and $CdSO_4$) and also in cadmium iodide. To explain these we may assume that the inactive molecules of these salts in reality combine with each other. Hittorf[2] was led

[1] This method has already been applied. Compare Raoult, *Ann. d. Ch. et d. Phys.* [6] **8**, 317 (1886). Paternò and Nasini, *Berichte d. deut. chem. Ges.*, 1886, p. 2527.

[2] Hittorf, *Pogg. Ann.*, **106**, 547 and 551 (1859); Wiedemann, *Elektricität*, **2**, 584.

to make this assumption for CdI_2 by the great change in the transport ratio; and if we look more closely at his table we find an unusually large alteration of the ratio for the three sulphates of the above series which he investigated, namely, $MgSO_4$, $CuSO_4$, and $ZnSO_4$. It is thus quite probable that this explanation is applicable to the salts mentioned. For the other salts we must assume that double molecules are only present in very small proportions. It remains now to indicate in a few words the grounds which have led previous authors to the assumption of a general occurrence of complex molecules in solutions.

Since, generally speaking, substances in the gaseous state consist of simple molecules, in accordance with Avogadro's Law, and since in the neighbourhood of the condensation point a small increase in density frequently occurs, which indicates a combination of these simple molecules, it was thought that in the change from the gaseous to the liquid state there would be a much greater amount of combination, that is, it was assumed that liquid molecules are in general not simple. I do not mean here to challenge the accuracy of this conclusion, but there is a great difference when the liquid considered is dissolved in another (*e.g.* HCl in water). For if we assume that on dilution the originally inactive molecules are transformed into active molecules, the ions being to a certain extent separated from each other, a process which is naturally accompanied by a great expenditure of energy, there is no difficulty in also making the assumption that the molecular complexes on mixing with water are likewise for the most part broken up, no great amount of work being thereby required. The fact that heat is absorbed on dilution has also been regarded as affording evidence for the existence of molecular complexes,[1] but, as we have seen, this might equally well be ascribed to the activation of the molecules. Some chemists, too, for the purpose of maintaining constant valency, assume the existence of molecular complexes in which the superfluous valencies satisfy each other.[2] The doctrine of constant valency is, however, so much disputed that one can scarcely use it as a basis for deduction. The conclusion arrived at in this way, that potassium chloride has the formula $(KCl)_3$, Lothar Meyer endeavours to support by drawing attention to the circumstance that KCl is much less volatile than $HgCl_2$ although it has a much smaller molecular weight. Apart altogether from the insecure basis on which such an argument is founded, the conclusion evidently applies only to pure substances, not to solutions. There are several other reasons adduced by Lothar Meyer for the existence of molecular complexes, for instance, that NaCl diffuses more slowly than HCl,[3] which, however, is to be attributed to the probably greater friction (according to electrolytic determinations) of Na than H against water. It is sufficient, however, to quote Meyer's own words: "However incomplete and uncertain all these various indications of methods for determining molecular weights in the liquid state may still be, yet they permit us to hope that in future it will be possible to determine the magnitude of the molecules. . . ."[4] van't Hoff's Law, however, now gives quite definite indications that in the vast majority of cases the number of molecular complexes in aqueous solution is negligible, although

[1] Ostwald, *Lehrb. d. allg. Chem.* [1st edition], 1, 811. L. Meyer, *Moderne Theorien der Chemie*, p. 319 (1880).

[2] L. Meyer, *loc. cit.*, p. 360. [3] *Ibid.*, p. 316.

[4] *Ibid.*, p. 321. van't Hoff's Law makes this possible, *vide supra*.

in a few cases, which are precisely those formerly used in support of the assumption of molecular complexes,[1] their existence is confirmed. It is not of course denied that such molecular complexes may exist in the case of other salts, particularly in concentrated solutions; but in solutions of the degree of dilution which Raoult examined they exist in general in such small quantity that they may be neglected in the above calculations without appreciable error.

Most of the properties of dilute salt solutions are of a so-called additive character. In other words, the numerical values of the properties may be regarded as the sum of the properties of the parts of the solution, *i.e.*, of the solvent and of the parts of the dissolved molecules, which in fact coincide with the ions. For example, the conductivity of a salt solution may be regarded as the sum of the conductivities of the solvent (generally zero), of the positive ion, and of the negative ion. In most cases this can be tested by comparing two salts (*e.g.* of K and Na) of one acid (*e.g.* HCl) with the corresponding salts of the same metals (K and Na) and another acid (*e.g.* HNO_3). Then the property of the first salt (KCl) minus the property of the second (NaCl) is equal to the property of the third salt (KNO_3) minus the property of the fourth ($NaNO_3$). This holds good in most cases for several properties, *e.g.* conductivity, lowering of the freezing point, refraction equivalent, heat of neutralisation, etc., which we shall briefly consider in the sequel, and finds its explanation in the almost complete dissociation of the great majority of salts into their ions. If a salt in aqueous solution is completely split into its ions, it follows naturally that the properties of this salt may in the main be expressed as the sum of the properties of the ions,

[1] Kohlrausch, *Wied. Ann.*, 6, p. 167 (1879).

since the ions are for the most part independent of each other, so that each ion has a characteristic value for the property, whatever be the oppositely charged ion with which it is associated. In the solutions actually investigated complete dissociation is probably never reached, and the above deduction is therefore not strictly accurate, but if we consider salts which are 80 or 90 per cent. dissociated, as are almost without exception the salts of strong bases with strong acids, we shall in general commit no very great errors by calculating the values of the properties on the assumption that the salts are completely decomposed into their ions. According to the foregoing table this applies also to the strong bases and acids $Ba(OH)_2$, $Sr(OH)_2$, $Ca(OH)_2$, LiOH, NaOH, KOH, TlOH and HCl, HBr, HI, HNO_3, $HClO_3$ and $HClO_4$. But there is another group of substances which have hitherto in most investigations played a subordinate part and are far removed from complete dissociation, even in dilute solution, as, for example, according to the table, $HgCl_2$ (and other Hg salts), CdI_2, $CdSO_4$, $FeSO_4$, $MgSO_4$, $CuSO_4$ and $Cu(C_2H_3O_2)_2$, the weak bases and acids, such as NH_3 and the various amines, H_3PO_4, H_2S, H_3BO_3, HCN, formic, acetic, butyric, tartaric, malic and lactic acids. The properties of these substances will not, as a rule, be of the same (additive) nature as those previously considered, as we shall find completely confirmed in the sequel. There exists naturally a number of transitional compounds between these two groups, as may be seen from the table. It should be mentioned here that in consideration of the almost universal occurrence of additive properties in substances of the first group, which are those by far the most frequently studied, several investigators have been led to the assumption of a certain complete dissociation of salts into their ions.[1] Since, however, from the purely chemical point of view no reason could be found why salt molecules should decompose in a perfectly definite way (into their ions), since in addition, for reasons which need not be discussed here, chemists have contested as long as possible the existence of unsaturated radicals (under which rubric ions are included), and since the foundations of such an assumption are somewhat uncertain,[2] the hypothesis of a complete dissociation has till now found little favour. The above table shows also that the reluctance of chemists to accept the complete dissociation demanded is not without a certain justification, inasmuch as at the dilutions actually employed the dissociation is never complete, being indeed for a large number of electrolytes (those of the second group) relatively insignificant.

From these observations we proceed to the special cases in which additive properties occur.

1. **Heat of neutralisation in dilute solutions.**— In the neutralisation of an acid by a base the energies of these two substances are disengaged as heat, but at the same time there is absorbed a certain quantity of heat arising from the energies of the water and salt (ions) which have been formed. We indicate by curved brackets the energies of the substances, for which it is immaterial so far as the deduction is concerned whether they occur as ions or not, and by square brackets those of the ions, the energies taken for them being always those in dilute solution. For example, on the provisional assumption of complete dissociation of the salts, we have for the neutralisation of NaOH by $\frac{1}{2}H_2SO_4$ (1)

[1] Valson, *Compt. rend.*, **73**, 441 (1871); **74**, 103 (1872). Favre and Valson, *Compt. rend.*, **75**, 1000 (1872). Raoult, *Ann. d. Chim. et d. Phys.* [6] **4**, 426.

[2] For Raoult's different hypotheses, compare *loc. cit.*, p. 401.

and HCl (2) and of KOH by $\frac{1}{2}H_2SO_4$ (3) and HCl (4) all in equivalent quantities, liberation of the following amounts of heat:—

$$(NaOH) + \tfrac{1}{2}(H_2SO_4) - (H_2O) - [Na] - \tfrac{1}{2}[SO_4] \quad (1)$$
$$(NaOH) + (HCl) - (H_2O) - [Na] - [Cl] \quad (2)$$
$$(KOH) + \tfrac{1}{2}(H_2SO_4) - (H_2O) - [K] - \tfrac{1}{2}[SO_4] \quad (3)$$
$$(KOH) + (HCl) - (H_2O) - [K] - [Cl] \quad (4)$$

We have obviously $(1) - (2) = (3) - (4)$ if we assume complete dissociation of the salts. As indicated above, this is approximately true for the instances which actually occur. It is all the more the case because the salts which are furthest from complete dissociation—here Na_2SO_4 and K_2SO_4—are dissociated approximately to the same extent, whereby the error occurring in the two members of the last equation is approximately the same, a circumstance which determines the occurrence of additive properties somewhat more frequently than one might expect. The following short table (p. 60) shows that on the neutralisation of strong bases and strong acids the additive properties are clearly apparent. For salts of weak bases with weak acids this is no longer the case, because they are probably partially decomposed by the water.

The numbers in brackets represent the difference between the heat evolution of the salt considered and that of the corresponding chloride. It will be seen if we neglect the last column that they are in each vertical column approximately constant. This is closely connected with the thermo-neutrality of salts; but since I have on a previous occasion treated this subject more particularly and emphasised the close connection with the Williamson-Clausius hypothesis,[1] I need not here enter into any detailed analysis.

[1] *Loc. cit.*, No. 14, p. 67.

Heats of Formation of some Salts in Dilute Solution.
(From the Data of Thomsen and Berthelot.)

	HCl, HBr, or HI.	HNO_3.	$C_2H_4O_2$.	CH_2O_2.
NaOH	13·7	13·7 (0·0)	13·3 (−0·4)	13·4 (−0·3)
KOH	13·7	13·8 (+0·1)	13·3 (−0·4)	13·4 (−0·3)
NH_3	12·4	12·5 (+0·1)	12·0 (−0·4)	11·9 (−0·5)
½ $Ca(OH)_2$	14·0	13·9 (−0·1)	13·4 (−0·6)	13·5 (−0·5)
½ $Ba(OH)_2$	13·8	13·9 (+0·1)	13·4 (−0·4)	13·5 (−0·3)
½ $Sr(OH)_2$	14·1	13·9 (−0·2)	13·3 (−0·8)	13·5 (−0·6)

	½ $(CO_2H)_2$.	½ H_2SO_4.	½ H_2S.	HCN.	½ CO_2.
NaOH	14·3 (+0·6)	15·8 (+2·1)	3·8 (−9·9)	2·9 (−10·8)	10·2 (−3·5)
KOH	14·3 (+0·6)	15·7 (+2·0)	3·8 (−9·9)	3·0 (−10·7)	10·1 (−3·6)
NH_3	12·7 (+0·3)	14·5 (+2·0)	3·1 (−9·3)	1·3 (−11·1)	5·3 (−7·1)
$Ca(OH)_2$	...	...	3·9 (−10·1)	...	...

2. **Specific volume and specific gravity of dilute solutions.**—If we add to a litre of water a small quantity of salt, whose ions may be regarded as completely independent, the volume is thereby changed. If x is the quantity of the one ion added and y that of the other ion, the volume as a first approximation will be $1 + ax + by$ litres where a and b are constants. Since the ions are dissociated from each other, the constant a of one ion will be independent of the nature of the other ion. In the same way the weight is $1 + cx + dy$ kilos where c and d are two other constants characteristic of the ions. Thus, for small quantities of x and y, the specific gravity is given by the formula

$$1 + (c - a)x + (b - d)y$$

where $(c - a)$ and $(b - d)$ are characteristic constants for the two ions. The specific gravity is thus for dilute solutions an additive property as Valson found.[1] But, according to Ostwald, "specific gravity is not a property to be used for the representation of stoichiometric laws,"[2] so we need not enter on a closer discussion of these results. The determination of the constants *a* and *b*, etc., would be promising, but so far has not been carried out.

In close connection with these phenomena is the change of volume on neutralisation. From similar considerations to those put forward for heat of neutralisation, it may be shown that the change of volume on neutralisation is an additive property. It is evident from the above table that all the K, Na, and NH_4 salts investigated are almost completely dissociated, as is still clearer from the later work of Ostwald, so that we may expect very good agreement for these salts. The differences in the change of volume on the formation of the corresponding salts of nineteen different acids appear as practically constant quantities.[3] Since the bases which form salts of the second group have not been investigated, no exceptions are yet known.

3. **Specific refraction of solutions.**—In mixtures of various substances the expression $P.\frac{n-1}{d}$ for the different components (where n is the refractive index, d the density and P the weight) may be summed to give the corresponding magnitude for the mixture. Consequently this magnitude, the refraction equivalent, must be an additive property for dissociated salts also. That this is in fact the case has been clearly shown by the researches of Gladstone. In this instance the

[1] Valson, *Compt. rend.*, **73**, 441 (1871). Ostwald, *Lehrbuch* [1st edition], **1**, 384.

[2] Ostwald, *loc. cit.*, p. 386.

[3] Ostwald, *loc. cit.*, p. 388.

potassium and sodium salts as well as the acids themselves have been investigated. The following table of molecular refraction equivalents has been extracted from Ostwald's *Lehrbuch*.[1]

	Potassium.	Sodium.	Hydrogen.	K – Na.	K – H.
Chloride . .	18·44	15·11	14·44	3·3	4·0
Bromide . .	25·34	21·70	20·63	3·6	4·7
Iodide . .	35·33	31·59	31·17	3·7	4·2
Nitrate . .	21·80	18·66	17·24	3·1	4·5
Sulphate .	30·55	...	22·45	...	2 × 4·1
Hydroxide .	12·82	9·21	5·95	3·6	6·8
Formate . .	19·93	16·03	13·40	3·9	6·5
Acetate . .	27·65	24·05	21·20	3·6	6·5
Tartrate . .	57·60	50·39	45·18	2 × 3·6	2 × 6·2

We see that the difference K – Na is everywhere much the same, as was to be expected from what we know of the degree of dissociation of the K and Na salts; the same also holds good for the difference K – H as long as the strong (dissociated) acids are considered. On the other hand the substances of the second group (the feebly dissociated acids) show quite different behaviour, the difference K – H being here much greater than for the first group.

4. **Capillarity.**—Valson[2] believed he had found additive properties in the capillary phenomena shown by salt solutions. We need not, however, consider the matter here, since it may be traced back to the circumstance that the specific gravity is an additive property as has been shown above.

5. **Conductivity.**—The credit of a great step in the development of the theory of electrolysis is due

[1] Ostwald, *Lehrbuch* [1st edition], **1**, 443.

[2] Valson, *Compt. rend.*, **74**, 103 (1872). Ostwald, *loc. cit.*, p. 492.

to F. Kohlrausch, who showed that conductivity is an additive property.[1] As we have already indicated in what sense this is to be understood, we may pass directly to the experimental data. For dilute solutions Kohlrausch gives in the memoir cited the following values:—

$K = 48$, $NH_4 = 47$, $Na = 31$, $Li = 21$, $Ag = 40$,
$H = 278$, $Cl = 49$, $Br = 53$, $I = 53$, $CN = 50$,
$OH = 141$, $F = 30$, $NO_3 = 46$, $ClO_3 = 40$,
$C_2H_3O_2 = 23$, $\frac{1}{2}Ba = 29$, $\frac{1}{2}Sr = 28$, $\frac{1}{2}Ca = 26$,
$\frac{1}{2}Mg = 23$, $\frac{1}{2}Zn = 20$, $\frac{1}{2}Cu = 29$.

These values, however, only hold good for the most highly dissociated substances (salts of monobasic acids, and the strong acids and bases). For the somewhat less dissociated sulphates and carbonates of the univalent metals (compare the previous table) he obtained much smaller values :

$K = 40$, $NH_4 = 37$, $Na = 22$, $Li = 11$, $Ag = 32$.
$H = 166$, $\frac{1}{2}SO_4 = 40$, $\frac{1}{2}CO_3 = 36$,

and for the least dissociated sulphates (those of the metals of the magnesium series) he obtained the still smaller values

$\frac{1}{2}Mg = 14$, $\frac{1}{2}Zn = 12$, $\frac{1}{2}Cu = 12$, $\frac{1}{2}SO_4 = 22$.

It is thus evident that Kohlrausch's Law is only applicable to the most highly dissociated salts, the less dissociated salts giving very different values. Since, however, the number of active molecules increases with rising dilution, so that at the most extreme dilution all salts have decomposed into only active (dissociated) molecules, it was to be expected that at high dilutions

[1] Kohlrausch, *Wied. Ann.*, **6**, 167 (1879); Wiedemann, *Elektricität*, **1**, 610; **2**, 955.

the salts would behave more regularly. I also showed from some examples that "too much importance should not be attributed to the anomalies of the salts (acetates and sulphates) of the metals of the magnesium series, since these anomalies disappear at high dilutions."[1] I likewise thought it was possible to carry out consistently the idea that conductivity is an additive property,[2] and attributed to the conductivity of hydrogen in all acids (even those which conduct worst of all, though their behaviour was *prima facie* at variance with this idea) a value completely independent of the nature of the acid, which could only be done with the help of the activity concept. The correctness of this view is more evident still from the later researches of Kohlrausch[3] and Ostwald.[4] In his last work on this subject Ostwald[4] tried to prove that without the assistance of the activity concept it is possible to apply generally the hypothesis that conductivity is an additive property, and this succeeds very well with the sodium, potassium, and lithium salts investigated by him, because these are as a rule very near complete dissociation, especially at very high dilution. The conclusion is still further supported by the fact that analogous salts of univalent metals, as being very closely related to each other, are at the same concentration about equally dissociated. If, however, we consider salts of less nearly related metals, we should arrive at quite different conclusions, as previous investigators clearly make evident. It is stated by Ostwald himself[5] that Kohlrausch's Law is not without modification applicable to acids, introduction of the activity

[1] *Loc. cit.*, No. 13, p. 41. [2] *Loc. cit.*, No. 14, p. 12.
[3] Kohlrausch, *Wied. Ann.*, **26**, 215 and 216 (1885).
[4] Ostwald, *Zeit. physik. Chem.*, **1**, 74 and 97 (1887).
[5] *Loc. cit.*, p. 79.

concept being necessary if it is to be carried through systematically. A closer investigation of copper acetate would lead to considerable difficulties, and still more would this be the case with mercury salts, for according to Grotrian's[1] investigations it appears as if these, even at the most extreme dilutions, yield a conductivity which is only a small fraction of that deduced from the law. It seems indeed that not all salts of the univalent metals are subject to this law, for Bouty[2] has found that potassium stibiotartrate, even in 0.001-normal solution, has only a fifth of the conducting power of KCl; according to Kohlrausch's Law it should have at least half. When, however, we introduce the concept of activity the law of Kohlrausch can be completely carried through, for in the above table the values of i calculated from this law by means of the activity concept show, even in the case of weak bases and acids as well as for $HgCl_2$ and $Cu(C_2H_3O_2)_2$, good agreement with the values of i calculated from Raoult's experiments.

6. **Lowering of the freezing-point.**—In one of his papers Raoult[3] indicates that the lowering of the freezing-point of water can be regarded as an additive property, as would naturally be the case according to our conception for the more dissociated salts in dilute solution. He gives the following values for the effect of the ions:—

Group I. Univalent electronegative ions (radicals) (Cl, Br, OH, NO_3, etc.) 20

" II. Bivalent electronegative ions (radicals) (SO_4, CrO_4, etc.) 11

[1] Grotrian, *Wied. Ann.*, **18**, 177 (1883).

[2] Bouty, *Ann. d. Chim. et d. Phys.* [6] **3**, 472 (1884).

[3] Raoult, *Ibid.*, [6] **4**, 616 (1885).

Group III. Univalent electropositive ions (radicals) (H, K, Na, NH_4, etc.). . . . 15

" IV. Bi- or multivalent[1] electropositive ions (radicals) (Ba, Mg, Al_2, etc.). . . 8

There are, however, very many exceptions, marked by unusually small dissociation, even in the most dilute solutions, as the following table shows :—

	Calculated.	Found.
Weak acids . . .	35	19·0
Cupric acetate . . .	48	31·1
Potassium stibiotartrate .	41	18·4
Mercuric chloride . .	48	20·4
Lead acetate . . .	48	22·2
Aluminium acetate . .	128	84·0
Ferric acetate . . .	128	58·1
Platinic chloride . . .	88	29·0

[1] According to the ideas developed above, all ions should have the same value 18·5. Raoult has artificially forced into the general law of the additive nature of freezing-point depressions, ions of the less dissociated substances like $MgSO_4$ by giving them much smaller values, namely, 8 and 11. The possibility of consistently attributing small values to the multivalent ions depends on the circumstance that in general the dissociation of salts is less as the valency of their ions is greater, as I have previously emphasised (*loc. cit.*, No. 13, p. 69 ; No. 14, p. 5) : "The inactivity (complexity) of a salt solution is the greater, the more easily the constituents of the salt (acid and base) form double compounds (acid or basic salts)." This result is completely supported by the later work of Ostwald (*Zeit. physikal. Chem.*, **1**, pp. 105 to 109). It is evident that if the correct value of 18.5 is attributed to the multivalent ions, the salts compounded from them would form pronounced exceptions. (Probably a similar view might rightly be adopted for other additive properties.) Thus, although Raoult by artificial devices has forced these little dissociated salts into the framework of his law, he has not succeeded with all salts, as is shown in the text.

From measurements of electrical conductivity we know that the molecules of the first four substances are very little dissociated; the others are closely related to them, so that we may expect the same to be true of them also, although their electrolytic properties have not yet been studied. From our point of view all these substances, in this case as in those previously discussed, are not to be regarded as exceptions, but as obeying precisely the same laws as those other substances previously regarded as normal.

In close connection with the lowering of the freezing-point there are several other properties of salt solutions which, as Guldberg[1] and van't Hoff[2] have shown, are proportional to the freezing-point depression. All these properties (lowering of vapour pressure, osmotic pressure, isotonic coefficient) are then to be looked upon as additive, as has indeed been shown by de Vries[3] for the isotonic coefficient. Since, however, all these properties can be deduced from the freezing-point depression, I do not think it necessary to discuss their details in this paper.

[1] Guldberg, *Compt. rend.*, **70**, 1349 (1870).

[2] van't Hoff, *loc. cit.*, p. 20.

[3] de Vries, "Eine Methode zur Analyse der Turgorkraft," *Pringsheims Jahrbücher*, **14**, 579 (1884); van't Hoff, *loc. cit.*, p. 26.

MENDELEEV, DMITRI IVANOVICH. (b. Tobolsk, Siberia [now Tyumen Oblast, R.S.F.S.R.], Russia, 8 February 1834; d. St. Petersburg [now Leningrad], Russia, 2 February 1907)

When he was sixteen years old, Mendeleev entered the Main Pedagogical Institute in St. Petersburg, graduating in 1855. His early researches in chemistry dealt with the interconnections between crystalline form, chemical composition, and specific volumes, as well as the structure of silicon compounds. During 1859–1860, he left Russia to collaborate with Robert Bunsen at the University of Heidelberg. Upon returning to Petersburg, he established his own laboratory, did some teaching, and was named to the chair of chemistry at the University of St. Petersburg in 1867. He resigned in 1890 in the wake of student disorders, having been sympathetic with their demands. His longtime interest in studying and improving the Russian petroleum and mining industries resulted in various governmental appointments, and, from 1893 until his death, Mendeleev was director of the Central Board of Weights and Measures.

As a young chemist Mendeleev adopted the chemical system of Gerhardt, including the theory of types, and he supported the new system of atomic weights discussed at the Karlsruhe Congress. His commitment to Daltonian chemistry and a natural system of chemical classification predisposed him against electrical theories of chemistry, including Arrhenius's ionization theory and J.J. Thomson's electron theory.

Mendeleev is rightly well-known for working out the periodic classification of the chemical elements. Using Gerhardt's values for atomic weights and Wurtz's theory of "atomicity," Mendeleev arranged the various groups of elements in the order of their atomic weights, demonstrating a regular progression in the differences between the atomic weights of the elements arranged in vertical columns. His predictions of three undiscovered elements and their properties became major evidence in favor of his periodic system when Lecoq de Boisbaudran discovered "gallium," which had the properties of Mendeleev's "eka-aluminum," in 1875. (In 1879 scandium, or "eka-boron," was discovered; and, in 1886, germanium, or "eka-silicon.") Mendeleev's 1889 lecture to the London Chemical Society described the origins and recent triumphs of his "periodic law," a law based on the conception of real chemical atoms, characterized by Newtonian mass or weight. This idea he now took to be strongly confirmed.

LXIII.—*The Periodic Law of the Chemical Elements.*

By Professor MENDELÉEFF.

(FARADAY LECTURE delivered before the Fellows of the Chemical Society in the Theatre of the Royal Institution, on Tuesday, June 4th, 1889.)

THE high honour bestowed by the Chemical Society in inviting me to pay a tribute to the world-famed name of Faraday by delivering this lecture has induced me to take for its subject the Periodic Law of the Elements—this being a generalisation in chemistry which has of late attracted much attention.

While science is pursuing a steady onward movement, it is convenient from time to time to cast a glance back on the route already traversed, and especially to consider the new conceptions which aim at discovering the general meaning of the stock of facts accumulated from day to day in our laboratories. Owing to the possession of laboratories, modern science now bears a new character, quite unknown not only to antiquity but even to the preceding century. Bacon's and Descartes' idea of submitting the mechanism of science simultaneously to experiment and reasoning has been fully realised in the case of chemistry, it having become not only possible but always customary to experiment. Under the all-penetrating control of experiment, a new theory, even if crude, is quickly strengthened, provided it be founded on a sufficient basis; the asperities are removed, it is amended by degrees, and soon loses the phantom light of a shadowy form or of one founded on mere prejudice; it is able to lead to logical conclusions and to submit to experimental proof. Willingly or not, in science we all must submit not to what seems to us attractive from one point of view or from another, but to what represents an agreement between theory and experiment; in other words, to demonstrated generalisation and to the approved experiment. Is it long since many refused to accept the generalisations involved in the law of Avogadro and Ampère, so widely extended by Gerhardt? We still may hear the voices of its opponents; they enjoy perfect freedom, but vainly will their voices rise so long as they do not use the language of demonstrated facts. The striking observations with the spectroscope which have permitted us to analyse the chemical constitution of distant worlds, seemed, at first, applicable to the task of determining the nature of the atoms themselves; but the working out of the idea in the laboratory soon demonstrated that the characters of spectra are determined—not directly by the atoms, but by the molecules into which the atoms are packed; and so it became evident that more verified facts must be collected before it will be possible to formulate new generalisations capable of taking their place beside those ordinary ones based upon the conception of simple bodies and atoms. But as the shade of the leaves and roots of living plants, together with the relics of a decayed vegetation, favour the growth of the seedling and serve to promote its luxurious development, in like manner sound generalisations—together with the relics of those which have proved to be untenable—promote scientific productivity, and ensure the luxurious growth of science under the influence of rays emanating from the centres of scientific energy. Such centres are scientific associations and societies. Before one of the oldest and most powerful of these I am about to take the liberty of passing in review the 20 years' life of a generalisation which is known under the name of the Periodic Law. It was in March, 1869, that I ventured to lay before the then youthful Russian Chemical Society the ideas upon the same subject, which I had expressed in my just written "Principles of Chemistry."

Without entering into details, I will give the conclusions I then arrived at, in the very words I used:—

"1. The elements, if arranged according to their atomic weights, exhibit an evident *periodicity* of properties.

"2. Elements which are similar as regards their chemical properties have atomic weights which are either of nearly the same value (*e.g.*, platinum, iridium, osmium) or which increase regularly (*e.g.*, potassium, rubidium, cæsium).

"3. The arrangement of the elements, or of groups of elements in the order of their atomic weights corresponds to their so-called *valencies* as well as, to some extent, to their distinctive chemical properties—as is apparent among other series in that of lithium, beryllium, barium, carbon, nitrogen, oxygen and iron.

"4. The elements which are the most widely diffused have *small* atomic weights.

"5. The *magnitude* of the atomic weight determines the character of the element just as the magnitude of the molecule determines the character of a compound body.

"6. We must expect the discovery of many yet *unknown* elements, for example, elements analogous to aluminium and silicon, whose atomic weight would be between 65 and 75.

"7. The atomic weight of an element may sometimes be amended by a knowledge of those of the contiguous elements. Thus, the atomic weight of tellurium must lie between 123 and 126, and cannot be 128.

"8. Certain characteristic properties of the elements can be foretold from their atomic weights.

"The aim of this communication will be fully attained if I succeed in drawing the attention of investigators to those relations which exist between the atomic weights of dissimilar elements, which, as far as I know, have hitherto been almost completely neglected. I believe that the solution of some of the most important problems of our science lies in researches of this kind."

To-day, 20 years after the above conclusions were formulated, they may still be considered as expressing the essence of the now well-known periodic law.

Reverting to the epoch terminating with the sixties, it is proper to indicate three series of data without the knowledge of which the periodic law could not have been discovered, and which rendered its appearance natural and intelligible.

In the first place, it was at that time that the numerical value of atomic weights became definitely known. Ten years earlier such knowledge did not exist, as may be gathered from the fact that in 1860 chemists from all parts of the world met at Karlsruhe in order to come to some agreement, if not with respect to views relating to atoms, at any rate as regards their definite representation. Many of those present probably remember how vain were the hopes of coming to an understanding, and how much ground was gained at that Congress by the followers of the unitary theory so brilliantly represented by Cannizzaro. I vividly remember the impression produced by his speeches, which admitted of no compromise, and seemed to advocate truth itself, based on the conceptions of Avogadro, Gerhardt and Regnault, which at that time were far from being generally recognised. And though no understanding could be arrived at, yet the objects of the meeting were attained, for the ideas of Cannizzaro proved, after a few years, to be the only ones which could stand criticism, and which represented an atom as—"the smallest portion of an element which enters into a molecule of its compound." Only such real atomic weights—not conventional ones—could afford a basis for generalisation. It is sufficient, by way of example, to indicate the following cases in which the relation is seen at once and is perfectly clear:—

K = 39	Rb = 85	Cs = 133
Ca = 40	Sr = 87	Ba = 137

whereas with the equivalents then in use—

K = 39	Rb = 85	Cs = 133
Ca = 20	Sr = 43·5	Ba = 68·5

the consecutiveness of change in atomic weight, which with the true values is so evident, completely disappears.

Secondly, it had become evident during the period 1860–70, and even during the preceding decade, that the relations between the atomic weights of analogous elements were governed by some general and simple laws. Cooke, Cremers, Gladstone, Gmelin, Lenssen, Pettenkofer, and especially Dumas, had already established many facts bearing on that view. Thus Dumas compared the following groups of analogous elements with organic radicles—

	Diff.		Diff.		Diff.		Diff.
		Mg = 12		P = 31		O = 8	
			8		44		8
Li = 7		Ca = 20		As = 75		S = 16	
	16		3 × 8		44		3 × 8
Na = 23		Sr = 44		Sb = 119		Se = 40	
	16		3 × 8		2 × 44		3 × 8
K = 39		Ba = 68		Bi = 207		Te = 64	

and pointed out some really striking relationships, such as the following:—

$$F = 19.$$
$$Cl = 35{\cdot}5 = 19 + 16{\cdot}5.$$
$$Br = 80 \quad = 19 + 2 \times 16{\cdot}5 + 28.$$
$$I = 127 = 2 \times 19 + 2 \times 16{\cdot}5 + 2 \times 28.$$

A. Strecker, in his work "Theorien und Experimente zur Bestimmung der Atomgewichte der Elemente" (Braunschweig, 1859), after summarising the data relating to the subject, and pointing out the remarkable series of equivalents—

Cr = 26·2 Mn = 27·6 Fe = 28 Ni = 29 Co = 30
Cu = 31·7 Zn = 32·5

remarks that: "It is hardly probable that all the above-mentioned relations between the atomic weights (or equivalents) of chemically analogous elements are merely accidental. We must, however, leave to the future the discovery of the *law* of the relations which appears in these figures." *

In such attempts at arrangement and in such views are to be recognised the real forerunners of the periodic law; the ground was prepared for it between 1860 and 1870, and that it was not expressed in a determinate form before the end of the decade, may, I suppose, be ascribed to the fact that only analogous elements had been compared. The idea of seeking for a relation between the atomic weights of all the elements was foreign to the ideas then current, so that neither the *vis tellurique* of De Chancourtois, nor the *law of octaves* of Newlands, could secure anybody's attention. And yet both De Chancourtois and Newlands, like Dumas and Strecker, more than Lenssen and Pettenkofer, had made an approach to the periodic law and had discovered its germs. The solution of the problem advanced but slowly, because the facts, and not the law, stood foremost in all attempts; and the law could not awaken a general interest so long as elements, having no apparent connection with each other, were included in the same octave, as for example:—

1st octave of Newlands..	H	F	Cl	Co & Ni	Br	Pd	I	Pt & Ir
7th Ditto....	O	S	Fe	Se	Rh & Ru	Te	Au	Os or Th

Analogies of the above order seemed quite accidental, and the more so as the octave contained occasionally 10 elements instead of eight, and when two such elements as Ba and V, Co and Ni, or Rh and Ru, occupied one place in the octave.† Nevertheless, the fruit was ripening, and I now see clearly that Strecker, De Chancourtois and Newlands stood foremost in the way towards the discovery of the periodic law, and that they merely wanted the boldness necessary to place the whole question at such a height that its reflection on the facts could be clearly seen.

A third circumstance which revealed the periodicity of chemical elements was the accumulation, by the end of the sixties, of new information respecting the rare elements, disclosing their many-sided relations to the other elements and to each other. The

* "Es ist wohl kaum anzunehmen, dass alle im Vorhergehenden hervorgehobenen Beziehungen zwischen den Atomgewichten (oder Aequivalenten) in chemischen Verhältnissen einander ähnliche Elemente bloss zufällig sind. Die Auffindung der in diesen Zahlen *gesetzlichen* Beziehungen müssen wir jedoch der Zukunft überlassen."

† To judge from J. A. R. Newlands' work, *On the Discovery of the Periodic Law*, London, 1884, p. 149; "On the Law of Octaves" (from the *Chemical News*, **12**, 83, August 18, 1865).

researches of Marignac on niobium, and those of Roscoe on vanadium were of special moment. The striking analogies between vanadium and phosphorus on the one hand, and between vanadium and chromium on the other, which became so apparent in the investigations connected with that element, naturally induced the comparison of V = 51 with Cr = 52, Nb = 94 with Mo = 96, and Ta = 192 with W = 194; while, on the other hand, P = 31 could be compared with S = 32, As = 75 with Se = 79, and Sb = 120 with Te = 125. From such approximations there remained but one step to the discovery of the law of periodicity.

The law of periodicity was thus a direct outcome of the stock of generalisations and established facts which had accumulated by the end of the decade 1860—1870: it is an embodiment of those data in a more or less systematic expression. Where, then, lies the secret of the special importance which has since been attached to the periodic law, and has raised it to the position of a generalisation which has already given to chemistry unexpected aid, and which promises to be far more fruitful in the future and to impress upon several branches of chemical research a peculiar and original stamp? The remaining part of my communication will be an attempt to answer this question.

In the first place we have the circumstance that, as soon as the law, made its appearance, it demanded a revision of many facts which were considered by chemists as fully established by existing experience. I shall return, later on, briefly to this subject, but I wish now to remind you that the periodic law, by insisting on the necessity for a revision of supposed facts, exposed itself at once to destruction in its very origin. Its first requirements, however, have been almost entirely satisfied during the last 20 years; the supposed facts have yielded to the law, thus proving that the law itself was a legitimate induction from the verified facts. But our inductions from data have often to do with such details of a science so rich in facts, that only generalisations which cover a wide range of important phenomena can attract general attention. What were the regions touched on by the periodic law? This is what we shall now consider.

The most important point to notice is, that periodic functions, used for the purpose of expressing changes which are dependent on variations of time and space, have been long known. They are familiar to the mind when we have to deal with motion in closed cycles, or with any kind of deviation from a stable position, such as occurs in pendulum-oscillations. A like periodic function became evident in the case of the elements, depending on the mass of the atom. The primary conception of the masses of bodies or of the masses of atoms belongs to a category which the present state of science forbids us to discuss, because as yet we have no means of dissecting or analysing the conception. All that was known of functions dependent on masses derived its origin from Galileo and Newton, and indicated that such functions either decrease or increase with the increase of mass, like the attraction of celestial bodies. The numerical expression of the phenomena was always found to be proportional to the mass, and in no case was an increase of mass followed by a recurrence of properties such as is disclosed by the periodic law of the elements. This constituted such a novelty in the study of the phenomena of nature that, although it did not lift the veil which conceals the true conception of mass, it nevertheless indicated that the explanation of that conception must be searched for in the masses of the atoms; the more so, as all masses are nothing but aggregations, or additions, of chemical atoms which would be best described as chemical individuals. Let me remark by the way that though the Latin word "individual" is merely a translation of the Greek word "atom," nevertheless history and custom have drawn so sharp a distinction between the two words, and the present chemical conception of atoms is nearer to that defined by the Latin word than by the Greek, although this latter also has acquired a special meaning which was unknown to the classics. The periodic law has shown that our chemical individuals display a harmonic periodicity of properties, dependent on their masses. Now, natural science has long been accustomed to deal with periodicities observed in nature, to seize them with the vice of mathematical analysis, to submit them to the rasp of experiment. And these instruments of scientific thought would surely, long since, have mastered the problem connected with the chemical elements, were it not for a new feature which was brought to light by the periodic law and which gave a peculiar and original character to the periodic function.

If we mark on an axis of abscissæ a series of lengths proportional to angles, and trace ordinates which are proportional to sines or other trigonometrical functions, we get periodic curves of a harmonic character. So it might seem, at first sight, that with the increase of atomic weights the function of the properties of the elements should also vary in the same harmonious way. But in this case there is no such continuous change as in the curves just referred to, because the periods do not contain the infinite number of points constituting a curve, but a *finite* number only of such points. An example will better illustrate this view. The atomic weights—

$$Ag = 108 \qquad Cd = 112 \qquad In = 113 \qquad Sn = 118$$
$$Sb = 120 \qquad Te = 125 \qquad I = 127$$

steadily increase, and their increase is accompanied by a modification of many properties which constitutes the essence of the periodic law.

Thus, for example, the densities of the above elements decrease steadily, being respectively—

10·5 8·6 7·4 7·2 6·7 6·4 4·9

while their oxides contain an increasing quantity of oxygen :—

Ag_2O Cd_2O_2 In_2O_3 Sn_2O_4 Sb_2O_5 Te_2O_6 I_2O_7

But to connect by a curve the summits of the ordinates expressing any of these properties would involve the rejection of Dalton's law of multiple proportions. Not only are there no intermediate elements between silver, which gives AgCl, and cadmium, which gives $CdCl_2$, but, according to the very essence of the periodic law there can be none; in fact a uniform curve would be inapplicable in such a case, as it would lead us to expect elements possessed of special properties at any point of the curve. The periods of the elements have thus a character very different from those which are so simply represented by geometers. They correspond to points, to numbers, to sudden changes of the masses, and not to a continuous evolution. In these sudden changes destitute of intermediate steps or positions, in the absence of elements intermediate between, say, silver and cadmium, or aluminium and silicon, we must recognise a problem to which no direct application of the analysis of the infinitely small can be made. Therefore, neither the trigonometrical functions proposed by Ridberg and Flavitzky, nor the pendulum-oscillations suggested by Crookes, nor the cubical curves of the Rev. Mr. Haughton, which have been proposed for expressing the periodic law, from the nature of the case, can represent the periods of the chemical elements. If geometrical analysis is to be applied to this subject it will require to be modified in a special manner. It must find the means of representing in a special way not only such long periods as that comprising,

K Ca Sc Ti V Cr Mn Fe Co Ni Cu Zn Ga G
As Se Br,

but short periods like the following:—

Na Mg Al Si P S Cl.

In the theory of numbers only do we find problems analogous to ours, and two attempts at expressing the atomic weights of the elements by algebraic formulæ seem to be deserving of attention, although neither of them can be considered as a complete theory, nor as promising finally to solve the problem of the periodic law. The attempt of E. J. Mills (1886) does not even aspire to attain this end. He considers that all atomic weights can be expressed by a logarithmic function,

$$15(n - 0{\cdot}9375^t),$$

in which the variables n and t are *whole numbers.* Thus, for oxygen, $n = 2$, and $t = 1$, whence its atomic weight is $= 15{\cdot}94$; in the case of chlorine, bromine, and iodine, n has respective values of 3, 6, and 9, while $t = 7$, 6, and 9; in the case of potassium, rubidium, and cæsium, $n = 4$, 6, and 9, and $t = 14$, 18, and 20.

Another attempt was made in 1888 by B. N. Tchitchérin. Its author places the problem of the periodic law in the first rank, but as yet he has investigated the alkaline metals only. Tchitchérin first noticed the simple relations existing between the atomic volumes of all alkaline metals; they can be expressed, according to his views, by the formula

$$A(2 - 0{\cdot}00535An),$$

where A is the atomic weight, and n is equal to 8 for lithium and sodium, to 4 for potassium, to 3 for rubidium, and to 2 for cæsium. If n remained equal to 8, during the increase of A, then the volume would become zero at $A = 46\frac{2}{3}$, and it would reach its maximum at $A = 23\frac{1}{3}$. The close approximation of the number $46\frac{2}{3}$ to the differences between the atomic weights of analogous elements (such as Cs — Rb, I — Br, and so on); the close correspondence of the number $23\frac{1}{3}$ to the atomic weight of sodium; the fact of n being necessarily a whole number, and several other aspects of the question, induce Tchitchérin to believe that they afford a clue to the understanding of the nature of the elements; we must, however, await the full development of his theory before pronouncing judgment on it. What we can at present only be certain of is this: that attempts like the two above named must be repeated and multiplied, because the periodic law has clearly shown that the masses of the atoms increase abruptly, by steps, which are clearly connected in some way with Dalton's law of multiple proportions; and because the periodicity of the elements finds expression in the transition from RX to RX_2, RX_3, RX_4, and so on till RX_8, at which point the energy of the combining forces being exhausted, the series begins anew from RX to RX_2, and so on.

While connecting by new bonds the theory of the chemical elements with Dalton's theory of multiple proportions, or atomic structure of bodies, the periodic law opened for natural philosophy a new and wide field for speculation. Kant said that there are in the world "two things which never cease to call for the admiration and reverence of man: the moral law within ourselves, and the stellar sky above us." But when we turn our thoughts towards the nature of the elements and the periodic law, we must add a third subject, namely, "the nature of the elementary individuals which we discover everywhere around us." Without them the stellar sky itself is inconceivable; and in the atoms we see at once their peculiar individualities, the infinite multiplicity of the individuals, and the submission of their seeming freedom to the general harmony of Nature.

Having thus indicated a new mystery of Nature, which does not yet yield to rational conception, the periodic law, together with the revelations of spectrum analysis, have contributed to again revive an old but remarkably long-lived hope—that of discovering, if not by experiment, at least, by a mental effort, the *primary matter*—which had its genesis in the minds of the Grecian philosophers, and has been transmitted, together with many other ideas of the classic period, to the heirs of their civilisation. Having grown, during the times of the alchemists up to the period when experimental proof was required, the idea has rendered good service; it induced those careful observations and experiments which later on called into being the works of Scheele, Lavoisier, Priestley and Cavendish. It then slumbered awhile, but was soon awakened by the attempts either to confirm or to refute the ideas of Prout as to the multiple proportion relationship of the atomic weights of all the elements. And once again the inductive or experimental method of studying Nature gained a direct advantage from the old Pythagorean idea: because atomic weights were determined with an accuracy formerly unknown. But again the idea could not stand the ordeal of experimental test, yet the prejudice remains and has not been uprooted, even by Stas; nay, it has gained a new vigour, for we see that all which is imperfectly worked out, new and unexplained, from the still scarcely studied rare metals to the hardly perceptible nebulæ, have been used to justify it. As soon as spectrum analysis appears as a new and powerful weapon of chemistry, the idea of a primary matter is immediately attached to it. From all sides we see attempts to constitute the imaginary substance *helium** the so much longed for primary matter. No attention is paid to the circumstance that the helium line is only seen in the spectrum of the solar protuberances, so that its universality in Nature remains as problematic as the primary matter itself; nor to the fact that the helium line is wanting amongst the Fraunhofer lines of the solar spectrum, and thus does not answer to the brilliant fundamental conception which gives its real force to spectrum analysis.

And finally, no notice is even taken of the indubitable fact that the brilliancies of the spectral lines of the simple bodies vary under different temperatures and pressures; so that all probabilities are in favour of the helium line simply belonging to some long since known element placed under such conditions of temperature, pressure, and gravity as have not yet been realised in our experiments. Again, the idea that the excellent investigations of Lockyer of the spectrum of

* That is, a body having a wave-length equal to 0·0005875 millimetre.

iron can be interpreted in favour of the compound nature of that element, evidently must have arisen from some misunderstanding. The spectrum of a compound body certainly does not appear as a sum of the spectra of its components; and therefore the observations of Lockyer can be considered precisely as a proof that iron undergoes no other changes at the temperature of the sun but those which it experiences in the voltaic arc—provided the spectrum of iron is preserved. As to the shifting of some of the lines of the spectrum of iron while the other lines maintain their positions, it can be explained, as shown by M. Kleiber (*Journal of the Russian Chemical and Physical Society*, 1885, 147), by the relative motion of the various strata of the sun's atmosphere, and by Zöllner's laws of the relative brilliancies of different lines of the spectrum. Moreover, it ought not to be forgotten that if iron were really proved to consist of two or more unknown elements, we simply should have an increase of the number of our elements—not a reduction, and still less a reduction of all of them to one single primary matter.

Feeling that spectrum analysis will not yield a support to the Pythagorean conception, its modern promoters are so bent upon its being confirmed by the periodic law, that the illustrious Berthelot, in his work *Les origines de l'Alchimie*, 1885, 313, has simply mixed up the fundamental idea of the law of periodicity with the ideas of Prout, the alchemists, and Democritus about primary matter.* But the periodic law, based as it is on the solid and wholesome ground of experimental research, has been evolved independently of any conception as to the nature of the elements; it does not in the least originate in the idea of an unique matter; and it has no historical connection with that relic of the torments of classical thought, and therefore it affords no more indication of the unity of matter or of the compound character of our elements, than the law of Avogadro, or the law of specific heats, or even the conclusions of spectrum analysis. None of the advocates of an unique matter have ever tried to explain the law from the standpoint of ideas taken from a remote antiquity when it was found convenient to admit the existence of many gods—and of an unique matter.

When we try to explain the origin of the idea of an unique primary matter, we easily trace that in the absence of inductions from experiment it derives its origin from the scientifically philosophical attempt at discovering some kind of unity in the immense diversity of individualities which we see around. In classical times

* He maintains (on p. 309) that the periodic law requires two new analogous elements, having atomic weights of 48 and 64, occupying positions between sulphur and selenium, although nothing of the kind results from any of the different readings of the law.

such a tendency could only be satisfied by conceptions about the immaterial world. As to the material world, our ancestors were compelled to resort to some hypothesis, and they adopted the idea of unity in the formative material, because they were not able to evolve the conception of any other possible unity in order to connect the multifarious relations of matter. Responding to the same legitimate scientific tendency, natural science has discovered throughout the universe a unity of plan, a unity of forces, and a unity of matter, and the convincing conclusions of modern science compel everyone to admit these kinds of unity. But while we admit unity in many things, we none the less must also explain the individuality and the apparent diversity which we cannot fail to trace everywhere. It has been said of old, "Give a fulcrum, and it will become easy to displace the earth." So also we must say, "Give anything that is individualised, and the apparent diversity will be easily understood." Otherwise, how could unity result in a multitude?

After a long and painstaking research, natural science has discovered the individualities of the chemical elements, and therefore it is now capable not only of analysing, but also of synthesising; it can understand and grasp the general and unity, as well as the individualised and the multitudinous. Unity and the general, like time and space, like force and motion, vary uniformly; the uniform admit of interpolations, revealing every intermediate phase. But the multitudinous, the individualised—like ourselves, like the chemical elements, like the members of a peculiar periodic function of elements, like Dalton's multiple proportions—is characterised in another way: we see in it—side by side with a connecting general principle—leaps, breaks of continuity, points which escape from the analysis of the infinitely small—a complete absence of intermediate links. Chemistry has found an answer to the question as to the causes of multitudes; and while retaining the conception of many elements, all submitted to the discipline of a general law, it offers an escape from the Indian Nirvana—the absorption in the universal, replacing it by the individualised. However, the place for individuality is so limited by the all-grasping, all-powerful universal, that it is merely a fulcrum for the understanding of multitude in unity.

Having touched upon the metaphysical bases of the conception of an unique matter which is supposed to enter into the composition of all bodies, I think it necessary to dwell upon another theory, akin to the above conception,—the theory of the compound character of the elements now admitted by some,—and especially upon one particular circumstance which being related to the periodic law is considered to be an argument in favour of that hypothesis.

Dr. Pelopidas, in 1883, made a communication to the Russian Chemical and Physical Society on the periodicity of the hydrocarbon radicles, pointing out the remarkable parallelism which was to be noticed in the change of properties of hydrocarbon radicles and elements when classed in groups. Professor Carnelley, in 1886, developed a similar parallelism. The idea of M. Pelopidas will be easily understood if we consider the series of hydrocarbon radicles which contain, say, 6 atoms of carbon :—

I	II	III	IV	V	VI	VII	VIII
C_6H_{13}	C_6H_{12}	C_6H_{11}	C_6H_{10}	C_6H_9	C_6H_8	C_6H_7	C_6H_6

The first of these radicles, like the elements of the Ist group, combines with Cl, OH, and so on, and gives the derivatives of hexyl alcohol, $C_6H_{13}(OH)$; but, in proportion as the number of hydrogen atoms decreases, the capacity of the radicles of combining with, say, the halogens increases. C_6H_{12} already combines with 2 atoms of chlorine; C_6H_{11} with 3 atoms, and so on. The last members of the series comprise the radicles of acids; thus C_6H_8, which belongs to the VIth group, gives, like sulphur, a bibasic acid, $C_6H_8O_2(OH)_2$, which is homologous with oxalic acid. The parallelism can be traced still further—because C_6H_5 appears as a monovalent radicle of benzene—and with it begins a new series of aromatic derivatives, so analogous to the derivatives of the fat series. Let me also mention another example from among those which have been given by M. Pelopidas. Starting from the alkaline radicle of monomethylammonium, $N(CH_3)H_3$, or NCH_6, which presents many analogies with the alkaline metals of the Ist group, he arrives, by successively diminishing the number of the atoms of hydrogen, at a seventh group which contains cyanogen, CN, which has long since been compared to the halogens of the VIIth group.

The most important consequence which, in my opinion, can be drawn from the above comparison is, that the periodic law, so apparent in the elements, has a wider application than might appear at first sight; it opens up a new vista of chemical evolutions. But, while admitting the fullest parallelism between the periodicity of the elements and that of the compound radicles, we must not forget that in the periods of the hydrocarbon radicles we have a *decrease* of mass as we pass from the representatives of the first group to the next; while in the periods of the elements the mass *increases* during the progression. It thus becomes evident that we cannot speak of an identity of periodicity in both cases, unless we put aside the ideas of mass and attraction, which are the real corner-stones of the whole of natural science and even enter into those very conceptions of simple bodies which came to light a full hundred years later than the immortal principles of Newton.*

From the foregoing, as well as from the failures of so many attempts at finding in experiment and speculation a proof of the compound character of the elements and of the existence of primordial matter, it is evident, in my opinion, that this theory must be classed amongst mere utopias. But utopias can only be combatted by freedom of opinion, by experiment, and by new utopias. In the republic of scientific theories freedom of opinions is guaranteed. It is precisely that freedom which permits me to criticise openly the widely diffused idea as to the unity of matter in the elements. Experiments and attempts at confirming that idea have been so numerous that it really would be instructive to have them all collected together, if only to serve as a warning against the repetition of old failures. And, now, as to new utopias which may be helpful in the struggle against the old ones, I do not think it quite useless to mention a *phantasy* of one of my students who imagined that the weight of bodies does not depend upon their mass, but upon the character of the motion of their atoms. The atoms, according to this new utopian, may all be homogeneous or heterogeneous, we know not which; we know them in motion only, and that motion they maintain with the same persistence as the stellar bodies maintain theirs. The weights of atoms differ only in consequence of their various modes and quantity of motion; the heaviest atoms may be much simpler than the lighter ones; thus an atom of mercury may be simpler than an atom of hydrogen—the manner in which it moves causes it to be heavier. My interlocutor even suggested that the view which attributes the greater complexity to the lighter elements finds confirmation in the fact that the hydrocarbon radicles mentioned by Pelopidas, while becoming lighter as they lose hydrogen, change their properties periodically in the same manner as the elements change theirs according as the atoms grow heavier.

The French proverb, *La critique est facile mais l'art est difficile*, however, may well be reversed in the case of all such ideal views, as it is much easier to formulate than to criticise them. Arising from the virgin soil of newly established facts, the knowledge relating to the elements, to their masses, and to the periodic changes of their properties, has given a motive for the formation of utopian hypotheses, probably because they could not be foreseen by the aid of any of the

* It is noteworthy that the year in which Lavoisier was born (1743)—the author of the idea of elements and of the indestructibility of matter—is later by exactly one century than the year in which the author of the theory of gravitation and mass was born (1643 N.S.). The affiliation of the ideas of Lavoisier and those of Newton is beyond doubt.

various metaphysical systems, and exist, like the idea of gravitation, as an independent outcome of natural science, requiring the acknowledgement of general laws, when these have been established with the same degree of persistency as is indispensable for the acceptance of a thoroughly established fact. Two centuries have elapsed since the theory of gravitation was enunciated, and although we do not understand its cause, we still must regard gravitation as a fundamental conception of natural philosophy, a conception which has enabled us to perceive much more than the metaphysicians did or could with their seeming omniscience. A hundred years later the conception of the elements arose; it made chemistry what it now is; and yet we have advanced as little in our comprehension of simple bodies since the times of Lavoisier and Dalton as we have in our understanding of gravitation. The periodic law of the elements is only 20 years old: it is not surprising therefore that, knowing nothing about the causes of gravitation and mass, or about the nature of the elements, we do not comprehend the *rationale* of the periodic law. It is only by collecting established laws, that is by working at the acquirement of truth, that we can hope gradually to lift the veil which conceals from us the causes of the mysteries of Nature and to discover their mutual dependency. Like the telescope and the microscope, laws founded on the basis of experiment are the instruments and means of enlarging our mental horizon.

In the remaining part of my communication I shall endeavour to show, and as briefly as possible, in how far the periodic law contributes to enlarge our range of vision. Before the promulgation of this law the chemical elements were mere fragmentary, incidental facts in Nature; there was no special reason to expect the discovery of new elements, and the new ones which were discovered from time to time appeared to be possessed of quite novel properties. The law of periodicity first enabled us to perceive undiscovered elements at a distance which formerly was inaccessible to chemical vision; and long ere they were discovered new elements appeared before our eyes possessed of a number of well-defined properties. We now know three cases of elements whose existence and properties were foreseen by the instrumentality of the periodic law. I need but mention the brilliant discovery of *gallium*, which proved to correspond to eka-aluminium of the periodic law, by Lecoq de Boisbaudran; of *scandium*, corresponding to eka-boron, by Nilson; and of *germanium*, which proved to correspond in all respects to eka-silicium, by Winckler. When, in 1871, I described to the Russian Chemical Society the properties, clearly defined by the periodic law, which such elements ought to possess, I never hoped that I should live to mention their discovery to the Chemical Society of Great Britain as a confirmation of the exactitude and the generality of the periodic law. Now, that I have had the happiness of doing so, I unhesitatingly say that although greatly enlarging our vision, even now the periodic law needs further improvements in order that it may become a trustworthy instrument in further discoveries.*

I will venture to allude to some other matters which chemistry has discerned by means of its new instrument, and which it could not have made out without a knowledge of the law of periodicity, and I will confine myself to simple bodies and to oxides.

Before the periodic law was formulated the atomic weights of the elements were purely empirical numbers, so that the magnitude of the equivalent, and the atomicity or the value in substitution possessed by an atom, could only be tested by critically examining the methods of determination, but never directly by considering the numerical values themselves; in short, we were compelled to move in the dark, to submit to the facts, instead of being masters of them. I need not recount the methods which permitted the periodic law at last to master the facts relating to atomic weights, and I would merely call to mind that it compelled us to modify the valencies of *indium* and *cerium*, and to assign to their compounds a different molecular composition. Determinations of the specific heats of these two metals fully confirmed the change. The trivalency of *yttrium*, which makes us now represent its oxide as Y_2O_3 instead of as YO, was also foreseen (in 1870) by the periodic law, and it now has become so probable that Cleve, and all other subsequent investigators of the rare metals, have not only adopted it but have also applied it without any new demonstration to bodies so imperfectly known as those of the cerite and gadolinite group, especially since Hildebrand determined the specific heats of lanthanum and didymium and confirmed the expectations suggested by the periodic law. But here, especially in the case of didymium, we meet with a series of difficulties long since foreseen through the periodic law, but only now becoming

* I foresee some more new elements, but not with the same certitude as before. I shall give one example, and yet I do not see it quite distinctly. In the series which contains Hg = 204, Pb = 206, and Bi = 208, we can guess the existence (at the place VI—11) of an element analogous to tellurium, which we can describe as dvi-tellurium, Dt having an atomic weight of 212, and the property of forming the oxide DtO_3. If this element really exists, it ought in the free state to be an easily fusible, crystalline, non-volatile metal of a grey colour, having a density of about 9·3, capable of giving a dioxide, DtO_2, equally endowed with feeble acid and basic properties. This dioxide must give on active oxidation an unstable higher oxide, DtO_3, which should resemble in its properties PbO_2 and Bi_2O_5. Dvi-tellurium hydride, if it be found to exist, will be a less stable compound than even H_2Te. The compounds of dvi-tellurium will be easily reduced, and it will form characteristic definite alloys with other metals.

evident, and chiefly arising from the relative rarity and insufficient knowledge of the elements which usually accompany didymium.

Passing to the results obtained in the case of the rare elements *beryllium*, *scandium* and *thorium*, it is found that these have many points of contact with periodic law. Although Avdéeff long since proposed the magnesia formula to represent beryllium oxide, yet there was so much to be said in favour of the alumina formula, on account of the specific heat of the metals and the isomorphism of the two oxides, that it became generally adopted and seemed to be well established. The periodic law, however, as Brauner repeatedly insisted (*Berichte*, 1878, 872; 1881, 53) was against the formula Be_2O_3; it required the magnesium formula BeO, that is, an atomic weight of 9, because there was no place in the system for an element like beryllium having an atomic weight of 13·5. This divergence of opinion lasted for years, and I often heard that the question as to the atomic weight of beryllium threatened to disturb the generality of the periodic law, or, at any rate, to require some important modifications of it. Many forces were operating in the controversy regarding beryllium, evidently because a much more important question was at issue than merely that involved in the discussion of the atomic weight of a relatively rare element; and during the controversy the periodic law became better understood, and the mutual relations of the elements became more apparent than ever before. It is most remarkable that the victory of the periodic law was won by the researches of the very observers who previously had discovered a number of facts in support of the trivalency of beryllium. Applying the higher law of Avogadro, Nilson and Petterson have finally shown that the density of the vapour of the beryllium chloride, $BeCl_2$, obliges us to regard beryllium as bivalent in conformity with the periodic law.* I consider the confirmation of Avdéeff's and Brauner's view as important in the

* Let me mention another proof of the bivalency of beryllium which may have passed unnoticed, as it was published in the Russian chemical literature. Having remarked (in 1884) that the density of such solutions of chlorides of metals, MCl_n, as contain 200 mols. of water (or a large and constant amount of water) regularly increases as the molecular weight of the dissolved salt increases, I proposed to one of our young chemists, M. Burdakoff, that he should investigate the beryllium chloride. If its molecule be $BeCl_2$ its weight must be = 80; and in such a case it must be heavier than the molecule of KCl = 74·5, and lighter than that of MgCl = 93. On the contrary, if beryllium chloride is a trichloride, BCl_3 = 120, its molecule must be heavier than that of $CaCl_2$ = 111, and lighter than that of $MnCl_2$ = 126. Experiment has shown the correctness of the former formula, the solution $BeCl_2 + 200H_2O$ having (at 15°/4°) a density of 1·0138, this being a higher density than that of the solution $KCl + 200H_2O$ (= 1·0121), and lower than that of $MgCl_2 + 200H_2O$ (= 1·0203). The bivalency of beryllium was thus confirmed in the case both of the dissolved and the vaporised chloride.

history of the periodic law as the discovery of scandium, which, in Nilson's hands, confirmed the existence of the eka-boron.

The circumstance that *thorium* proved to be quadrivalent, and Th = 232, in accordance with the views of Chydenius and the requirements of the periodic law, passed almost unnoticed, and was accepted without opposition, and yet both thorium and uranium are of great importance in the periodic system, as they are its last members, and have the highest atomic weights of all the highest elements.

The alteration of the atomic weight of *uranium* from U = 120 into U = 240 attracted more attention, the change having been made on account of the periodic law, and for no other reason. Now that Roscoe, Rammelsberg, Zimmermann, and several others have admitted the various claims of the periodic law in the case of uranium, its high atomic weight is received without objection, and it endows that element with a special interest.

While thus demonstrating the necessity of modifying the atomic weights of several insufficiently known elements, the periodic law enabled us also to detect errors in the determination of the atomic weights of several elements whose valencies and true position among other elements were already well known. Three such cases are especially noteworthy: those of tellurium, titanium and platinum. Berzelius had determined the atomic weight of *tellurium* to be 128, while the periodic law claimed for it an atomic weight below that of iodine, which had been fixed by Stas at 126·5, and which was certainly not higher than 127. Brauner then undertook the investigation, and he has shown that the true atomic weight of tellurium is lower than that of iodine, being near to 125. For *titanium* the extensive researches of Thorpe have confirmed the atomic weight of Ti = 48, indicated by the law, and already foreseen by Rose, but contradicted by the analyses of Pierre and several other chemists. An equally brilliant confirmation of the expectations based on the periodic law has been given in the case of the series osmium, iridium, platinum, and gold. At the time of the promulgation of the periodic law the determinations of Berzelius, Rose, and many others gave the following figures:—

Os = 200; Ir = 197; Pt = 198; Au = 196.

The expectations of the periodic law* have been confirmed, first, by new determinations of the atomic weight of *platinum* (by Seubert, Dittmar and M'Arthur), which proved to be near to 196 (taking O = 16, as proposed by Marignac, Brauner, and others); secondly,

* I pointed them out in the *Liebig's Annalen*, Supplement Band viii, 1871, p. 211.

by Seubert having proved that the atomic weight of *osmium* is really lower than that of platinum, and that it is near to 191; and thirdly, by the investigations of Krüss, and Thorpe and Laurie proving that the atomic weight of *gold* exceeds that of platinum, and approximates to 197. The atomic weights which were thus found to require correction were precisely those which the periodic law had indicated as affected with errors; and it has been proved therefore that the periodic law affords a means of testing experimental results. If we succeed in discovering the exact character of the periodical relationships between the increments in atomic weights of allied elements discussed by Ridberg in 1885, and again by Bazaroff in 1887, we may expect that our instrument will give us the means of still more closely controlling the experimental data relating to atomic weights.

Let me next call to mind that, while disclosing the variation of chemical properties,* the periodic law has also enabled us to systematically discuss many of the physical properties of elementary bodies, and to show that these properties are also subject to the law of periodicity. At the Moscow Congress of Russian Naturalists in August, 1869, I dwelt upon the relations which existed between density and the atomic weight of the elements. The following year Professor Lothar Meyer, in his well-known paper,† studied the same subject in more detail, and thus contributed to spread information about the periodic law. Later on, Carnelley, Laurie, L. Meyer, Roberts-Austen, and several others applied the periodic system to represent the order in the changes of the magnetic properties of the elements, their melting points, the heats of formation of their haloid compounds, and even of such mechanical properties as the coefficient of elasticity, the breaking stress, &c., &c. These deductions, which have received further support in the discovery of new elements endowed not only with chemical but even with physical properties which were foreseen by the law of periodicity, are well known; so I need not dwell upon the subject, and may pass to the consideration of oxides.‡

* Thus, in the typical small period of

Li, Be, B, C, N, O, F,

we see at once the progression from the alkaline metals to the acid non-metals, such as are the halogens.

† *Liebig's Annalen*, Erz. Bd. vii, 1870.

‡ A distinct periodicity can also be discovered in the spectra of the elements. Thus the researches of Hartley, Ciamician, and others have disclosed, first, the homology of the spectra of analogous elements; secondly, that the alkaline metals have simpler spectra than the metals of the following groups; and thirdly, that there is a certain likeness between the complicated spectra of manganese and iron on the one hand, and the no less complicated spectra of chlorine and bromine on

In indicating that the gradual increase of the power of elements of combining with oxygen is accompanied by a corresponding decrease in their power of combining with hydrogen, the periodic law has shown that there is a limit of oxidation, just as there is a well-known limit to the capacity of elements for combining with hydrogen. A single atom of an element combines with at most four atoms of either hydrogen or oxygen: and while CH_4 and SiH_4 represent the highest hydrides, so RuO_4 and OsO_4 are the highest oxides. We are thus led to recognise types of oxides, just as we have had to recognise types of hydrides.*

The periodic law has demonstrated that the maximum extent to which different non-metals enter into combination with oxygen is determined by the extent to which they combine with hydrogen, and that the sum of the number of equivalents of both must be equal to 8. Thus chlorine, which combines with 1 atom, or 1 equivalent of hydrogen, cannot fix more than 7 equivalents of oxygen, giving Cl_2O_7: while sulphur, which fixes 2 equivalents of hydrogen, cannot combine with more than 6 equivalents or 3 atoms of oxygen. It thus becomes evident that we cannot recognise as a fundamental property of the elements the atomic valencies deduced from their hydrides; and that we must modify, to a certain extent, the theory of atomicity if we desire to raise it to the dignity of a general principle capable of affording an insight into the constitution of all compound molecules. In other words, it is only to carbon, which is quadrivalent with regard both to oxygen and hydrogen, that we can apply the theory of constant valency and of bond, by means of which so many still endeavour to explain the structure of compound molecules. But I should go too far if I ventured to explain in detail the conclusions which can be drawn from the above considerations. Still, I think it necessary to dwell upon one particular fact which must be explained from the point of view of the periodic law in order to clear the way to its extension in that particular direction.

the other hand, and their likeness corresponds to the degree of analogy between those elements which is indicated by the periodic law.

* Formerly it was supposed that, being a bivalent element, oxygen can enter into any grouping of the atoms, and there was no limit foreseen as to extent to which it could further enter into combination. We could not explain why bivalent sulphur, which forms compounds such as

$$S<^{O}_{O}> \quad \text{and} \quad S<^{O}_{O}>O,$$

could not also form oxides such as—

$$S<^{O-O}_{O-O}> \quad \text{or} \quad S<^{O-O}_{O-O}>O,$$

while other elements, as, for instance, chlorine, form compounds such as—

$$Cl-O-O-O-O-K.$$

The higher oxides yielding salts the formation of which was foreseen by the periodic system—for instance, in the short series beginning with sodium—

$$Na_2O,\ MgO,\ Al_2O_3,\ SiO_2,\ P_2O_5,\ SO_3,\ Cl_2O_7,$$

must be clearly distinguished from the higher degrees of oxidation which correspond to hydrogen peroxide and bear the true character of peroxides. Peroxides such as Na_2O_2, BaO_2, and the like have long been known. Similar peroxides have also recently become known in the case of chromium, sulphur, titanium, and many other elements, and I have sometimes heard it said that discoveries of this kind weaken the conclusions of the periodic law in so far as it concerns the oxides. I do not think so in the least, and I may remark, in the first place, that all these peroxides are endowed with certain properties—obviously common to all of them, which distinguish them from the actual, higher, salt-forming oxides, especially their easy decomposition by means of simple contact agencies; their incapacity of forming salts of the common type; and their capacity of combining with other peroxides (like the faculty which hydrogen peroxide possesses of combining with barium peroxide, discovered by Schoene). Again, we remark that some groups are especially characterised by their capacity of generating peroxides. Such is, for instance, the case in the VIth group, where we find the well-known peroxides of sulphur, chromium, and uranium; so that further investigation of peroxides will probably establish a new periodic function, foreshadowing that molybdenum and wolfram will assume peroxide forms with comparative readiness. To appreciate the constitution of such peroxides, it is enough to notice that the peroxide form of sulphur (so-called persulphuric acid) stands in the same relation to sulphuric acid as hydrogen peroxide stands to water:—

$$H(OH),\ \text{or}\ H_2O,\ \text{responds to}\ (OH)(OH),\ \text{or}\ H_2O_2,$$

and so also—

$$H(HSO_4),\ \text{or}\ H_2SO_4\ \text{responds to}\ (HSO_4)(HSO_4),\ \text{or}\ H_2S_2O_8.$$

Similar relations are seen everywhere, and they correspond to the principle of substitutions which I long since endeavoured to represent as one of the chemical generalisations called into life by the periodic law. So also sulphuric acid, if considered with reference to hydroxyl, and represented as follows—

$$HO(SO_2OH),$$

has its corresponding compound in dithionic acid—

$$(SO_2OH)(SO_2OH),\ \text{or}\ H_2S_2O_6.$$

Therefore, also, phosphoric acid, $HO(POH_2O_2)$, has, in the same sense, its corresponding compound in the subphosphoric acid of Saltzer:—

$$(POH_2O_2)(POH_2O_2), \text{ or } H_4P_2O_6;$$

and we must suppose that the peroxide compound corresponding to phosphoric acid, if it be discovered, will have the following structure:—

$$(H_2PO_4)_2 \text{ or } H_4P_2O_8 = 2H_2O + 2PO_3.*$$

As far as is known at present, the highest form of peroxides is met with in the peroxide of uranium, UO_4, prepared by Fairley;† while OsO_4 is the highest oxide giving salts. The line of argument which is inspired by the periodic law, so far from being weakened by the discovery of peroxides, is thus actually strengthened, and we must hope that a further exploration of the region under consideration will confirm the applicability to chemistry generally of the principles deduced from the periodic law.

Permit me now to conclude my rapid sketch of the oxygen compounds by the observation that the periodic law is especially brought into evidence in the case of the oxides which constitute the immense majority of bodies at our disposal on the surface of the earth.

The oxides are evidently subject to the law, both as regards their chemical and their physical properties, especially if we take into account the cases of polymerism which are so obvious when comparing CO_2 with Si_nO_{2n}. In order to prove this I give the densities s and the specific volumes v of the higher oxides of two short periods. To render comparison easier, the oxides are all represented as of the form R_2O_n. In the column headed Δ the differences are given between the volume of the oxygen compound and that of the parent element, divided by n, that is, by the number of atoms of oxygen in the compound:—‡

* In this sense, oxalic acid, $(COOH)_2$, also corresponds to carbonic acid, $OH(COOH)$, in the same way that dithionic acid corresponds to sulphuric acid, and subphosphoric acid to phosphoric; therefore, if a peroxide, corresponding to carbonic acid, be obtained, it will have the structure of $(HCO_3)_2$, or $H_2C_2O_6 = H_2O + C_2O_5$. So also lead must have a real peroxide, Pb_2O_5.

† The compounds of uranium prepared by Fairley seem to me especially instructive in understanding the peroxides. By the action of hydrogen peroxide on uranium oxide, UO_3, a peroxide of uranium, UO_44H_2O, is obtained ($U = 240$) if the solution be acid; but if hydrogen peroxide act on uranium oxide in the presence of caustic soda, a crystalline deposit is obtained, which has the composition $Na_4UO_84H_2O$, and evidently is a combination of sodium peroxide, Na_2O_2, with uranium peroxide, UO_4. It is possible that the former peroxide, UO_44H_2O, contains the elements of hydrogen peroxide and uranium peroxide, U_2O_7, or even $U(OH)_6H_2O_2$, like the peroxide of tin recently discovered by Spring, which has the constitution $Sn_2O_5H_2O_2$.

‡ Δ thus represents the average increase of volume for each atom of oxygen con-

	s.	v.	Δ.		s.	v.	Δ.
Na_2O	2·6	24	−22	K_2O	2·7	35	−55
Mg_2O_2	3·6	22	−3	Ca_2O	3·15	36	−7
Al_2O_3	4·0	26	+1·3	Sc_2O_3	3·86	35	0
Si_2O_4	2·65	45	5·2	Li_2O_4	4·2	38	+5
P_2O_5	2·39	59	6·2	V_2O_5	3·49	52	6·7
S_2O_6	1·96	82	8·7	Cr_2O_6	2·74	73	9·5

I have nothing to add to these figures, except that like relations appear in other periods as well. The above relations were precisely those which made it possible for me to be certain that the relative density of eka-silicon oxide would be about 4·7; germanium oxide, actually obtained by Winckler, proved, in fact, to have the relative density 4·703.

The foregoing account is far from being an exhaustive one of all that has already been discovered by means of the periodic law telescope in the boundless realms of chemical evolution. Still less is it an exhaustive account of all that may yet be seen, but I trust that the little which I have said will account for the philosophical interest attached in chemistry to this law. Although but a recent scientific generalisation, it has already stood the test of laboratory verification and appears as an instrument of thought which has not yet been compelled to undergo modification; but it needs not only new applications, but also improvements, further development, and plenty of fresh energy. All this will surely come, seeing that such an assembly of men of science as the Chemical Society of Great Britain has expressed the desire to have the history of the periodic law described in a lecture dedicated to the glorious name of Faraday.

OSTWALD, FRIEDRICH WILHELM. (b. Riga, Latvia, Russia, 2 September 1853; d. Leipzig, Germany, 4 April 1932)

After attending the Realgymnasium in Riga, Ostwald studied at the University of Dorpat (now Tartu) and completed his doctorate in 1878. He was professor of chemistry at the Riga Polytechnic Institute from 1881 to 1887, when he was appointed, at the University of Leipzig, to the only chair of physical chemistry then existing in Germany. In 1887 he founded, with Van't Hoff as coeditor, the *Zeitschrift für physikalische Chemie*, and, in 1898, he became the director of the new physical chemistry institute there. He retired in 1906. Ostwald, in addition to his chemical work, investigated the theory of colors and supported the establishment of a new international language, Esperanto. In 1889 he began editing a series of scientific classics known as *Ostwalds Klassiker der Exakten Wissenschaften*, which now includes 256 volumes.

Ostwald's first chemical researches concerned chemical affinities; from these studies he went on to investigate electrolytic dissociation, electrical conductivity, mass action, reaction velocities, and catalysis. He was awarded the Nobel Prize in chemistry in 1909 for the work on catalysis, which included studies of crystallization from supersaturated solutions and the relationship between rate of catalytic activity and acid strength. That a catalyst causes no change in overall energy relations and does not alter the thermodynamically stable equilibrium position of a reaction was an important argument made by him. However, it is for his role in the ionization theory and the general development of physical chemistry that he is best known. At the same time his writing of two of the principal chemistry textbooks of the late nineteenth century, the *Lehrbuch der allgemeine Chemie* (1885–1887) and the *Grundriss der allgemeinen Chemie* (1889) also contributed to his influence and reputation.

Despite his support of Arrhenius's ideas on ionization, Ostwald was not an advocate of atomism nor of molecular-kinetic theory. Indeed, it was only after the work of Jean Perrin and Thé Svedberg on Brownian motion that Ostwald conceded, in 1909, that there was good experimental evidence for the physical reality of atoms and molecules. He argued that the principal laws associated with chemical atomism could be deduced from chemical dynamics, a view naturally developed from "energetics" as a system of scientific explanation superior to a materialistic and mass-dependent mechanics. In 1895 at the annual meeting of the German Society of Scientists and Physicians, Ostwald and his Dresden colleague Georg Helm defended the view that energy had displaced matter, or mass, as the fundamental basis of physical theory. Ostwald's delivered paper was challenged at the Lübeck meeting and in subsequent journal publications by Ludwig Boltzmann, Felix Nernst, and others.

Science Progress.

No. 24. February, 1896. Vol. IV.

EMANCIPATION FROM SCIENTIFIC MATERIALISM.[1]

IT has at all periods been a source of complaint that so little unanimity should prevail with regard to the most important and fundamental of human problems. In our own times, however, the grievance concerning one of the greatest of these questions has almost disappeared. For, although there still exist many and varied contradictions, nevertheless it may be asserted that scarcely in any age has there been so comparatively close an agreement regarding our conception of the outer world of phenomena as exists in this present scientific century of ours. From mathematician to practising doctor, every scientifically thinking man, if called upon to express his opinion as to the "inner structure" of the universe, would sum up his ideas in the conception that things consisted of atoms in motion, and that these atoms and their mutual forces were the final realities underlying all phenomena. We read and hear, with countless repetition, the statement that the only intelligent explanation of the physical world is to be found in a "Mechanics of the Atoms"; matter and motion appear as the final principles to which natural phenomena in all

[1] A paper read at the third general sitting of the Assembly of the Society of German Scientists and Physicians at Lübeck, 20th September, 1895. By Wilhelm Ostwald, Professor in Chemistry at the University of Leipzig. Authorised translation by F. G. Donnan and F. B. Kenrick.

their variety must be referred. This conception we may term scientific materialism.

I here propose to state my conviction that this so generally accepted view is untenable; that this mechanical idea of the universe does not fulfil the purpose for which it was designed, and that it is inconsistent with undoubted and generally known and recognised truths. The conclusion to be drawn from this is obvious. The scientifically untenable view must be abandoned and its place filled, if possible, by a new and a better one. The natural question as to whether such another and better conception can be found, I think I can answer in the affirmative. My remarks consequently divide themselves naturally into two parts: a destructive part and a constructive part. As in all cases, so here, destruction is easier than construction; the inefficiency of the customary mechanical treatment is more easy to demonstrate than the efficiency of the new one, which I would characterise as the "energetical" view. I may remark at once that the new view has already had an opportunity of proving its worth on quiet reflection, and by impartial research in particularly favourable regions of experimental science. Although this cannot prove its correctness, it still gives the new conception a claim to notice.

It will not perhaps be superfluous to state at the outset that I am dealing to-day exclusively with a question of natural science. I draw aside on principle from all conclusions of an ethical or religious nature which may be deduced from the result of this discussion. I do this not because I undervalue the significance of such conclusions, but because my arguments have been founded, independently of such considerations, on the firm ground of the exact sciences. For the tillage of this soil also, it may be said that he who putteth his hand to the plough and looketh back is not fit for this kingdom. The scientific investigator of nature is bound to set forth his results neither to pain nor to please, and we may trust ourselves to the power of Truth which, though perhaps temporarily, can never lead far from the right way him who seeks her earnestly.

I do not fail to recognise that my undertaking places me in opposition to the views of men who have done great things for science, and to whom we all look up in wonder and admiration. But do not charge me with presumption if I oppose myself to such men in a matter of so great importance. You do not call it presumption when the sailor on duty at the mast-head turns the great ship from her course by his cry of "Breakers ahead!" albeit he is but an insignificant member of her crew. His duty is to announce what he sees, and did he fail to do this he would prove untrue to this duty. It is in this sense a duty which I am to-day discharging. Nevertheless, no one is bound to alter his scientific course in answer to my cry of "Breakers ahead!" Every one is at liberty to test whether it is a reality that stands before my eyes, or whether a mirage deceives my vision. I believe however that the special nature of my scientific duties gives me for the moment a clearer insight into certain phenomena than may be had from other points of view, and for this reason I could not but regard it as wrong, were I for extraneous motives to leave unsaid what I have seen.

In order to find our way clearly through the infinite variety of the world of phenomena, we always make use of the same scientific method, namely, grouping together things that are similar and in variety seeking the non-varying. In this way our gradual mastery over the infinitely various phenomena of the outer world is acquired and ever more effective means of co-ordination are continually developed. From the simple *list* we proceed to the *system*, from this again to the *law of nature*, and the most comprehensive form of the latter condenses into a *general conception*. We perceive that the phenomena of the actual world, limitless as is their variety, form nevertheless only certain perfectly definite and particular cases of the theoretically conceivable possibilities. The significance of a Law of Nature consists in the determination of the *actual* among the *possible* cases, and the form to which each can be referred is the finding of an *invariant*, that is to say, a quantity which remains unchanged even when all the other determining elements vary within the possible limits imposed by the law. Thus we perceive that the historical development of scientific conceptions is ever associated with the discovery and working out of such invariants; in them we behold the mile-stones which mark the track traversed by human knowledge.

Such an invariant of universal significance was found in the idea of mass. This gives not only the constants of the laws of Astronomy, but also appears no less invariable in the case of the most deep-going changes to which we can subject the objects of the outer world, namely, chemical changes. Accordingly this idea appeared to be excellently suited to form the central point of scientific law. Of course it was too poor in connotation to serve alone for the representation of the various phenomena, and had therefore to be correspondingly extended. This was effected by fusing with the simple mechanical idea the series of properties which are associated with the property of mass and are proportional to it. In this manner arose the conception of matter, in which everything was summed up that was associated in our sense-impressions with mass, and which always accompanied it, as, for example, weight, extension in space, chemical properties, etc., and the *physical law* of the conservation of mass passed into the *metaphysical axiom* of the conservation of matter.

It is important to observe that, with this extension, a great many hypothetical elements crept into a conception originally free from hypothesis. For example, in the light of this conception a chemical reaction had, contrary to appearance, to be so considered that none of the matter affected by the chemical change could possibly disappear and be replaced by new matter with new properties. On the contrary, the view required the assumption that when, for instance, all the perceptible properties of iron and oxygen had disappeared in iron oxide, nevertheless iron and oxygen still existed in the body produced and had only assumed other properties. At present we are so accustomed to such an idea that it is difficult for us to conceive its strangeness, indeed its absurdity. But if we consider for a moment that all we know of a given substance consists in the knowledge of its properties it is evident that the assertion that a certain substance continues to exist without possessing these properties is not very far removed from pure nonsense. As a matter of fact, this purely formal assumption serves simply to unite the general facts of chemistry, in particular the stöchiometric laws of mass, with the arbitrary notion of an intrinsically unvarying matter.

But even the so extended idea of matter, together with the necessary attendant suppositions, is not sufficient to embrace all phenomena, not even indeed in the inorganic world. We think of matter as something in itself intrinsically motionless and unvarying; hence in order to make the representation of an ever-changing world possible, we must supplement this conception with another independent one which gives expression to this changeableness. A conception of this sort was, in the most successful manner, developed by Galileo, the founder of scientific physics, in the idea of force as the constant cause of motion. Galileo had, in fact, discovered an invariant of great importance for the varying phenomena of free and indirect fall; and with the assumption of the intrinsically constant force of gravitation, whose effects continuously accumulate, the complete representation of these phenomena became possible. The significance of this conception became evident when Newton, with his idea that the same force acted between the heavenly bodies, brought the whole star-world under the sway of science. It was this step forward in particular which gave rise to the conviction that just as the astronomical so also all other physical phenomena must be capable of representation by this means. When, moreover, at the beginning of our century the efforts of numerous, and especially French, astronomers had shown that Newton's law of gravitation could not only represent the motions of the heavenly bodies to a first approximation, but withstood also the far severer test of a second approximation in their ability to express with equal certainty and accuracy the small deviations from the typical motions (*i.e.*, the perturbations), the confidence in the sufficiency of this theory must have been extraordinarily strengthened. What was then more likely than the expectation that the theory which had shown itself so perfectly adapted to represent the motions of the great bodies of the Universe, must be the correct, nay, the only means of bringing the events of the little world of atoms under the sway of science? So arose the mechanical conception of nature according to which all phenomena (at first of inorganic nature) were to be finally referred simply to the motions of atoms according to the same laws which had been recognised to hold for heavenly bodies. That this conception should be immediately carried over from the region of inorganic to that of animated nature, was a necessary consequence when it was once perceived that the same laws which hold in the former, claim here also their inviolable rights. This conception of the universe found its classical expression in Laplace's idea of the "Universe-Formula" by means of which every past and future event was to be capable of deduction by rigid analysis applied to mechanical laws. For this purpose an intellect was to be required which though far beyond the human mind in power was not essentially of a different nature.

We do not generally notice in what an extraordinary degree this widespread view is hypothetical, nay, metaphysical; on the contrary we usually regard it as the most exact expression for the actual relations. Nevertheless it must be remarked that a confirmation of the natural deduction from this theory, namely, that all non-mechanical phenomena such as heat, radiation, electricity, magnetism and chemical action are actually mechanical, has in no single case been obtained. In no single case has the attempt to represent the actual relations by means of a mechanical system so far succeeded that nothing remained over to explain. I grant that for many individual phenomena, the mechanical analogues have been given with more or less success. But all attempts to completely represent the whole of the known facts in any department by means of some such mechanical analogue have resulted without exception in some unexplainable contradiction between what really happens, and what we should expect from our mechanical model. This contradiction may long remain hidden; but the history of science teaches us that it sooner or later makes its inevitable appearance, and that all we can say with complete certainty regarding such mechanical similes or analogues—usually termed mechanical theories of the phenomena in question—is that they will doubtless on some occasion fail.

The history of optical theories offers a striking example of these facts. So long as the whole of optics embraced only reflection and refraction, its phenomena could be represented by the mechanical system proposed by Newton, according to which light consisted of small particles which were shot out in straight lines from the radiating object, and obeyed the laws of motion for perfectly elastic bodies. The fact that another mechanical view, the vibration theory of Huygens and Euler, yielded just as much in these respects might well have awakened doubts as to the exclusive ability of the earlier hypothesis to meet the requirements of the case, but it was not able to usurp the latter's scientific position. When, however, the phenomena of interference and polarisation were discovered, Newton's mechanical analogue showed itself to be quite unsuitable, and the other, namely, the *vibration theory*, was considered established, since from its assumptions the chief points at least of the new phenomena could be deduced.

The life of the vibration theory as a *mechanical* theory has also had its bounds, for in our own time it too has been carried to its grave without drum or fife, and its place taken by the electro-magnetic theory of light. A post-mortem examination reveals clearly the cause of death; it has resulted from the failure of the mechanical parts. The hypothetical ether whose task it was to "beat" had to fulfil its duty under particularly difficult conditions. For the phenomena of polarisation demanded unconditionally that the vibrations should be transversal. Now this presupposes a solid body, and the calculations of Lord Kelvin have finally shown that a medium with such properties as were thus required of the ether could not possibly be stable, and, as a necessary consequence, can have no physical existence. Doubtless the idea of sparing the now generally received electro-magnetic theory of light a similar fate prompted the immortal Hertz, to whom this theory owes so much, when he expressly declined to see in it anything beyond a system of six differential equations. That the evolution of the theory should end in this point is a far more convincing argument than any I could adduce against the permanent value of the theoretical methods previously followed on mechanical lines.

But you may urge the fruitfulness of these theories. Yes, they were fruitful in so far as they contained correct elements, just as they were harmful on account of the false ones. Which were the right and which were the wrong elements, however, was only revealed after long and dearly bought experience.

The result of our remarks up to the present is in the first place purely negative; we have learned how *not* to proceed, and it may appear of little use to bring forward such negative results. However we may already note one gain here which many of you will not consider worthless. In this way we discover the possibility of critically correcting a view which in its own time created no small sensation, and caused many of those interested great trouble. I refer to the widely known views first expounded twenty-three years ago by the celebrated Berlin Professor of Physiology, Emil du Bois-Reymond, at the Leipzig meeting of the German Association of Natural Philosophers, and later in some other much-read papers; views dealing with the prospects of our future knowledge of nature and culminating in the much-discussed *ignorabimus*. In the long controversy which this speech gave rise to du Bois-Reymond has, so far as I can see, victoriously withstood all attacks, and naturally so, since all his opponents have proceeded from the same premises which led him to his *ignorabimus*, and his conclusions stand as firm as the basis upon which he built them. This basis, which meanwhile had been called in question by no one, is the mechanical conception of the universe, namely, the assumption that the resolving of phenomena into a system of moving particles is the goal to be aimed at in our explanation of nature. Should, however, this foundation fall—and we have seen that it must—then with it goes the *ignorabimus*, and science is once more free to move onward.

I do not believe that this result will be received with astonishment, for judging by my own experience, no investigator of nature has seriously believed in the *ignorabimus*, although it was not clear where the weak point of the argument lay. Hence the result gained by the rejection of the mechanical conception of the Universe, namely, the banishment of that menacing spectre, may well be of some value to many a thinker unable to find a flaw in the resistless logic of du Bois-Reymond's argument.

What has here been set forth for the sake of clearness with respect to special discussions such as the foregoing, has, however, a far wider significance. The doing away with the mechanical construction of the universe goes down to the very foundations of the whole materialistic conception of things, taking the word materialistic in its scientific sense. If it appears a vain undertaking, ending with every serious attempt in final failure, to give a mechanical representation of the known phenomena of physics, we are driven to the conclusion that similar attempts in the incomparably more complicated phenomena of organic life will be still less likely to succeed. The same fundamental contradictions occur here also, and the assertion that all the phenomena of nature can be primarily referred to mechanical ones cannot even be designated here as a practical working-hypothesis; it is simply incorrect.

This error appears more clearly when viewed in the light of the following fact. The equations of mechanics all possess the property that they still hold good when the sign of the quantity denoting *time* is changed. That is to say, theoretically perfect mechanical processes can take place just as well backwards as forwards. In a purely mechanical world there would be, therefore, no Before and no After, in the sense of our world; the tree could return again to the sapling and the sapling to the seed, the butterfly transform itself once more into the caterpillar, and the old man become again a babe. For the fact that this does not occur, the mechanical conception of the world has no explanation to offer, and can have none on account of the already mentioned property of mechanical equations. The evident irreversibility of actual natural phenomena proves, therefore, the existence of processes which cannot be represented by mechanical equations, and with this statement the judgment on scientific materialism is passed.

We must accordingly, and this appears to follow with absolute certainty from these considerations, give up all hope of getting a clear idea of the physical world by referring phenomena to an atomistic mechanics. But, perhaps one of you will say, what means shall we have left of picturing to ourselves what really occurs in nature when the conception of atoms in motion is abolished? To such a question I would answer: Thou shalt not make unto thyself any image or likeness. Our task is not to view the world in a more or less bedimmed and crooked mirror, but as directly as the nature of our minds will permit. To co-ordinate realities, *i.e.*, definite and measurable quantities, so that when certain of them are given the others can be deduced, is the problem set before science, and this problem cannot be solved by assuming as substratum any hypothetical analogue, but only by the determination of the mutual relations existing between measurable magnitudes.

Undoubtedly this way is long and tiring, yet it is the only permissible one. But we need not tread this path with bitter self-renunciation hoping that it will finally lead our grandchildren to the longed-for summit. No, it is we who are the fortunate ones, and the most hopeful bequest which the departing century can bestow on the one that is just dawning is the replacement of the mechanical by the "energetical" conception of the universe.

I consider it of the greatest importance to state here that all this is by no means a novelty, a production of to-day. No, for half a century it has, though unrecognised, been in our possession. Here indeed if anywhere we may fitly apply the words: "Geheimnissvoll offenbar" (mysteriously revealed). Daily could we read it and we understood it not.

When Julius Robert Mayer fifty-three years ago first discovered the equivalency of the various natural forces, or as we now say, the various forms of energy, he had already taken an important step in the critical direction. But according to an ever-recurring law in collective thinking, a new idea is never accepted in the pure and unsullied form in which it is offered. He who has not inwardly experienced its development, but who has received the knowledge from without, seeks above all to adapt that which is new as well as possible to his previous notions. In this way the new idea is marred, and, even if not actually perverted, nevertheless robbed of its best power. Indeed so active is this peculiarity of thought that it does not even leave free the discoverer himself. Copernicus' powerful intellect sufficed indeed to cause sun and earth to change places in their motions, but failed to conceive the motions of the other planets in their simplicity; for these he retained the traditional theory of the epicycles. We see the same thing in Mayer's case. The task of the succeeding generation consisted, as is almost always the case, not simply in reaping the results of the new doctrine, but rather in separating piece by piece the arbitrary and extraneous additions until finally the fundamental idea should appear again in its pure simplicity.

We observe also in our case a similar development. When J. R. Mayer had set forth the law of equivalency his idea of equivalent transmutability of the different forms of energy was in its simplicity too strange to be directly accepted. Indeed the three scientists to whom we are mostly indebted for the working out of this law—Helmholtz, William Thomson and Clausius—all three believed that the law could be "explained" by assuming that the different forms of energy were fundamentally the same, namely, *mechanical* in nature. In this way what appeared to be the most urgent need was satisfied, namely, a direct connection with the dominating mechanical conception of nature. But a fundamental part of the new thought was thereby lost.

Half a century has been necessary to mature the conviction that this hypothetical addition to the law of energy is in reality no deepening of insight but a renunciation of its most important aspect: its freedom from every arbitrary hypothesis. It was not, however, the recognition of this general fact which actually brought about this advance, namely, the rejection (as far as it has gone) of the mechanical explanation, but rather the final failure of all attempts to interpret satisfactorily by the mechanical treatment the phenomena connected with the remaining forms of energy.

But you will be impatient to learn how it is to be possible by such an abstract idea as energy to form a conception of the universe which shall be comparable, in clearness and intuitiveness, with the mechanical one. The answer ought not to be difficult. What do we know of the physical world? Obviously only what is vouchsafed us through our organs of sense. But under what conditions are these organs set in action? Turn the matter as we may, the only principle we find common to all is this: *The Organs of perception react in response to differences of energy between them and the surroundings.* In a world in which the temperature was everywhere that of our bodies we should have no experience of heat, just as we have no perception of the constant atmospherical pressure under which we live. Not till we have produced a space of different pressure do we become aware of its existence.

Well and good, this you will be ready to admit. But you will not be so ready to abandon matter, for energy must of course have a *carrier.* But I ask in return: Why? Since our total knowledge of the outer world consists in its energy relations, what right have we to assume in this very outer world the existence of something of which we have had no experience? But energy, it has been urged, is only something thought of, an abstraction, while matter is a reality; exactly the reverse, I reply. Matter is a thing of thought which we have constructed for ourselves (rather imperfectly) in order to express that which is lasting in the changeableness of phenomena. Now that we begin to grasp that the Actual, *i.e.*, that which *acts* upon us, is energy alone, we must inquire in what relation the two conceptions stand to one another, and the result is undoubtedly that the predicate of reality can be affirmed of Energy only.

This decisive aspect of the new conception will perhaps stand out more clearly if I sketch the development of the idea in question in a short historical *resumé*. We have already seen that progress in science is characterised by the discovery of ever more general invariants, and I have also indicated how the first of these invariable quantities, mass, was extended to matter, *i.e.*, mass accompanied by volume, weight, and chemical properties. This conception was, however, obviously insufficient to embrace phenomena in their continual changeableness, and from Galileo's time onwards the idea of force has been added in order to suit this phase of nature. But force lacked the property of constancy, and after *vis viva* and work had been found to be mechanical functions exhibiting the properties of partial invariants, Mayer discovered in energy the most general invariant which rules the whole dominion of physical forces.

In accordance with this historical development, matter and energy continued to exist side by side, and all that could be said about their mutual relation was that they occurred for the most part together or that matter was the carrier or holder of energy.

But are matter and energy things essentially different from one another, as perhaps Body and Soul? Or is not rather all that we know and say of matter already contained in the idea of energy, so that we can represent with this latter quantity the totality of phenomena? According to my conviction there can be no doubt as to the answer. Hidden in the conception of matter are the following ideas: first of all, mass, that is, the capacity factor of kinetic energy; further, the occupation of space or volume energy; again, weight or that particular kind of distance energy which appears as gravitation; and finally chemical properties, *i.e.*, chemical energy. It is always a question of energy alone, and if we imagine the various kinds of energy removed from matter there remains *nothing*, not even the space it occupied; for space makes itself known only through the expenditure of energy which the penetration into it requires. Matter is therefore nothing but a group of various forms of energy co-ordinated in space, and all that we try to say of matter is really said of these energies.

What I am endeavouring to lay down is so important that you will pardon my venturing to approach the subject from another quarter. Allow me to use for this purpose the most drastic illustration I can find. Imagine that you receive a blow from a stick. What do you feel, the stick or its energy?

The only possible answer is: The energy. For the stick is the most harmless thing in the world as long as it is not wielded. But we can also knock ourselves against a stationary stick. Certainly. What we perceive, as already stated, are *differences* of energy conditions relative to our sense organs, and it is consequently immaterial whether the stick moves towards us or we towards the stick. If both we and the stick are moving in the same direction with the same velocity, the latter has no further existence for our sense of feeling, for it can no longer come in contact with us and effect an exchange of energy.

These considerations show, I hope, that *all* that we have until now been able to express by the ideas of Matter and Force—and indeed much more besides—may actually be expressed by the idea of energy. It is a question simply of transferring to this conception those properties and laws which were formerly ascribed to matter and force. We gain further the enormous advantage of doing away with the contradictions which were attendant on the former method of treatment, and to which I alluded in the earlier part of my paper. By making no assumption as to the relation between the different forms of energy, except that given by the law of conservation, we leave ourselves free to study objectively the various properties of these forms of energy. We can, further, by rational consideration and arrangement of these properties, establish a system which shall represent explicitly not only the similarities, but also the differences in these forms of energy, and which will therefore lead us much further scientifically than would be possible when slurring over their differences through the hypothetical assumption of their "inner" identity. A good illustration of the meaning I intend to convey may be found in the kinetic hypothesis of gases which at present enjoys almost universal recognition. According to this hypothesis the pressure of a gas arises from the blows delivered by the moving particles. Now pressure is a quantity which possesses no special direction: a gas exerts pressure in all directions equally; but a blow is caused by a moving object, and this motion has a definite direction. Consequently one of these quantities cannot be referred directly to the other. The kinetic hypothesis gets round this difficulty by the assumption that the blows occur uniformly in all directions, whereby the vector-property really possessed by the blow is artificially done away with. In this case the artificial adaptation of the properties of the different energies is successful, but in other cases it is not completely possible. The factors of electrical energy, potential and electrical quantity, form a case in point—quantities which I propose to call *polar*, *i.e.*, they are characterised not only by a numerical value but also by a sign in such a way that the sum of two equal values with opposite signs is equal to zero and not to their double value. In mechanics the purely polar quantities are unknown, and this is why all attempts to set up even a partially workable mechanical hypothesis for electrical phenomena must essentially fail. Should we succeed in contriving a mechanical magnitude with polar properties (as is perhaps not impossible, and certainly worthy of careful consideration) we should have the means wherewith to picture to ourselves mechanically at least a few phases of electrical phenomena. We may say with certainty that it will be a question of a few only, and that the imperfection of all mechanical hypotheses will show itself here again and will prevent the complete carrying out of the idea.

But even if the laws of natural phenomena may really be reduced to the laws of the corresponding forms of energy, what advantage do we gain therefrom? First, the very considerable one, that a science free from hypothesis becomes possible. We seek no longer forces whose existence we cannot prove between atoms which we cannot observe, but we judge a process by the kind and quantity of the vanishing and appearing energies. *These* we are able to measure, and all that it is necessary to know may be expressed in this manner. What an enormous general advantage this is, will be clear to every one whose scientific conscience has suffered under the continual amalgamation of facts and hypothesis which the physics and chemistry of to-day offer us as rational science. Energetics offers us a means of fulfilling in its true sense the demand of Kirchhoff so oft misunderstood, namely, the substitution of the description of phenomena for the so-called explanation of nature. With this freedom from hypothesis appears a methodical unity which, it may be unhesitatingly affirmed, has up till now never been attained. I have already referred to the philosophical significance of this unity of principle in the conception of natural phenomena; although obvious from the very nature of the case, I may still call attention in particular to the enormous advantage with regard to the teaching and understanding of science which accrues from this philosophical unification. To illustrate by an example: we may assert that all equations without exception which connect two or more different classes of phenomena must necessarily be equations between energy values, others are altogether impossible. This follows as a necessary result from the fact that besides the intuitive ideas of space and time, energy is the only quantity which is common to each and every class of phenomena; and therefore energy quantities are the only ones which can possibly be equated between these classes.

I must, unfortunately, refrain from entering into the

many relations—some already known, some new—which may by this principle be directly written down, but which formerly could be deduced only by more or less clumsy calculations. I must also forbear from discussing the new phases which have been exhibited in the light of general energetical considerations by the other earlier laws of thermodynamics—the most extended branch of Energetics. All these things must indeed be so if what I said concerning the importance of this new way of considering nature is well founded.

There is a final question which I would not like to pass untouched. When we once succeed in grasping in its pure entirety a fruitful and important truth we are only too apt to look upon it as all-comprehensive. This mistake we see daily committed in science, and the conception to the combating of which I have devoted the half of the time allowed me has arisen from exactly such a cause. We shall forthwith have to ask ourselves the question: Is energy, necessary and useful as it is for the understanding of nature, also *sufficient* for this purpose? Or are there phenomena which cannot be completely represented by the known laws of energy?

I do not think I can better justify the responsibility which I have by these remarks incurred towards you than by emphasising the fact that this question must be answered in the *negative.* Immense as are the advantages possessed by the energetical conception of the universe over the mechanical or materialistic, it seems to me, nevertheless, that already certain points may be noted which are not covered by the known laws of energy, and which therefore point to the existence of principles which extend beyond these. Energetics will remain beside these new laws; but it will not be in the future, as we must to-day consider it, the most comprehensive principle ruling natural phenomena; it will perhaps appear as a special case of a still more general relation, of the nature of which we can at present have scarcely an inkling.

I am not afraid of having lowered by what I have said the value of that mental progress which we were discussing; I think I have rather raised it. For once again are we met by the fact that Science may never recognise bounds to her progress, and that amidst the struggles for some new possession her eyes must not become blinded to the fact that beyond the territory she is striving to conquer still wider plains extend which later must also be subdued. Let us have done with the time when the smoke and dust of battle confined our vision to the narrow limits of the combat. To-day that is no longer allowed; to-day we fire with smokeless powder—or rather ought to do so—and have, therefore, with the possibility, also the duty not to fall into the errors of past periods.

WILHELM OSTWALD.

BOLTZMANN, LUDWIG. (b. Vienna, Austria, 20 February 1844; d. Duino, near Trieste, 5 September 1906).

Boltzmann took his doctorate in 1867 at the University of Vienna and became an assistant in the Physical Institute. He held professorships at Graz (1875–1891), Munich (1891–1895), Leipzig (1900–1902), and Vienna (1895–1900 and 1902–1906). Among his students and collaborators was Svante Arrhenius, who worked with Boltzmann at Graz during Arrhenius's *Wanderjahren* of 1884–1890. During 1900–1902 Boltzmann taught at Leipzig with

Ostwald at the same time that the two men argued radically different views on the value and validity of the physical hypothesis of atomism. Boltzmann committed suicide in 1906, shortly after returning to Vienna, where he had replaced Ernst Mach in a professorship in the philosophy of science and in theoretical physics.

Boltzmann's interest in atomistics and kinetic theory was due in part to his studies at Vienna with Josef Loschmidt, who published, in 1865, the first reliable estimate of molecular dimensions, using the kinetic theory of Clausius and Maxwell. Boltzmann developed the statistical approach of Maxwell in papers on kinetic theory and thermal equilibrium. His greatest achievement demonstrated systematically how thermodynamic entropy is related to the statistical distribution of molecular particles, and how increasing entropy corresponds to the increasing randomness of molecular distribution. Boltzmann's controversial result that a function (H or E) always decreases for nonequilibrium systems is now known as the "H-theorem" and has been shown to be valid only under certain special conditions. In 1877 he introduced the probabilistic relation $S = k \log W$ for a description of entropy (S), where W is the number of possible molecular configurations corresponding to a given macroscopic state of the system.

Throughout his career Boltzmann concerned himself with the mathematical and physical implications of an atomic view of matter. Crucial in many discussions of the validity of molecular mechanics and kinetic theory were paradoxes dealing with the irreversibility of entropy processes and the reversibility of mechanical systems. After challenging Ostwald's defense of energetics at the 1895 Lübeck meeting, Boltzmann wrote several essays renewing arguments for the molecular-kinetic theory, including "On the Necessity of Atomic Theories in Physics." Here he reviewed the epistemological, methodological, and experimental advantages of using the atomistic hypothesis despite some of its seemingly paradoxical drawbacks.

ON THE NECESSITY OF ATOMIC THEORIES IN PHYSICS.[1]

IN addition to the atomic hypothesis now in vogue, there is a second method employed in theoretical physics which seeks to represent the facts of some given and very narrowly circumscribed domain by means of differential equations. We shall call it *mathematico-physical phenomenology.* Since this method offers a new constructive representation of the facts, and since it is of advantage to possess as many such representations as possible, unquestionably the new doctrine is qualified to take rank beside the atomistic hypothesis in its present form as a method of great value and utility. Another form of phenomenology, which I should like to term the *energic phenomenology,* will claim our consideration later.

Now, the opinion has grown rife that our representations of nature as drawn from phenomenological considerations enjoy intrinsic advantages over our representations as obtained by the atomistic method. It is my wont to avoid general philosophical discussions of this character, where there are no practical consequences at stake, for the reason that they do not admit of the precise formulation that special questions do, and consequently their solution remains largely a matter of taste. But I have the impression that the doctrine of atomism has, for the rather untenable reason above adduced, recently been fast losing ground, and I have therefore felt constrained to do what is in my power to avert the

[1]Translated by T. J. McCormack from the *Proceedings of the Imperial Academy of Sciences of Vienna,* Vol. CV., No. 8, Section of Mathematics, Astronomy, Physics.

injury that I believe would result to science if phenomenology should be made a dogma, as atomism once was.

To avoid misunderstandings, I shall declare it at the outset to be the purpose of the following discussion to answer certain quite definite questions. And since the services that atomism has performed for science cannot possibly be a subject of doubt, these questions may be formulated as follows:

(1) Has not the atomic theory in its present shape distinct advantages over the phenomenologic theory now current?

(2) Is there any likelihood that there shall develop from phenomenology in the near future a theory possessing precisely the advantages which constitute the distinctive features of the atomistic theory?

(3) Granting the possibility of the atomic theory's being eventually discarded, does not the equal possibility exist that gradually phenomenology may become absorbed in the atomic theory?

(4) Finally, would it not be a positive detriment to science, if the reigning views of atomism did not continue to be cultivated with the same ardor as the views of phenomenology?

I may state at once that the outcome of the following reflexions will be to answer the foregoing questions in favor of atomism.

The differential equations of mathematico-physical phenomenology are obviously nothing but rules or directions for constructing and combining numbers and geometric concepts. But these numbers and concepts are themselves nothing but mental constructs from which the phenomena of nature may be predicted;[1] and such is precisely the case also with the constructs of atomism, so that for my part I can discover not the slightest difference between them in this regard. Indeed, to my mind and feeling no direct description of a comprehensive domain of facts, no single and exclusive mental representation of it, at all, is possible. It is accordingly wrong to say with Ostwald, "Thou shalt make unto thee no mental image or likeness whatsoever," but one may merely

[1] Compare Mach's *Principien der Wärmelehre*, Leipsic, J. A. Barth, 1896, p. 363. Mach's writings have contributed much toward shaping my philosophy.

say, "Thou shalt give to such images the fewest possible arbitrary features."

Mathematico-physical phenomenology often combines with its preference for the differential equations a certain disparagement of atomism. But it is my conviction that there is a vicious circle running through the reasoning that the differential equations transcend the facts in less degree than the most generalised form of the atomistic conception. Of course, if we are antecedently of the opinion that our sense-perceptions are mentally representable by the simile of a *continuum*, the differential equations will not, and the atomistic conceptions will, transcend our preconceived opinion. But quite different the case if we are wont to think atomistically: then the situation is reversed, and the conception of a *continuum* appears to transcend the facts.

Let us analyse, for example, the import of Fourier's equation for the conduction of heat, which has become classic in this regard. This equation is merely the statement of a rule, consisting of two parts:

(1) Conceive in the interior of a body (or, more generally, regularly aranged in some correspondingly bounded three-dimensional manifold) a large number of tiny things (call them elementary particles), each of which at the outset has any temperature you please. After the lapse of a very small interval of time (or on a very slight increase of a fourth variable), let the temperature of each particle be the arithmetic mean of the temperatures which the immediately surrounding bodies previously had.[1] After a like interval of time, repeat this process; and so on.

(2) Conceive both the elementary particles and the minute intervals of time to diminish constantly in size and their number to increase in a corresponding ratio until thermometric results are reached of sufficient accuracy to render the influence of further diminution inappreciable.

Likewise, the definite integrals that give the solution of the differential equations, can in general be evaluated only by mechan-

[1] Mach, *loc. cit.*, p. 118.

ical quadratures, or, in other words, themselves first require decomposition into a finite number of parts.

Let no one fancy he has attained a clear comprehension of what a continuum is from the mere utterance of the word *continuum* or merely by writing down a differential equation. On closer scrutiny, the differential equation will be found to be merely a statement of the fact that at the outset a finite number of things is to be held in the mind; this is the condition precedent of the entire process; not until afterwards is the number of the things increased until further increase is without influence on the result. Of what advantage is it now to suppress the postulate requiring us to conceive large numbers of individual things just after we have in our explanation of a differential equation defined by means of that postulate the value expressed by that equation? I may be pardoned the rather hackneyed expression if I say that the person who imagines he has cast off the thrall of atomism by differential equations does not see the woods for the trees. And as for explaining differential equations by complicated geometrical or other physical notions, this would be an out-and-out presentation of the subject in the light of an analogy, instead of a direct description. In point of fact, we are unable to distinguish the adjacent parts. But a constructive image in which we were unable to distinguish the adjacent parts at the very beginning, would be indistinct; we could not perform the prescribed numerical operations with it.

In declaring, therefore, a differential equation, or a formula containing definite integrals, to be the most appropriate representation of the facts, I am surrendering myself to an illusion if I fancy that in so doing I have eliminated the atomistic conception from my mental representation, without which conception the notion of limit is meaningless; I am rather merely making the additional assertion that to whatever degree of refinement our means of observation may be pushed, there will be no noticeable differences ever discovered between the limiting values and the facts.

Does not that constructive representation which assumes a very large but finite number of elementary particles, transcend the facts far less than such an implication? Is not the situation completely reversed? Whereas formerly the assumption of atoms of a definite size was regarded as a crude and arbitrary theory passing unnecessarily beyond the facts, now that assumption appears the more natural one, and the assertion that there will never be differences discovered between the facts and the limits for the reason that such have never hitherto been discovered (perhaps not once in all cases), adds to the representation something new and undemonstrated. Why the representation is made clearer, or more probable, by this adscititious and supplemental assertion is incomprehensible to me.[1] Atomism appears to be inseparable from the notion of a continuum. Laplace, Poisson, Cauchy, and the rest evidently proceeded from atomistic conceptions for the reason that inquirers were at that time more perfectly aware of the fact that differential equations were only symbols for atomistic conceptions, and that they therefore felt more vividly the need of simplifying these conceptions.

Like the equation for the conduction of heat, the fundamental equation of elasticity can in general be solved only by first conceiving a finite number of elementary particles which act on one another according to certain simple laws, and by subsequently seeking the limit after augmenting the number. This limit is thus again the real definition of the fundamental equations, and that representation of the facts which assumes at the outset a large but finite number of things, again appears the simpler.

We may obtain thus, by investing the atoms in question with as many properties as are necessary to describe a small province of facts in the simplest manner, a special atomistic system for every province of facts,[2] which, whilst it is in my opinion as little

[1] Our sensations of sight correspond to the excitations of a finite number of nerve-fibers, and are thence doubtless better represented by a mosaic than by a continuous surface. The same holds true of the other sensations. Is it not more probable, therefore, that the models representing complexes of sensations are more fitly composed of discrete parts?

[2] If Hertz is consistent, he can assign to his statement that a definite system of differential equations constitutes his theory of electro-magnetic phenomena, no other meaning than that he thinks these phenomena under the image of two heterogeneous intellectual entities filling space and having both the characteristics of vectors, the variation of which in time, here with reference to intensity and direc-

a direct description as that which is commonly called atomism, is yet a representation of the facts as devoid as can be of adscititious elements.

Now, phenomenology endeavors to combine all these atomistic constructs without previously simplifying them, in order to de-

---

tion, is, as in the conduction of heat, determined solely by the immediate environment, but is dependent on the same in a more complicated though readily assignable manner. We have given in this an atomistic theory of electro-magnetism, which contains as few adscititious elements as possible. The demand that this domain shall be explained mechanically, is but another expression of the need of simplifying the construct representing it and of rendering it homogeneous with the constructs representing the facts of the other departments. It is to this lack of homogeneity, which does not impress one on merely comparing the differential equations superficially, and to the probability that simpler constructs exist, that people doubtless wish to give expression when they say we do not know what electricity is.

The ordinary equations of the theory of elasticity, when additionally involving the displacements u, v, w, and the elastic forces X_x, X_y . . . ., represent (if the meaning of the notion of limits be recalled) pretty complicated rules for the variation of the coördinates $x+u$, $y+v$, $z+w$ of ordinary points, and the simultaneous variation of vector-atoms. Even the equations which arise after elimination of the elastic forces, are in need of some reduction before they yield the customary atomistic construct of elastic phenomena. To obtain the latter, certain combinations and decompositions of the equations or of the constructs representing them have to be performed, just as in mechanics forces have to be composed and decomposed in order to obtain as simple a description as possible.

Likewise, differential coefficients with respect to time impliedly require that in our mental construct of nature time shall be conceived at the outset as decomposed into very small, finite portions (time-atoms). If I relinquished, therefore, as unsubstantiated by experience, the view that deviations can never be discovered from the limit to which the construct approaches as the time-atoms are diminished in magnitude, I should be forced to the contention that even the laws of the mechanics of a material point are only approximately correct. Just to give some slight idea of the variety of the constructs at our command, I shall sketch forth here one special example of the representations in question.

Conceive in space (or better, in a tridimensional manifold) a large number of spheres in contact. The arrangement of these spheres varies, in accordance with a law A still to be determined, very slightly from one time-atom to another, but nevertheless by a finite amount. The variously formed interstices between the spheres take the place of the atoms of the old construct, and the law A is to be so chosen that the variations of the interstices in time shall furnish a theory of the universe. If it were possible to discover such a construct, which exhibited more homogeneity than the usual atomism, its title would be established by this fact alone. The conception of atoms as material points, and of forces as functions of their distance, is thus in all likelihood a provisional view, which for lack of a better must still be retained.

scribe the facts as they actually are, that is, to bring into conformity with them all the conceptions contained in these various atomistic theories. But, forasmuch as they introduce a multitude of notions gathered from many narrow provinces of facts, not to speak of a host of differential equations of which each, despite varied analogies, furnishes its own quota of idiosyncracy, it is to be antecedently expected that the description will assume an extremely complicated form. In point of fact, it appears that enormously complicated equations, utterly lacking in synoptic features, are requisite even when phenomenology attempts to describe the concatenation of some very few provinces of phenomena where the processes are virtually stationary (say, elastic deformation accompanied with heating and magnetisation, etc.). Furthermore, hypothetical features, features transcending the facts, must perforce be introduced when it is attempted to describe, for example, the dissociation of gases with Gibbs, or that of electrolysis with Planck.

Then there is the additional consideration that all the concepts of phenomenology have been derived from approximately stationary phenomena, and will not hold for turbulent motions. For example, we may define the temperature of a quiescent body by means of a thermometer inserted in it. If the body move as a whole, the thermometer may move with it. But if every element of volume of the body have a different motion, the definition is rendered nugatory, and it is then probable, or rather possible, that the different forms of energy (as to what is heat and what visible motion) can no longer be sharply distinguished.

If this be borne in mind, and account be also taken of the complications which the phenomenological equations assume even in the few instances where the concatenation of several provinces of phenomena is to be described, we shall obtain some conception of the difficulties encompassing the description by this method of turbulent phenomena in general, especially such as are accompanied with chemical changes, that is to say, of describing them without previously harmonising by arbitrary simplifications the various atomistic theories appertaining to the several provinces.

A special phenomenology, which I shall denominate the *energic*, in its widest acceptation seeks to bring some sort of harmony into the atomistic theories appertaining to the different fields, by ascertaining what features are common to them. Two classes of such features are known. To the first class belong certain general theorems, like the conservation of energy, the principle of entropy, etc., general integral theorems as I might term them, which hold good in all provinces. The second class consists of analogies which pervade a great variety of provinces. These are largely nothing more than a similarity of form which certain equations always assume at a certain stage of approximation, whilst in more minute details the analogies would frequently appear to fail.[1] Yet, in spite of the enormous importance of the integral theorems, which is owing to their universal validity and the high degree of certainty that flows therefrom, and in spite of the importance of the analogies, which is due to the advantages in computation and the new points of view they offer, both integral theorems and analogies furnish a small fraction only of the entire range and compass of the facts. Inquirers were compelled, therefore, in order even to describe exactly single domains, to introduce additionally so many special constructs (the natural history of the domain), that they could not obtain, as I believe I have shown elsewhere, a single unequivocal and comprehensive description of any one department of stationary phenomena by this method, let alone a synoptic view of all, or, worst of all, of turbulent, phenomena. The question whether comprehensive representations of nature will ever be obtained along this line, has therefore a purely academic value.

To reach this latter goal, the current atomistic theory is seeking to bring the fundamental features of the various phenomenological atomisms into better conformity with one another, by arbitrarily extending and curtailing the properties of the atoms required in each province, and thus rendering them suitable for the simul-

[1] Approximate proportionality of slight changes of the function to changes of the argument, the remaining of the first and second differential quotients with approximately constant coefficients, linearity with respect to small quantities and consequently superposition. So, too, the analogies in the behavior of the different forms of energy appear in part to have such purely algebraical sources.

taneous description of several provinces.[1] It decomposes, so to speak, the properties of the atoms required for a single province, into components (see p. 70, footnote), rendering them applicable to several provinces. Naturally, as in the case of the decomposition of forces into components, this is not possible without some violence and transcension of the facts.[2] In compensation, however, it enjoys the advantage of offering a simple and synoptic representation of a far greater mass of facts.

Whereas phenomenology stands in need of distinct and largely unallied constructs even for the mechanics of the movements of the centre of gravity and of rigid bodies, for elasticity, hydrodynamics, etc., the present atomic theory forms a perfectly apposite representation of all mechanical phenomena, and in view of the compactness of this province it is hardly to be expected that phenomena will yet be discovered that do not fit into the frame of the system. The atomic theory also embraces the phenomena of heat.

---

[1] I do not mean to say by this that the phenomenologic equations have always preceded the advances made by modern atomism in point of time. As a fact, most of the phenomenologic equations were reached by the consideration of specialised forms of atoms taken from another province (mechanics), and did not acquire until subsequently, by being stripped of these considerations, the character of phenomenologic equations. This circumstance cannot surprise us when it is known that these equations in reality always involve the supposition of atomistic constructs, and it can therefore only tell in favor of atomism.

[2] One property of this sort arbitrarily introduced into our conception of atoms is their unchangeability. The criticism that we are concerned here with an unwarranted generalisation of the merely transitory unchangeableness of solid bodies, would be justified if we attempted, as was formerly done, to prove the atoms unalterable, *à priori*. We have introduced this property into our conception in order to render it capable of representing in aggregate as many phenomena as possible, just as we introduce into the equation for the conduction of heat the first differential coefficient with respect to the coördinates, in order that it may conform to the facts. We are ready to relinquish unalterability in all cases where another assumption would more fitly represent the facts. In reality, the vector-atoms of the ether mentioned in the footnote on page 70 would not be unalterable with the time.

The unalterability of the atoms belongs thus to those conceptions which have proved themselves of great service, although the metaphysical considerations by which they have been reached cannot stand before unbiassed criticism. But it is precisely owing to this serviceability that we are forced to grant at least the probability that radiant energy so called admits of representations by means of constructs similar to matter (that the luminous ether is a substance).

The fact that this latter circumstance cannot be definitely demonstrated, is due to the difficulty of calculating the molecular motions. In any event, all essential facts may be rediscovered in our theory, which has also proved itself extremely useful in the presentation of crystallographic facts, of the constant proportions of masses in chemical combinations,[1] of chemical isomerisms, of the relations between rotation of the plane of polarisation and chemical constitution, etc., etc.

Then, again, the atomic theory is susceptible of still greater development. We are at liberty to conceive among the atoms individuals endowed with any sort of properties, as, for instance, the vector-atoms, which, as we saw in the footnote on page 70, yielded for the time being the simplest description of electromagnetic phenomena.[2]

As to the turbulent phenomena which are as yet entirely inaccessible to phenomenology, the attitude of atomism toward these is admittedly one of definite preconceptions; yet in compensation for this bias, it offers valuable clues for the most likely modes of presentation of the phenomena, nay in many cases even enables us to predict them. Thus the theory of gases is able to predict the course of all mechanical and thermal phenomena in gases, even in

---

[1] No chemical combination is produced instantaneously; each is propagated in space with a finite, though great, velocity. Applying therefore the analysis of the notion of continuity above presented, Mach's and Ostwald's theory of chemism will be found to assert that in each instance a elementary particles of one substance and b of another disappear, while c of a new substance make their appearance. The difference between this and the current conceptions of chemistry is manifestly not very essential. Nothing material would be altered if the limit as usually found should be made to represent the facts.

[2] If by a mechanical explanation of nature be understood one reposing on the present laws of mechanics, it must be declared to be altogether problematical whether the atomic theory of the future will be a mechanical explanation in the sense defined. Only in so far as it shall be under the constant necessity of supplying the simplest possible laws for the time-variation of numerous individual entities distributed in a manifold of probably three dimensions, may it be termed, at least in the traditional sense, a mechanical theory. For example, should no simpler description of electromagnetic phenomena be actually forthcoming, then the vector-atoms mentioned in the text would have to be retained. Whether the laws according to which these vary with the time are to be termed mechanical or not, depends wholly on our own taste.

the case of turbulent motions, and so affords indications as to how temperature, pressure, etc., are to be defined for these phenomena. Now this is precisely the chief end and purpose of science so to shape the constructs representing one group of facts that from them may be predicted the behavior of other similar facts. It is understood, of course, that the prediction is afterwards to be checked by experiment. Probably it will be only in part confirmed. There is then hope of so modifying and supplementing the constructs that they will conform also to the new facts; which is tantamount to making new discoveries with respect to the constitution of the atoms.

Unquestionably, the demand is justified that no more adscititious elements (and even these are to be as general as possible) shall be added to the construct than are absolutely necessary to describe extensive provinces of phenomena, that we shall always be ready to modify the construct, nay, even to bear in mind the possibility that it might be advisable to substitute an entirely new and utterly different one in its place. And for the very reason that the construction of the new construct would have to be effected on the basis of the old special phenomenologic constructs that have remained intact, therefore these also are to be carefully cultivated by the side of atomism.

In conclusion, I am inclined to go so far even as to assert that it is of the very nature of a construct, that it should introduce certain arbitrary features in its attempts at representation, and that strictly speaking one transcends experience every time one infers from a construct conforming to a given set of facts a single new additional fact. It is mathematically certain that, in order to represent all the facts, we should not be permitted to put in place of Fourier's equation for the conduction of heat an entirely different equation, identical with Fourier's only in regard to facts previously observed, with the result that on the first new observation made we should have to alter our construct utterly and with it our entire conception of the interchange of heat among the smallest particles. All bodies hitherto investigated, for instance, might accidentally exhibit just that particular group of regularities with the absence of which Fourier's equation would be rendered nugatory.

Just as Fourier transferred to the smallest particles (atoms) the law of specific heat and the fact that the interchange of heat between bodies in contact is proportional to the difference of the temperatures, so the theory of gases transfers to these particles the general laws of mechanics and the fact that bodies in close contact displace one another, but at some distance do not do so. With these smallest particles we cannot, as we see, dispense, when considering bodies of any size.

So, too, the assumption that the same atoms suffice for the description of both the liquid and gaseous aggregate states, appears to me, in view of the continuity of these two states, to be well founded, and conforms admirably with the requirement of simplicity in our description of nature. Granting the reasonableness of the two foregoing assumptions, we cannot escape the conclusion that the atoms are impressed with relative motions invisible to the eye, which absorb visible kinetic energy and the perceptibility of which by certain nerves is certainly not improbable (special mechanical theory of heat); and, further, that in highly rarefied bodies they usually describe nearly rectilinear paths (kinetic theory of gases). The construct by which we represent mechanical phenomena would only be made more complicated, if not self-contradictory, by abandonment of these conclusions. The additional assumption that the molecular motions do not cease whilst the induced visible motions are transformed gradually into molecular motions, also conforms perfectly to established mechanical laws.

All the consequences of the special mechanical theory of heat, howsoever disparate be the domains to which they pertain, have been confirmed by experience; indeed, I may say that they have been attuned in their minutest details to the rhythmical pulsations of nature.[1]

---

[1] Of many instances I will mention but the explanation of the three states of aggregation and their transitions from one to another; further, the agreement of the concept of entropy with the mathematical expression of the probability or disorder of a motion. The assertion that a moving system of very many minute bodies

Fourier's assumptions with regard to the conduction of heat are of course of so extremely simple a nature, and the facts calculable from them conformed so perfectly to verified observations, that the assertion of his assumption and equations not being absolutely trustworthy, as first approximations, appears rather of the nature of casuistry. But I do not find it at all strange that the requirements of the situation should be satisfied by such extremely simple and plausible assumptions, for the province is very arbitrarily limited; or that subsequently facts greatly different from the kind already verified should be derivable from them.

If there should ever be forthcoming so comprehensive a theory as the present atomism, one likewise which reposed on so distinct and unassailable a foundation as Fourier's theory of the conduction of heat, then indeed this were an ideal. Whether this result is to be attained by the subsequent combination of the previously unsimplified phenomenologic equations, or whether it is to be realised by the eventual asymptotic approach of current atomic conceptions to the plane of evidence of Fourier's theory by continual adaptation and constant confirmation by experience,—this appears to me to be as yet entirely undetermined.[1] For, though the existing ob-

---

tends, apart from unobservably few exceptions, towards a condition for which an assignable mathematical expression measuring the probability of the state is a maximum, appears to me to be superior, after all, to the almost tautological assertion that the system tends towards the most stable condition. As for the rest, Mach is correct in his conjecture (*loc. cit.*, p. 381) that at the time I composed a popular lecture on this topic I was unacquainted with the works cited by him as treating of the tendency toward stability, all of which but one were published years after my lecture, and all of them after the original papers of which my lecture was but a popular presentation.

If the principle of energy were the only support of the special mechanical theory of heat, and the explanation of that principle its only purpose, then the theory would be superfluous after the principle had been fully explained. We saw, however, that there are many other reasons speaking in favor of it, and that it furnishes a constructive representation of many other phenomena.

The theory of the electric fluids was from the outset artificial after an entirely different manner, and had been recognised by many inquirers from the start as provisional.

[1] Considerable remoulding and adapting (cf. Mach, *loc. cit.*, p. 380) are still to be done in the case of both theories. Fourier's equations for the conduction of heat

$$\frac{du}{dt} = k\,\Delta u$$

servations going to show that molecular motions in liquids and gases may be observed directly, be not regarded as conclusive, yet the possibility of future conclusive observations cannot be gainsaid. It accordingly seems to me utterly beside the mark to assert positively that such constructive representations of nature as the special mechanical theory of heat or the atomic theory of chemism and crystallisation shall ever vanish perforce from science. The sole question that can be raised is whether the precipitate zeal exhibited in the cultivation of such constructs, or the excessive caution that recommends total abstinence from them, be of greater disadvantage to science.

The extent to which physics, chemistry, and crystallography have profited from atomism in point of palpableness and perspicuity, is well known; that it has also been a clog, and hence in many cases has appeared as superfluous ballast, especially at the time when it far less perfectly conformed to phenomena than at present, and was more impregnated with metaphysical considerations, shall not be denied. Synoptic control will not be lost, nor certitude in the least impaired, if we rigorously sunder the phenomenology of the results that have been established with the greatest possible certainty from the atomic hypotheses that serve the purposes of unification, and continue to cultivate each with equal zeal, as alike indispensable. But we should not lay exclusive emphasis on the advantages of phenomenology and assert that this latter doctrine is one day to supplant the atomism now current.

---

is, if k be constant, unquestionably wrong. That with k variable it should necessarily take the form

$$h\frac{du}{dt}=\frac{d}{dx}\left(k\frac{du}{dx}\right)+\frac{d}{dy}\left(k\frac{du}{dy}\right)+\frac{d}{dz}\left(k\frac{du}{dz}\right)$$

has hardly been sufficiently confirmed by experience. And it does not represent at all the effect on the distribution of heat of the compressions and dilatations inseparably incident to non-stationary conduction, or the direct action of the heated volume-elements on other more remote elements in a diathermanous body by thermal radiation (and who knows but all bodies are diathermanous for certain rays transmitting energy and necessarily therefore also heat). If it be said these phenomena are not part of the process of pure thermal conduction, it may be replied that a pure process of this character is then nothing but a metaphysical hypostatised concept.

Even though the possibility exist of combining the constructs of phenomenology into a comprehensive theory differently from the manner pursued by atomism, still the following points must be admitted:

(1) This theory can never be a mere inventory, in the sense of having a definite symbol attached to each individual fact: it would in such a case be as difficult to command the situation as it would be to have experience of all the facts individually and severally. Like the present atomism, therefore, it can be nothing more than a body of directions for the construction of a theory and view of the world.

(2) If we will but free ourselves from illusion regarding the meaning of a differential equation, or of any continuously extended magnitude generally, we cannot fail to see that the phenomenologic view of the world is itself necessarily an atomistic view, viz., a direction to think by certain definite laws the time-changes of an immensely large number of things arranged in a manifold of probably three dimensions. The things in question may of course be either homogeneous or heterogeneous, variable or invariable. By assuming a large finite number or the limit of a continuously increasing number, our construct could correctly represent all the phenomena.

Assuming the possibility of an all-embracing theory of the world, every feature of which has the same evidence as Fourier's theory of the conduction of heat, it is still a question whether it can be more easily reached by the phenomenologic method or through the perfection and empiric confirmation of the constructs of atomism. It would be even permissible to assume that several representations of the world were possible, all of which possessed the ideal traits.

L. Boltzmann.

Leipsic.

THOMSON, JOSEPH JOHN. (b. Cheetham Hill, near Manchester, England, 18 December 1856; d. Cambridge, England, 20 August 1940)

Thomson took an engineering degree at Owens College, Manchester and studied at Trinity College, Cambridge University from 1876 to 1880, where he became a fellow in 1881. Despite his youth, he succeeded Lord Rayleigh in 1884 as Cavendish Professor of Physics, directing the Cavendish Laboratory and teaching physics at Trinity College. He also accepted a professorship of Natural Philosophy at the Royal Institution in 1905. In

1918, the year after he accepted the mastership of Trinity, Thomson resigned the Cavendish chair in favor of Rutherford. During Thomson's tenure at the Cavendish Laboratory, he received the 1906 Nobel Prize in physics and helped train a generation of Nobel Prize winners.

Thomson's first notable paper, which won the 1882 Adams Prize, was an essay on vortex rings, based on Kelvin's 1867 theory of the vortex atom. Thomson's early work in electromagnetic theory included studies of the applicability of Lagrangian functions, the development of the idea that electricity flows in metals much as electricity flows in electrolytes, and the discovery of the electromagnetic mass possessed by moving electrified bodies as a result of their charge. Once appointed to the Cavendish, he began a program of researches on gas discharge, using an electrolytic analogy for interpretation of some of his results. In 1893 he published *Recent Researches in Electricity and Magnetism*, which was widely regarded as a sequel to Maxwell's 1873 *Treatise on Electricity and Magnetism.*

The research that established Thomson's international reputation was his work done in the 1890s on cathode rays, which some physicists (mostly Germans) then regarded as an electromagnetic radiation, and others (mostly English) took to be charged particles. In 1897 J. J. Thomson published the "Cathode Rays" paper in which he reported the successful measurement of the charge-to-mass ratio of negatively charged cathode "rays" and the result that this ratio was at least one thousand times smaller than the ratio for ions in electrolysis. Demonstrating that the ratio was independent of the gas and that the charge was identical to that for negatively charged ions, he announced, in a Friday evening lecture at the Royal Institution, the discovery of a particle that is smaller than the hydrogen atom and is a universal constituent of matter. He immediately applied his result on the "corpuscle" (or "electron") to the elucidation of the structure of atoms and their chemical reactions, suggesting, in 1904, an electron model for the hydrogen atom. His later work on the deflection of "positive rays" resulted in methods for determining the atomic weights of gas particles and led to the development, by Francis Aston, of the mass spectroscope.

THE

LONDON, EDINBURGH, AND DUBLIN

PHILOSOPHICAL MAGAZINE

AND

JOURNAL OF SCIENCE.

[FIFTH SERIES.]

OCTOBER 1897.

XL. *Cathode Rays.* *By* J. J. Thomson, *M.A.*, *F.R.S.*, *Cavendish Professor of Experimental Physics, Cambridge**.

THE experiments† discussed in this paper were undertaken in the hope of gaining some information as to the nature of the Cathode Rays. The most diverse opinions are held as to these rays; according to the almost unanimous opinion of German physicists they are due to some process in the æther to which—inasmuch as in a uniform magnetic field their course is circular and not rectilinear—no phenomenon hitherto observed is analogous: another view of these rays is that, so far from being wholly ætherial, they are in fact wholly material, and that they mark the paths of particles of matter charged with negative electricity. It would seem at first sight that it ought not to be difficult to discriminate between views so different, yet experience shows that this is not the case, as amongst the physicists who have most deeply studied the subject can be found supporters of either theory.

The electrified-particle theory has for purposes of research a great advantage over the ætherial theory, since it is definite and its consequences can be predicted; with the ætherial theory it is impossible to predict what will happen under any given circumstances, as on this theory we are dealing with hitherto

* Communicated by the Author.

† Some of these experiments have already been described in a paper read before the Cambridge Philosophical Society (Proceedings, vol. ix. 1897), and in a Friday Evening Discourse at the Royal Institution ('Electrician,' May 21, 1897).

unobserved phenomena in the æther, of whose laws we are ignorant.

The following experiments were made to test some of the consequences of the electrified-particle theory.

Charge carried by the Cathode Rays.

If these rays are negatively electrified particles, then when they enter an enclosure they ought to carry into it a charge of negative electricity. This has been proved to be the case by Perrin, who placed in front of a plane cathode two coaxial metallic cylinders which were insulated from each other: the outer of these cylinders was connected with the earth, the inner with a gold-leaf electroscope. These cylinders were closed except for two small holes, one in each cylinder, placed so that the cathode rays could pass through them into the inside of the inner cylinder. Perrin found that when the rays passed into the inner cylinder the electroscope received a charge of negative electricity, while no charge went to the electroscope when the rays were deflected by a magnet so as no longer to pass through the hole.

This experiment proves that something charged with negative electricity is shot off from the cathode, travelling at right angles to it, and that this something is deflected by a magnet; it is open, however, to the objection that it does not prove that the cause of the electrification in the electroscope has anything to do with the cathode rays. Now the supporters of the ætherial theory do not deny that electrified particles are shot off from the cathode; they deny, however, that these charged particles have any more to do with the cathode rays than a rifle-ball has with the flash when a rifle is fired. I have therefore repeated Perrin's experiment in a form which is not open to this objection. The arrangement used was as follows:—Two coaxial cylinders (fig. 1) with slits in them are placed in a bulb connected with the discharge-tube; the cathode rays from the cathode A pass into the bulb through a slit in a metal plug fitted into the neck of the tube; this plug is connected with the anode and is put to earth. The cathode rays thus do not fall upon the cylinders unless they are deflected by a magnet. The outer cylinder is connected with the earth, the inner with the electrometer. When the cathode rays (whose path was traced by the phosphorescence on the glass) did not fall on the slit, the electrical charge sent to the electrometer when the induction-coil producing the rays was set in action was small and irregular; when, however, the rays were bent by a magnet so as to fall on the slit there was a large charge of negative electricity sent to the electrometer. I was surprised at the magnitude of the charge; on some occasions enough negative electricity went through the narrow slit into the inner cylinder in one second to alter the potential of a capacity of 1·5 microfarads by 20 volts. If the rays were so

Fig. 1.

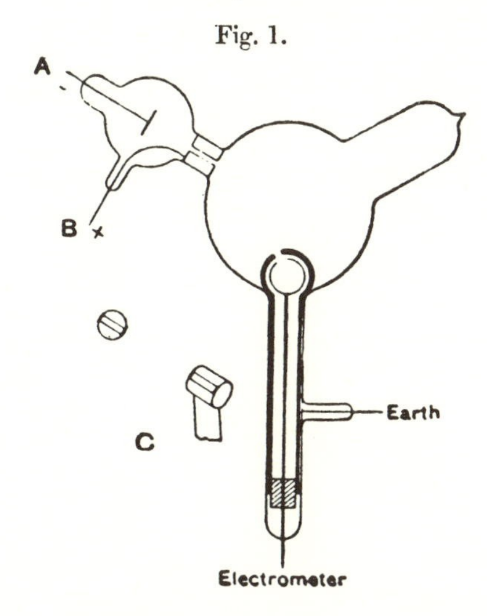

much bent by the magnet that they overshot the slits in the cylinder, the charge passing into the cylinder fell again to a very small fraction of its value when the aim was true. Thus this experiment shows that however we twist and deflect the cathode rays by magnetic forces, the negative electrification follows the same path as the rays, and that this negative electrification is indissolubly connected with the cathode rays.

When the rays are turned by the magnet so as to pass through the slit into the inner cylinder, the deflexion of the electrometer connected with this cylinder increases up to a certain value, and then remains stationary although the rays continue to pour into the cylinder. This is due to the fact that the gas in the bulb becomes a conductor of electricity when the cathode rays pass through it, and thus, though the inner cylinder is perfectly insulated when the rays are not passing, yet as soon as the rays pass through the bulb the air between the inner cylinder and the outer one becomes a conductor, and the electricity escapes from the inner cylinder to the earth. Thus the charge within the inner cylinder does not go on continually increasing; the cylinder settles down into a state of equilibrium in which the rate at which it gains negative electricity from the rays is equal to the rate at which it loses it by conduction through the air. If the inner cylinder has initially a positive charge it rapidly loses that charge and acquires a negative one; while if the initial charge is a negative one, the cylinder will leak if the initial negative potential is numerically greater than the equilibrium value.

Deflexion of the Cathode Rays by an Electrostatic Field.

An objection very generally urged against the view that the cathode rays are negatively electrified particles, is that hitherto no deflexion of the rays has been observed under a small electrostatic force, and though the rays are deflected when they pass near electrodes connected with sources of large differences of potential, such as induction-coils or electrical machines, the deflexion in this case is regarded by the supporters of the ætherial theory as due to the discharge passing between the electrodes, and not primarily to the electrostatic field. Hertz made the rays travel between two parallel plates of metal placed inside the discharge-tube, but found that they were not deflected when the plates were connected with a battery of storage-cells; on repeating this experiment I at first got the same result, but subsequent experiments showed that the absence of deflexion is due to the conductivity conferred on the rarefied gas by the cathode rays. On measuring this conductivity it was found that it diminished very rapidly as the exhaustion increased; it seemed then that on trying Hertz's experiment at very high exhaustions there might be a chance of detecting the deflexion of the cathode rays by an electrostatic force.

The apparatus used is represented in fig. 2.

Fig. 2.

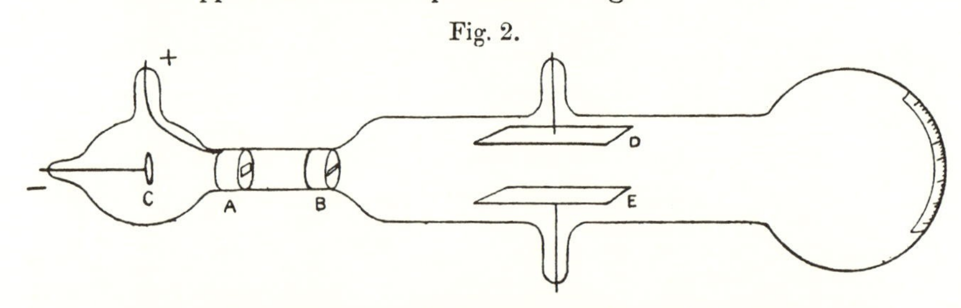

The rays from the cathode C pass through a slit in the anode A, which is a metal plug fitting tightly into the tube and connected with the earth; after passing through a second slit in another earth-connected metal plug B, they travel between two parallel aluminium plates about 5 cm. long by 2 broad and at a distance of 1·5 cm. apart; they then fall on the end of the tube and produce a narrow well-defined phosphorescent patch. A scale pasted on the outside of the tube serves to measure the deflexion of this patch.

At high exhaustions the rays were deflected when the two aluminium plates were connected with the terminals of a battery of small storage-cells; the rays were depressed when the upper plate was connected with the negative pole of the battery, the lower with the positive, and raised when the upper plate was connected with the positive, the lower with the negative pole. The deflexion was proportional to the difference of potential between the plates, and I could detect the deflexion when the potential-difference was as small as two volts. It was only when the vacuum was a good one that the deflexion took place, but that the absence of deflexion is due to the conductivity of the medium is shown by what takes place when the vacuum has just arrived at the stage at which the deflexion begins. At this stage there is a deflexion of the rays when the plates are first connected with the terminals of the battery, but if this connexion is maintained the patch of phosphorescence gradually creeps back to its undeflected position. This is just what would happen if the space between the plates were a conductor, though a very bad one, for then the positive and negative ions between the plates would slowly diffuse, until the positive plate became coated with negative ions, the negative plate with positive ones; thus the electric intensity between the plates would vanish and the cathode rays be free from electrostatic force. Another illustration of this is afforded by what happens when the pressure is low enough to show the deflexion and a large difference of potential, say 200 volts, is established between the plates; under these circumstances there is a large deflexion of the cathode rays, but the medium under the large electromotive force breaks down every now and then and a bright discharge passes between the plates; when this occurs the phosphorescent patch produced by the cathode rays jumps back to its undeflected position. When the cathode rays are deflected by the electrostatic field, the phosphorescent band breaks up into several bright bands separated by comparatively dark spaces; the phenomena are exactly analogous to those observed by Birkeland when the cathode rays are deflected by a magnet, and called by him the magnetic spectrum.

A series of measurements of the deflexion of the rays by the electrostatic force under various circumstances will be found later on in the part of the paper which deals with the velocity of the rays and the ratio of the mass of the electrified particles to the charge carried by them. It may, however, be mentioned here that the deflexion gets smaller as the pressure diminishes, and when in consequence the potential-difference in the tube in the neighbourhood of the cathode increases.

Conductivity of a Gas through which Cathode Rays are passing.

The conductivity of the gas was investigated by means of the apparatus shown in fig. 2. The upper plate D was connected with one terminal of a battery of small storage-cells, the other terminal of which was connected with the earth; the other plate E was connected with one of the coatings of a condenser of one microfarad capacity, the other coating of which was to earth; one pair of quadrants of an electrometer was also connected with E, the other pair of quadrants being to earth. When the cathode rays are passing between the plates the two pairs of quadrants of the electrometer are first connected with each other, and then the connexion between them was broken. If the space between the plates were a non-conductor, the potential of the pair of quadrants not connected with the earth would remain zero and the needle of the electrometer would not move; if, however, the space between the plates were a conductor, then the potential of the lower plate would approach that of the upper, and the needle of the electrometer would be deflected. There is always a deflexion of the electrometer, showing that a current passes between the plates. The magnitude of the current depends very greatly upon the pressure of the gas; so much so, indeed, that it is difficult to obtain consistent readings in consequence of the changes which always occur in the pressure when the discharge passes through the tube.

We shall first take the case when the pressure is only just low enough to allow the phosphorescent patch to appear at the end of the tube; in this case the relation between the current between the plates and the initial difference of potential is represented by the curve shown in fig. 3. In this

Fig. 3.

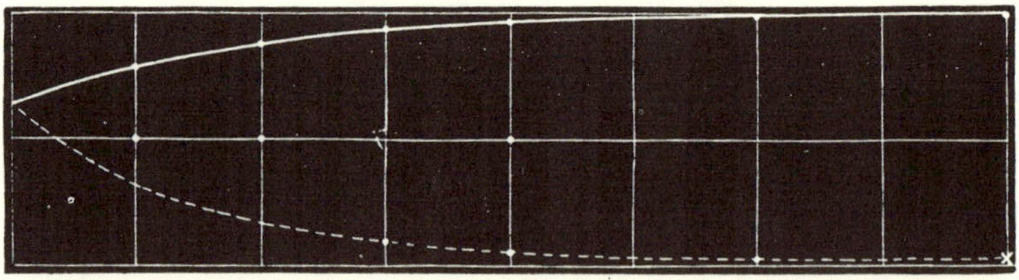

figure the abscissæ represent the initial difference of potential between the plates, each division representing two volts, and the ordinates the rise in potential of the lower plate in one minute each division again representing two volts. The quantity of electricity which has passed between the plates in one minute is the quantity required to raise 1 microfarad to the potential-difference shown by the curve. The upper and lower curve relates to the case when the upper plate is connected with the negative and positive pole respectively of the battery.

Even when there is no initial difference of potential between the plates the lower plate acquires a negative charge from the impact on it of some of the cathode rays.

We see from the curve that the current between the plates soon reaches a value where it is only slightly affected by an increase in the potential-difference between the plates; this is a feature common to conduction through gases traversed by Röntgen rays, by uranium rays, by ultra-violet light, and, as we now see, by cathode rays. The rate of leak is not greatly different whether the upper plate be initially positively or negatively electrified.

The current between the plates only lasts for a short time; it ceases long before the potential of the lower plate approaches that of the upper. Thus, for example, when the potential of the upper plate was about 400 volts above that of the earth, the potential of the lower plate never rose above 6 volts: similarly, if the upper plate were connected with the negative pole of the battery, the fall in potential of the lower plate was very small in comparison with the potential-difference between the upper plate and the earth.

These results are what we should expect if the gas between the plates and the plug B (fig. 2) were a very much better conductor than the gas between the plates, for the lower plate will be in a steady state when the current coming to it from the upper plate is equal to the current going from it to the plug: now if the conductivity of the gas between the plate and the plug is much greater than that between the plates, a small difference of potential between the lower plate and the plug will be consistent with a large potential-difference between the plates.

So far we have been considering the case when the pressure is as high as is consistent with the cathode rays reaching the end of the tube; we shall now go to the other extreme and consider the case when the pressure is as low as is consistent with the passage of a discharge through the bulb. In this case, when the plates are not connected with the battery we get a negative charge communicated to the lower plate, but only very slowly in comparison with the effect in the previous case. When the upper plate is connected with the negative pole of a battery, this current to the lower plate is only slightly increased even when the difference of potential is as much as 400 volts: a small potential-difference of about

20 volts seems slightly to decrease the rate of leak. Potential-differences much exceeding 400 volts cannot be used, as though the dielectric between the plates is able to sustain them for some little time, yet after a time an intensely bright arc flashes across between the plates and liberates so much gas as to spoil the vacuum. The lines in the spectrum of this glare are chiefly mercury lines; its passage leaves very peculiar markings on the aluminium plates.

If the upper plate was charged positively, then the negative charge communicated to the lower plate was diminished, and stopped when the potential-difference between the plates was about 20 volts; but at the lowest pressure, however great (up to 400 volts) the potential-difference, there was no leak of positive electricity to the lower plate at all comparable with the leak of negative electricity to this plate when the two plates were disconnected from the battery. In fact at this very low pressure all the facts are consistent with the view that the effects are due to the negatively electrified particles travelling along the cathode rays, the rest of the gas possessing little conductivity. Some experiments were made with a tube similar to that shown in fig. 2, with the exception that the second plug B was absent, so that a much greater number of cathode rays passed between the plates. When the upper plate was connected with the positive pole of the battery a luminous discharge with well-marked striations passed between the upper plate and the earth-connected plug through which the cathode rays were streaming; this occurred even though the potential-difference between the plate and the plug did not exceed 20 volts. Thus it seems that if we supply cathode rays from an external source to the cathode a small potential-difference is sufficient to produce the characteristic discharge through a gas.

Magnetic Deflexion of the Cathode Rays in Different Gases.

The deflexion of the cathode rays by the magnetic field was studied with the aid of the apparatus shown in fig. 4. The cathode was placed in a side-tube fastened on to a bell-jar; the opening between this tube and the bell-jar was closed by a metallic plug with a slit in it; this plug was connected with the earth and was used as the anode. The cathode rays passed through the slit in this plug into the bell-jar, passing in front of a vertical plate of glass ruled into small squares. The bell-jar was placed between two large parallel coils arranged as a Helmholtz galvanometer. The course of the rays was determined by taking photographs of the bell-jar when the cathode rays were passing through it; the divisions on the plate enabled the path of the rays to be determined. Under the action of the magnetic field the narrow beam of cathode rays spreads out into a broad fan-shaped luminosity in the gas. The luminosity in this fan is not uniformly

Fig. 4.

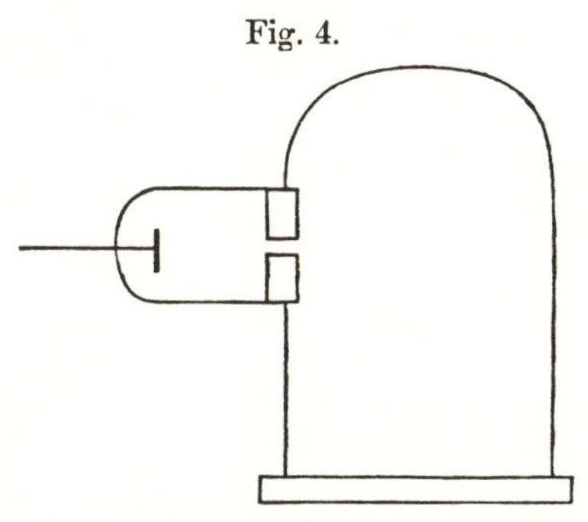

distributed, but is condensed along certain lines. The phosphorescence on the glass is also not uniformly distributed; it is much spread out, showing that the beam consists of rays which are not all deflected to the same extent by the magnet. The luminosity on the glass is crossed by bands along which the luminosity is very much greater than in the adjacent parts. These bright and dark bands are called by Birkeland, who first observed them, the magnetic spectrum. The brightest spots on the glass are by no means always the terminations of the brightest streaks of luminosity in the gas; in fact, in some cases a very bright spot on the glass is not connected with the cathode by any appreciable luminosity, though there may be plenty of luminosity in other parts of the gas. One very interesting point brought out by the photographs is that in a given magnetic field, and with a given mean potential-difference between the terminals, the path of the rays is independent of the nature of the gas. Photographs were taken of the discharge in hydrogen, air, carbonic acid, methyl iodide, *i. e.*, in gases whose densities range from 1 to 70, and yet, not only were the paths of the most deflected rays the same in all cases, but even the details, such as the distribution of the bright and dark spaces, were the same; in fact, the photographs could hardly be distinguished from each other. It is to be noted that the pressures were not the same; the pressures in the different gases were adjusted so that the mean potential-differences between the cathode and the anode were the same in all the gases. When the pressure of a gas is lowered, the potential-difference between the terminals increases, and the deflexion of the rays produced by a magnet diminishes, or at any rate the deflexion of the rays when the phosphorescence is a maximum diminishes. If an air-break is inserted an effect of the same kind is produced.

In the experiments with different gases, the pressures were as high as was consistent with the appearance of the phosphorescence on the glass, so as to ensure having as much as possible of the gas under consideration in the tube.

As the cathode rays carry a charge of negative electricity, are deflected by an electrostatic force as if they were negatively electrified, and are acted on by a magnetic force in just the way in which this force would act on a negatively electrified body moving along the path of these rays, I can see no escape from the conclusion that they are charges of negative electricity carried by particles of matter. The question next arises, What are these particles? are they atoms, or molecules, or matter in a still finer state of subdivision? To throw some light on this point, I have made a series of measurements of the ratio of the mass of these particles to the charge carried by it. To determine this quantity, I have used two independent methods. The first of these is as follows:—Suppose we consider a bundle of homogeneous cathode rays. Let m be the mass of each of the particles, e the charge carried by it. Let N be the number of particles passing across any section of the beam in a given time; then Q the quantity of electricity carried by these particles is given by the equation

$$Ne = Q.$$

We can measure Q if we receive the cathode rays in the inside of a vessel connected with an electrometer. When these rays strike against a solid body, the temperature of the body is raised; the kinetic energy of the moving particles being converted into heat; if we suppose that all this energy is converted into heat, then if we measure the increase in the temperature of a body of known thermal capacity caused by the impact of these rays, we can determine W, the kinetic energy of the particles, and if v is the velocity of the particles,

$$\tfrac{1}{2}Nmv^2 = W.$$

If ρ is the radius of curvature of the path of these rays in a uniform magnetic field H, then

$$\frac{mv}{e} = H\rho = I,$$

where I is written for $H\rho$ for the sake of brevity. From these equations we get

$$\frac{1}{2}\frac{m}{e}v^2=\frac{W}{Q}.$$

$$v=\frac{2W}{QI},$$

$$\frac{m}{e}=\frac{I^2Q}{2W}.$$

Thus, if we know the values of Q, W, and I, we can deduce the values of v and m/e.

To measure these quantities, I have used tubes of three different types. The first I tried is like that represented in fig. 2, except that the plates E and D are absent, and two coaxial cylinders are fastened to the end of the tube. The rays from the cathode C fall on the metal plug B, which is connected with the earth, and serves for the anode; a horizontal slit is cut in this plug. The cathode rays pass through this slit, and then strike against the two coaxial cylinders at the end of the tube; slits are cut in these cylinders, so that the cathode rays pass into the inside of the inner cylinder. The outer cylinder is connected with the earth, the inner cylinder, which is insulated from the outer one, is connected with an electrometer, the deflexion of which measures Q, the quantity of electricity brought into the inner cylinder by the rays. A thermo-electric couple is placed behind the slit in the inner cylinder; this couple is made of very thin strips of iron and copper fastened to very fine iron and copper wires. These wires passed through the cylinders, being insulated from them, and through the glass to the outside of the tube, where they were connected with a low-resistance galvanometer, the deflexion of which gave data for calculating the rise of temperature of the junction produced by the impact against it of the cathode rays. The strips of iron and copper were large enough to ensure that every cathode ray which entered the inner cylinder struck against the junction. In some of the tubes the strips of iron and copper were placed end to end, so that some of the rays struck against the iron, and others against the copper; in others, the strip of one metal was placed in front of the other; no difference, however, could be detected between the results got with these two arrangements. The strips of iron and copper were weighed, and the thermal capacity of the junction calculated. In one set of junctions this capacity was 5×10^{-3}, in another 3×10^{-3}. If we assume that the cathode rays which strike against the junction give their energy up to it, the deflexion of the galvanometer gives us W or $\frac{1}{2}Nmv^2$.

The value of I, *i. e.*, Hρ, where ρ is the curvature of the path of the rays in a magnetic field of strength H was found as follows :—The tube was fixed between two large circular coils placed parallel to each other, and separated by a distance equal to the radius of either; these coils produce a uniform magnetic field, the strength of which is got by measuring with an ammeter the strength of the current passing through them. The cathode rays are thus in a uniform field, so that their path is circular. Suppose that the rays, when deflected by a magnet, strike against the glass of the tube at E

Fig. 5.

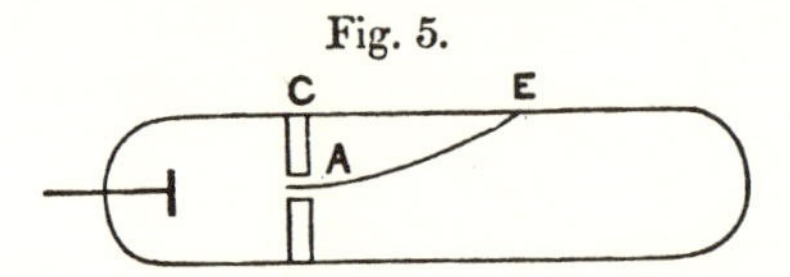

(fig. 5), then, if ρ is the radius of the circular path of the rays,

$$2\rho=\frac{\mathrm{CE}^2}{\mathrm{AC}}+\mathrm{AC}\ ;$$

thus, if we measure CE and AC we have the means of determining the radius of curvature of the path of the rays.

The determination of ρ is rendered to some extent uncertain, in consequence of the pencil of rays spreading out under the action of the magnetic field, so that the phosphorescent patch at E is several millimetres long ; thus values of ρ differing appreciably from each other will be got by taking E at different points of this phosphorescent patch. Part of this patch was, however, generally considerably brighter than the rest ; when this was the case, E was taken as the brightest point ; when such a point of maximum brightness did not exist, the middle of the patch was taken for E. The uncertainty in the value of ρ thus introduced amounted sometimes to about 20 per cent.; by this I mean that if we took E first at one extremity of the patch and then at the other, we should get values of ρ differing by this amount.

The measurement of Q, the quantity of electricity which enters the inner cylinder, is complicated by the cathode rays making the gas through which they pass a conductor, so that though the insulation of the inner cylinder was perfect when the rays were off, it was not so when they were passing through the space between the cylinders ; this caused some of the charge communicated to the inner cylinder to leak away so that the actual charge given to the cylinder by the cathode rays was larger than that indicated by the electrometer.

To make the error from this cause as small as possible, the inner cylinder was connected to the largest capacity available, 1·5 microfarad, and the rays were only kept on for a short time, about 1 or 2 seconds, so that the alteration in potential of the inner cylinder was not large, ranging in the various experiments from about ·5 to 5 volts. Another reason why it is necessary to limit the duration of the rays to as short a time as possible, is to avoid the correction for the loss of heat from the thermo-electric junction by conduction along the wires; the rise in temperature of the junction was of the order 2° C.; a series of experiments showed that with the same tube and the same gaseous pressure Q and W were proportional to each other when the rays were not kept on too long.

Tubes of this kind gave satisfactory results, the chief drawback being that sometimes in consequence of the charging up of the glass of the tube, a secondary discharge started from the cylinder to the walls of the tube, and the cylinders were surrounded by glow; when this glow appeared, the readings were very irregular; the glow could, however, be got rid of by pumping and letting the tube rest for some time. The results got with this tube are given in the Table under the heading Tube 1.

The second type of tube was like that used for photographing the path of the rays (fig. 4); double cylinders with a thermo-electric junction like those used in the previous tube were placed in the line of fire of the rays, the inside of the bell-jar was lined with copper gauze connected with the earth. This tube gave very satisfactory results; we were never troubled with any glow round the cylinders, and the readings were most concordant; the only drawback was that as some of the connexions had to be made with sealing-wax, it was not possible to get the highest exhaustions with this tube, so that the range of pressure for this tube is less than that for tube 1. The results got with this tube are given in the Table under the heading Tube 2.

The third type of tube was similar to the first, except that the openings in the two cylinders were made very much smaller; in this tube the slits in the cylinders were replaced by small holes, about 1·5 millim. in diameter. In consequence of the smallness of the openings, the magnitude of the effects was very much reduced; in order to get measurable results it was necessary to reduce the capacity of the condenser in connexion with the inner cylinder to ·15 microfarad, and to make the galvanometer exceedingly sensitive, as the rise in temperature of the thermo-electric junction was in these experiments only about ·5° C. on the average. The results obtained in this tube are given in the Table under the heading Tube 3.

The results of a series of measurements with these tubes are given in the following Table :—

Gas.	Value of W/Q.	I.	m/e	v.
Tube 1.				
Air	$4{\cdot}6 \times 10^{11}$	230	$\cdot57\times10^{-7}$	4×10^{9}
Air	$1{\cdot}8 \times 10^{12}$	350	$\cdot34\times10^{-7}$	1×10^{10}
Air	$6{\cdot}1 \times 10^{11}$	230	$\cdot43\times10^{-7}$	$5{\cdot}4\times10^{9}$
Air	$2{\cdot}5 \times 10^{12}$	400	$\cdot32\times10^{-7}$	$1{\cdot}2\times10^{10}$
Air	$5{\cdot}5 \times 10^{11}$	230	$\cdot48\times10^{-7}$	$4{\cdot}8\times10^{9}$
Air	1×10^{12}	285	$\cdot4 \times10^{-7}$	7×10^{9}
Air	1×10^{12}	285	$\cdot4 \times10^{-7}$	7×10^{9}
Hydrogen	6×10^{12}	205	$\cdot35\times10^{-7}$	6×10^{9}
Hydrogen	$2{\cdot}1 \times 10^{12}$	460	$\cdot5 \times10^{-7}$	$9{\cdot}2\times10^{9}$
Carbonic acid	$8{\cdot}4 \times 10^{11}$	260	$\cdot4 \times10^{-7}$	$7{\cdot}5\times10^{9}$
Carbonic acid	$1{\cdot}47\times10^{12}$	340	$\cdot4 \times10^{-7}$	$8{\cdot}5\times10^{9}$
Carbonic acid	$3{\cdot}0 \times 10^{12}$	480	$\cdot39\times10^{-7}$	$1{\cdot}3\times10^{10}$
Tube 2.				
Air	$2{\cdot}8 \times 10^{11}$	175	$\cdot53\times10^{-7}$	$3{\cdot}3\times10^{9}$
Air	$4{\cdot}4 \times 10^{11}$	195	$\cdot47\times10^{-7}$	$4{\cdot}1\times10^{9}$
Air	$3{\cdot}5 \times 10^{11}$	181	$\cdot47\times10^{-7}$	$3{\cdot}8\times10^{9}$
Hydrogen	$2{\cdot}8 \times 10^{11}$	175	$\cdot53\times10^{-7}$	$3{\cdot}3\times10^{9}$
Air	$2{\cdot}5 \times 10^{11}$	160	$\cdot51\times10^{-7}$	$3{\cdot}1\times10^{9}$
Carbonic acid	2×10^{11}	148	$\cdot54\times10^{-7}$	$2{\cdot}5\times10^{9}$
Air	$1{\cdot}8 \times 10^{11}$	151	$\cdot63\times10^{-7}$	$2{\cdot}3\times10^{9}$
Hydrogen	$2{\cdot}8 \times 10^{11}$	175	$\cdot53\times10^{-7}$	$3{\cdot}3\times10^{9}$
Hydrogen	$4{\cdot}4 \times 10^{11}$	201	$\cdot46\times10^{-7}$	$4{\cdot}4\times10^{9}$
Air	$2{\cdot}5 \times 10^{11}$	176	$\cdot61\times10^{-7}$	$2{\cdot}8\times10^{9}$
Air	$4{\cdot}2 \times 10^{11}$	200	$\cdot48\times10^{-7}$	$4{\cdot}1\times10^{9}$
Tube 3.				
Air	$2{\cdot}5 \times 10^{11}$	220	$\cdot9\times10^{-7}$	$2{\cdot}4\times10^{9}$
Air	$3{\cdot}5 \times 10^{11}$	225	$\cdot7\times10^{-7}$	$3{\cdot}2\times10^{9}$
Hydrogen	3×10^{11}	250	$1{\cdot}0\times10^{-7}$	$2{\cdot}5\times10^{9}$

It will be noticed that the value of m/e is considerably greater for Tube 3, where the opening is a small hole, than for Tubes 1 and 2, where the opening is a slit of much greater area. I am of opinion that the values of m/e got from Tubes 1 and 2 are too small, in consequence of the leakage from the inner cylinder to the outer by the gas being rendered a conductor by the passage of the cathode rays.

It will be seen from these tables that the value of m/e is independent of the nature of the gas. Thus, for the first tube the mean for air is $\cdot 40 \times 10^{-7}$, for hydrogen $\cdot 42 \times 10^{-7}$, and for carbonic acid gas $\cdot 4 \times 10^{-7}$; for the second tube the mean for air is $\cdot 52 \times 10^{-7}$, for hydrogen $\cdot 50 \times 10^{-7}$, and for carbonic acid gas $\cdot 54 \times 10^{-7}$.

Experiments were tried with electrodes made of iron instead of aluminium; this altered the appearance of the discharge and the value of v at the same pressure, the values of m/e were, however, the same in the two tubes; the effect produced by different metals on the appearance of the discharge will be described later on.

In all the preceding experiments, the cathode rays were first deflected from the cylinder by a magnet, and it was then found that there was no deflexion either of the electrometer or the galvanometer, so that the deflexions observed were entirely due to the cathode rays; when the glow mentioned previously surrounded the cylinders there was a deflexion of the electrometer even when the cathode rays were deflected from the cylinder.

Before proceeding to discuss the results of these measurements I shall describe another method of measuring the quantities m/e and v of an entirely different kind from the preceding; this method is based upon the deflexion of the cathode rays in an electrostatic field. If we measure the deflexion experienced by the rays when traversing a given length under a uniform electric intensity, and the deflexion of the rays when they traverse a given distance under a uniform magnetic field, we can find the values of m/e and v in the following way:—

Let the space passed over by the rays under a uniform electric intensity F be l, the time taken for the rays to traverse this space is l/v, the velocity in the direction of F is therefore

$$\frac{\mathrm{F}e}{m}\frac{l}{v},$$

so that θ, the angle through which the rays are deflected when they leave the electric field and enter a region free from electric force, is given by the equation

$$\theta=\frac{Fe}{m}\frac{l}{v^2}.$$

If, instead of the electric intensity, the rays are acted on by a magnetic force H at right angles to the rays, and extending across the distance l, the velocity at right angles to the original path of the rays is

$$\frac{Hev}{m}\frac{l}{v},$$

so that ϕ, the angle through which the rays are deflected when they leave the magnetic field, is given by the equation

$$\phi=\frac{He}{m}\frac{l}{v}.$$

From these equations we get

$$v=\frac{\phi}{\theta}\frac{F}{H}$$

and

$$\frac{m}{e}=\frac{H^2\theta \, . \, l}{F\phi^2}.$$

In the actual experiments H was adjusted so that $\phi=\theta$; in this case the equations become

$$v=\frac{F}{H},$$

$$\frac{m}{e}=\frac{H^2 l}{F\theta}.$$

The apparatus used to measure v and m/e by this means is that represented in fig. 2. The electric field was produced by connecting the two aluminium plates to the terminals of a battery of storage-cells. The phosphorescent patch at the end of the tube was deflected, and the deflexion measured by a scale pasted to the end of the tube. As it was necessary to darken the room to see the phosphorescent patch, a needle coated with luminous paint was placed so that by a screw it could be moved up and down the scale; this needle could be seen when the room was darkened, and it was moved until it coincided with the phosphorescent patch. Thus, when light was admitted, the deflexion of the phosphorescent patch could be measured.

The magnetic field was produced by placing outside the tube two coils whose diameter was equal to the length of the plates; the coils were placed so that they covered the space occupied by the plates, the distance between the coils was equal to the radius of either. The mean value of the magnetic force over the length l was determined in the following way: a narrow coil C whose length was l, connected with a ballistic galvanometer, was placed between the coils; the plane of the windings of C was parallel to the planes of the coils; the cross section of the coil was a rectangle 5 cm. by 1 cm. A given current was sent through the outer coils and the kick α of the galvanometer observed when this current was reversed. The coil C was then placed at the centre of two very large coils, so as to be in a field of uniform magnetic force: the current through the large coils was reversed and the kick β of the galvanometer again observed; by comparing α and β we can get the mean value of the magnetic force over a length l; this was found to be

$$60 \times \iota,$$

where ι is the current flowing through the coils.

A series of experiments was made to see if the electrostatic deflexion was proportional to the electric intensity between the plates; this was found to be the case. In the following experiments the current through the coils was adjusted so that the electrostatic deflexion was the same as the magnetic:—

Gas.	θ.	H.	F.	l.	m/e.	v.
Air	8/110	5·5	$1{\cdot}5\times10^{10}$	5	$1{\cdot}3\times10^{-7}$	$2{\cdot}8\times10^{9}$
Air	9·5/110	5·4	$1{\cdot}5\times10^{10}$	5	$1{\cdot}1\times10^{-7}$	$2{\cdot}8\times10^{9}$
Air	13/110	6·6	$1{\cdot}5\times10^{10}$	5	$1{\cdot}2\times10^{-7}$	$2{\cdot}3\times10^{9}$
Hydrogen	9/110	6·3	$1{\cdot}5\times10^{10}$	5	$1{\cdot}5\times10^{-7}$	$2{\cdot}5\times10^{9}$
Carbonic acid...	11/110	6·9	$1{\cdot}5\times10^{10}$	5	$1{\cdot}5\times10^{-7}$	$2{\cdot}2\times10^{9}$
Air	6/110	5	$1{\cdot}8\times10^{10}$	5	$1{\cdot}3\times10^{-7}$	$3{\cdot}6\times10^{9}$
Air	7/110	3·6	1×10^{10}	5	$1{\cdot}1\times10^{-7}$	$2{\cdot}8\times10^{9}$

The cathode in the first five experiments was aluminium, in the last two experiments it was made of platinum; in the last experiment Sir William Crookes's method of getting rid of the mercury vapour by inserting tubes of pounded sulphur, sulphur iodide, and copper filings between the bulb and the pump was adopted. In the calculation of m/e and v no allowance has been made for the magnetic force due to the coil in the region outside the plates; in this region the magnetic force will be in the opposite direction to that between the plates, and will tend to bend the cathode rays in the opposite direction: thus the effective value of H will be smaller than the value used in the equations, so that the values of m/e are larger, and those of v less than they would be if this correction were applied. This method of determining the values of m/e and v is much less laborious and probably more accurate than the former method; it cannot, however, be used over so wide a range of pressures.

From these determinations we see that the value of m/e is independent of the nature of the gas, and that its value 10^{-7} is very small compared with the value 10^{-4}, which is the smallest value of this quantity previously known, and which is the value for the hydrogen ion in electrolysis.

Thus for the carriers of the electricity in the cathode rays m/e is very small compared with its value in electrolysis. The smallness of m/e may be due to the smallness of m or the largeness of e, or to a combination of these two. That the carriers of the charges in the cathode rays are small compared with ordinary molecules is shown, I think, by Lenard's results as to the rate at which the brightness of the phosphorescence produced by these rays diminishes with the length of path travelled by the ray. If we regard this phosphorescence as due to the impact of the charged particles, the distance through which the rays must travel before the phosphorescence fades to a given fraction (say $1/e$, where $e=2{\cdot}71$) of its original intensity, will be some moderate multiple of the mean free path. Now Lenard found that this distance depends solely upon the density of the medium, and not upon its chemical nature or physical state. In air at atmospheric pressure the distance was about half a centimetre, and this must be comparable with the mean free path of the carriers through air at atmospheric pressure. But the mean free path of the molecules of air is a quantity of quite a different order. The carrier, then, must be small compared with ordinary molecules.

The two fundamental points about these carriers seem to me to be (1) that these carriers are the same whatever the gas through which the discharge passes, (2) that the mean free paths depend upon nothing but the density of the medium traversed by these rays.

It might be supposed that the independence of the mass of the carriers of the gas through which the discharge passes was due to the mass concerned being the quasi mass which a charged body possesses in virtue of the electric field set up in its neighbourhood ; moving the body involves the production of a varying electric field, and, therefore, of a certain amount of energy which is proportional to the square of the velocity. This causes the charged body to behave as if its mass were increased by a quantity, which for a charged sphere is $\frac{1}{5} e^2/\mu a$ ('Recent Researches in Electricity and Magnetism'), where e is the charge and a the radius of the sphere. If we assume that it is this mass which we are concerned with in the cathode rays, since m/e would vary as e/a, it affords no clue to the explanation of either of the properties (1 and 2) of these rays. This is not by any means the only objection to this hypothesis, which I only mention to show that it has not been overlooked.

The explanation which seems to me to account in the most simple and straightforward manner for the facts is founded on a view of the constitution of the chemical elements which has been favourably entertained by many chemists : this view is that the atoms of the different chemical elements are different aggregations of atoms of the same kind. In the form in which this hypothesis was enunciated by Prout, the atoms of the different elements were hydrogen atoms ; in this precise form the hypothesis is not tenable, but if we substitute for hydrogen some unknown primordial substance X, there is nothing known which is inconsistent with this hypothesis, which is one that has been recently supported by Sir Norman Lockyer for reasons derived from the study of the stellar spectra.

If, in the very intense electric field in the neighbourhood of the cathode, the molecules of the gas are dissociated and are split up, not into the ordinary chemical atoms, but into these primordial atoms, which we shall for brevity call corpuscles ; and if these corpuscles are charged with electricity and projected from the cathode by the electric field, they would behave exactly like the cathode rays. They would evidently give a value of m/e which is independent of the nature of the gas and its pressure, for the carriers are the same whatever the gas may be ; again, the mean free paths of these corpuscles would depend solely upon the density of the medium through which they pass. For the molecules of the medium are composed of a number of such corpuscles separated by considerable spaces; now the collision between a single corpuscle and the molecule will not be between the corpuscles and the molecule as a whole, but between this corpuscle and the individual corpuscles which form the molecule ; thus the number of collisions the particle makes as it moves through a crowd of these molecules will be proportional, not to the number of the molecules in the crowd, but to the number of the individual corpuscles. The mean free path is inversely proportional to the number of collisions in unit time, and so is inversely proportional to the number of corpuscles in unit volume; now as these corpuscles are all of the same mass, the number of corpuscles in unit volume will be proportional to the mass of unit volume, that is the mean free path will be inversely proportional to the density of the gas. We see, too, that so long as the distance between neighbouring corpuscles is large compared with the linear dimensions of a corpuscle the mean free path will be independent of the way they are arranged, provided the number in unit volume remains constant, that is the mean free path will depend only on the density of the medium traversed by the corpuscles, and will be independent of its chemical nature and physical state : this from Lenard's very remarkable measurements of the absorption of the cathode rays by various media, must be a property possessed by the carriers of the charges in the cathode rays.

Thus on this view we have in the cathode rays matter in a new state, a state in which the subdivision of matter is carried very much further than in the ordinary gaseous state : a state in which all matter—that is, matter derived from different sources such as hydrogen, oxygen, &c.—is of one and the same kind; this matter being the substance from which all the chemical elements are built up.

With appliances of ordinary magnitude, the quantity of matter produced by means of the dissociation at the cathode is so small as to almost to preclude the possibility of any direct chemical investigation of its properties. Thus the coil I used would, I calculate, if kept going uninterruptedly night and day for a year, produce only about one three-millionth part of a gramme of this substance.

The smallness of the value of m/e is, I think, due to the largeness of e as well as the smallness of m. There seems to me to be some evidence that the charges carried by the corpuscles in the atom are large compared with those carried by the ions of an electrolyte. In the molecule of HCl, for example, I picture the components of the hydrogen atoms as held together by a great number of tubes of electrostatic force; the components of the chlorine atom are similarly held together, while only one stray tube binds the hydrogen atom to the chlorine atom. The reason for attributing this high charge to the constituents of the atom is derived from the values of the specific inductive capacity of gases : we may imagine that the specific inductive capacity of a gas is due to the setting in the electric field of the electric doublet formed by the two oppositely electrified atoms which form the molecule of the gas. The measurements of the specific inductive capacity show, however, that this is very approximately an additive quantity: that is, that we can assign a certain value to each element, and find the specific inductive capacity of HCl by adding the value for hydrogen to the value for chlorine; the value of H_2O by adding twice the value for hydrogen to the value for oxygen, and so on. Now the electrical moment of the doublet formed by a positive charge on one atom of the molecule and a negative charge on the other atom would not be an additive property; if, however, each atom had a definite electrical moment, and this were large compared with the electrical moment of the two atoms in the molecule, then the electrical moment of any compound, and hence its specific inductive capacity, would be an additive property. For the electrical moment of the atom, however, to be large compared with that of the molecule, the charge on the corpuscles would have to be very large compared with those on the ion.

If we regard the chemical atom as an aggregation of a number of primordial atoms, the problem of finding the configurations of stable equilibrium for a number of equal particles acting on each other according to some law of force —whether that of Boscovich, where the force between them is a repulsion when they are separated by less than a certain critical distance, and an attraction when they are separated by a greater distance, or even the simpler case of a number of mutually repellent particles held together by a central force —is of great interest in connexion with the relation between the properties of an element and its atomic weight. Unfortunately the equations which determine the stability of such a collection of particles increase so rapidly in complexity with the number of particles that a general mathematical investigation is scarcely possible. We can, however, obtain a good deal of insight into the general laws which govern such configurations by the use of models, the simplest of which is the floating magnets of Professor Mayer. In this model the magnets arrange themselves in equilibrium under their mutual repulsions and a central attraction caused by the pole of a large magnet placed above the floating magnets.

A study of the forms taken by these magnets seems to me to be suggestive in relation to the periodic law. Mayer showed that when the number of floating magnets did not exceed 5 they arranged themselves at the corners of a regular polygon— 5 at the corners of a pentagon, 4 at the corners of a square, and so on. When the number exceeds 5, however, this law no longer holds: thus 6 magnets do not arrange themselves at the corners of a hexagon, but divide into two systems, consisting of 1 in the middle surrounded by 5 at the corners of a pentagon. For 8 we have two in the inside and 6 outside; this arrangement in two systems, an inner and an outer, lasts up to 18 magnets. After this we have three systems: an inner, a middle, and an outer; for a still larger number of magnets we have four systems, and so on.

Mayer found the arrangement of magnets was as follows:—

1.	2.	3.	4.	5.
1.5	2.6	3.7	4.8	5.9
1.6	2.7	3.8	4.9	
1.7				
1.5.9	2.7.10	3.7.10	4.8.12	5.9.12
1.6.9	2.8.10	3.7.11	4.8.13	5.9.13
1.6.10	2.7.11	3.8.10	4.9.12	
1.6.11		3.8.11	4.9.13	
		3.8.12		
		3.8.13		
1.5.9.12	2.7.10.15	3.7.12.13	4.9.13.14	
1.5.9.13	2.7.12.14	3.7.12.14	4.9.13.15	
1.6.9.12		3.7.13.14	4.9.14.15	
1.6.10.12		3.7.13.15		
1.6.10.13				
1.6.11.12				
1.6.11.13				
1.6.11.14				
1.6.11.15				
1.7.12.14				

where, for example, 1.6.10.12 means an arrangement with one magnet in the middle, then a ring of six, then a ring of ten, and a ring of twelve outside.

Now suppose that a certain property is associated with two magnets forming a group by themselves; we should have this property with 2 magnets, again with 8 and 9, again with 19 and 20, and again with 34, 35, and so on. If we regard the system of magnets as a model of an atom, the number of magnets being proportional to the atomic weight, we should have this property occurring in elements of atomic weight 2, (8, 9), 19, 20, (34, 35). Again, any property conferred by three magnets forming a system by themselves would occur with atomic weights 3, 10, and 11; 20, 21, 22, 23, and 24; 35, 36, 37 and 39; in fact, we should have something quite analogous to the periodic law, the first series corresponding to the arrangement of the magnets in a single group, the second series to the arrangement in two groups, the third series in three groups, and so on.

Velocity of the Cathode Rays.

The velocity of the cathode rays is variable, depending upon the potential-difference between the cathode and anode, which is a function of the pressure of the gas—the velocity increases as the exhaustion improves; the measurements given above show, however, that at all the pressures at which experiments were made the velocity exceeded 10^9 cm./sec. This velocity is much greater than the value 2×10^7 which I previously obtained (Phil. Mag. Oct. 1894) by measuring directly the interval which separated the appearance of luminosity at two places on the walls of the tube situated at different distances from the cathode.

In my earlier experiments the pressure was higher than in the experiments described in this paper, so that the velocity of the cathode rays would on this account be less. The difference between the two results is, however, too great to be wholly explained in this way, and I attribute the difference to the glass requiring to be bombarded by the rays for a finite time before becoming phosphorescent, this time depending upon the intensity of the bombardment. As this time diminishes with the intensity of bombardment, the appearance of phosphorescence at the piece of glass most removed from the cathode would be delayed beyond the time taken for the rays to pass from one place to the other by the difference in time taken by the glass to become luminous; the apparent velocity measured in this way would thus be less than the true velocity. In the former experiments endeavours were made to diminish this effect by making the rays strike the glass at the greater distance from the cathode less obliquely than they struck the glass nearer to the cathode; the obliquity was adjusted until the brightness of the phosphorescence was approximately equal in the two cases. In view, however, of the discrepancy between the results obtained in this way and those obtained by the later method, I think that it was not successful in eliminating the lag caused by the finite time required by the gas to light up.

Experiments with Electrodes of Different Materials.

In the experiments described in this paper the electrodes were generally made of aluminium. Some experiments, however, were made with iron and platinum electrodes.

Though the value of m/e came out the same whatever the material of the electrode, the appearance of the discharge varied greatly; and as the measurements showed, the potential-difference between the cathode and anode depended greatly upon the metal used for the electrode ; the pressure being the same in all cases.

To test this point further I used a tube like that shown in fig. 6, where *a, b, c* are cathodes made of different metals, the anodes being in all cases platinum wires. The cathodes were disks of aluminium, iron, lead, tin, copper, mercury, sodium amalgam, and silver chloride ; the potential-difference

Fig. 6.

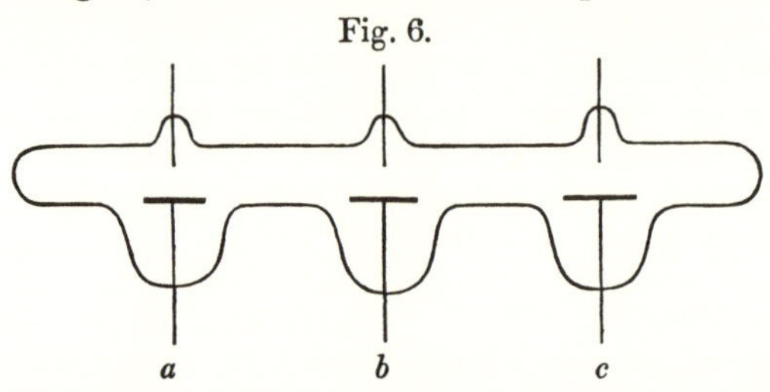

between the cathode and anode was measured by Lord Kelvin's vertical voltmeter, and also by measuring the length of spark in air which, when placed in parallel with the anode and cathode, seemed to allow the discharge to go as often through the spark-gap as through the tube. With this arrangement the pressures were the same for all the cathodes. The potential-difference between the anode and cathode and the equivalent spark-length depended greatly upon the nature of the cathode. The extent of the variation in potential may be estimated from the following table :—

Cathode.	Mean Potential-Difference between Cathode and Anode.
Aluminium	1800 volts.
Lead	2100 ,,
Tin	2400 ,,
Copper	2600 ,,
Iron......................	2900 ,,

The potential-difference when the cathode was made of sodium amalgam or silver chloride was less even than that of aluminium.

The order of many of the metals changed about very capriciously, experiments made at intervals of a few minutes frequently giving quite different results. From the abrupt way in which these changes take place I am inclined to think that gas absorbed by the electrode has considerable influence on the passage of the discharge.

I have much pleasure in thanking Mr. Everitt for the assistance he has given me in the preceding investigation.

Cambridge, Aug. 7, 1897.

POINCARÉ, HENRI. (b. Nancy, France, 29 April 1854; d. Paris, France, 17 July 1912)

Poincaré entered the Ecole Polytechnique in 1873 and studied afterwards at the Ecole des Mines. He completed his doctoral thesis in mathematics at the Paris Sciences Faculty in 1879, taught at the Caen Sciences Faculty during 1879–1881, and then at the Paris Sciences Faculty until his death in 1912. He was the most distinguished French mathematician at the turn of the century, with an esteemed reputation in astronomy as well as in mathematical physics. His work in the philosophy of science came to be labelled "conventionalism," falling within the positivist and phenomenalist tradition, and his book *Science and Hypothesis* (1902) has become a classic in the philosophy of science.

In mathematics, Poincaré's principal contributions lie in the theory of functions, algebraic topology, and the theory of differential equations. Before he was twenty years old, he discovered the automorphic or "fuchsian" functions of one complex variable. This allowed him to make transformations of non-Euclidean geometry into properties of Euclidean figures; it also enabled him to resolve a problem involving a system of curvilinear triangles which he had not been able to solve through traditional Euclidean geometry. In celestial mechanics Poincaré dealt with the theory of orbits and investigated the three-body problem.

Poincaré was the first Parisian scientist to introduce Maxwell's electromagnetic theory in his courses, and his interest in Lorentz's electron theory led to mathematical results similar to those of Lorentz and Einstein in electrodynamics and the special theory of relativity. He was strongly interested in the new radiations reported in the 1890s, and it was his suggestion that phosphorescent materials might emit X rays that started Henri Becquerel on his path to the discovery of radioactivity in phosphorescent uranium salts. Like Einstein, Poincaré was concerned about reconciling classical mechanics, molecular-kinetic theory, thermodynamics, and electrodynamics. Consistent with his epistemological view that conceptual elements which remain true in scientific theories are relations between things and not ideas of the things themselves, Poincaré regarded the choice between matter and energy, or atoms and equivalents, as a matter of convention. Yet he was an influential voice for the usefulness and heuristic value of the molecular-kinetic theory, as expressed in his 1900 speech "Relations between Experimental Physics and Mathematical Physics" given at the International Congress of Physics.

RELATIONS BETWEEN EXPERIMENTAL PHYSICS AND MATHEMATICAL PHYSICS.[1]

RÔLE OF EXPERIMENT AND OF GENERALISATION.

EXPERIMENT is the sole source of truth: this alone can teach us something new; this alone can give us certainty. These two points no one may question.

But if experiment is all, what place is there for mathematical physics? What has experimental physics to do with such an auxiliary which seems useless and even perhaps dangerous? Nevertheless mathematical physics exists, and has been of undeniable service; this fact needs explanation.

Observation is not sufficient; use must be made of our observations, and for that generalisation is necessary. This has always been done; but man profiting from past errors, has observed more and more and generalised less and less. Each century has scoffed at the preceding, accusing it of generalising too boldly and too naïvely. Descartes commiserated the Ionians; Descartes in his turn makes us smile; without doubt our sons will some day laugh at us. Is there no way to get at the gist of the matter at once and escape the raillery that we foresee? May we not be content with experiment alone?

No, that is impossible and would be misunderstanding completely the true character of science. The savant must work with

---

[1] Written in 1900 and delivered before the International Congress of Physics in Paris. Translated by George K. Burgess, Docteur de l'Université de Paris, Instructor in Physics, University of California.

method; science is made of facts as a house of stones; but an accumulation of facts is no more a science than a pile of stones a house. Above all the scientist must foresee. Carlyle says: "The fact alone matters; John Lackland passed by here, that is what is admirable, here is a reality for the which I would give all the theories in the world." Carlyle was a compatriot of Bacon; like him he desired to proclaim the cult *for the God of Things as they are*, but Bacon would not have said that. It is the language of the historian. Most likely the physicist would have said: "John Lackland passed by here; never mind, for he will not pass this way again."

We all know that there are good experiments and poor ones. The latter accumulate in vain; whether there are a hundred or a thousand, a single piece of work by a real master, a Pasteur for instance, suffices to make them fall into obscurity. This, Bacon would have well understood; is it not he who invented the expression *experimentum crucis?* But Carlyle would not have understood it. A fact is a fact; a student has read a number on his thermometer, taking no precautions; no matter, he has read it, and if it is the fact only that counts, this is a reality of the same degree as the wanderings of John Lackland. What then is a good experiment? It is one which teaches us something more than an isolated fact; it aids us to predict, and enables us to generalise.

Without generalisation, prediction is impossible. The circumstances under which one has operated will never be simultaneously reproduced. The observed fact can never be realised again; the only thing that can be affirmed, is that under analogous circumstances, an analogous fact will be produced. In order to predict it is necessary to invoke analogy, that is, to generalise.

However timid one may be, it is necessary to interpolate; experiment gives us only a certain number of isolated points, which must be united by a continuous line; this is a true generalisation. But one does more, the curve so traced will pass between and near these points; but not through them. So that one is not limited to the generalisation of the experiment, he corrects it; and the physicist who would abstain from these corrections and content himself solely with experiment would be forced to announce the most extraordinary laws. The detached facts are not enough; that is why we must have Science ordered, or better, organised.

It is often said that we must experiment with no preconceived idea. That is not possible; not only would this render sterile every experiment, but even if we wanted to do so, it could not be done. Every one has within him his idea of the world, which cannot be so easily put aside. For example, we have to make use of language, which is made up necessarily of preconceived ideas. Such ideas unconsciously held are the most dangerous of all.

Shall we say that if we cause to intervene others of which we have full consciousness, we shall but aggravate the evil? I do not think so; I believe rather that they will act as mutual counterweights, I was going to say antidotes, that in general will accord poorly and even conflict with each other forcing us to look at things from different aspects. This is enough to free us: he who can choose his master is no longer a slave.

Thus, thanks to generalisation, each observed fact enables us to predict a great number of others; but we must not forget that the first alone is certain and the others merely probable. However well founded a prediction may seem, we are never *absolutely* sure that experiment will not prove it false, if we undertake to verify it. But the probability of truth is often so great that practically we may be content with it. Better is it to predict without certainty than never to have predicted at all.

We should never disdain to make a verification when the occasion presents itself. But every experiment is long and difficult, the workers are few, and the quantity of facts that we need to predict is immense; beside this mass, the number of direct verifications that we can make will ever be a negligible quantity. Of this little that we may directly reach, we must select the better part; it is necessary that each experiment should allow the greatest possible number of predictions having the highest degree of probability. The problem is, so to speak, to increase the efficiency of the scientific machine.

Allow me to compare science to a library which must increase indefinitely; the librarian has at his disposal for purchases but limited funds, which must not be wasted. It is experimental physics that is charged with the buying; she alone can enrich the library. As for mathematical physics, her mission is to make the catalogue; if this is well made, the library will not be richer; but it may aid the reader to make use of these riches. Also by showing the librarian the gaps in his collections, it will aid him to make judicious use of his funds; which is the more important as the funds are quite inadequate.

Such is the rôle of mathematical physics; she should direct generalisation so as to augment what I have just called the efficiency of Science. By what means she accomplishes this, and how she may do so without danger, that is what we shall examine.

THE UNITY OF NATURE.

We observe in the first place that every generalisation supposes in a certain measure the belief in the unity and in the simplicity of nature. In the case of unity there can be no difficulty. If the different parts of the universe were not as the organs of the same body, they would not react on each other, they would ignore each other mutually; and we in particular could know but one part. Consequently we have not to ask ourselves if nature is one, but how she is one.

As to the second point, all is not so clear. It is not certain that nature is simple. Can we without danger act as if she were so?

There was a time when the simplicity of Mariotte's law was an argument presented in favor of its exactness, when Fresnel himself, after having said, in conversation with Laplace, that nature did not occupy herself with analytical difficulties, was obliged to explain his words, so as not to offend the current public opinion. To-day ideas have changed much; nevertheless those who do not believe that natural laws must be simple, are still often obliged to act as if they so believed. They cannot separate themselves entirely from this appearance without rendering impossible all generalisation and consequently all science.

It is clear that any fact soever may be generalised in an infinite number of ways, and one must choose among them. The choice will be determined by considerations of simplicity. Take the case of interpolation. We draw a line as regularly as possible among the points given by observation. Why do we avoid the discordant points, the too sharp inflections? Why do we not describe a curve having the most capricious zigzags? It is because we know beforehand, or we think we know, that the law to be expressed cannot be as complicated as that. Jupiter's mass may be deduced either from the movements of his satellites, from the perturbations of the greater planets, or from those of the lesser planets. If the averages of the determinations obtained by these methods are taken, we find three numbers nearly but not quite identical. This result might be interpreted by supposing that the gravitation constant is not the same in the three cases; the observations would be certainly much better represented. Why do we reject this interpretation? Not because it is absurd but that it is uselessly complicated. It will not be accepted until it is forced upon us, and that day is not yet.

To resume, every law is reputed simple until proved otherwise.

This custom is forced upon physicists by the reasons that I have indicated; but how justify it in the presence of discoveries that daily show us new details richer and more complex? How reconcile it even with the unity of nature? For if all things are interdependent, the relations in which so many different objects intermingle cannot be simple.

If we study the history of science, we see produced two phenomena that are, so to speak, the inverse of each other: on the one hand there is a simplicity hidden under complex appearances, on the other hand an apparent simplicity conceals extremely complex realities.

What is more complicated than the troubled movements of the planets, what more simple than Newton's law? There, nature playing, as Fresnel said, with the analytical difficulties, employs but simple means and engenders by their combination I know not what tangled snarl. Here is a case of hidden simplicity,—one which must be unravelled.

Examples of the other kind abound. In the kinetic theory of gases, we consider the molecules animated with great velocities, whose paths, deformed by incessant impacts, have the most capricious shapes, and cross space in all directions. The observable result is the simple law of Mariotte; each individual fact was complicated; the law of great numbers has re-established simplicity in the mean. Here the simplicity is only apparent, and the coarseness of our senses alone prevents us from perceiving the complexity.

Many phenomena obey a law of proportionality; but why? Because in these phenomena there is something which is very small. The simple law observed is then but a translation of this general analytical rule, according to which the infinitely small increment of a function is proportional to the increment of the variable. Since in reality the increments are not infinitely small, but very small, the proportionality law is but approximate and the simplicity is but apparent. What I have said applies to the law of the superposition of small movements, whose use is so fruitful and which is the basis of optics.

And Newton's law itself? Its simplicity, so long hidden, is perhaps only apparent. Who knows if it is not due to some complicated mechanism, to the impact of some subtle matter animated with irregular movements, and if it has not become simple merely by the play of averages and of large numbers? In any case it is difficult not to suppose that the true law contains supplementary terms, which may become sensible at small distances. If in astronomy they are negligible in comparison with Newton's expression, and if the law becomes thus simplified, this is merely on account of the enormity of the celestial distances.

Without doubt, if our means of investigation became more and more penetrating, we should discover the simple within the complex, then the complex from the simple, then again the simple within the complex, and so on, without being able to predict which would be the last term. It is necessary to stop somewhere, and for science to be possible, we must stop where we have found simplicity. That is the only foundation upon which we can construct the edifice of our generalisations. But, the simplicity being only apparent, will this foundation be solid enough? That is what is to be studied.

For this, let us see what rôle our generalisations play in the belief in simplicity. We have verified a simple law in a considerable number of particular cases; we refuse to admit that this occurrence, so often repeated, is a result of mere chance, and we conclude that the law must be true in the general case.

Kepler finds that the positions of a planet observed by Tycho are all on the same ellipse. He has not for a single instant the thought that, by a singular chance, Tycho never regarded the heavens but at the moment when the true trajectory of the planet happened to cut this ellipse.

What does it matter then if the simplicity is real, or if it conceals a complex truth? Whether it be due to the influence of large numbers which level individual differences, or to the greatness or smallness of certain quantities which allow of neglecting certain terms, in no case is it due to chance. This simplicity, real or apparent, has always a cause. We may then reason in the same way at all times, and if a simple law has been observed in several particular cases, we may legitimately suppose that it will still be true in analogous cases. To refuse to so consider the matter would be to attribute an inadmissible rôle to chance.

Nevertheless there is a difference. If the simplicity was real and profound, it would bear the test of the increasing precision of our methods of measurement; if then we believe nature to be profoundly simple, we must conclude that it is an approximate and not a rigorous simplicity. This was formerly done; but this is what we no longer have the right to do.

The simplicity of Kepler's laws, for example, is only apparent. This does not prevent their being applied, almost exactly, to all systems analogous to the solar system, but it prevents their being rigorously exact.

THE RÔLE OF HYPOTHESIS.

Every generalisation is a hypothesis; the hypothesis has then a necessary rôle that no one has ever contested. But it should always, as soon and as often as possible, be submitted to verification. It is evident, that if it does not stand this test, it must be thrown aside without regret. This is what is usually done, but sometimes with impatience.

This impatience, however, is not justifiable; the physicist who has just renounced one of his hypotheses should be glad, on the contrary, for he has just found an unhoped-for occasion of discovery. His hypothesis, I imagine, had not been lightly adopted; it took account of all the known factors which seemed to be able to intervene in the phenomenon. If the verification is not made, it is because there is something unexpected, something extraordinary; we are on the point of finding something unknown.

Has the hypothesis so rejected been sterile? Far from it. One may even say that it has rendered more service than a true hypothesis; not only has it been the occasion of a decisive experiment, but if the experiment had been made by chance, without the existence of the hypothesis, nothing would have been inferred; nothing extraordinary would have been seen; merely one fact more would have been catalogued without deducing the least consequence.

Now under what conditions is the use of hypothesis without danger?

The firm purpose to submit all to experiment does not suffice; there are still hypotheses that are dangerous; they are in the first place and above all those that are tacit and unconscious. Since we make them without knowing it, we are powerless to abandon them. Here again is a service that mathematical physics may render. By the precision proper to it, we are obliged to formulate all the hypotheses that we should make without this aid, but without being aware of their existence.

Note, besides, that it is important not to multiply our hypotheses too fast, but to make them only one after another. If we construct a theory founded on multiple hypotheses, and if experiment condemns it, which among our premises must we change? It is impossible to know. And conversely, if the experiment succeeds, are we to think all the hypotheses verified at once? Have several unknowns been determined with a single equation?

Care must also be taken to distinguish between the several kinds of hypotheses. First there are those that are quite natural and without which we could hardly do. It is difficult not to suppose that the influence of very distant bodies is quite negligible, that small movements obey a linear law, that the effect is a continuous function of the cause. I will say as much for the conditions imposed by symmetry. All these hypotheses form, so to speak, the common foundation of all theories in mathematical physics. They are the last that should be abandoned.

There is a second category of hypotheses that I will qualify as indifferent. In the greater number of questions, the analyst supposes at the outset of his calculations, either that matter is continuous, or inversely that it is made up of atoms. By either method his results will be the same. If he chooses the latter, and experiment confirms his results, will he think he has demonstrated, for example, the real existence of atoms?

Into optical theories two vectors are introduced, which are regarded, the one as a velocity, the other as a vortex. This is again an indifferent hypothesis, since the same conclusions would have been reached with contrary suppositions; the success of the experiment cannot prove that the first vector is a velocity; it proves but one thing, namely that it is a vector; this is really the only hypothesis that was introduced in the premises. To give it that concrete appearance that the weakness of our intellects requires, it was necessary to consider it either as a velocity or as a vortex; likewise it was necessary to represent it by a letter, as x or y; but the result, whatever it be, will not prove that we were right or wrong to regard it as a velocity; no more can it be proved correct or not to call it x and not y.

These indifferent hypotheses are never dangerous, provided their character is not misunderstood. They may be useful, either as artifices for calculation, or to sustain our comprehension by concrete images, to fix our ideas, as we say. There is then no reason to proscribe them.

The hypotheses of the third category are veritable generalisations. They are the ones that experiment will confirm or prove false. Verified or condemned, they will always be fruitful. But, for the reasons that I have given, this holds only if they are not too numerous.

ORIGIN OF MATHEMATICAL PHYSICS.

Let us go farther and study at close range the conditions which have brought about the development of mathematical physics. We recognise at once that savants have always tried to resolve the complex phenomenon given directly by experiment into a very great number of elementary phenomena. And this in three different ways:

First, with respect to time, instead of embracing in its entirety the progressive development of a phenomenon, we seek simply to join each instant to the one immediately preceding; it is admitted that the actual state of the world depends only on the immediate past, without being influenced by the memory of a more remote past. Thanks to this postulate, instead of studying directly the whole succession of phenomena, it is possible to write its *differential equation* representing a single epoch; for Newton's laws Kepler's are substituted.

Next, we seek to decompose the phenomenon in space. What experiment gives us, is a confused collection of facts spread over a field of considerable extent; the task is to discern the elementary phenomenon, which is localised in a very small region of space.

A few examples will perhaps make my meaning clearer. If one wished to study in all its complexity the distribution of temperature in a solid which is cooling, it would be impossible to do so. All becomes simple if we reflect that a point in the solid cannot impart heat to a distant point, but only to the nearest, and it is only gradually that the flow of heat will be able to reach other portions of the solid. The elementary phenomenon is the exchange of heat between two contiguous points; it is strictly localised, and it is relatively simple, if it be admitted, as is natural, that it is not influenced by the temperature of molecules whose distance is sensible.

I bend a rod; it will assume a very complicated form whose direct study would be impossible; but I shall be able to attack the problem, if I observe that the flexure is only the resultant of the deformations of the very small elements of the rod, and that the deformation of each of these elements depends only on the forces which are directly applied to it and in nowise on those which may act upon the other elements.

In all these examples, which may be increased indefinitely, it is admitted that there is no action at a distance or at great distances. This is a hypothesis; it is not always true, as the law of gravitation proves; it must then be submitted to verification. If it is confirmed, even approximately, it is precious, for it is going to permit the use of mathematical physics by successive approximations at least. If it does not stand the test, something analogous must be sought, for there are still other ways to reach the elementary phenomenon. If several bodies act simultaneously, it may happen that their actions are independent and may be added together, either as vectors or as scalar quantities. The elementary phenomenon is then the action of an isolated body. Or perhaps one has to do with small movements, or more generally with small variations, which obey the well-known law of superposition. The observed movement will then be decomposed into simple movements; for example, a sound into its harmonics, white light into its monochromatic components.

When we have discerned in what direction to seek the elementary phenomenon, by what means may we reach it?

It will often happen that to predict it, or rather to predict what is useful for us, it will not be necessary to know the mechanism; the law of great numbers will suffice. Consider the example of the propagation of heat; each molecule radiates towards its neighbors, according to a law which we have no need of knowing; if we make any supposition in this regard it will be an indifferent hypothesis and consequently useless and unverifiable. And, indeed, by the action of averages and thanks to the symmetry of the medium, all differences are razed, and whatever the hypothesis, the result is always the same.

The same circumstances are present in the theories of elasticity and capillarity; the neighboring molecules attract and repel each other, we have no need to know according to what law; it suffices that this attraction is sensible at small distances only, that the molecules are very numerous, that the medium is symmetrical, and we have but to let the law of great numbers act.

Here again the simplicity of the elementary phenomenon was hidden beneath the complexity of the observable resultant phenomenon; but in its turn, this simplicity was only apparent and concealed a very complex mechanism.

The best way to reach the elementary phenomenon would be evidently by experiment. It would be necessary by experimental artifices, to dissociate the complex beam that nature offers to our researches and study with care its elements as purified as possible; for example, natural white light can be decomposed into monochromatic lights by means of a prism and into polarised lights by means of a polariser.

Unfortunately this is neither always possible nor sufficient, and it is sometimes necessary for the mind to anticipate the experiment. I will cite but a single example which has always appealed to me.

If I decompose white light, I can isolate a small portion of the spectrum, but however small it may be, it always conserves a certain width. Similarly, the natural lights called *monochromatic* give us a very fine line, although not infinitely fine. One might suppose that in studying experimentally the properties of these natural lights, operating with finer and finer spectral beams, and passing at last to the limit, one would come to know the properties of a light rigorously monochromatic. This would not be so. Imagine that two beams start from the same source, that they are polarised in two planes at right angles, afterwards brought into the same plane of polarisation, and that one tries to make them interfere. If the light were *rigorously* monochromatic they would interfere, but with our nearly monochromatic lights there would be no interference, and this however narrow the beam; it would be necessary in order to have it otherwise that the beam be several million times narrower than the finest known. Here then the passage to the limit would have deceived us; the intellect has outstripped experiment, and if this has been successfully done, it is because the former was guided by the instinct of simplicity.

A knowledge of the elementary fact permits us to put the problem into the form of an equation; it only remains to deduce from this by combination the complex observable and verifiable fact. This is what is called *integration*; it is the mathematician's affair.

It may be asked why, in the physical sciences, a generalisation readily takes the mathematical form. The reason is now easy to see; it is not merely that one has to express numerical laws; it is because the observable phenomenon is due to the superposition of a great number of elementary phenomena *all similar to each other*; in this way the differential equations are quite readily introduced.

It is not sufficient that each elementary phenomenon obeys simple laws, it is necessary that all to be combined obey the same law. It is only then that the intervention of mathematics may be useful; mathematics teaches us, in fact, to combine like with like. Its goal is to divine the result of a combination, without passing through all the intermediate steps each time. If we have to repeat several times the same operation, it enables us to avoid this repetition by informing us beforehand of the result by a sort of induction. In such cases all these operations must be similar to each other, otherwise we should have to go step by step, and mathematics would become useless.

It is thus of the approximate homogeneity of matter studied by the physicist that mathematical physics could be born. In the natural sciences, we do not find these conditions: homogeneity, relative independence of distant parts, simplicity of the elementary part; and that is why naturalists are obliged to make use of other modes of generalisation.

SIGNIFICATION OF PHYSICAL THEORIES.

Men of the world are struck to see how transient are scientific theories. After several years of prosperity, they see them successively abandoned; they see ruins pile on ruins; they predict that the theories current to-day will, after a brief delay, in their turn succumb, and they conclude that such theories are absolutely in vain. It is what they call the *bankruptcy of science.*

Their scepticism is superficial; they take no account whatever of the object and rôle of scientific theories, otherwise they would understand that the ruins are still good for something. No theory seemed so well established as Fresnel's which attributed light to movements of the ether. However, that of Maxwell is to-day preferred. Does this mean that the work of Fresnel has been in vain? No, for Fresnel's goal was not to know whether there really is an ether, whether or not it is formed of atoms, whether these atoms move in such or such a way; it was to predict optical phenomena. As for that, Fresnel's theory enables us to do this to-day as well as it did before Maxwell. The differential equations are always true; they may always be integrated by the same methods and the results of this integration ever preserve their value.

Let no one say that we thus reduce physical theories to simple practical recipes; these equations express actual relations, and if the equations remain true, it is because these relations preserve their reality. They teach us, now as before, that there is such and such a relation between this thing and that; only, something which we called *movement* before, we now call *electric current.* But these names were only images substituted for the real objects that nature will forever hide from us. The true relations between these real objects are the only reality that we can reach, and the sole condition is that the same relations shall exist between these objects as between the images we are forced to put in their place. If these relations are known to us, what matters it if we judge it convenient to replace one image by another?

That a given periodic phenomenon (an electrical oscillation for instance) is really due to the vibration of a given atom which, behaving like a pendulum, is displaced in such or such a way, all this is neither certain nor interesting. But that there is between the electrical oscillation, the movement of the pendulum, and all periodic movements an intimate relationship which corresponds to a profound reality; that this relationship, this similitude, or better this parallelism is continued in the details; that it is a consequence of more general principles, as the conservation of energy and least action,—this we may affirm; this is the truth that will remain forever the same in all the guises in which we may see fit to dress it.

Numerous theories of dispersion have been proposed. The first were imperfect and contained but little truth. Then came Helmholtz's, which was modified in various ways; and its author himself has imagined another based on Maxwell's principles. But the remarkable thing is, that all the scientists who have followed Helmholtz reach the same equations, from seemingly widely separated starting-points. I venture to say that these theories are all true at once, not merely because they allow us to predict the same phenomena, but because they express a true relation, that between absorption and anomalous dispersion. In the premises of these theories, that which is true is common to all; it is the affirmation of such or such a relation between certain things that some call by one name some by another.

The kinetic theory of gases has given rise to many objections, to which reply would be difficult, if there had been any claim that it contained absolute truth. But all these objections cannot refute its past usefulness, particularly in revealing to us the one true relation, otherwise profoundly hidden, between gaseous and osmotic pressures. In this sense it may be said to be true.

When a physicist finds a contradiction between two theories which are equally dear to him, he sometimes says: Let us not be troubled but let us hold fast to the two ends of the chain that the intermediate links be not lost to us. This argument of the embarrassed theologian would be ridiculous if we are to attribute to physical theories the sense given them by men of the world. In case of contradiction, one of them at least should then be considered false. It is no longer so if we will seek in them what is to be sought. It may be they both express true relations and that there is contradiction only in the images with which we have dressed reality.

To those who find that we restrict too much the domain accessible to the scientist, I reply: These questions which we prohibit you from studying and which you so regret, are not only insoluble, they are also illusory and void of sense.

Your philosopher claims that all physics can be explained by the mutual impact of atoms. If he means that the same relations obtain among physical phenomena as among the mutual impacts of a great number of billiard-balls, nothing better, this is verifiable, it is perhaps true. But he means to say something more; and we think we understand him because we think we know what an impact is in itself. Why? simply because we have often seen a game of billiards. Are we to understand that God, in contemplating his work, feels the same sensations as we in the presence of a billiard match? If we do not wish to give to his assertion this fantastic meaning, if also we do not wish to give it the one I previously mentioned, then it has no meaning whatever.

Hypotheses of this nature have only a symbolic sense. The scientist should not banish them any more than a poet banishes metaphor; but he should know what they are worth. They may be useful to give satisfaction to the mind, and they will not be harmful provided they are but indifferent hypotheses.

These considerations show us why certain theories that were thought to be abandoned and definitely condemned by experiment, are suddenly revived from their ashes and recommence a new life. It is because they express true relations, and had not ceased to do so, when for some reason or other we thought it necessary to enunciate the same relations in another language. They had thus kept a sort of latent life.

Hardly fifteen years ago, was there anything more ridiculous, more quaintly old-fashioned, than the fluids of Coulomb? But nevertheless here they reappear under the name *electrons*. In what do these molecules electrified in a permanent way differ from the electric molecules of Coulomb? True, in the electrons the electricity is supported by a little, though very little, matter; in other words, they have mass. But Coulomb did not gainsay mass to his fluids; or if he did, it was reluctantly. It would be rash to affirm that the belief in electrons will not also undergo its eclipse; but it was not less curious to remark this unexpected renaissance.

But the most striking example is Carnot's principle. Carnot established it, starting from false hypotheses. When it was perceived that heat is not indestructible but may be converted into work, his ideas were completely abandoned; later Clausius returned to them and caused them to triumph definitively. Carnot's theory, in its primitive form, expressed, besides true relations, other inexact relations, *débris* of old ideas; but the presence of the latter did not alter the reality of the others. Clausius had but to separate them as one cuts away dead branches.

The result was the second law of thermodynamics. The relations were always the same, although these relations did not hold, in appearance at least, between the same objects. This sufficed to preserve for the principle its value. Nor have the reasonings of Carnot perished by reason of this; they were applied to matter infected with error; but their form (that is to say, their essential part) remained correct.

What I have said throws light at the same time on the rôle of general principles like the principles of least action and the conservation of energy. These principles have a very great value; they were obtained in seeking what was common in the statements of numerous physical laws; they thus represent the quintessence of innumerable observations. However, from their very generality results a consequence to which I have called attention in the preface to my *Course on Thermodynamics;* it is that they are of necessity verified. Since we cannot give energy a general definition, the principle of the conservation of energy signifies simply that there is a *something* that remains constant. Whatever new notions of the world future experiments may give us, we are certain beforehand that there is something which will remain constant, and which we may call *energy*.

Does this mean that the principle has no sense and vanishes

into a tautology? Not at all; it means that the different things we call *energy* are joined by a true relationship. But even if this principle has a meaning, it may be false; perhaps we have no right to deduce applications from it indefinitely, and yet it is sure beforehand to be verified in the strict sense of the word. How then shall we be warned when it has reached the full development that we may legitimately give it? Simply when it ceases to be useful, or when we may no longer use it to correctly predict new phenomena. We shall be sure in such cases that the relation affirmed is no longer true; for otherwise it would be fruitful; experiment, without directly contradicting a new extension of the principle, will nevertheless have condemned it.

PHYSICS AND MECHANISM.

Most theorists have a constant predilection for explanations borrowed from mechanics or dynamics. Some would be satisfied if they could account for all phenomena by the movement of molecules attracting one another according to certain laws. Others are more exacting, they would suppress attractions at a distance; their molecules would follow rectilinear paths from which they could only be deviated by impacts. Still others, as Hertz, suppress also the forces, but suppose their molecules submitted to geometrical connections analogous, for example, to those of articulated systems; they thus wish to reduce dynamics to a sort of kinematics. All, in a word, wish to bend nature into a certain form, lacking which their minds cannot be satisfied. Is nature flexible enough for this?

I have already put the question in the preface to my work: *Electricity and Optics*. I have shown that every time the principles of energy and of least action are satisfied, not only is there always a mechanical explanation possible, but there is always an infinity of them. Thanks to a well-known theorem on articulated systems due to Koenigs, it may be shown that everything may be explained in an infinite number of ways by connections after the manner of Hertz, or else by central forces. Without doubt, it might be just as easily demonstrated that everything may be explained by simple impacts.

For this, bear in mind, it is not sufficient to be content with ordinary matter, which comes in contact with our senses and whose movements we observe directly. Ordinary matter may be conceived either as formed of atoms whose inner movements escape us, the displacement of the whole being alone accessible to our senses, or one of those subtle fluids may be imagined which, under the name *ether* or other names, have always played such an important rôle in physical theories.

Often one goes farther and regards the ether as the only primitive matter, or as the only true matter. The more moderate consider ordinary matter as condensed ether, which is in no way startling; but others reduce still further its importance and see in matter only the geometrical locus of the singularities in the ether. Thus, for Kelvin, what we call *matter* is but the locus of the points at which the ether is animated by vortex motions; for Riemann, it was the locus of the points at which ether is constantly destroyed; for more recent writers, Wiechert or Larmor, it is the locus of the points at which the ether has undergone a sort of torsion of a very particular kind. Taking any one of these points of view, the question arises in my mind, by what right do we apply to the ether, under pretext that it is true matter, the mechanical properties observed in ordinary matter, which is but false matter?

The ancient fluids, caloric, electricity, etc., were abandoned when it was seen that heat is not indestructible. But they were abandoned also for another reason. In materialising them, their individuality, so to speak, was emphasised, gaps were opened between them. It was necessary to fill in these gaps when the sentiment of the unity of nature became stronger, and when the intimate relations binding all parts were perceived. In multiplying the fluids, not only did the ancient physicists create unnecessary entities, but they broke down real ties. It is not sufficient that a theory does not affirm false relations, neither must it hide true relations.

Does our ether actually exist?

We know whence comes our belief in the ether. If light takes several years to reach us from a star, it is no longer upon the star nor yet upon the earth; but it must be somewhere, and supported by some material agency.

The same idea can be expressed in a more mathematical and abstract form. What we note are changes undergone by material molecules; we see, for example, that our photographic plate experiences the consequences of phenomena of which the incandescent mass of a star was the theatre several years ago. Now, in ordinary mechanics, the state of the system studied depends only on its state at the moment immediately preceding; the system satisfies certain differential equations. On the other hand, if we did not believe in the ether, the state of the material universe would depend not only upon the state immediately preceding, but also upon much more ancient states; the system would satisfy equations of finite differences. It is to obviate this transgression of the general mechanical laws that we have invented the ether.

This would oblige us to fill the interplanetary space with ether, but not to make it penetrate into the midst of material media. Fizeau's experiment goes farther. By the interference of rays that have passed through water or air in motion, it seems to show us two different media penetrating each other and yet moving with respect to each other. We all but touch the ether.

Situations may be conceived in which we can touch it closer still. Suppose Newton's principle of the equality of action and reaction is not true if applied to matter *only* and that this is demonstrated. The geometrical sum of all the forces applied to all the material molecules would no longer be zero. It would be necessary, if we did not wish to change the whole science of mechanics, to introduce the ether, in order that the action that matter here apparently undergoes should be counterbalanced by the reaction of matter on something.

Or again, suppose we discover that optical and electrical phenomena are influenced by the movement of the earth. It would follow that these phenomena could reveal to us not only the relative movements of material bodies, but also what would seem to be their absolute movements. It would again be necessary to have an ether, in order that these so-called absolute movements should no take place with respect to empty space, but with respect to somet thing concrete.

Will this ever be accomplished? I do not cherish the hope, and I will say shortly why. And yet, it is not so absurd since others have entertained it. For example, if the theory of Lorentz were true, Newton's principle would not apply to matter *alone*, and the difference would not be very far from being accessible to experiment. On the other hand, many experiments have been made on the influence of the earth's movement. The results have always been negative. But if these experiments have been undertaken, it is because we were not sure beforehand, and indeed according to the reigning theories, the compensation should be only approximate, and we should expect to see improved methods give positive results.

I think that such an experiment is illusory; it was none the less interesting to show that a success of this kind would open in a certain sense a new world.

And now allow me to digress slightly; I must explain why I do not believe, in spite of Lorentz, that more exact observations will ever make evident anything else than relative displacements of material bodies. Experiments have been made that should have disclosed the terms of the first order; the results were negative; can that have been by chance? No one has admitted it; a general explanation was sought, and Lorentz found it; he showed that the first order terms should cancel each other, but not the second order terms. Then more precise experiments were made, which were also negative; neither could this be a result of chance; an explanation was necessary and was found; they are always found; hypotheses are what we lack the least.

But this is not enough; who does not think this leaves too important a rôle to chance? Would it not be also a chance that this singular concurrence would cause a certain circumstance to destroy the terms of the first order, and that a totally different circumstance should cause those of the second order to vanish? No, it is necessary to find the same explanation for the two cases, and everything tends to show that this explanation would serve just as well for the higher order terms, and that the mutual destruction of these terms will be rigorous and absolute.

ACTUAL STATE OF THE SCIENCE.

In the history of the development of physics two opposite tendencies are to be distinguished. On the one hand, at each instant new relations are discovered between objects which seemed destined to remain forever separated; scattered facts cease to be strangers to each other; they tend to arrange themselves into an imposing synthesis. Science marches towards unity and simplicity.

On the other hand, observation reveals every day new phenomena; they must wait for their place a long time; and sometimes to make one, a corner of the edifice must be demolished. In the known phenomena themselves, where our crude senses indicate unity, we perceive details more varied from day to day; what we thought to be simple becomes complex and science seems to march towards diversity and complication.

Of these two opposite tendencies each of which seems to triumph in turn, which will win? If the first, science is possible; but nothing proves this *a priori*, and possibly after vain efforts to bend nature in spite of herself to our ideal of unity, submerged by the ever-mounting flood of our new riches, we shall be compelled to renounce classifying them, abandon our ideal, and reduce science to the recording of innumerable recipes.

We cannot reply to this question. All that we can do is to observe the science of to-day and to compare it with that of yesterday. From this examination we may doubtless draw some conjectures.

A half century ago, hopes were high. The discovery of the conservation of energy and of its transformations had just revealed the unity of force. It showed also that the phenomena of heat could be explained by molecular movements. The nature of these movements was not exactly known, yet no one doubted but that it soon would be. For light, the work seemed completely done. As concerns electricity, the advancement was less great. Electricity had just annexed magnetism. This was a considerable step towards unity, and a definite one. But in what way was electricity to enter in its turn into the general unity, how was it to be included in the universal mechanism? No one had any idea. The possibility of this reduction was not doubted by any one; they had faith. Finally, as to what concerns the molecular properties of material bodies, the reduction seemed still easier; but all the details were hazy. In a word, the hopes were vast, they were strong, but they were vague.

To-day what do we see? In the first place, a step in advance, an immense progress. The relations between electricity and light are now known; the three domains of light, electricity, and magnetism, formerly separated, are but one now; and this combination seems definite. This conquest, nevertheless, has cost us some sacrifices. Optical phenomena enter as particular cases in electrical phenomena; as long as the former remained isolated, it was easy to explain them by movements thought to be known in all their details; that was easy. But now an explanation, to be acceptable, must be readily applicable to the whole electrical domain. This often causes difficulty.

The most satisfactory theory we have, is that of Lorentz; it is unquestionably the one that best explains the known facts, the one that sheds light on the greatest number of true relations, the one in which are to be found the most traces of definite construction. Nevertheless, it still possesses a serious fault, as I have above shown; it is in contradiction with Newton's principle of the equality of action and reaction; or rather, in the eyes of Lorentz, this principle is not applicable to matter alone; in order to be true, it must take account of the actions exerted by the ether on matter, and of the reaction of matter upon the ether. At present it seems most probable that things do not happen in this way.

However this may be, thanks to Lorentz, the results of Fizeau on the optics of moving bodies, the laws of normal and anomalous dispersion and of absorption have been connected together and with the other properties of the ether by bonds that doubtless will not break. Look at the ease with which the Zeeman effect found its place, and even helped to classify the magnetic rotation of Faraday which had remained rebellious to Maxwell's efforts. This facility proves that Lorentz's theory is not an artificial assemblage destined to give way. Probably it should be modified, but not destroyed.

Lorentz had no other ambition than to include in a single whole all the optics and electrodynamics of moving bodies; he made no pretense to give a mechanical explanation. Larmor goes farther; keeping of Lorentz's theory what is essential, he grafts on it MacCullagh's ideas on the direction of the movement of the ether. However ingenious this effort may be, the fault in Lorentz's theory remains, and is even aggravated. According to Lorentz, we do not know what the movements of the ether are; thanks to this ignorance, we might suppose them such as compensated those of matter and re-established the equality of action and reaction. With Larmor, we know the movements of the ether, and we can demonstrate that the compensation does not take place.

If Larmor has to my mind failed, does that mean that a mechanical explanation is impossible? Far from it: I said above that as long as a phenomenon obeys the two principles of energy and least action, it permits of an infinite number of mechanical explanations. It is the same for optical and electrical phenomena.

But that does not suffice: for a mechanical explanation to be good, it must be simple; in order to choose it from among all those that are possible, there must be other reasons than the necessity to make a choice. Well, a theory which satisfies this condition and which consequently might be useful, we do not posses as yet. Are we to complain? That would be to forget the end sought, which is not the mechanism, but the true and sole aim of unity.

We should then bridle our ambition; let us not seek to formulate a mechanical explanation; let us be content to show that we may always find one if we so wish. In this we have succeeded; the principle of the conservation of energy has always been confirmed; a second principle has been joined to this, that of least action, put in the form appropriate to physics. This also has always been verified, at least as far as concerns reversible phenomena, which obey Lagrange's equations, that is to say, the most general laws of physics.

The irreversible phenomena are much more rebellious. They also, however, are being arranged and tend to enter the unity: the light which illuminates them has come from Carnot's principle. For a long time thermodynamics was confined to the study of the dilatation of bodies and their change of state. Later it became bolder and enlarged its domain considerably. We owe to it the theories of the voltaic cell and thermo-electric phenomena; there is not in all physics a corner that it has not explored, and it has even attacked chemistry. Everywhere the same laws reign; everywhere under a diversity of appearances Carnot's principle reappears; everywhere also appears that eminently abstract concept of entropy, which is as universal as energy, and like it seems to conceal a reality. Radiant heat seemed to escape it; but recently that too has been brought under the same laws.

In this way new analogies are revealed, which may often be pursued in detail; electric resistance resembles the viscosity of liquids; hysteresis resembles rather the friction of solids. In all cases, friction appears to be the type imitated by the most diverse irreversible phenomena, and this relationship is real and profound.

A strictly mechanical explanation of these phenomena has also been sought. Such is hardly possible. To find it, it has been necessary to suppose that the irreversibility is but an appearance, that the elementary phenomena are reversible and obey the known laws of dynamics. But the elements are extremely numerous and blend more and more, so that to our crude eyes all appears to tend towards uniformity, that is to say, all seems to march in the same direction without hope of return. The apparent irreversibility is thus but an effect of the law of great numbers. Only a being of infinitely subtle senses, as the imaginary demon of Maxwell, could untangle this snarl and turn the world about.

This conception, which is connected with the kinetic theory of gases, has cost great effort, and has been on the whole not very fruitful; it may become so. This is not the place to examine if it leads to contradictions, and if it conforms well to the true nature of things.

Let us notice, however, the original ideas of Gouy on the Brownian movement. According to this *savant*, this singular movement does not obey Carnot's principle. The particles that it sets moving about are smaller than the meshes of this tightly drawn net; they should then be ready to unravel them and in that way turn the world about. One may imagine he sees Maxwell's demon at work.

To resume, phenomena long known are better and better classified; but new phenomena come to claim their place; and most of them, as the Zeemann effect, find it at once.

But we have the cathode rays, the X-rays, the uranium and radium radiations. There is a whole world that none suspect. How many unexpected guests to find a place for! No one can yet predict the place that they will occupy. But I do not think they will destroy the general unity, I think rather they will complete it. On the one hand, indeed, the new radiations seem to be connected with the phenomena of luminescence; not only do they excite fluorescence, but they arise sometimes under the same conditions as it. Neither are they without relationship with the causes producing the spark discharge under the action of ultra-violet light.

Finally, and above all, it is believed that in all these phenomena there exist ions,—animated, it is true, with far greater velocities than in electrolytes.

All this is very vague, but it will become clearer.

Phosphorescence and the action of light on a spark were regions quite isolated and consequently somewhat neglected by investigators. It is to be hoped that now a new path may be made which will facilitate their communication with the rest of science.

Not only do we discover new phenomena, but in those that we think we know, unlooked-for aspects are revealed. In the free ether, the laws preserve their majestic simplicity; but matter, properly so called, seems more and more complex; all that is said of it is but approximate and at each instant our formulæ require new terms.

Nevertheless the ranks are not broken; the relations that we have recognised between objects that we believed simple, still remain between the same objects when recognised in their complexity, and that alone is important. Our equations become more and more complicated, it is true, so as to embrace more closely the complexities of nature; but nothing is changed in the relations which permit these equations to be derived from one another. In a word, the *form* of these equations persists.

Take for example the laws of reflexion; Fresnel established them by a simple and attractive theory, which experiment seemed to confirm. Subsequently, more precise researches proved that this verification was only approximate; they showed everywhere traces of elliptical polarisation. But, thanks to the aid given us by the first approximation, the cause of these anomalies was soon found in the presence of a transition layer; and Fresnel's theory has remained in all its essentials.

It would seem, nevertheless, that all these relations would never have been noted if the complexity of the objects they joined had been known beforehand. Long ago it was said: If Tycho had had instruments ten times as precise, we should never have had either Kepler, Newton, or Astronomy. It is a misfortune for a science to be born too late, when the means of observation have become too perfect. This is what is happening to-day with physical chemistry; the founders are hampered in their estimates by the third and fourth decimals; happily they are men of robust faith.

As the properties of matter are better known, we see that continuity reigns. From the work of Andrews and Van der Waals, we see how the transition from the liquid to the gaseous state is made, and that it is not brusque. Similarly there is no gap between the liquid and solid states, and we note in their work by the side of articles on the rigidity of liquids memoirs on the flow of solids.

With this tendency simplicity without doubt is lost; such and such an effect was represented by several straight lines; it is necessary now to join these lines by curves more or less complicated. In return unity is gained. These separated categories quiet the mind but do not satisfy it.

Finally the methods of physics have invaded a new domain, that of chemistry; physical chemistry is born. It is still quite young, but we see that already it has allowed us to connect such phenomena as electrolysis, osmosis, and the movements of ions.

From this rapid exposition, what do we conclude?

Taking all things into account, unity has become more nearly realised; this has not been as quickly done as was hoped fifty years ago, and the way predicted has not always been followed; but, on the whole, much ground has been gained.

H. POINCARÉ.

PARIS, 1900.

SODDY, FREDERICK (b. Eastbourne, England, 2 September 1877; d. Brighton, England, 22 September 1956)

After studying at Eastbourne College and Aberystwyth, Soddy attended Merton College, Oxford from 1895 to 1898 and carried out independent chemical research for two years before becoming a demonstrator at McGill University in Montreal, Canada. In 1903 he returned to England to work with William Ramsay in London. From 1904–1914 he taught chemistry at the University of Glasgow; from 1914–1919, at Aberdeen; and from 1919–1936, at Oxford. During 1901–1903 he developed the disintegration theory of radioactivity with Ernest Rutherford, and, in 1903, Soddy and William Ramsay confirmed, spectroscopically, that helium is a product of the spontaneous disintegration or transmutation of the radium atom.

In 1910 Soddy first suggested the idea of a chemical isotope, introducing the term "isotope" in 1913. The idea followed from the chemical inseparability of various radioelements and natural elements, and Soddy concluded that elements of different atomic weights may in fact have the same chemical properties. It was for his researches on the origin and nature of isotopes that Soddy was awarded the Nobel Prize in chemistry in 1921.

Soddy's earliest important work in chemistry was done in collaboration with Ernest Rutherford at McGill University. They sorted out the various components of radiation emitted by radium and other naturally radioactive elements, concentrating on the alpha radiations which they demonstrated to be positively charged particles. Their conclusions that radioactivity is the result of a spontaneously disintegrating atom seemed to be a new chemical "alchemy." Their theory immediately provided force to J. J. Thomson's electron theory and to the idea that chemical and physical atoms are not indivisible and indestructible bits of matter, but are structured with smaller particles characterized by electrical charge and mass. This structure helps explain the mechanism of chemical reactions.

XLI. *The Cause and Nature of Radioactivity.*—Part I. *By* E. RUTHERFORD, *M.A., D.Sc., Macdonald Professor o Physics, and* F. SODDY, *B.A. (Oxon.), Demonstrator in Chemistry, McGill University, Montreal* *.

CONTENTS.

I. *Introduction.*

THE following papers give the results of a detailed investigation of the radioactivity of thorium compounds which has thrown light on the questions connected

* Communicated by the Authors. Accounts of these researches, during the progress of the investigation, have already been given to the London Chemical Society.

with the source and maintenance of the energy dissipated by radioactive substances. Radioactivity is shown to be accompanied by chemical changes in which new types of matter are being continuously produced. These reaction products are at first radioactive, the activity diminishing regularly from the moment of formation. Their continuous production maintains the radioactivity of the matter producing them at a definite equilibrium-value. The conclusion is drawn that these chemical changes must be sub-atomic in character.

The present researches had as their starting-point the facts that had come to light with regard to thorium radioactivity (Rutherford, Phil. Mag. 1900, vol. xlix. pp. 1 & 161). Besides being radioactive in the same sense as the uranium compounds, the compounds of thorium continuously emit into the surrounding atmosphere a gas which possesses the property of temporary radioactivity. This "emanation," as it has been named, is the source of rays, which ionize gases and darken the photographic film *.

The most striking property of the thorium emanation is its power of exciting radioactivity on all surfaces with which it comes into contact. A substance after being exposed for some time in the presence of the emanation behaves as if it were covered with an invisible layer of an intensely active material. If the thoria is exposed in a strong electric field, the excited radioactivity is entirely confined to the negatively charged surface. In this way it is possible to concentrate the excited radioactivity on a very small area. The excited radioactivity can be removed by rubbing or by the action of acids, as, for example, sulphuric, hydrochloric, and hydrofluoric acids. If the acids be then evaporated, the radioactivity remains on the dish.

The emanating power of thorium compounds is independent of the surrounding atmosphere, and the excited activity it produces is independent of the nature of the substance on which it is manifested. These properties made it appear that both phenomena were caused by minute quantities of special kinds of matter in the radioactive state, produced by the thorium compound.

The next consideration in regard to these examples of radioactivity, is that the activity in each case diminishes regularly with the lapse of time, the intensity of radiation at each instant being proportional to the amount of energy remaining to be radiated. For the emanation a period of

* If thorium oxide be exposed to a white heat its power of giving an emanation is to a large extent destroyed. Thoria that has been so treated is referred to throughout as "de-emanated."

one minute, and for the excited activity a period of eleven hours, causes the activity to fall to half its value.

These actions—(1) the production of radioactive material, and (2) the dissipation of its available energy by radiation—which are exhibited by thorium compounds in the secondary effects of emanating power and excited radioactivity, are in reality taking place in all manifestations of radioactivity. The constant radioactivity of the radioactive elements is the result of an equilibrium between these two opposing processes.

II. *The Experimental Methods of investigating Radioactivity.*

Two methods are used for the measurement of radioactivity, the electrical and the photographic. The photographic method is of a qualitative rather than a quantitative character; its effects are cumulative with time, and as a rule long exposures are necessary when the radioactivity of a feeble agent like thoria is to be demonstrated. In addition, Russell has shown that the darkening of a photographic plate is brought about also by agents of a totally different character from those under consideration, and, moreover, under very general conditions. Sir William Crookes (Proc. Roy. Soc. (1900) lxvi. p. 409) has sounded a timely note of warning against putting too much confidence in the indications of the photographic method of measuring radioactivity. The uncertainty of an effect produced by cumulative action over long periods of time quite precludes its use for work of anything but a qualitative character.

But the most important objection to the photographic method is that certain types of rays from radioactive substances, which ionize gases strongly, produce little if any effect on the sensitive film. In the case of uranium, these protographically inactive rays form by far the greatest part of the total radiation, and much of the previous work on uranium by the photographic method must be interpreted differently (Soddy, Proc. Chem. Soc. 1902, p. 121).

On the other hand, it is possible to compare intensities of radiation by the electrical method with greater rapidity and with an error not exceeding 1 or 2 per cent. These methods are based on the property generally possessed by all radiations of the kind in question, of rendering a gas capable of discharging both positive and negative electricity. These, as will be shown, are capable of great refinement and certainty. An ordinary quadrant electrometer is capable of detecting and measuring a difference of potential of at least 10^{-2} volts. With special instruments, this sensitiveness may be increased a hundredfold. An average value for the capacity of the electrometer and connexions is 3×10^{-5} microfarads; and when this is charged up to 10^{-2} volts, a quantity of electricity corresponding to 3×10^{-13} coulombs is stored up. Now in the electrolysis of water one gram of hydrogen carries a charge of 10^5 coulombs. Assuming, for the sake of example, that the conduction of electricity in gases is analogous to that in liquids, this amount of electricity corresponds to the transport of a mass of 3×10^{-18} grams of hydrogen; that is, a quantity of the order of 10^{-12} times that detected by the balance. For a more delicate instrument, this amount would produce a large effect.

The examples of radium in pitchblende and of the thorium-excited radioactivity make it certain that comparatively large ionization effects are produced by quantities of matter beyond the range of the balance or spectroscope.

The electrometer also affords the means of recognizing and differentiating between the emanations and radiations of different chemical substances. By the rate of decay the emanation from thorium, for example, can be instantly distinguished from that produced by radium; and although a difference in the rate of decay does not of itself argue a fundamental difference of nature, the identity of the rate of decay furnishes at least strong presumption of identity of nature.

Radiations, on the other hand, can be compared by means of their penetration powers (Rutherford, Phil. Mag. 1899, vol. xlvii. p. 122). If the rays from various radioactive substances are made to pass through successive layers of aluminium-foil, each additional layer of foil cuts down the radiation to a fraction of its former value, and a curve can be plotted with the thickness of metal penetrated as abscissæ, and the intensity of the rays after penetration as ordinates, expressing at a glance the penetration power of the rays under examination. The curves so obtained are quite different for different radioactive substances. The radiations from uranium, radium, thorium, each give distinct and characteristic curves, whilst that of the last-named again is quite different from that given by the excited radioactivity produced by the thorium emanation. It has been recently found (Rutherford and Grier, *Phys. Zeit.* 1902, p. 385) that thorium compounds, in addition to a type of easily absorbed Röntgen-rays, non-deviable in the magnetic field, emit also rays of a very penetrating character deviable in the magnetic field. The latter are therefore similar to cathode-rays, which are known to consist of material particles travelling with a velocity approaching that of light. But thorium, in comparison with uranium and radium, emits a much smaller proportion of deviable radiation. The determination of the proportion between the deviable and non-deviable rays affords a new means of investigating thorium radioactivity.

The electrometer thus supplies the study of radioactivity with methods of quantitative and qualitative investigation, and there is therefore no reason why the cause and nature of the phenomenon should not be the subject of chemical investigation.

Fig. 1 shows the general arrangement. From 0·5 to 0·1 gram of the compound to be tested, reduced to fine powder, is uniformly sifted over a platinum plate 36 sq. cms. in area.

Fig. 1.

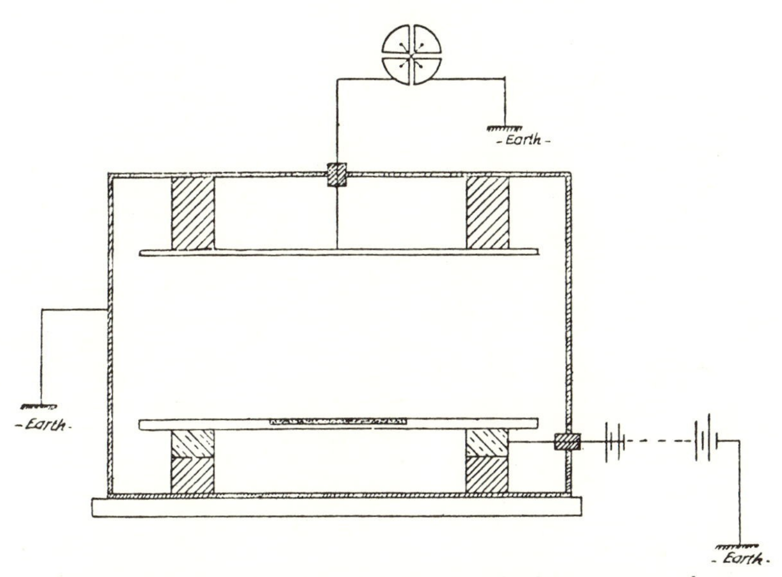

This plate was placed on a large metal plate connected to one pole of a battery of 300 volts, the other pole of which was earthed. An insulated parallel plate was placed about 6 cm. above it, and the whole apparatus inclosed in a metal box connected to earth, to prevent electrostatic disturbance. The shaded portions in the figure represented insulators. A door was made in the apparatus so that the plate could be rapidly placed in position or removed. Both pairs of quadrants are first connected to earth. On connecting the one pair with the apparatus, the deflexion of the needle from zero increases uniformly with time, and the time taken to pass over 100 divisions of the scale is taken by a stop-watch. The *rate* of movement is a measure of the ionization-current between the plates. The ratio of the currents for different substances is a comparative measure of their radioactivity.

With this apparatus ·5 gr. of thorium oxide produces a current of $1·1 \times 10^{-11}$ amperes, which, with the electrometer used, working at average sensitiveness, corresponds to 100 divisions of the scale in 36 seconds. In certain cases a special modification of the Dolezalek electrometer was employed which is 100 times more sensitive. With this instrument the radioactivity of 1 milligram of thoria produces a measurable effect. If the substance gives off an emanation, the current between the plates increases with time. Under these conditions, when the thorium compound is exposed in thin layers with a maximum of radiating surface, all but one or two per cent. of the total effect is due to the straight-line radiation. Even when the effect due to the emanation has attained a maximum, this constitutes a very small fraction of the whole. This effect, however, may to a large extent be eliminated by taking the current between the electrodes immediately after the material is placed in the testing-apparatus. It may be completely eliminated by passing a current of air between the electrodes to remove the emanation as fast as it is formed.

The current between the plates observed with the electrometer at first increases with the voltage, but a stage is very soon reached when there is a very small increase for a large additional voltage. A P.D. of 300 volts was sufficient to obtain the maximum current, so that all the ions reached the electrodes before any appreciable recombination occurred.

It must, however, at once be pointed out that it is difficult to make any absolute measure of radioactivity. The radiation from thorium is half absorbed by a thickness of aluminium of ·0004 cm.; and since thorium oxide is far denser than aluminium, it is probable that the radiation in this case is confined to a surface-layer only ·0001 cm. deep. It is obvious that different preparations, each containing the same percentage of thorium but with different densities and different states of division, will not give the same intensity of radiation. In comparing two different specimens of the same compound, it is important that the final steps in their preparation should be the same in each case. As a rule absolute measurements of this kind have been avoided. It is possible, however, to trace with great accuracy the *change* of radioactivity of any preparation with time by leaving it undisturbed on its original plate, and comparing it with a similarly undisturbed constant comparison sample. Most of the investigations have been carried out by this method.

III. *The Separation of a Radioactive Constituent from Thorium Compounds.*

During an investigation of the emanating power of thorium compounds, to be described later, evidence was obtained of the separation of an intensely radioactive constituent by chemical methods. It had been noticed that in certain cases thorium hydroxide, precipitated from dilute solutions of thorium nitrate by ammonia, possessed an abnormally low emanating power. This led naturally to an examination being made of the filtrates and washings obtained during the process. It was found that the filtrates invariably possessed emanating power, although from the nature of their production they are chemically free from thorium. If the filtrate is evaporated to dryness, and the ammonium salts removed by ignition, the small residues obtained exhibit radioactivity also, to an extent very much greater than that possessed by the same weight of thorium. As a rule these residues were of the order of one-thousandth part by weight of the thorium salt originally taken, and were many hundred, in some cases over a thousand, times more active than an equal weight of thoria. The separation of an active constituent from thorium by this method is not all dependent on the purity of the salt used. By the kindness of Dr. Knöfler, of Berlin, who, in the friendliest manner, presented us with a large specimen of his purest thorium nitrate, we were enabled to test this point. This specimen, which had been purified by a great many processes, did not contain any of the impurities found in the commercial salt before used. But its radioactivity and emanating power were at least as great, and the residues from the filtrates after precipitation by ammonia were no less active than those before obtained. These residues are free from thorium, or at most contain only the merest traces, and when redissolved in nitric acid do not appear to give any characteristic reaction.

An examination of the penetrating power of the rays from the radioactive residue, showed that the radiations emitted were in every respect identical with the ordinary thorium radiation. In another experiment the nature of the emanation from a similar intensely active thorium-free residue was submitted to examination. The rate of decay was quite indistinguishable from that of the ordinary thorium emanation; that is, substances chemically free from thorium have been prepared possessing thorium radioactivity in an intense degree.

The thorium hydroxide which had been submitted to the above process was found to be less than half as radioactive as the same weight of thorium oxide. It thus appeared that a constituent responsible for the radioactivity of thorium had been obtained, which possessed distinct chemical properties and an activity of the order of at least a thousand times as great as the material from which it had been separated.

Sir William Crookes (Proc. Roy. Soc. 1900, lxvi. p. 409) succeeded in separating a radioactive constituent of great activity and distinct chemical nature from uranium, and gave the name UrX to this substance. For the present, until more is known of its real nature, it will be convenient to name the active constituent of thorium ThX, similarly. Like UrX, however, ThX does not answer to any definite analytical reactions, but makes its appearance with precipitates formed in its solution even when no question of insolubility is involved. This accords with the view that it is present in infinitesimal quantity, and possesses correspondingly great activity. Even in the case of the most active preparations, these probably are composed of some ThX associated with accidental admixtures large in proportion.

These results receive confirmation from observations made on a different method of separating ThX. The experiment was tried of washing thoria with water repeatedly, and seeing if the radioactivity was thereby affected. In this way it was found that the filtered washings, on concentration, deposited small amounts of material with an activity often of the order of a thousand times greater than that of the original sample. In one experiment, 290 grams of thoria were shaken for a long time with nine quantities, each of 2 litres of distilled water. The first washing, containing thorium sulphate present as an impurity, was rejected, the rest concentrated to different stages and filtered at each stage. One of the residues so obtained weighed 6·4 mg., and was equivalent in radioactivity to 11·3 grams of the original thoria, and was therefore no less than 1800 times more radioactive. It was examined chemically, and gave, after conversion into sulphate, the characteristic reaction of thorium sulphate, being precipitated from its solution in cold water by warming. *No other substance than thorium could be detected by chemical analysis*, although of course the quantity was too small for a minute examination. The penetrating power of the radiation from this substance again established its identity with the ordinary thorium radiation.

In another experiment, a small quantity of thoria was shaken many times with large quantities of water. In this case, the radioactivity of the residue was examined and found to be about 20 per cent. less radioactive than the original sample.

The influence of Time on the activity of Thorium and ThX.—The preparations employed in our previous experiments were allowed to stand over during the Christmas vacation. On examining them about three weeks later it was found that the thorium hydroxide, which had originally possessed only about 36 per cent. of its normal activity, had almost completely recovered the usual value. The active residues, on the other hand, prepared by both methods, had almost completely lost their original activity. The chemical separation effected was thus not permanent in character. At this time M. Becquerel's paper (*Comptes Rendus*, cxxxiii. p. 977, Dec. 9th, 1901) came to hand, in which he shows that the same phenomena of recovery and decay are presented by uranium after it has been partially separated from its active constituent by chemical treatment.

A long series of observations was at once started to determine—

(1) The rate of recovery of the activity of thorium rendered less active by removal of ThX ;

(2) The rate of decay of the activity of the separated ThX ;

in order to see how the two processes were connected. The results led to the view that may at once be stated. The radioactivity of thorium at any time is the resultant of two opposing processes—

(1) The production of fresh radioactive material at a constant rate by the thorium compound ;

(2) The decay of the radiating power of the active material with time.

The normal or constant radioactivity possessed by thorium is an equilibrium value, where the rate of increase of radioactivity due to the production of fresh active material is balanced by the rate of decay of radioactivity of that already formed. It is the purpose of the present paper to substantiate and develope this hypothesis.

IV. *The Rates of Recovery and Decay of Thorium Radioactivity.*

A quantity of the pure thorium nitrate was separated from ThX in the manner described by several precipitations with ammonia. The radioactivity of the hydroxide so obtained was tested at regular intervals to determine the rate of recovery of its activity. For this purpose the original specimen of ·5 gram was left undisturbed throughout the whole series of measurements on the plate over which it had been sifted, and was compared always with ·5 gram of ordinary de-emanated thorium oxide spread similarly on a second plate and also left undisturbed. The emanation from the hydroxide was prevented from interfering with the results by a speeial arrangement for drawing a current of air over it during the measurements.

The active filtrate from the preparation was concentrated and made up to 100 c.c. volume. One quarter was evaporated to dryness and the ammonium nitrate expelled by ignition in a platinum dish, and the radioactivity of the residue tested at the same intervals as the hydroxide to determine the rate of decay of its activity. The comparison in this case was a standard sample of uranium oxide kept undisturbed on a metal plate, which repeated work has shown to be a perfectly constant source of radiation. The remainder of the filtrate was used for other experiments.

The following table gives an example of one of a numerous series of observations made with different preparations at different times. The maximum value obtained by the hydroxide and the original value of the ThX are taken as 100 :—

Time in days.	Activity of Hydroxide.	Activity of ThX.
0	44	100
1	37	117
2	48	100
3	54	88
4	62	72
5	68	
6	71	53
8	78	
9	...	29·5
10	83	25·2
13	...	15·2
15	...	11·1
17	96·5	
21	99	
28	100	

Fig. 2 shows the curves obtained by plotting the radioactivities as ordinates, and the time in days as abscissæ. Curve II. illustrates the rate of recovery of the activity of thorium, curve I. the rate of decay of activity of ThX. It will be seen that neither of the curves is regular for the first two days. The activity of the hydroxide at first actually

Fig. 2.

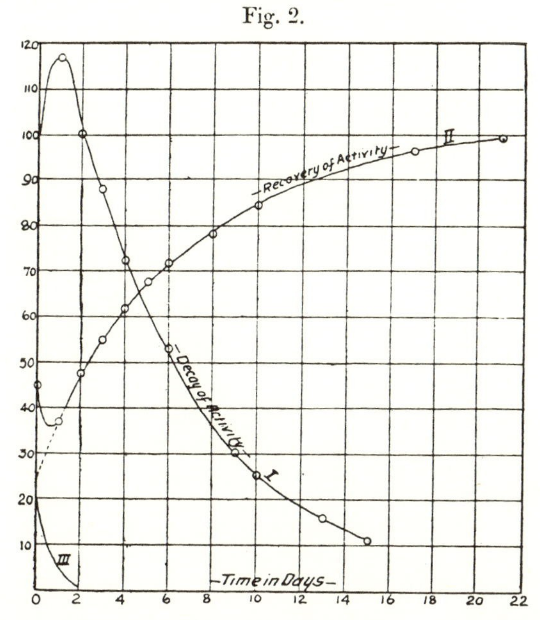

diminished and was at the same value after two days as when first prepared. The activity of the ThX, on the other hand, at first increases and does not begin to fall below the original value till after the lapse of two days (compare section IX.). These results cannot be ascribed to errors of measurement, for they have been regularly observed whenever similar preparations have been tested. The activity of the residue obtained from thorium oxide by the second method of washing decayed very similarly to that of ThX, as shown by the above curve.

If for present purposes the initial periods of the curve are disregarded and the later portions only considered, it will be seen at once that the time taken for the hydroxide to recover one half of its lost activity is about equal to the time taken by the ThX to lose half its activity, viz., in each case about 4 days, and speaking generally the percentage proportion of the lost activity regained by the hydroxide over any given interval is approximately equal to the percentage proportion of the activity lost by the ThX during the same interval. If the recovery curve is produced backwards in the normal direction to cut the vertical axis, it will be seen to do so at a minimum of about 25 per cent., and the above result holds even more accurately if the recovery is assumed to start from this constant minimum, as, indeed, it has been shown to do under suitable conditions (section IX., fig. 4).

This is brought out by fig. 3, which represents the recovery

Fig. 3.

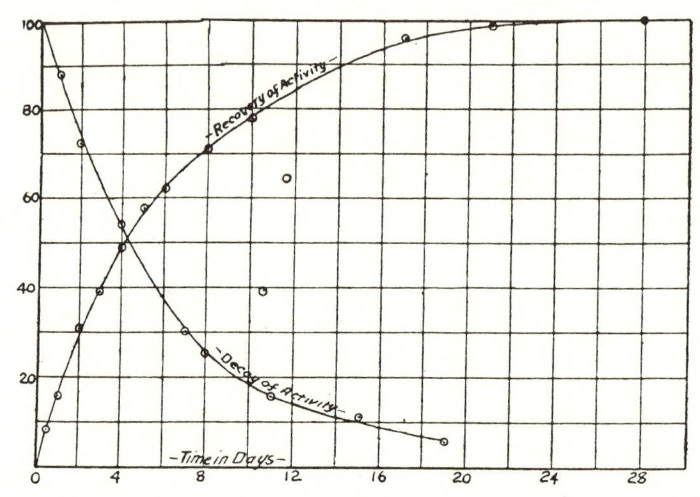

curve of thorium in which the percentage amounts of activity recovered, reckoned from this 25 per cent. minimum, are plotted as ordinates. In the same figure the decay curve after the second day is shown on the same scale.

The activity of ThX decreases very approximately in a geometrical progression with the time, *i. e.* if I_0 represent the initial activity and I_t the activity after time t,

$$\frac{I_t}{I_0} = e^{-\lambda t}, \quad . \quad . \quad . \quad . \quad . \quad . \quad . \quad . \quad (1)$$

where λ is a constant and e the base of natural logarithms.

The experimental curve obtained with the hydroxide for the rate of rise of its activity from a minimum to a maximum value will therefore be approximately expressed by the equation

$$\frac{I_t}{I_0} = 1 - e^{-\lambda t}, \quad . \quad . \quad . \quad . \quad . \quad . \quad (2)$$

where I_0 represents the amount of activity recovered when the maximum is reached, and I_t the activity recovered after time t, λ *being the same constant as before.*

Now this last equation has been theoretically developed in other places (compare Rutherford, Phil. Mag. 1900, pp. 10 and 181) to express the rise of activity to a constant maximum of a system consisting of radiating particles in which

(1) The rate of supply of fresh radiating particles is constant.

(2) The activity of each particle dies down geometrically with the time according to equation (1).

It therefore follows that if the initial irregularities of the curves are disregarded and the residual activity of thorium is assumed to possess a *constant* value, the experimental curve obtained for the recovery of activity will be explained if two processes are supposed to be taking place :

(1) That the active constituent ThX is being produced at a constant rate;

(2) That the activity of the ThX decays geometrically with time.

Without at first going into the difficult questions connected with the initial irregularities and the residual activity, the main result that follows from the curves given can be put to experimental test very simply. The primary conception is that the major part of the radioactivity of thorium is not due to the thorium at all, but to the presence of a non-thorium substance in minute amount which is being continuously produced.

V. *Chemical Properties of ThX.*

The fact that thorium on precipitation from its solutions by ammonia leaves the major part of its activity in the filtrate does not of itself prove that a material *constituent* responsible for this activity has been chemically separated. It is possible that the matter constituting the non-thorium part of the solution is rendered temporarily radioactive by its association with thorium, and this property is retained through the processes of precipitation, evaporation, and ignition, and manifests itself finally on the residue remaining.

This view, however, can be shown to be quite untenable, for upon it any precipitate capable of removing thorium completely from its solution should yield active residues similar to those obtained from ammonia. Quite the reverse, however, holds.

When thorium nitrate is precipitated by sodium or ammonium carbonate, the residue from the filtrate by evaporation and ignition is free from activity, and the thorium carbonate possesses the normal value for its activity.

The same holds true when oxalic acid is used as the

precipitant. This reagent even in strongly acid solution precipitates almost all of the thorium. When the filtrate is rendered alkaline by ammonia, filtered, evaporated, and ignited, the residue obtained is inactive.

In the case where sodium phosphate is used as the precipitant in ordinary acid solution, the part that comes down is more or less free from ThX. On making the solution alkaline with ammonia, the remainder of the thorium is precipitated as phosphate, and carries with it the whole of the active constituent, so that the residue from the filtrate is again inactive.

In fact ammonia is the only reagent of those tried capable of separating ThX from thorium.

The result of Sir William Crookes with uranium, which we have confirmed working with the electrical method, may be here mentioned. UrX is completely precipitated by ammonia together with uranium, and the residue obtained by the evaporation of the filtrate is quite inactive.

There can thus be no question that both ThX and UrX are distinct types of matter with definite chemical properties. Any hypothesis that attempts to account for the recovery of activity of thorium and uranium with time must of necessity start from this primary conception.

VI. *The Continuous Production of ThX.*

If the recovery of the activity of thorium with time is due to the production of ThX, it should be possible to obtain experimental evidence of the process. The first point to be ascertained is how far the removal of ThX by the method given reduces the total radioactivity of thorium. A preliminary trial showed that the most favourable conditions for the separation are by precipitating in hot dilute solutions by dilute ammonia. A quantity of 5 grams of thorium nitrate, as obtained from the maker, was so precipitated by ammonia, the precipitate being redissolved in nitric acid and reprecipitated under the same conditions successively *without lapse of time.*

The removal of ThX was followed by measuring the activity of the residues obtained from the successive filtrates. The activity of the ThX from the first filtrate was equivalent to 4·25 grams of thoria, from the second to 0·33 gram, and from the third to 0·07 gram. It will be seen that by two precipitations practically the whole of the ThX is removed. The radioactivity of the separated hydroxide was 48 per cent. of that of the standard de-emanated sample of thoria.

Rate of production of ThX.—A quantity of thorium nitrate solution that had been freed from ThX about a month before, was again subjected to the same process. The activity of the residue from the filtrate in an experiment in which 10 grams of this nitrate had been employed was equivalent to 8·3 grams of thorium oxide. This experiment was performed on the same day as the one recorded above, in which 5 grams of new nitrate had been employed, and it will be seen that there is no difference in the activity of the filtrate in the two cases. In one month the activity of the ThX in a thorium compound again possesses its maximum value.

If a period of 24 hours is allowed to elapse between the successive precipitations, the activity of the ThX formed during that time corresponds to about one-sixth of the maximum activity of the total thorium employed. In three hours the activity of the amount produced is about one-thirtieth. The rate of production of ThX worked out from those figures well agrees with the form of the curve obtained for the recovery of activity of thorium, if the latter is taken to express the continuous production of ThX at a constant rate and the diminution of the activity of the product in geometrical progression with the time.

By using the sensitive electrometer, the course of production of ThX can be followed after extremely short intervals. Working with 10 grams of thorium nitrate, the amount produced in the minimum time taken to carry out the successive precipitations is as much as can be conveniently measured. If any interval is allowed to lapse the effect is beyond the range of the instrument, unless the sensitiveness is reduced to a fraction of its ordinary value by the introduction of capacities into the system. Capacities of ·01 and ·02 microfarad, which reduce the sensitiveness to less than one two-hundredth of the normal, were frequently employed in dealing with these active residues.

The process of the production of ThX is continuous, and no alteration was observed in the amount produced in a given time after repeated separations. In an experiment carried out for another purpose (section IX.) after 23 successive precipitations extending over 9 days, the amount formed during the last interval was as far as could be judged no less than what occurred at the beginning of the process.

The phenomenon of radioactivity, by means of the electrometer as its measuring instrument, thus enables us to detect and measure changes occurring in matter after a few minutes interval, which have never yet been detected by the balance or suspected of taking place.

VII. *Influence of Conditions on the Changes occurring in Thorium.*

It has been shown that in thorium compounds the decay of radioactivity with time is balanced by a continuous production of fresh active material. The change which produces this material must be chemical in nature, for the products of the action are different in chemical properties from the thorium from which they are produced. The first step in the study of the nature of this change is to examine the effects of conditions upon its rate.

Effect of conditions on the rate of decay.—Since the activity of the products affords the means of measuring the amount of change, the influence of conditions on the rate of decay must be first found. It was observed that, like all other types of temporary radioactivity, the rate of decay is unaltered by any known agency. It is unaffected by ignition and chemical treatment, and the material responsible for it can be dissolved in acids and re-obtained by the evaporation of the solution, without affecting the activity. The following experiment shows that the activity decays at the same rate in solutions as in the solid state. The remainder of the solution that had been used to determine the decay curve of ThX (fig. 2) was allowed to stand, and at the end of 12 days a second quarter was evaporated to dryness and ignited, and its activity compared with that of the first which had been left since evaporation upon its original platinum dish. The activities of the two specimens so compared with each other were the same, showing that in spite of the very different conditions the two fractions had decayed at equal rates. After 19 days a third quarter was evaporated, and the activity, now very small, was indistinguishable from that of the fraction first evaporated. Re-solution of the residues after the activity had decayed does not at all regenerate it. The activity of ThX thus decays at a rate independent of the chemical and physical condition of the molecule.

Thus the rate of recovery of activity under different conditions in thorium compounds affords a direct measure of the rate of production of ThX under these conditions. The following experiments were performed :—

One part of thorium hydroxide newly separated from ThX was sealed up in a vacuum obtained by a good Töpler pump, and the other part exposed to air. On comparing the samples 12 days later no difference could be detected between them either in their radioactivity or emanating power.

In the next experiment a quantity of hydroxide freed from

ThX was divided into two equal parts; one was exposed for 20 hours to the heat of a Bunsen burner in a platinum crucible, and then compared with the other. No difference in the activities was observed. In a second experiment, one half was ignited for 20 minutes on the blast, and then compared with the other with the same result. The difference of temperature and the conversion of thorium hydroxide into oxide thus exercised no influence on the activity.

Some experiments that were designed to test in as drastic a manner as possible the effect of the chemical condition of the molecule on the rate of production of ThX brought to light small differences, but these are almost certainly to be accounted for in another way. It will be shown later (section IX.) that about 21 per cent. of the normal radioactivity of thorium oxide under ordinary conditions consists of a secondary activity excited on the mass of the material. This portion is of course a variable, and since it is divided among the total amount of matter present, the conditions of aggregation, &c., will affect the value of this part. This effect of excited radioactivity in thorium makes a certain answer to the question difficult, and on this account the conclusion that the rate of production of ThX is independent of the molecular conditions is not final. The following experiment, however, makes it extremely probable.

A quantity of thorium nitrate as obtained from the maker was converted into oxide in a platinum crucible by treatment with sulphuric acid and ignition to a white heat. The de-emanated oxide so obtained was spread on a plate, and any change in radioactivity with time, which under these circumstances could certainly be detected, was looked for during the first week from preparation. None whatever was observed, whereas if the rate of production of ThX in thorium nitrate is different from that in the oxide, the equilibrium point, at which the decay and increase of activity balance each other, will be altered in consequence. There should have therefore occurred a logarithmic rise or fall from the old to the new value. As, however, the radioactivity remained constant, it appears very probable that the changes involved are independent of the molecular condition.

It will be seen that the assumption is here made that the proportion of excited radioactivity in the two compounds is the same, and for this reason compounds were chosen which possess but low emanating power. (Compare section IX. last paragraph.)

Uranium is a far simpler example of a radioactive element than thorium, as the phenomena of excited radioactivity and emanating power are here absent. The separation of UrX and the recovery of the activity of the uranium with time appear, however, analogous to these processes in thorium, and the rate of recovery and decay of uranium activity are at present under investigation. It is proposed to test the influence of conditions on the rate of change more thoroughly in the case of uranium, as here secondary changes do not interfere.

VIII. *The Cause and Nature of Radioactivity.*

The foregoing conclusions enable a great generalization to be made in the subject of radioactivity. Energy considerations require that the intensity of radiation from any source should die down with time unless there is a constant supply of energy to replace that dissipated. This has been found to hold true in the case of all known types of radioactivity with the exception of the "naturally" radioactive elements—to take the best established cases, thorium, uranium, and radium. It will be shown later that the radioactivity of the emanation produced by thorium compounds decays geometrically with the time under all conditions, and is not affected by the most drastic chemical and physical treatment. The same has been shown by one of us (Phil. Mag. 1900, p. 161) to hold for the excited radioactivity produced by the thorium emanation. This decays at the same rate whether on the wire on which it is originally deposited, or in solution of hydrochloric or nitric acid. The excited radioactivity produced by the radium emanation appears analogous. All these examples satisfy energy considerations. In the case of the three naturally occurring radioactive elements, however, it is obvious that there must be a continuous replacement of the dissipated energy, and no satisfactory explanation has yet been put forward.

The nature of the process becomes clear in the light of the foregoing results. The material constituent responsible for the radioactivity, when it is separated from the thorium which produces it, then behaves in the same way as the other types of radioactivity cited. Its activity decays geometrically with the time, and the rate of decay is independent of the molecular conditions. The normal radioactivity is, however, maintained at a constant value by a chemical change which produces fresh radioactive material at a rate also independent of the conditions. The energy required to maintain the radiations will be accounted for if we suppose that the energy of the system after the change has occurred is less than it was before.

The work of Crookes and Becquerel on the separation of UrX and the recovery of the activity of the uranium with time, makes it appear extremely probable that the same explanation holds true for this element. The work of M. and Mme. Curie, the discoverers of radium, goes to show that this body easily suffers a temporary decrease of its activity by chemical treatment, the normal value being regained after the lapse of time, and this can be well interpreted on the new view. All known types of radioactivity can thus be brought under the same category.

IX. *The Initial Portions of the Curves of Decay and Recovery.*

The curves of the recovery and decay of the activities of thorium and ThX with time suggested the explanation that the radioactivity of thorium was being maintained by the production of ThX at a constant rate. Before this can be considered rigidly established, two outstanding points remain to be cleared up. 1. What is the meaning of the early portion of the curves? The recovery curve drops before it rises, and the decay curve rises before it drops. 2. Why does not the removal of ThX render thorium completely inactive? A large proportion of the original radioactivity is not affected by the removal of ThX.

A study of the curves (fig. 2) shows that in each case a double action is probably at work. It may be supposed that the normal decay and recovery are taking place, but are being masked by a simultaneous rise and decay from other causes. From what is known of thorium radioactivity, it was surmised that an action might be taking place similar to that effected by the emanation of exciting radioactivity on surrounding inactive matter. It will be shown later that the ThX, and not thorium, is the cause of the emanating power of thorium compounds. On this view, the residual activity of thorium might consist in whole or in part of a secondary or excited radioactivity produced on the whole mass of the thorium compound by its association with the ThX. The drop in the recovery-curve on this view would be due to the decay of this excited radioactivity proceeding simultaneously with, and at first reversing the effect of the regeneration of ThX. The rise of the decay-curve would be the increase due to the ThX exciting activity on the matter with which it is associated, the increase from this cause being greater than the decrease due to the decay of the activity of the ThX. It is easy to put this hypothesis to experimental test. If the ThX is removed from the thorium as soon as it is formed over a sufficient period, the former will be prevented from exciting activity on the latter, and that already excited will decay spontaneously. The experiment was therefore performed. A quantity of nitrate was precipitated as hydroxide in the usual way to remove ThX, the precipitate redissolved in nitric acid, and again precipitated after a certain interval. From time to time a portion of the hydroxide was removed and its radioactivity tested. In this way the thorium was precipitated in all 23 times in a period of 9 days, and the radioactivity reduced to a constant minimum. The following table shows the results:—

	Activity of Hydroxide. per cent.
After first precipitation	46
After precipitations at three intervals of 24 hours	39
At three more intervals each of 24 hours, and three more each of 8 hours	22
At three more each of 8 hours	24
At six more each of 4 hours	25

The constant minimum thus attained—about 25 per cent. of the original activity—is thus about 21 per cent. below that obtained by two successive precipitations without interval, which has been shown to remove all the ThX separable by the process. The rate of recovery of this 23 times precipitated hydroxide was then measured (fig. 4). It will be

Fig. 4.

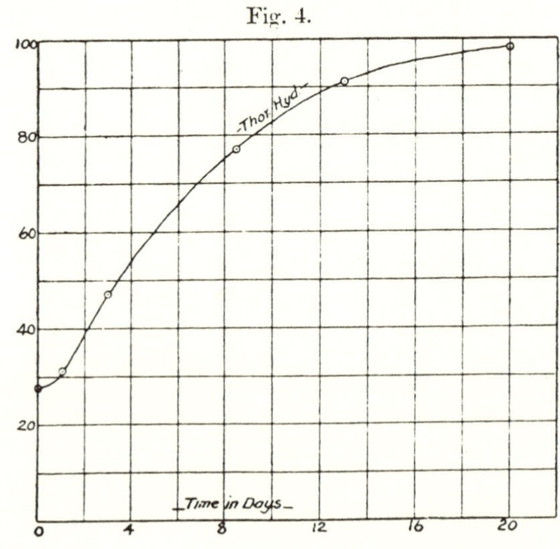

seen that it is now quite normal, and the initial drop characteristic of the ordinary curve is quite absent. It is in fact almost identical with the ordinary curve (fig. 2) that has been produced back to cut the vertical axis, and there is thus no doubt that there is a residual activity of thorium unconnected apparently with ThX, and constituting about one fourth of the whole.

The decay-curves of several of the fractions of ThX separated in this experiment after varying intervals of time were taken for the first few days. All of them showed the initial rise of about 15 per cent. at the end of 18 hours, and then a normal decay to zero. The position is thus proved that the initial irregularities are caused by the secondary radiation excited by ThX upon the surrounding matter. By suitably choosing the conditions the recovery-curve can be made to rise normally from a constant minimum, and the decay-curve be shown to consist of two curves, the first the rate of production of excited radioactivity, and the second the rate of decay of the activity as a whole.

So far nothing has been stated as to whether the excited radioactivity which contributes about 21 per cent. of the total activity of thorium is the same or different from the known type produced by the thorium emanation. All that has been assumed is that it should follow the same general law; *i. e.* the effect will increase with the time of action of the exciting cause, and decrease with time after the cause is removed. If the rate of rise of the excited activity be worked out from the curves given (fig. 5) it will be found to agree with that of the ordinary excited activity, *i. e.* it rises to half value in about 12 hours. Curve 1 is the observed decay-curve for ThX; curve 2 is the theoretical curve, assuming that it decreases geometrically with time and falls to half value in four days. Curve 3 is obtained by plotting the difference between these two, and therefore constitutes the curve of excited activity. Curve 4 is the experimental curve obtained for the rise of the excited radioactivity from the thorium emanation when the exciting cause is constant. But the exciting cause (ThX) in the present case is not constant, but is itself falling to half value in 4 days, and hence the difference curve, at first almost on the other, drops away from it as time goes on, and finally decays to zero. There is thus no reason to doubt that the effect is the same as that produced by the thorium emanation, which is itself a secondary effect of ThX. Curve 3 (fig. 2) represents a similar difference curve for *the decay* of excited activity, plotted from *the recovery curve* of thorium.

Since this effect of excited activity is caused by the emanation, it seemed reasonable to suppose that it will be greater, the less the emanation succeeds in escaping in the radioactive state, and therefore that de-emanated compounds should

Fig. 5.

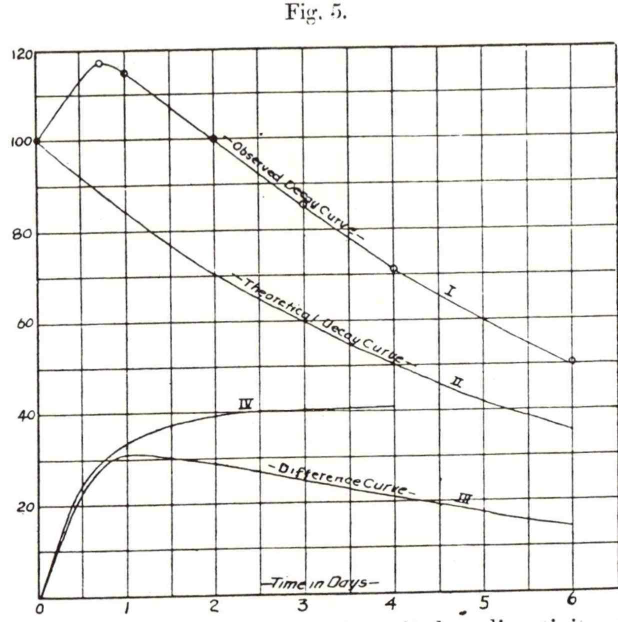

possess a greater proportion of excited radioactivity than those with high emanating power. This conclusion was tested by converting a specimen of thorium carbonate with an emanating power five times that of ordinary thoria, into oxide and de-emanating by intense ignition. The energy that before escaped in the form of emanation is now, all but a few per cent., prevented from escaping. The radioactivity of the oxide so prepared rose in the first three days about thirty per cent. of its original amount, and there thus seem to be grounds for the view that the excited radioactivity will contribute a much greater effect in a non-emanating thorium compound than in one possessing great emanating power.

Additional confirmation of this view is to be found in the nature of the radiations emitted by the two classes of compounds (Section XI.).

X. *The Non-separable Radioactivity of Thorium.*

It has not yet been found possible by any means to free thorium from its residual activity, and the place of this part in the scheme of radioactivity of thorium remains to be considered. Disregarding the view that it is a separate phenomenon, and not connected with the major part of the activity, two hypotheses can be brought forward capable of experimental test, and in accordance with the views advanced on the nature of radioactivity, to account for the existence of this part. First, if there was a second type of excited activity produced by ThX similar to that known, but with a very slow rate of decay, it would account for the existence of the non-separable activity. If this is true it will not be found possible to free thorium from this activity by chemical means, but the continuous removal of ThX over a very long period would, as in the above case, cause its spontaneous decay.

Secondly, if the change which gives rise to ThX produces a second type of matter at the same time, *i. e.* if it is of the type of a decomposition rather than a depolymerization, the second type would also in all probability be radioactive, and would cause the residual activity. On this view the second type of matter should also be amenable to separation by chemical means, although it is certain from the failure of the methods already tried that it resembles thorium much more closely than ThX. But until it is separated from the thorium producing it, its activity will not decay spontaneously. Thus what has already been shown to hold for ThX will be true for the second constituent if methods are found to remove it from the thorium.

It has been shown (Soddy, *loc. cit.*) that uranium also possesses a non-separable radioactivity extremely analogous to that possessed by thorium, and whatever view is taken of the one will in all probability hold also for the other. This consideration makes the second hypothesis, that the residual activity is caused by a second non-thorium type of matter produced in the original change, the more probable of the two.

XI. *The Nature of the Radiations from Thorium and ThX.*

From the view of radioactivity put forward it necessarily follows that the total radioactivity of thorium is altered neither in character nor amount by chemical treatment. With regard to the first, the amount of activity, it has been pointed out that the intensity of radiations *emitted* do not furnish alone a measure of the activity. The absorption in the mass of material must be considered also. The radiations of thorium oxide are derived from a very dense powder; those from ThX, on the other hand, have only to penetrate a very thin film of material. The difficulty can be overcome to some extent by taking for the comparison the radioactivity of a thin film of a soluble thorium salt produced by evaporating a solution to dryness over a large metal plate. Compared in this way, the radioactivity of ThX when first separated almost exactly equals the activity of the nitrate from which it is produced, while the hydroxide retains about two-fifths of this amount. The total activity of the products is therefore greater than that of the original salt; but this is to be expected, for it is certain that more absorption takes place in the nitrate than in the products into which it is separated.

Similar difficulties stand in the way of an answer to the second question, whether the nature of the radiations is affected by chemical treatment, for it has been experimentally observed that the penetrating power of these radiations decreases with the thickness of material traversed. The character of the radiations from ThX and thorium have, however, been compared by the method of penetration power. A large number of comparisons justifies the view that the character of thorium radioactivity is unaltered by chemical treatment and the separation of ThX, although the different types are unequally distributed among the separated products.

Determinations of the proportion of rays deviable by the magnetic field in thorium and ThX throws fresh light on the question. The general result is that ThX gives out both deviable and non-deviable rays, and the same applies to the excited activity produced by ThX. But in the experiment in which the excited radiation was allowed to spontaneously decay, by removing ThX as formed, the thorium compound obtained after 23 precipitations was found to be quite free from deviable radiation. This is one of the most striking resemblances between the non-separable radioactivities of uranium and thorium, and warrants the question whether the primary radiation of ThX is not, like that of UrX, composed entirely of cathode-rays. There is, however, no means of deciding this point owing to the excited radiation which always accompanies the primary radiation of ThX, and which itself comprises both types of rays.

Finally, it may be mentioned that the proportion of deviable and non-deviable radiation is different for different compounds of thorium. The nitrate and ignited oxide, compounds which hardly possess any emanating power, have a higher proportion of deviable radiation than compounds with great emanating power. This is indirect evidence of the correctness of the view already put forward (Section IX.), that when the emanation is prevented from escaping it augments the proportion of excited radioactivity of the compound.

XII. *Summary of Results.*

The foregoing experimental results may be briefly summarized. The major part of the radioactivity of thorium—ordinarily about 54 per cent.—is due to a non-thorium type of matter, ThX, possessing distinct chemical properties, which is temporarily radioactive, its activity falling to half value in about four days. The constant radioactivity of thorium is maintained by the production of this material at a constant rate. Both the rate of production of the new material and the rate of decay of its activity appear to be independent of the physical and chemical condition of the system.

The ThX further possesses the property of exciting radioactivity on surrounding inactive matter, and about 21 per cent. of the total activity under ordinary circumstances is derived from this source. Its rate of decay and other considerations make it appear probable that it is the same as the excited radioactivity produced by the thorium emanation, which is in turn produced by ThX. There is evidence that, if from any cause the emanation is prevented from escaping in the radioactive state, the energy of its radiation goes to augment the proportion of excited radioactivity in the compound.

Thorium can be freed by suitable means from both ThX and the excited radioactivity which the latter produces, and then possesses an activity about 25 per cent. of its original value, below which it has not been reduced. This residual radiation consists entirely of rays non-deviable by the magnetic field, whereas the other two components comprise both deviable and non-deviable radiation. Most probably this residual activity is caused by a second non-thorium type of matter produced in the same change as ThX, and it should therefore prove possible to separate it by chemical methods.

XIII. *General Theoretical Considerations.*

Turning from the experimental results to their theoretical interpretation, it is necessary to first consider the generally accepted view of the nature of radioactivity. It is well established that this property is the function of the atom and not of the molecule. Uranium and thorium, to take the most definite cases, possess the property in whatever molecular condition they occur, and the former also in the elementary state. So far as the radioactivity of different compounds of different density and states of division can be compared together, the intensity of the radiation appears to depend only on the quantity of active element present. It is not at all dependent on the source from which the element is derived, or the process of purification to which it has been subjected, provided sufficient time is allowed for the equilibrium point to be reached. It is not possible to explain the phenomena by the existence of impurities associated with the radioactive elements, even if any advantage could be derived from the assumption. For these impurities must necessarily be present always to the same extent in different specimens derived from the most widely different sources, and, moreover, they must persist *in unaltered amount* after the most refined processes of purification. This is contrary to the accepted meaning of the term impurity.

All the most prominent workers in this subject are agreed in considering radioactivity an atomic phenomenon. M. and Mme. Curie, the pioneers in the chemistry of the subject, have recently put forward their views (*Comptes Rendus*, cxxxiv. 1902, p. 85). They state that this idea underlies their whole work from the beginning and created their methods of research. M. Becquerel, the original discoverer of the property for uranium, in his announcement of the recovery of the activity of the same element after the active constituent had been removed by chemical treatment, points out the significance of the fact that uranium is giving out cathode-rays. These, according to the hypothesis of Sir William Crookes and Prof. J. J. Thomson, are *material* particles of mass one thousandth of the hydrogen atom.

Since, therefore, radioactivity is at once an atomic phenomenon and accompanied by chemical changes in which new types of matter are produced, these changes must be occurring within the atom, and the radioactive elements must be undergoing spontaneous transformation, The results that have so far been obtained, which indicate that the velocity of this reaction is unaffected by the conditions, makes it clear that the changes in question are different in character from any that have been before dealt with in chemistry. It is apparent that we are dealing with phenomena outside the sphere of known atomic forces. Radioactivity may therefore be considered as a manifestation of subatomic chemical change.

The changes brought to knowledge by radioactivity, although undeniably material and chemical in nature, are of a different order of magnitude from any that have before been dealt with in chemistry. The course of the production of new matter which can be recognized by the electrometer, by means of the property of radioactivity, after the lapse of a few hours or even minutes, might conceivably require geological epochs to attain to quantities recognized by the balance. However the well-defined chemical properties of both ThX and UrX are not in accordance with the view that the actual amounts involved are of this extreme order of minuteness. On the other hand, the existence of radioactive elements at all in the earth's crust is an *à priori* argument against the magnitude of the change being anything but small.

Radioactivity as a new property of matter capable of exact quantitative determination thus possesses an interest apart from the peculiar properties and powers which the radiations themselves exhibit. Mme. Curie, who isolated from pitchblende a new substance, radium, which possessed distinct chemical properties and spectroscopic lines, used the property as a means of chemical analysis. An exact parallel is to be found in Bunsen's discovery and separation of cæsium and rubidium by means of the spectroscope.

The present results show that radioactivity can also be used to follow *chemical changes occurring in matter*. The properties of matter that fulfil the necessary conditions for the study of chemical change without disturbance to the reacting system are few in number. It seems not unreasonable to hope, in the light of the foregoing results, that radioactivity, being such a property, affords the means of obtaining information of the processes occurring within the chemical atom, in the same way as the rotation of the plane of polarization and other physical properties have been used in chemistry for the investigation of the course of molecular change.

EINSTEIN, ALBERT. (b. Ulm, Germany, 14 March 1879; d. Princeton, New Jersey, 18 April 1955)

Einstein attended *Gymnasia* in Munich and in Aarau, Switzerland, and then studied physics and mathematics at the Eidgenössische Technische Hochschule (Polytechnic) in Zurich. While working in the Swiss Patent Office in Berne, he wrote his 1905 doctoral thesis, "A New Determination of Molecular Dimensions," and other papers that have become famous as his classical papers on Brownian motion, special relativity, and the photoelectric effect. In 1909 Einstein became associate professor of physics at the University of Zurich; in 1911 he was made professor at the German University in Prague, and, in 1912, professor at Zurich Polytechnic. From 1914 to 1933 he was the director of the Kaiser Wilhelm Institute for Physics and professor at the University of Berlin, lecturing regularly in Leiden, as well, during the later years. When Hitler became Chancellor in 1933 Einstein left Germany to accept a position at the newly established Institute for Advanced Study in Princeton. He received the Nobel Prize in physics in 1921.

Einstein is best known for his work on the electrodynamics of moving bodies, first developed as a restricted theory in 1905 and then extended to a general framework in which the effects of acceleration are identical to the effects of gravitation. The 1919 solar eclipse measurements of the bending of starlight in the vicinity of the sun's gravitational field provided dramatic confirmation for Einstein's general theory of relativity; he soon achieved a popular reputation equalled by no modern scientist. His commitment to a unification of the laws of electromagnetism and gravitation (unified field theory) dominated his later work in physics and cosmology. He never accepted the validity of the new quantum mechanics, which denied the possibility, in principle, of a causal, nonprobabilistic set of laws for physical theory.

Einstein's earliest scientific work dealt with electromagnetism, thermodynamics, and a statistical approach to the entropy function. His 1905 paper "On the Movement of Small Particles Suspended in a Stationary Liquid Demanded by the Molecular-Kinetic Theory of Heat" provided crucial support for the physical validity of Avogadro's hypothesis and molecular-kinetic theory. The photoelectric-effect paper, "On a Heuristic Point of View about the Creation and Conversion of Light," demonstrated the necessity of incorporating the atomistic (or quantum) idea into the electromagnetic theory of light. Here Einstein demonstrated that the mathematical description for the entropy of black-body radiation in a closed volume is identical to that for a gas in the same volume. By analogy, then, electromagnetic radiation may be treated as a dynamic collection of particles, as is the case for a gas, where the energy of the electromagnetic or light particles is proportional to the frequency of radiation. Einstein's revolutionary paper made clear that the "atomistic" or discontinuous nature of matter is characteristic of energy as well, generalizing Planck's recent work on the existence of discontinuities, or "quanta" of energy in black-body radiation.

3 On a Heuristic Point of View about the Creation and Conversion of Light†

A. EINSTEIN

THERE exists an essential formal difference between the theoretical pictures physicists have drawn of gases and other ponderable bodies and Maxwell's theory of electromagnetic processes in so-called empty space. Whereas we assume the state of a body to be completely determined by the positions and velocities of an, albeit very large, still finite number of atoms and electrons, we use for the determination of the electromagnetic state in space continuous spatial functions, so that a finite number of variables cannot be considered to be sufficient to fix completely the electromagnetic state in space. According to Maxwell's theory, the energy must be considered to be a continuous function in space for all purely electromagnetic phenomena, thus also for light, while according to the present-day ideas of physicists the energy of a ponderable body can be written as a sum over the atoms and electrons. The energy of a ponderable body cannot be split into arbitrarily many, arbitrarily small parts, while the energy of a light ray, emitted by a point source of light is according to Maxwell's theory (or in general according to any wave theory) of light distributed continuously over an ever increasing volume.

The wave theory of light which operates with continuous functions in space has been excellently justified for the representation of purely optical phenomena and it is unlikely ever to be replaced by another theory. One should, however, bear in mind that optical observations refer to time averages and not to

† *Ann. Physik* **17**, 132 (1905).

instantaneous values and notwithstanding the complete experimental verification of the theory of diffraction, reflexion, refraction, dispersion, and so on, it is quite conceivable that a theory of light involving the use of continuous functions in space will lead to contradictions with experience, if it is applied to the phenomena of the creation and conversion of light.

In fact, it seems to me that the observations on "black-body radiation", photoluminescence, the production of cathode rays by ultraviolet light and other phenomena involving the emission or conversion of light can be better understood on the assumption that the energy of light is distributed discontinuously in space. According to the assumption considered here, when a light ray starting from a point is propagated, the energy is not continuously distributed over an ever increasing volume, but it consists of a finite number of energy quanta, localised in space, which move without being divided and which can be absorbed or emitted only as a whole.

In the following, I shall communicate the train of thought and the facts which led me to this conclusion, in the hope that the point of view to be given may turn out to be useful for some research workers in their investigations.

1. On a Difficulty in the Theory of "Black-body Radiation"

To begin with, we take the point of view of Maxwell's theory and electron theory and consider the following case. Let there be in a volume completely surrounded by reflecting walls, a number of gas molecules and electrons moving freely and exerting upon one another conservative forces when they approach each other, that is, colliding with one another as gas molecules according to the kinetic theory of gases.† Let there further be a number of electrons which are bound to points in space, which are far from one

† This assumption is equivalent to the preposition that the average kinetic energies of gas molecules and electrons are equal to one another in temperature equilibrium. It is well known that Mr. Drude has theoretically derived in this way the relation between the thermal and electrical conductivities of metals.

another, by forces proportional to the distance from those points and in the direction towards those points. These electrons are also assumed to be interacting conservatively with the free molecules and electrons as soon as the latter come close to them. We call the electrons bound to points in space "resonators"; they emit and absorb electromagnetic waves with definite periods.

According to present-day ideas on the emission of light, the radiation in the volume considered—which can be found for the case of dynamic equilibrium on the basis of the Maxwell theory—must be identical with the "black-body radiation"—at least provided we assume that resonators are present of all frequencies to be considered.

For the time being, we neglect the radiation emitted and absorbed by the resonators and look for the condition for dynamic equilibrium corresponding to the interaction (collisions) between molecules and electrons. Kinetic gas theory gives for this the condition that the average kinetic energy of a resonator electron must equal the average kinetic energy corresponding to the translational motion of a gas molecule. If we decompose the motion of a resonator electron into three mutually perpendicular directions of oscillation, we find for the average value $\bar{E}$ of the energy of such a linear oscillatory motion

$$\bar{E} = \frac{R}{N} T,$$

where R is the gas constant, N the number of "real molecules" in a gramme equivalent and T the absolute temperature. This follows as the energy $\bar{E}$ is equal to $\frac{2}{3}$ of the kinetic energy of a free molecules of a monatomic gas since the time averages of the kinetic and the potential energy of a resonator are equal to one another. If, for some reason—in our case because of radiation effects—one manages to make the time average of a resonator larger or smaller than $\bar{E}$, collisions with the free electrons and molecules will lead to an energy transfer to or from the gas which has a non-vanishing average. Thus, for the case considered by us, dynamic equilibrium will be possible only if each resonator has the average energy $\bar{E}$.

We can now use a similar argument for the interaction between the resonators and the radiation which is present in space. Mr. Planck[1] has derived for this case the condition for dynamic equilibrium under the assumption that one can consider the radiation as the most random process imaginable.† He found

$$\bar{E}_\nu = \frac{L^3}{8\pi\nu^2}\rho_\nu,$$

where $\bar{E}_\nu$ is the average energy of a resonator with eigenfrequency ν (per oscillating component), L the velocity of light, ν the frequency and $\rho_\nu d\nu$ the energy per unit volume of that part of the radiation which has frequencies between ν and $\nu+d\nu$.

If the radiation energy of frequency ν is not to be either decreased or increased steadily, we must have

$$\frac{R}{N}T = \bar{E} = \bar{E}_\nu = \frac{L^3}{8\pi\nu^2}\rho_\nu,$$

† One can formulate this assumption as follows. We expand the z-component of the electrical force (Z) at a given point in space between the time $t=0$ and $t=T$ (where T indicates a time which is large compared to all oscillation periods considered) in a Fourier series

$$Z = \sum_{\nu=1}^{\infty} A_\nu \sin\left(2\pi\nu\frac{t}{T}+\alpha_\nu\right),$$

where $A_\nu \geqq 0$ and $0 \leqq \alpha_\nu \leqq 2\pi$. For the same point in space, one considers to have made such an expansion arbitrarily often with arbitrarily chosen initial times. In that case, we have for the frequency of different combinations of values for the quantities A_ν and α_ν (statistical) probabilities dW of the form

$$dW = f(A_1, A_2, \ldots, \alpha_1, \alpha_2, \ldots)\, dA_1\, dA_2 \ldots d\alpha_1\, d\alpha_2 \ldots.$$

Radiation is now the most random process imaginable, if

$$f(A_1, A_2, \ldots, \alpha_1, \alpha_2, \ldots) = F_1(A_1)F_2(A_2)\ldots f_1(\alpha_1)f_2(\alpha_2)\ldots,$$

that is, when the probability for a given value of one of the A or the α is independent of the values of the other A and α. The more closely the condition is satisfied that the separate pairs of quantities A_ν and α_ν depend on the emission and absorption processes of *special* groups of resonators, the more definitely can we thus say in the case treated by us that the radiation can be considered to be the most random imaginable one.

$$\rho_\nu = \frac{R}{N}\frac{8\pi\nu^2}{L^3}T.$$

This relation, which we found as the condition for dynamic equilibrium does not only lack agreement with experiment, but it also shows that in our picture there can be no question of a definite distribution of energy between aether and matter. The greater we choose the range of frequencies of the resonators, the greater becomes the radiation energy in space and in the limit we get

$$\int_0^\infty \rho_\nu \, d\nu = \frac{R}{N}\frac{8\pi}{L^3}T \int_0^\infty \nu^2 \, d\nu = \infty.$$

2. On Planck's Determination of Elementary Quanta

We shall show in the following that determination of elementary quanta given by Mr. Planck is, to a certain extent, independent of the theory of "black-body radiation" constructed by him.

Planck's formula[2] for ρ_ν which agrees with all experiments up to the present is

$$\rho_\nu = \frac{\alpha\nu^3}{e^{\beta\nu/T} - 1},$$

where $\alpha = 6{\cdot}10 \times 10^{-56}$, $\beta = 4{\cdot}866 \times 10^{-11}$.

For large values of T/ν, that is, for long wavelengths and high radiation densities, this formula has the following limiting form

$$\rho_\nu = \frac{\alpha}{\beta}\nu^2 T.$$

One sees that this formula agrees with the one derived in section 1 from Maxwell theory and electron theory. By equating the coefficients in the two formulae, we get

$$\frac{R}{N}\frac{8\pi}{L^3} = \frac{\alpha}{\beta}$$

or
$$N = \frac{\beta}{\alpha} \frac{8\pi R}{L^3} = 6{\cdot}17 \times 10^{23},$$

that is, one hydrogen atom weighs $1/N = 1{\cdot}62 \times 10^{-24}$ g. This is exactly the value found by Mr. Planck, which agrees satisfactorily with values of this quantity found by different means.

We thus reach the conclusion: the higher the energy density and the longer the wavelengths of radiation, the more usable is the theoretical basis used by us; for short wavelengths and low radiation densities, however, the basis fails completely.

In the following, we shall consider "black-body radiation", basing ourselves upon experience without using a picture of the creation and propagation of the radiation.

3. On the Entropy of the Radiation

The following considerations are contained in a famous paper by Mr. W. Wien and are only mentioned here for the sake of completeness.

Consider radiation which takes up a volume v. We assume that the observable properties of this radiation are completely determined if we give the radiation energy $\rho(\nu)$ for all frequencies.† As we may assume that radiations of different frequencies can be separated without work or heat, we can write the entropy of the radiation in the form

$$S = v \int_0^\infty \phi(\rho, \nu)\, d\nu,$$

where ϕ is a function of the variables ρ and ν. One can reduce ϕ to a function of one variable only by formulating the statement that the entropy of radiation between reflecting walls is not changed by an adiabatic compression. We do not want to go into this, but at once investigate how one can obtain the function ϕ from the radiation law of a black body.

† This is an arbitrary assumption. Of course, one sticks to this simplest assumption until experiments force us to give it up.

In the case of "black-body radiation", ρ is such a function of ν that the entropy is a maximum for a given energy, that is,

$$\delta \int_0^\infty \phi(\rho, \nu)\, d\nu = 0,$$

if

$$\delta \int_0^\infty \rho\, d\nu = 0.$$

From this it follows that for any choice of $\delta\rho$ as function of ν

$$\int_0^\infty \left(\frac{\partial \phi}{\partial \rho} - \lambda\right) \delta\rho\, d\nu = 0,$$

where λ is independent of ν. In the case of black-body radiation, $\partial\phi/\partial\rho$ is thus independent of ν.

If the temperature of a black-body radiation in a volume $v = 1$ increases by dT, we have the equation

$$dS = \int_{\nu=0}^{\nu=\infty} \frac{\partial \phi}{\partial \rho}\, d\rho\, d\nu,$$

or, as $\partial\phi/\partial\rho$ is independent of ν:

$$dS = \frac{\partial \phi}{\partial \rho}\, dE.$$

As dE is equal to the heat transferred and as the process is reversible, we have also

$$dS = \frac{1}{T}\, dE.$$

Through comparing, we get

$$\frac{\partial \phi}{\partial \rho} = \frac{1}{T}.$$

This is the black-body radiation law. One can thus from the function ϕ obtain the black-body radiation law and conversely from the latter the function ϕ through integration, bearing in mind that ϕ vanishes for $\rho = 0$.

4. Limiting Law for the Entropy of Monochromatic Radiation for Low Radiation Density

From the observation made so far on "black-body radiation", it is clear that the law

$$\rho = \alpha \nu^3 \, e^{-\beta\nu/T}$$

put forward originally for "black-body radiation" by Mr. W. Wien is not exactly valid. However, for large values of ν/T, it is in complete agreement with experiment. We shall base our calculations on this formula, though bearing in mind that our results are valid only within certain limits.

First of all, we get from this equation

$$\frac{1}{T} = -\frac{1}{\beta\nu} \ln \frac{\rho}{\alpha\nu^3},$$

and then, if we use the relation found in the preceding section

$$\phi(\rho, \nu) = -\frac{\rho}{\beta\nu}\left[\ln \frac{\rho}{\alpha\nu^3} - 1\right].$$

Let there now be radiation of energy E with a frequency between ν and $\nu + d\nu$ and let the volume of the radiation be v. The entropy of this radiation is

$$S = v\phi(\rho, \nu)\, d\nu = -\frac{E}{\beta\nu}\left[\ln \frac{E}{v\alpha\nu^3\, d\nu} - 1\right].$$

If we restrict ourselves to investigating the dependence of the entropy on the volume occupied by the radiation, and if we denote the entropy of the radiation by S_0 if it occupies a volume v_0, we get

$$S - S_0 = \frac{E}{\beta\nu} \ln \frac{v}{v_0}.$$

This equation shows that the entropy of a monochromatic radiation of sufficiently small density varies with volume according to the same rules as the entropy of a perfect gas or of a dilute solution. The equation just found will in the following be interpreted on the basis of the principle, introduced by Mr. Boltzmann into physics, according to which the entropy of a system is a function of the probability of its state.

5. Molecular–Theoretical Investigation of the Volume-dependence of the Entropy of Gases and Dilute Solutions

When calculating the entropy in molecular gas theory one often uses the word "probability" in a sense which is not the same as the definition of probability given in probability theory. Especially, often "cases of equal probability" are fixed by hypothesis under circumstances where the theoretical model used is sufficiently definite to deduce probabilities rather than fixing them by hypothesis. I shall show in a separate paper that when considering thermal phenomena it is completely sufficient to use the so-called "statistical probability", and I hope thus to do away with a logical difficulty which is hampering the consistent application of Boltzmann's principle. At the moment, however, I shall give its general formulation and the application to very special cases.

If it makes sense to talk about the probability of a state of a system and if, furthermore, any increase of entropy can be considered as a transition to a more probable state, the entropy S_1 of a system will be a function of the probability W_1 of its instantaneous state. If, therefore, one has two systems which do not interact with one another, one can write

$$S_1 = \phi_1(W_1), \quad S_2 = \phi_2(W_2).$$

If one considers these two systems as a single system of entropy S and probability W we have

$$S = S_1 + S_2 = \phi(W) \quad \text{and} \quad W = W_1 . W_2.$$

This last relation states that the states of the two systems are independent.

From these equations it follows that

$$\phi(W_1 \,.\, W_2) = \phi_1(W_1)+\phi_2(W_2),$$

and hence finally

$$\phi_1(W_1) = C \ln W_1 + \text{const},$$

$$\phi_2(W_2) = C \ln W_2 + \text{const},$$

$$\phi(W) = C \ln W + \text{const}.$$

The quantity C is thus a universal constant; it follows from kinetic gas theory that it has the value R/N where the constants R and N have the same meaning as above. If S_0 is the entropy of a certain initial state of the system considered and W the relative probability of a state with entropy S, we have in general

$$S - S_0 = \frac{R}{N} \ln W.$$

We now consider the following special case. Let us consider a number, n, moving points (e.g., molecules) in a volume v_0. Apart from those, there may be in this space arbitrarily many other moving points of some kind or other. We do not make any assumptions about the laws according to which the points considered move in space, except that as far as their motion is concerned no part of space—and no direction—is preferred above others. The number of the (first-mentioned) points which we are considering be moreover so small that we can neglect their mutual interaction.

There corresponds a certain entropy S_0 to the system under consideration, which may be, for instance, a perfect gas or a dilute solution. Consider now the case where a part v of the volume v_0 contains all n moving points while otherwise nothing is changed in the system. This state clearly corresponds to a different value, S_1 of the entropy, and we shall now use Boltzmann's principle to determine the entropy difference.

We ask: how large is the probability of this state relative to the original state? Or: how large is the probability that at an arbitrary moment all n points moving independently of one another in a given volume v_0 are (accidentally) in the volume v?

One gets clearly for this probability, which is a "statistical probability":

$$W = \left(\frac{v}{v_0}\right)^n;$$

one obtains from this, applying Boltzmann's principle:

$$S - S_0 = R\frac{n}{N}\ln\frac{v}{v_0}.$$

It must be noted that it is unnecessary to make any assumptions about the laws, according to which the molecules move, to derive this equation from which one can easily derive thermodynamically the Boyle–Gay–Lussac law and the same law for the osmotic pressure.†

6. Interpretation of the Expression for the Volume-dependence of the Entropy of Monochromatic Radiation according to Boltzmann's Principle

In Section 4, we found for the volume-dependence of the entropy of monochromatic radiation the expression

$$S - S_0 = \frac{E}{\beta\nu}\ln\frac{v}{v_0}.$$

If we write this equation in the form

$$S - S_0 = \frac{R}{N}\ln\left[\left(\frac{v}{v_0}\right)^{NE/R\beta\nu}\right],$$

and compare it with the general formula which expresses

† If E is the energy of the system, we have

$$-d(E-TS) = p\,dv = T\,dS = RT\frac{n}{N}\frac{dv}{v}$$

or

$$pv = R\frac{n}{N}T.$$

Boltzmann's principle,

$$S - S_0 = \frac{R}{N} \ln W,$$

we arrive at the following conclusion:

If monochromatic radiation of frequency ν and energy E is enclosed (by reflecting walls) in a volume v_0, the probability that at an arbitrary time the total radiation energy is in a part v of the volume v_0 will be

$$W = \left(\frac{v}{v_0}\right)^{NE/R\beta\nu}.$$

From this we then conclude:

Monochromatic radiation of low density behaves—as long as Wien's radiation formula is valid—in a thermodynamic sense, as if it consisted of mutually independent energy quanta of magnitude $R\beta\nu/N$.

We now wish to compare the average magnitude of the "black-body" energy quanta with the average kinetic energy of the translational motion of a molecule at the same temperature. The latter is $\frac{3}{2}RT/N$, while we get from Wien's formula for the average magnitude of the energy quantum

$$\frac{\int_0^\infty \alpha\nu^3 e^{-\beta\nu/T}\, d\nu}{\int_0^\infty \frac{N}{R\beta\nu}\alpha\nu^3 e^{-\beta\nu/T}\, d\nu} = 3\frac{R}{N}T.$$

If monochromatic radiation—of sufficiently low density—behaves, as far as the volume-dependence of its entropy is concerned, as a discontinuous medium consisting of energy quanta of magnitude $R\beta\nu/N$, it is plausible to investigate whether the laws on creation and transformation of light are also such as if light consisted of such energy quanta. This question will be considered in the following.

7. On Stokes' Rule

Consider monochromatic light which is changed by photoluminescence to light of a different frequency; in accordance with the result we have just obtained, we assume that both the original and the changed light consist of energy quanta of magnitude $(R/N)\beta\nu$, where ν is the corresponding frequency. We must then interpret the transformation process as follows. Each initial energy quantum of frequency ν_1 is absorbed and is—at least when the distribution density of the initial energy quanta is sufficiently low—by itself responsible for the creation of a light quantum of frequency ν_2; possibly in the absorption of the initial light quantum at the same time also light quanta of frequencies ν_3, ν_4, ... as well as energy of a different kind (e.g. heat) may be generated. It is immaterial through what intermediate processes the final result is brought about. Unless we can consider the photoluminescing substance as a continuous source of energy, the energy of a final light quantum can, according to the energy conservation law, not be larger than that of an initial light quantum; we must thus have the condition

$$\frac{R}{N}\beta\nu_2 \leqq \frac{R}{N}\beta\nu_1, \quad \text{or} \quad \nu_2 \leqq \nu_1$$

This is the well-known Stokes' rule.

We must emphasise that according to our ideas the intensity of light produced must—other things being equal—be proportional to the incident light intensity for weak illumination, as every initial quantum will cause one elementary process of the kind indicated above, independent of the action of the other incident energy quanta. Especially, there will be no lower limit for the intensity of the incident light below which the light would be unable to produce photoluminescence.

According to the above ideas about the phenomena deviations from Stokes' rule are imaginable in the following cases:

1. When the number of the energy quanta per unit volume

involved in transformations is so large that an energy quantum of the light produced may obtain its energy from several initial energy quanta.

2. When the initial (or final) light energetically does not have the properties characteristic for "black-body radiation" according to Wien's law; for instance, when the initial light is produced by a body of so high a temperature that Wien's law no longer holds for the wavelengths considered.

This last possibility needs particular attention. According to the ideas developed here, it is not excluded that a "non-Wienian radiation", even highly-diluted, behaves energetically differently than a "black-body radiation" in the region where Wien's law is valid.

8. On the Production of Cathode Rays by Illumination of Solids

The usual idea that the energy of light is continuously distributed over the space through which it travels meets with especially great difficulties when one tries to explain photo-electric phenomena, as was shown in the pioneering paper by Mr. Lenard.[3]

According to the idea that the incident light consists of energy quanta with an energy $R\beta\nu/N$, one can picture the production of cathode rays by light as follows. Energy quanta penetrate into a surface layer of the body, and their energy is at least partly transformed into electron kinetic energy. The simplest picture is that a light quantum transfers all of its energy to a single electron; we shall assume that that happens. We must, however, not exclude the possibility that electrons only receive part of the energy from light quanta. An electron obtaining kinetic energy inside the body will have lost part of its kinetic energy when it has reached the surface. Moreover, we must assume that each electron on leaving the body must produce work P, which is characteristic for the body. Electrons which are excited at the surface and at right angles to it will leave the body with the greatest normal velocity. The kinetic energy of such electrons is

$$\frac{R}{N}\beta\nu - P$$

If the body is charged to a positive potential Π and surrounded by zero potential conductors, and if Π is just able to prevent the loss of electricity by the body, we must have

$$\Pi\varepsilon = \frac{R}{N}\beta\nu - P,$$

where ε is the electrical mass of the electron, or

$$\Pi E = R\beta\nu - P',$$

where E is the charge of a gram equivalent of a single-valued ion and P' is the potential of that amount of negative electricity with respect to the body.†

If we put $E = 9{\cdot}6 \times 10^3$, $\Pi \times 10^{-8}$ is the potential in Volts which the body assumes when it is irradiated in a vacuum.

To see now whether the relation derived here agrees, as to order of magnitude, with experiments, we put $P' = 0$, $\nu = 1{\cdot}03 \times 10^{15}$ (corresponding to the ultraviolet limit of the solar spectrum) and $\beta = 4{\cdot}866 \times 10^{-11}$. We obtain $\Pi \times 10^7 = 4{\cdot}3$ Volt, a result which agrees, as to order of magnitude, with Mr. Lenard's results.[3]

If the formula derived here is correct, Π must be, if drawn in Cartesian coordinates as a function of the frequency of the incident light, a straight line, the slope of which is independent of the nature of the substance studied.

As far as I can see, our ideas are not in contradiction to the properties of the photoelectric action observed by Mr. Lenard. If every energy quantum of the incident light transfers its energy to electrons independently of all other quanta, the velocity distribution of the electrons, that is, the quality of the resulting cathode radiation, will be independent of the intensity of the incident light; on the other hand, ceteris paribus, the number of

† If one assumes that it takes a certain amount of work to free a single electron by light from a neutral molecule, one has no need to change this relation; one only must consider P' to be the sum of two terms.

electrons leaving the body should be proportional to the intensity of the incident light.[3]

As far as the necessary limitations of these rules are concerned, we could make remarks similar to those about the necessary deviations from the Stokes rule.

In the preceding, we assumed that the energy of at least part of the energy quanta of the incident light was always transferred completely to a single electron. If one does not make this obvious assumption, one obtains instead of the earlier equation the following one

$$\Pi E + P' \leqq R\beta\nu.$$

For cathode-luminescence, which is the inverse process of the one just considered, we get by a similar argument

$$\Pi E + P' \geqq R\beta\nu.$$

For the substances investigated by Mr. Lenard, ΠE is always considerably larger than $R\beta\nu$, as the voltage which the cathode rays must traverse to produce even visible light is, in some cases a few hundred, in other cases thousands of volts.[3] We must thus assume that the kinetic energy of an electron is used to produce many light energy quanta.

9. On the Ionisation of Gases by Ultraviolet Light

We must assume that when a gas is ionised by ultraviolet light, always one absorbed light energy quantum is used to ionise just one gas molecule. From this follows first of all that the ionisation energy (that is, the energy theoretically necessary for the ionisation) of a molecule cannot be larger than the energy of an effective, absorbed light energy quantum. If J denotes the (theoretical) ionisation energy per gram equivalent, we must have

$$R\beta\nu \geqq J.$$

According to Lenard's measurements, the largest effective wavelength for air is about $1{\cdot}9 \times 10^{-5}$ cm, or

$$R\beta\nu = 6{\cdot}4 \times 10^{12}\,\text{erg} \geqq J.$$

An upper limit for the ionisation energy can also be obtained from ionisation voltages in dilute gases. According to J. Stark[4] the smallest measured ionisation voltage (for platinum anodes) in air is about 10 Volt.† We have thus an upper limit of $9{\cdot}6 \times 10^{12}$ for J which is about equal to the observed one. There is still another consequence, the verification of which by experiment seems to me to be very important. If each light energy quantum which is absorbed ionises a molecule, the following relation should exist between the absorbed light intensity L and the number j of moles ionised by this light:

$$j = \frac{L}{R\beta\nu}.$$

This relation should, if our ideas correspond to reality, be valid for any gas which—for the corresponding frequency—does not show an appreciable absorption which is not accompanied by ionisation.

† In the interior of the gas, the ionisation voltage for negative ions is anyhow five times larger.

References

1. M. PLANCK, *Ann. Physik* **1**, 99 (1900).
2. M. PLANCK, *Ann. Physik* **4**, 561 (1901); this paper contains essentially the results of the two papers, reprinted as 1 and 2 in the present volume.
3. P. LENARD, *Ann. Physik* **8**, 149 (1902).
4. J. STARK, *Die Elektrizität in Gasen*, Leipzig, 1902, p. 57.

DUHEM, PIERRE-MAURICE-MARIE (b. Paris, France, 10 June 1861; d. Cabrespine, France, 14 September 1916)

Duhem entered the Collège Stanislas in Paris in 1872 and then continued his education at the Ecole Normale Supérieure. After preparing a doctoral thesis in thermodynamics which was rejected in 1884, he successfully defended a thesis on the theory of magnetism in 1888. He taught at the Sciences Faculties in Lille (1887-1893), Rennes (1893-1894), and Bordeaux (1894-1916). At Bordeaux he held the first chair of theoretical physics in France.

Duhem's reputation rests on his contributions in the history and philosophy of science, as well as in physics, although he turned down a position in the history of science at the Collège de France on the grounds that he was a physicist.

Duhem's principal work in physics lay in thermodynamics, physical chemistry, hydrodynamics, and elasticity. His 1884 thesis was grounded in the thermodynamics and energy functions of J. W. Gibbs and Hermann von Helmholtz, and it included a refutation of Marcellin Berthelot's principle of maximum work. The criticism of Berthelot, along with Duhem's militant royalism and ultramontane Catholicism, were crucial factors blocking his advancement in Parisian scientific institutions. Duhem's approach in thermodynamics was axiomatic and deductive. He did not employ the hypotheses of atoms and molecules, but attempted to establish a generalized system of thermodynamics, or energetics, from which Newtonian mechanics could be derived as a special case. He used the concept of thermodynamic potential to deduce experimental results for homogeneous and heterogeneous systems, and to elucidate the properties of systems in real and apparent equilibrium.

During the period from 1892 to 1906, Duhem published many papers, including book reviews, which form an important corpus in the philosophy of science. His historical interests lay mainly in the history of chemistry and mechanics; the three-volume *Etudes sur Léonard de Vinci* was published in 1906–1913 and the ten-volume *Système de Monde* in 1913–1959. Duhem was uncompromising in his opposition to physical atomism and to naive mechanical models in physics and chemistry. He argued that the aim of science is a system of scientific laws which form part of a natural classification of phenomena; and in describing the relationships among postulates, theorems, laws, and evidence, Duhem made the startling and influential argument that there can be no crucial experiments to decide for or against a single hypothesis. His book *La Théorie physique, son objet et sa structure* (1906) remains a classic in the philosophy of science that still is a starting point for discussion of the aims and methods of science.

PHYSICAL THEORY AND NATURAL CLASSIFICATION

1. *What Is the True Nature of a Physical Theory and the Operations Constituting It?*

WHILE we regard a physical theory as a hypothetical explanation of material reality, we make it dependent on metaphysics. In that way, far from giving it a form to which the greatest number of minds can give their assent, we limit its acceptance to those who acknowledge the philosophy it insists on. But even they cannot be entirely satisfied with this theory since it does not draw all its principles from the metaphysical doctrine from which it is claimed to be derived.

These thoughts, discussed in the preceding chapter, lead us quite naturally to ask the following two questions:

Could we not assign an aim to physical theory that would render it *autonomous*? Based on principles which do not arise from any metaphysical doctrine, physical theory might be judged in its own terms without including the opinions of physicists who depend on the philosophical schools to which they may belong.

Could we not conceive a method which might be *sufficient* for the construction of a physical theory? Consistent with its own definition the theory would employ no principle and have no recourse to any procedure which it could not legitimately use.

We intend to concentrate on this aim and this method, and to study both.

Let us posit right now a definition of physical theory; the sequel of this book will clarify it and will develop its complete content: A physical theory is not an explanation. It is a system of mathematical propositions, deduced from a small number of principles, which aim to represent as simply, as completely, and as exactly as possible a set of experimental laws.

In order to start making this definition somewhat more precise, let us characterize the four successive operations through which a physical theory is formed:

1. Among the physical properties which we set ourselves to repre-

sent we select those we regard as simple properties, so that the others will supposedly be groupings or combinations of them. We make them correspond to a certain group of mathematical symbols, numbers, and magnitudes, through appropriate methods of measurement. These mathematical symbols have no connection of an intrinsic nature with the properties they represent; they bear to the latter only the relation of sign to thing signified. Through methods of measurement we can make each state of a physical property correspond to a value of the representative symbol, and vice versa.

2. We connect the different sorts of magnitudes, thus introduced, by means of a small number of propositions which will serve as principles in our deductions. These principles may be called "hypotheses" in the etymological sense of the word for they are truly the grounds on which the theory will be built; but they do not claim in any manner to state real relations among the real properties of bodies. These hypotheses may then be formulated in an arbitrary way. The only absolutely impassable barrier which limits this arbitrariness is logical contradiction either among the terms of the same hypothesis or among the various hypotheses of the same theory.

3. The diverse principles or hypotheses of a theory are combined together according to the rules of mathematical analysis. The requirements of algebraic logic are the only ones which the theorist has to satisfy in the course of this development. The magnitudes on which his calculations bear are not claimed to be physical realities, and the principles he employs in his deductions are not given as stating real relations among those realities; therefore it matters little whether the operations he performs do or do not correspond to real or conceivable physical transformations. All that one has the right to demand of him is that his syllogisms be valid and his calculations accurate.

4. The various consequences thus drawn from the hypotheses may be translated into as many judgments bearing on the physical properties of the bodies. The methods appropriate for defining and measuring these physical properties are like the vocabulary and key permitting one to make this translation. These judgments are compared with the experimental laws which the theory is intended to represent. If they agree with these laws to the degree of approximation corresponding to the measuring procedures employed, the theory has attained its goal, and is said to be a good theory; if not, it is a bad theory, and it must be modified or rejected.

Thus a true theory is not a theory which gives an explanation of physical appearances in conformity with reality; it is a theory which represents in a satisfactory manner a group of experimental laws. A false theory is not an attempt at an explanation based on assumptions contrary to reality; it is a group of propositions which do not agree with the experimental laws. *Agreement with experiment is the sole criterion of truth for a physical theory.*

The definition we have just outlined distinguishes four fundamental operations in a physical theory: (1) the definition and measurement of physical magnitudes; (2) the selection of hypotheses; (3) the mathematical development of the theory; (4) the comparison of the theory with experiment.

Each one of these operations will occupy us in detail as we proceed with this book, for each of them presents difficulties calling for minute analysis. But right now it is possible for us to answer a few questions and to refute a few objections raised by the present definition of physical theory.

2. *What Is the Utility of a Physical Theory? Theory Considered as an Economy of Thought*

And first, of what use is such a theory?

Concerning the very nature of things, or the realities hidden under the phenomena we are studying, a theory conceived on the plan we have just drawn teaches us absolutely nothing, and does not claim to teach us anything. Of what use is it, then? What do physicists gain by replacing the laws which experimental method furnishes directly with a system of mathematical propositions representing those laws?

First of all, instead of a great number of laws offering themselves as independent of one another, each having to be learnt and remembered on its own account, physical theory substitutes a very small number of propositions, viz., fundamental hypotheses. The hypotheses once known, mathematical deduction permits us with complete confidence to call to mind all the physical laws without omission or repetition. Such condensing of a multitude of laws into a small number of principles affords enormous relief to the human mind, which might not be able without such an artifice to store up the new wealth it acquires daily.

The reduction of physical laws to theories thus contributes to that "intellectual economy" in which Ernst Mach sees the goal and directing principle of science.[1]

[1] E. Mach, "Die ökonomische Natur der physikalischen Forschung," *Populär-wissenschaftliche Vorlesungen* (3rd ed.; Leipzig, 1903), Ch. XIII, p. 215.

The experimental law itself already represented a first intellectual economy. The human mind had been facing an enormous number of concrete facts, each complicated by a multitude of details of all sorts; no man could have embraced and retained a knowledge of all these facts; none could have communicated this knowledge to his fellows. Abstraction entered the scene. It brought about the removal of everything private or individual from these facts, extracting from their total only what was general in them or common to them, and in place of this cumbersome mass of facts it has substituted a single proposition, occupying little of one's memory and easy to convey through instruction: it has formulated a physical law.

"Thus, instead of noting individual cases of light-refraction, we can mentally reconstruct all present and future cases, if we know that the incident ray, the refracted ray, and the perpendicular lie in the same plane and that $\sin i/\sin r = n$. Here, instead of the numberless cases of refraction in different combinations of matter and under all different angles of incidence, we have simply to note the rule above stated and the values of n—which is much easier. The economical purpose here is unmistakable."[2]

The economy achieved by the substitution of the law for the concrete facts is redoubled by the mind when it condenses experimental laws into theories. What the law of refraction is to the innumerable facts of refraction, optical theory is to the infinitely varied laws of light phenomena.

Among the effects of light only a very small number had been reduced to laws by the ancients; the only laws of optics they knew were the law of the rectilinear propagation of light and the laws of reflection. This meager contingent was reinforced in Descartes' time by the law of refraction. An optics so slim could do without theory; it was easy to study and teach each law by itself.

Today, on the contrary, how can a physicist who wishes to study

(Translator's note: Translated by T. J. McCormack, "The Economical Nature of Physical Research," Mach's *Popular Scientific Lectures* [3rd ed.; La Salle, Ill.: Open Court, 1907], Ch. XIII.)

See also E. Mach, *La Mécanique; exposé historique et critique de son développement* (Paris, 1904), Ch. IV, Sec. 4: "La Science comme économie de la pensée," p. 449. (Translator's note: Translated from the German 2nd ed. by T. J. McCormack, *The Science of Mechanics: a Critical and Historical Account of Its Development* [Open Court, 1902], Ch. IV, Sec. iv: "The Economy of Science," pp. 481-494.)

[2] E. Mach, *La Mécanique* . . . , p. 453. (Translator's note: Translated in *The Science of Mechanics* . . . , p. 485.)

optics, as we know it, acquire even a superficial knowledge of this enormous domain without the aid of a theory? Consider the effects of simple refraction, of double refraction by uniaxial or biaxial crystals, of reflection on isotropic or crystalline media, of interference, of diffraction, of polarization by reflection and by simple or double refraction, of chromatic polarization, of rotary polarization, etc. Each one of these large categories of phenomena may occasion the statement of a large number of experimental laws whose number and complication would frighten the most capable and retentive memory.

Optical theory supervenes, takes possession of these laws, and condenses them into a small number of principles. From these principles we can always, through regular and sure calculation, extract the law we wish to use. It is no longer necessary, therefore, to keep watch over the knowledge of all these laws; the knowledge of the principles on which they rest is sufficient.

This example enables us to take firm hold of the way the physical sciences progress. The experimenter constantly brings to light facts hitherto unsuspected and formulates new laws, and the theorist constantly makes it possible to store up these acquisitions by imagining more condensed representations, more economical systems. The development of physics incites a continual struggle between "nature that does not tire of providing" and reason that does not wish "to tire of conceiving."

3. *Theory Considered as Classification*

Theory is not solely an economical representation of experimental laws; it is also a *classification* of these laws.

Experimental physics supplies us with laws all lumped together and, so to speak, on the same plane, without partitioning them into groups of laws united by a kind of family tie. Very often quite accidental causes or rather superficial analogies have led observers in their research to bring together different laws. Newton put into the same work the laws of the dispersion of light crossing a prism and the laws of the colors adorning a soap bubble, simply because of the colors that strike the eye in these two sorts of phenomena.

On the other hand, theory, by developing the numerous ramifications of the deductive reasoning which connects principles to experimental laws, establishes an order and a classification among these laws. It brings some laws together, closely arranged in the same group; it separates some of the others by placing them in two groups very far apart. Theory gives, so to speak, the table of contents and the chapter headings under which the science to be studied will be methodically divided, and it indicates the laws which are to be arranged under each of these chapters.

Thus, alongside the laws which govern the spectrum formed by a prism it arranges the laws governing the colors of the rainbow; but the laws according to which the colors of Newton's rings are ordered go elsewhere to join the laws of fringes discovered by Young and Fresnel; still in another category, the elegant coloration analyzed by Grimaldi is considered related to the diffraction spectra produced by Fraunhofer. The laws of all these phenomena, whose striking colors lead to their confusion in the eyes of the simple observer, are, thanks to the efforts of the theorist, classified and ordered.

These classifications make knowledge convenient to use and safe to apply. Consider those utility cabinets where tools for the same purpose lie side by side, and where partitions logically separate instruments not designed for the same task: the worker's hand quickly grasps, without fumbling or mistake, the tool needed. Thanks to theory, the physicist finds with certitude, and without omitting anything useful or using anything superfluous, the laws which may help him solve a given problem.

Order, wherever it reigns, brings beauty with it. Theory not only renders the group of physical laws it represents easier to handle, more convenient, and more useful, but also more beautiful.

It is impossible to follow the march of one of the great theories of physics, to see it unroll majestically its regular deductions starting from initial hypotheses, to see its consequences represent a multitude of experimental laws down to the smallest detail, without being charmed by the beauty of such a construction, without feeling keenly that such a creation of the human mind is truly a work of art.

4. *A Theory Tends to Be Transformed into a Natural Classification*[3]

This esthetic emotion is not the only reaction that is produced by a theory arriving at a high degree of perfection. It persuades us also to see a natural classification in a theory.

Now first, what is a natural classification? For example, what

[3] We have already noted natural classification as the ideal form toward which physical theory tends in "L'Ecole anglaise et les théories physiques," Art. 6, *Revue des questions scientifiques*, October 1893.

does a naturalist mean in proposing a natural classification of vertebrates?

The classification he has imagined is a group of intellectual operations not referring to concrete individuals but to abstractions, species; these species are arranged in groups, the more particular under the more general. In order to form these groups the naturalist considers the diverse organs—vertebral column, cranium, heart, digestive tube, lungs, swim-bladder—not in the particular and concrete forms they assume in each individual, but in the abstract, general, schematic forms which fit all the species of the same group. Among these organs thus transfigured by abstraction he establishes comparisons, and notes analogies and differences; for example, he declares the swim-bladder of fish analogous to the lung of vertebrates. These homologies are purely ideal connections, not referring to real organs but to generalized and simplified conceptions formed in the mind of the naturalist; the classification is only a synoptic table which summarizes all these comparisons.

When the zoologist asserts that such a classification is natural, he means that those ideal connections established by his reason among abstract conceptions correspond to real relations among the associated creatures brought together and embodied in his abstractions. For example, he means that the more or less striking resemblances which he has noted among various species are the index of a more or less close blood-relationship, properly speaking, among the individuals composing these species; that the cascades through which he translates the subordination of classes, of orders, of families, and of genera reproduce the genealogical tree in which the various vertebrates are branched out from the same trunk and root. These relations of real family affiliation can be established only by comparative anatomy; to grasp them in themselves and put them in evidence is the business of physiology and of paleontology. However, when he contemplates the order which his methods of comparison introduce into the confused multitude of animals, the anatomist cannot assert these relations, the proof of which transcends his methods. And if physiology and paleontology should someday demonstrate to him that the relationship imagined by him cannot be, that the evolutionist hypothesis is controverted, he would continue to believe that the plan drawn by his classification depicts real relations among animals; he would admit being deceived about the nature of these relations but not about their existence.

The neat way in which each experimental law finds its place in the classification created by the physicist and the brilliant clarity imparted to this group of laws so perfectly ordered persuade us in an overwhelming manner that such a classification is not purely artificial, that such an order does not result from a purely arbitrary grouping imposed on laws by an ingenious organizer. Without being able to explain our conviction, but also without being able to get rid of it, we see in the exact ordering of this system the mark by which a natural classification is recognized. Without claiming to explain the reality hiding under the phenomena whose laws we group, we feel that the groupings established by our theory correspond to real affinities among the things themselves.

The physicist who sees in every theory an explanation is convinced that he has grasped in light vibration the proper and intimate basis of the quality which our senses reveal in the form of light and color; he believes in an ether, a body whose parts are excited by this vibration into a rapid to-and-fro motion.

Of course, we do not share these illusions. When, in the course of an optical theory, we talk about luminous vibration, we no longer think of a real to-and-fro motion of a real body; we imagine only an abstract magnitude, i.e., a pure, geometrical expression. It is a periodically variable length which helps us state the hypotheses of optics, and to regain by regular calculations the experimental laws governing light. This vibration is to our mind a *representation*, and not an *explanation*.

But when, after much groping, we succeed in formulating with the aid of this vibration a body of fundamental hypotheses, when we see in the plan drawn by these hypotheses a vast domain of optics, hitherto encumbered by so many details in so confused a way, become ordered and organized, it is impossible for us to believe that this order and this organization are not the reflected image of a real order and organization; that the phenomena which are brought together by the theory, e.g., interference bands and colorations of thin layers, are not in truth slightly different manifestations of the same property of light; and that phenomena separated by the theory, e.g., the spectra of diffraction and of dispersion, do not have good reasons for being in fact essentially different.

Thus, physical theory never gives us the explanation of experimental laws; it never reveals realities hiding under the sensible appearances; but the more complete it becomes, the more we apprehend that the logical order in which theory orders experimental laws is the reflection of an ontological order, the more we suspect that the relations it establishes among the data of observation correspond to real relations among things,[4] and the more we feel that theory tends to be a natural classification.

The physicist cannot take account of this conviction. The method at his disposal is limited to the data of observation. It therefore cannot prove that the order established among experimental laws reflects an order transcending experience; which is all the more reason why his method cannot suspect the nature of the real relations corresponding to the relations established by theory.

But while the physicist is powerless to justify this conviction, he is nonetheless powerless to rid his reason of it. In vain is he filled with the idea that his theories have no power to grasp reality, and that they serve only to give experimental laws a summary and classificatory representation. He cannot compel himself to believe that a system capable of ordering so simply and so easily a vast number of laws, so disparate at first encounter, should be a purely artificial system. Yielding to an intuition which Pascal would have recognized as one of those reasons of the heart "that reason does not know," he asserts his faith in a real order reflected in his theories more clearly and more faithfully as time goes on.

Thus the analysis of the methods by which physical theories are constructed proves to us with complete evidence that these theories cannot be offered as explanations of experimental laws; and, on the other hand, an act of faith, as incapable of being justified by this analysis as of being frustrated by it, assures us that these theories are not a purely artificial system, but a natural classification. And so, we may here apply that profound thought of Pascal: "We have an impotence to prove, which cannot be conquered by any dogmatism; we have an idea of truth which cannot be conquered by any Pyrrhonian skepticism."

5. *Theory Anticipating Experiment*

There is one circumstance which shows with particular clarity our belief in the natural character of a theoretical classification; this circumstance is present when we ask of a theory that it tell us the results of an experiment before it has occurred, when we give it the bold injunction: "Be a prophet for us."

A considerable group of experimental laws had been established by investigators; the theorist has proposed to condense the laws into a very small number of hypotheses, and has succeeded in doing so;

[4] Cf. H. Poincaré, *La Science et l'Hypothèse* (Paris, 1903), p. 190. (Translator's note: Translated by Bruce Halsted, "Science and Hypothesis" in *Foundations of Science* [Lancaster, Pa.: Science Press, 1905].)

each one of the experimental laws is correctly represented by a consequence of these hypotheses.

But the consequences that can be drawn from these hypotheses are unlimited in number; we can, then, draw some consequences which do not correspond to any of the experimental laws previously known, and which simply represent possible experimental laws.

Among these consequences, some refer to circumstances realizable in practice, and these are particularly interesting, for they can be submitted to test by facts. If they represent exactly the experimental laws governing these facts, the value of the theory will be augmented, and the domain governed by the theory will annex new laws. If, on the contrary, there is among these consequences one which is sharply in disagreement with the facts whose law was to be represented by the theory, the latter will have to be more or less modified, or perhaps completely rejected.

Now, on the occasion when we confront the predictions of the theory with reality, suppose we have to bet for or against the theory; on which side shall we lay our wager?

If the theory is a purely artificial system, if we see in the hypotheses on which it rests statements skillfully worked out so that they represent the experimental laws already known, but if the theory fails to hint at any reflection of the real relations among the invisible realities, we shall think that such a theory will fail to confirm a new law. That, in the space left free among the drawers adjusted for other laws, the hitherto unknown law should find a drawer already made into which it may be fitted exactly would be a marvelous feat of chance. It would be folly for us to risk a bet on this sort of expectation.

If, on the contrary, we recognize in the theory a natural classification, if we feel that its principles express profound and real relations among things, we shall not be surprised to see its consequences anticipating experience and stimulating the discovery of new laws; we shall bet fearlessly in its favor.

The highest test, therefore, of our holding a classification as a natural one is to ask it to indicate in advance things which the future alone will reveal. And when the experiment is made and confirms the predictions obtained from our theory, we feel strengthened in our conviction that the relations established by our reason among abstract notions truly correspond to relations among things.

Thus, modern chemical symbolism, by making use of developed formulas, establishes a classification in which diverse compounds are ordered. The wonderful order this classification brings about in the tremendous arsenal of chemistry already assures us that the classification is not a purely artificial system. The relations of analogy and derivation by substitution it establishes among diverse compounds have meaning only in our mind; yet, we are convinced that they correspond to kindred relations among substances themselves, whose nature remains deeply hidden but whose reality does not seem doubtful. Nevertheless, for this conviction to change into overwhelming certainty, we must see the theory write in advance the formulas of a multitude of bodies and, yielding to these indications, synthesis must bring to light a large number of substances whose composition and several properties we should know even before they exist.

Just as the syntheses announced in advance sanction chemical notation as a natural classification, so physical theory will prove that it is the reflection of a real order by anticipating observation.

Now the history of physics provides us with many examples of this clairvoyant guesswork; many a time has a theory forecast laws not yet observed, even laws which appear improbable, stimulating the experimenter to discover them and guiding him toward that discovery.

The Académie des Sciences had set, as the subject for the physics prize that was to be awarded in the public meeting of March 1819, the general examination of the phenomena of the diffraction of light. Two memoirs were presented, and one by Fresnel was awarded the prize, the commission of judges consisting of Biot, Arago, Laplace, Gay-Lussac, and Poisson.

From the principles put forward by Fresnel, Poisson deduced through an elegant analysis the following strange consequence: If a small, opaque, and circular screen intercepts the rays emitted by a point source of light, there should exist behind the screen, on the very axis of this screen, points which are not only bright, but which shine exactly as though the screen were not interposed between them and the source of light.

Such a corollary, so contrary, it seems, to the most obvious experimental certainties, appeared to be a very good ground for rejecting the theory of diffraction proposed by Fresnel. Arago had confidence in the natural character arising from the clairvoyance of this theory. He tested it, and observation gave results which agreed absolutely with the improbable predictions from calculation.[5]

[5] *Oeuvres complètes d'Augustin Fresnel*, 3 vols. (Paris, 1866-1870), I, 236, 365, 368.

Thus physical theory, as we have defined it, gives to a vast group of experimental laws a condensed representation, favorable to intellectual economy.

It classifies these laws and, by classifying, renders them more easily and safely utilizable. At the same time, putting order into the whole, it adds to their beauty.

It assumes, while being completed, the characteristics of a natural classification. The groups it establishes permit hints as to the real affinities of things.

This characteristic of natural classification is marked, above all, by the fruitfulness of the theory which anticipates experimental laws not yet observed, and promotes their discovery.

That sufficiently justifies the search for physical theories, which cannot be called a vain and idle task even though it does not pursue the explanation of phenomena.

PRIMARY QUALITIES

1. *On the Excessive Multiplication of Primary Qualities*

FROM THE MIDST of the empirically given physical world we shall detach the qualities which appear to us the ones that should be regarded as primary. We shall not try to explain these qualities or to reduce them to other more hidden attributes. We shall accept them just as our means of observation make us acquainted with them, whether they appear to us in the form of quantities or are given to us as perceived qualities; in either case we shall regard them as irreducible notions, as the very *elements* which are to constitute our theories. But we shall correlate these properties, qualitative or quantitative, with corresponding mathematical symbols which will allow us in reasoning about them to borrow the language of algebra.

Will this manner of proceeding commit us to the abuse for which the promoters of Renaissance science harshly scolded Scholastic physics and for which they rigorously and definitively brought it to justice?

Undoubtedly the scientists or scholars to whom we owe modern physics could not pardon the Scholastic philosophers for being averse to discussion of natural laws in mathematical language: "If we know anything," cried Gassendi, "we know it by mathematics; but those people have no concern for the true and legitimate science of things! They cling to trivialities!"[1]

But this was not the grievance most often and most sharply brought against the Scholastic doctors by the reformers of physics. Above all, their charge was that the Scholastics invented a new quality every time they looked at a new phenomenon, attributing to a special virtue each effect they had neither studied nor analyzed, and imagining they had given an explanation when they had only given a name. They had thus transformed science into a vain and pretentious jargon.

"This manner of philosophizing," Galileo used to say, "has to my

[1] P. Gassendi, *Exercitationes paradoxicae adversus Aristotelicos* (Grenoble, 1624), Problem I.

mind a great analogy with the manner one of my friends had in painting; he would write on the canvas with chalk: 'Here, I want a fountain with Diana and her nymphs, as well as some hunting dogs; there, a hunter with a stag's head; in the distance, a little woods, a field, a hill'; then he left to the artist the trouble of painting all these things and went away convinced that he had painted the metamorphosis of Acteon when he had only given some names."[2] And Leibniz compared the method followed in physics by the philosophers who on every occasion introduced new forms and new qualities with one "who would be content to say that a clock has the clocklike quality derived from its form without considering in what the latter consists."[3]

A laziness of the mind, which finds it convenient to be paid with words, and an intellectual dishonesty, which finds it profitable to pay others with words, are vices that are widespread in mankind. Certainly the Scholastic physicists, so prompt in endowing the form of each body with all the virtues their vague and superficial systems proclaimed, were often deeply tainted with these vices. But the philosophy that admits qualitative properties does not have a sad monopoly of these faults, for we find them as well among the followers of schools that pride themselves on reducing everything to quantity.

Gassendi, for example, was a convinced atomist; for him every observable quality was but an appearance; in reality there was nothing but the atoms, their shapes, groupings, and motions. But if we had asked him to explain essential physical qualities according to these principles—if we had asked him, "What is taste? What is odor? What is sound? What is light?"—how would he have answered us?

"In the very thing we call tasteful, taste does not seem to consist in anything else than in corpuscles of such a configuration that by penetrating the tongue or palate they affect the contexture of this organ and set it in motion in a manner that gives rise to the sensation we call taste.

"In reality, odor seems to be nothing else than certain corpuscles of such a configuration that when they are exhaled and penetrate the nostrils they conform to the contexture of these organs so as to give rise to the sensation we call olfaction or odor.

"Sound does not seem to be anything else than certain corpuscles

[2] Galileo, *Dialogo sopra i due massimi sistemi del mundo* (Florence, 1632), "Giornata terza."

[3] G. W. Leibniz, *Die philosophischen Schriften*, 7 vols., ed. C. I. Gerhardt (Berlin, 1875-1890), IV, 434.

which, configurated in a certain fashion and transmitted rapidly far from the sounding body, penetrate the ear, set it in motion, and cause the sensation called hearing.

"In a luminous body light does not seem to be anything else than very tenuous corpuscles configurated in a certain fashion and emitted by the luminous body with incredible velocity; they penetrate the organ of sight, and are apt to set it in motion and to create the sensation of vision."[4]

It was an Aristotelian, a *doctus bachelieurus* (learned doctor), who when asked:

> Demandabo causam et rationem quare
> Opium facit dormire?*

answered:

> Quiat est in eo
> Virtus dormitiva
> Cujus est natura
> Sensus assoupire.†

If this bachelor of science had given up Aristotle and had made himself an atomist, Molière would have undoubtedly met him at the philosophical lectures given at Gassendi's home, where the great writer of comedies often visited.

Moreover, the Cartesians would have been mistaken in shouting so triumphantly at the common ridicule into which they saw the Aristotelians and atomists fall. Pascal must have been thinking of one of these Cartesians when he wrote: "There are some who go to the absurd extreme of explaining a word by the same word. One of them, I know, defined light as follows: 'Light is a luminary motion of a luminous body,' as if we could understand the words 'luminary' and 'luminous' without understanding 'light.'"[5]

The allusion, in fact, was to Father Noël, a former teacher of Descartes at the school in La Flèche, who later became one of his fervent disciples, and who in a letter to Pascal on the vacuum had written: "Light, or rather illumination, is a luminary motion of the rays constituting lucid bodies which fill transparent bodies and are not moved luminarily except by other lucid bodies."

When one attributes light to a virtue of brightening, to luminous

[4] P. Gassendi, *Syntagma philosophicum* (Florence, 1727), I, v, Chs. IX, X, XI.

* Translator's note: "What is the cause and reason why opium causes one to sleep?"

† Translator's note: "Because there is in it a dormitive virtue whose nature is to cause the senses to become drowsy."

[5] B. Pascal, *De l'esprit géométrique.*

corpuscles, or to a luminary motion, he is an Aristotelian, an atomist, or a Cartesian, respectively; but if one boasts of having in that way added a particle to our knowledge concerning light, he does not have a sound mind. In all the schools we find people with false minds who imagine themselves to be filling a flask with a precious liqueur when they simply stick a fancy label on it; but all physical doctrines soundly interpreted agree in condemning this illusion. We should bend our efforts, therefore, to avoiding it.

2. *A Primary Quality Is a Quality Irreducible in Fact, Not by Law*

Moreover, our principles put us on guard against that travesty of thought which consists in putting into bodies as many distinct qualities, or almost as many, as there are diverse effects to be explained. We propose to give as simplified and as summary a representation of a group of physical laws as is possible; we aspire to achieve the most complete economy of thought realizable. It is therefore clear that for the construction of our theory we shall have to employ the least number of notions regarded as primitive and of qualities taken as simple. We shall push as far as it will go the method of analysis and reduction, a method which dissociates complex properties, especially those grasped by the senses, and reduces them to a small number of elementary properties.

How shall we know that our dissection has been pushed to the very end, and that the qualities at the end of our analysis cannot in turn be resolved into simpler qualities?

Physicists who tried to construct explanatory theories drew upon the philosophical precepts to which they had subjected themselves for touchstones and reagents to enable them to recognize whether the analysis of a quality had penetrated to the elements. For example, so long as an atomist had not reduced a physical effect to the size, configuration, and action of atoms and to the laws of impact, he knew that his task was not accomplished; so long as a Cartesian found something in a quality other than "bare extension and its modification," he was certain its true nature had not been reached.

If we, on our part, do not claim to explain the properties of bodies but only to give a condensed algebraic representation of them, if we do not proclaim in the construction of our theories any metaphysical principle but intend to make physics an autonomous doctrine, where shall we go for a criterion allowing us to declare that such and such a quality is truly simple and irreducible or that such and such a complex is destined for a more penetrating dissection?

When we regard a property as primary and elementary, we shall not in any way assert that this quality is by its nature simple and indecomposable; we shall declare that all our efforts to reduce this quality to others have failed and that it has been impossible for us to decompose it.

Every time, therefore, that a physicist ascertains a set of phenomena hitherto unobserved or discovers a group of laws apparently showing a new property, he will first investigate whether this property is not a combination, formerly unsuspected, of already known qualities accepted in prevailing theories. Only after he has failed in his many varied efforts, will he decide to regard this property as a new primary quality and introduce into his theories a new mathematical symbol.

"Every time an *exceptional* fact has been discovered," wrote H. Sainte-Claire Deville, describing the hesitations of his thought when he recognized the first phenomena of dissociation, "the first job, I shall say the first duty, practically imposed on the man of science has been to make every effort to cause the fact to come under the common rule by means of an explanation which sometimes requires more work and reflection than the discovery itself. When we succeed, we experience a very keen satisfaction in extending, so to speak, the domain of a physical law, and in increasing the simplicity and generality of a great classification. . . . But when an exceptional fact escapes every explanation, or at least resists every effort conscientiously made to subject it to common law, we must look for other facts which are analogous to it; when they are found they must be classified *provisionally* by means of the theory that has been formed."[6]

When Ampère discovered the mechanical action between two electrical wires, each connected to one of the two poles of a battery, the attractions and repulsions between electrical conductors had been known for a long time. The quality manifested in these attractions and repulsions had been analyzed; it had been represented by an appropriate mathematical symbol, the positive or negative charge of each material element. The use of this symbol had led Poisson to build a mathematical theory which represented most felicitously the experimental laws established by Coulomb.

[6] H. Sainte-Claire Deville, "Recherches sur la décomposition des corps par la chaleur et la dissociation," *Archives des Sciences physiques et naturelles* of the Bibliothèque Universelle, new period, IX (1860), 59.

Might not newly discovered laws be reduced to this quality, whose introduction into physics was an accomplished fact? Might not one explain the attractions and repulsions exerted between two wires in a closed circuit by admitting that certain charges are suitably distributed on the surface of these wires or within them, and that these charges attract or repel each other inversely with the square of the distance, according to the fundamental thesis underlying the theory of Coulomb and Poisson? It was legitimate for this question to have been asked and investigated by physicists; if some one of them had succeeded in giving an affirmative answer to it and reduced the laws of the actions observed by Ampère to the laws of electrostatics established by Coulomb, he would have given us an electrical theory free from the consideration of any primary quality other than the electric charge.

Attempts to reduce the laws of the forces Ampère had put in evidence to electrostatic actions were first of all multiplied. But Faraday cut short these attempts by showing that these forces could give rise to movements of continuous rotation; indeed, as soon as Ampère learned of the phenomenon discovered by the great English physicist, he understood its whole import. This phenomenon, he said, "proves that the action emanating from two conductors of electricity cannot be due to a special distribution of certain fluids at rest in these conductors, as are ordinary electrical attractions and repulsions."[7] "In fact, from the principle of the conservation of living force, which is a necessary consequence of the laws of motion, it follows necessarily that when the elementary forces, here the attractions and repulsions in inverse ratio to the squares of the distances, are expressed by simple functions of the mutual distances of the points between which they act, and if some of these points are constantly connected to one another and move only by virtue of these forces, the others remaining fixed, the first points cannot return to the same position relative to the second points with velocities greater than they had when they started from that position. Now, in the continuous motion impressed on a moving conductor by the action of a fixed one, all points of the former return to the same position with velocities increasing with each revolution, until the friction and resistance of the battery acid in which the end of the conductor is immersed puts an end to the increase

[7] A. M. Ampère, "Exposé sommaire des nouvelles expériences électrodynamiques," read before the Academy, April 8, 1822, *Journal de Physique*, xciv, 65.

of the speed of this conductor's rotation, which then becomes constant despite the friction and resistance.

"It is therefore completely proven that we cannot account for the phenomena produced by the action of two voltaic conductors by supposing that electrical molecules acting inversely with the square of the distance are distributed over the conducting wires."[8]

Strict necessity demanded that there be attributed to the various parts of a voltaic conductor a property not reducible to static electricity; it was necessary to recognize a new primary quality whose existence was to be expressed by saying that the wire is "traversed by a current." This electrical current appears to be bound in a certain direction or affected with a certain sense of direction. It shows a lesser or greater intensity which can by a choice of a scale be correlated with a smaller or larger number, a number to which we assign the name "intensity of electrical current." This intensity of electrical current, a mathematical symbol of a primary quality, allowed Ampère to develop his theory of electrodynamic phenomena, a theory which dispenses with the Frenchman's need to envy the Englishman's pride in the glory of Newton.

The physicist who asks a metaphysical doctrine for the principles with which to develop his theories acquires from that doctrine the marks by which he will recognize whether a quality is simple or complex, and these two words have an absolute sense for him. The physicist who seeks to make his theories autonomous and independent of any philosophical system attributes an entirely relative sense to the words "simple quality" or "primary property"; they designate for him simply a property that it has been impossible for him to resolve into other qualities.

The meaning that the chemists attribute to the words "simple body" or "element" has undergone an analogous transformation.

For an Aristotelian only the four elements fire, air, water, and earth deserved the name of a simple body; all other bodies were complex, and so long as they had not been dissociated to the point of separating out the four elements which could enter into their composition, analysis had not reached its end. Similarly, an alchemist knew that his spagyric art of decomposition had not attained the ultimate goal of his operations until he had separated out the salt, sulphur, and mercury whose union made up all mixtures. The alchemist and the Aristotelian both claimed to know the

[8] A. M. Ampère, *Théorie mathématique des phénomènes électrodynamiques uniquement déduite de l'expérience* (Paris, 1826). Reprinted in the edition published by Hermann (Paris, 1883), p. 96.

marks which characterize the truly simple body in an absolute manner.

Lavoisier and his school led chemists to adopt an entirely different idea of a simple body; it is not a body that a certain philosophical doctrine declares indecomposable, but a body that we have not been able to decompose, a body which has resisted every means of analysis used in laboratories.[9]

When the alchemist and the Aristotelian pronounced the word element, they were proudly asserting their claim to know the very nature of the materials which have gone into the construction of every body in the universe. In the mouth of the modern chemist the same word is a gesture of modesty, an admission of impotence; he is confessing that a body has victoriously resisted every effort made to reduce it.

Chemistry has been compensated for this modesty by its enormous fertility. Is it not legitimate to hope that a similar modesty will procure for theoretical physics the same gains?

3. *A Quality Is Never Primary, except Provisionally*

"We can, therefore, never be sure," Lavoisier said, "that what we regard as simple today will be so in fact. All that we can say is that such a substance is the present end-term at which chemical analysis has arrived, and that the substance cannot be subdivided further in the present state of our knowledge. It is presumable that earth-substances will soon cease to be counted among the simple substances. . . ."[10]

Indeed, in 1807 Humphry Davy transformed Lavoisier's guess into a demonstrated truth, and proved that potash and soda are the oxides of two metals which he called potassium and sodium. Since that time a great many bodies which had long resisted every attempt at analysis have been decomposed and are now excluded from the number of elements.

The title "element" which certain bodies bear is a quite provisional one; it is at the mercy of a more ingenious or more powerful analysis than those in use up to date, a means of analysis which will perhaps dissociate the substance regarded as simple into several distinct substances.

No less provisional is the title "primary quality." The quality

[9] The reader who desires to know the phases through which the idea of a simple body has passed may consult our book *Le Mixte et la Combinaison chimique. Essai sur l'évolution d'une idée.* (Paris, 1902), Part II, Ch. 1.

[10] A. L. Lavoisier, *Traité élémentaire de Chimie* (3rd ed.), I, 194.

which today cannot be reduced to any other physical property will cease tomorrow to be independent; tomorrow, perhaps, the progress of physics will make us recognize in the primary quality a combination of properties which some apparently very different effects have revealed to us for a long time.

The study of the phenomena of light leads to the consideration of a primary quality, light. A direction is given to this quality; its intensity, far from being fixed, varies periodically with enormous rapidity, repeating itself several hundred trillion times a second. A line whose length varies periodically with this extraordinary frequency furnishes a geometrical symbol appropriate for imagining light; the symbol, light vibration, will serve to deal with this quality by mathematical reasoning. Light vibration will be the essential element by means of which the theory of light will be built; its components will serve in writing some equations with partial derivatives and some boundary conditions, condensing and classifying with admirable order and brevity all the laws of the propagation of light, its partial or total reflection, its refraction, and its diffraction.

In another quarter, the analysis of the phenomena that insulating substances like sulphur, vulcanized rubber, and wax show in the presence of electrically charged bodies have led physicists to attribute to these dielectric bodies a certain property. After trying in vain to reduce this property to the electric charge, they had to decide to treat it as a primary quality with the name dielectric polarization. The latter has at each point of the insulating substance and at each instant not only a certain intensity but also a certain direction and a certain sense so that a line segment furnishes the mathematical symbol allowing one to speak about dielectric polarization in the language of mathematicians.

A bold extension of the electrodynamics formulated by Ampère furnished Maxwell with a theory of the variable state of dielectrics. This theory condenses and orders the laws of all the phenomena produced inside insulators where the dielectric polarization varies from one instant to the next. All these laws are summarized in a small number of equations, some of which are satisfied at every point of the same insulating body and the others at every point of the surface separating two distinct dielectrics.

The equations governing light vibration have all been established as though the dielectric polarization did not exist; the equations on which dielectric polarization depends have been discovered by a theory in which the word light is not even mentioned.

Now, see how a surprising convergence between these equations is established.

A dielectric polarization which varies periodically has to verify equations all of which are similar to the equations governing light vibration.

And not only do these equations have the same form but also the coefficients figuring in them have the same numerical value. Thus in a vacuum or in air, at first without any electric action polarizing a certain region, electric polarization once begun is propagated with a certain velocity; the equations of Maxwell allow one to determine this velocity by purely electrical procedures wherein nothing is borrowed from optics; numerous measurements agree that the value of this velocity is around 300,000 kilometers per second; this number is precisely equal to the velocity of light in air or in a vacuum, a velocity that four purely optical methods, distinct from one another, have taught us.

The conclusion imposed by this unexpected convergence is: Light is not a primary quality; light vibration is nothing else than a periodically variable dielectric polarization; the electromagnetic theory of light created by Maxwell has resolved a property we thought irreducible; it has derived it from a quality with which for many years there appeared to be no connection.

Thus the progress of theories may itself lead physicists to reduce the number of qualities that they had at first considered as primary, and to prove that two properties regarded as distinct are but two diverse aspects of the same property.

Must we conclude that the number of qualities admitted into our theories will diminish from day to day, that matter which is the subject of our theorizing will be less and less rich in essential attributes, and that it will tend towards a simplicity comparable to that of atomistic or Cartesian matter? I think that would be a rash conclusion. Undoubtedly, the very development of theory may from time to time produce the fusion of two distinct qualities, similar to that fusion of light and dielectric polarization established by the electromagnetic theory of light. But on the other hand, the constant progress of experimental physics frequently brings on the discovery of new categories of phenomena, and in order to classify these phenomena and group their laws, it is necessary to endow matter with new properties.

Which of these two contrary movements will prevail—the one which reduces qualities to other qualities and tends to simplify matter, or the one which discovers new properties and tends to complicate? It would be imprudent to formulate any long-term prediction about this subject. At least, it seems certain that in our time the second trend is much more powerful than the first and is leading our theories toward a more and more complex conception of matter, richer in attributes.

Besides, the analogy between the primary qualities of physics and the simple bodies of chemistry is here again a marked one. If perhaps the day will come when powerful methods of analysis will resolve the numerous bodies we today call simple into a small number of elements, there is no certain or probable sign allowing us to announce the dawn of that day yet. In our own day* chemistry is making progress in constantly discovering new simple bodies. For half a century, the rare earths have continued to furnish new recruits to an already long list of metals; gallium, germanium, scandium, etc. show us chemists proud to inscribe the name of their country on this list. In the air we breathe, a mixture of nitrogen and oxygen apparently so well known since Lavoisier, we see revealed a whole family of new gases: argon, helium, xenon, crypton. Finally, the study of new radiations, which will surely compel physics to enlarge the circle of its primary qualities, furnishes chemistry with hitherto unknown bodies: radium and perhaps polonium and actinium.

Most assuredly we have gone a long way from the beautifully simple bodies which Descartes dreamed up, those bodies which were reduced "simply to bare extension and its modification." Chemistry piles up a collection of about a hundred material bodies irreducible to one another, and to each of these physics associates a form capable of a multitude of diverse properties. Each of these two sciences strives to reduce the number of its elements as much as it can, and yet, in proportion to the progress each science makes, it sees this number grow.

* Translator's note: About 1900.

PERRIN, JEAN BAPTISTE. (b. Lille, France, 30 September 1870; d. New York, New York, 17 April 1942)

Perrin studied at the *lycée* in Lyon and at the Lycée Janson-de-Sailly in Paris before entering the Ecole Normale Supérieure in 1891. He completed a doctoral thesis on X rays and cathode rays in 1897 and began teaching a course in physical chemistry at the Paris Faculty of Sciences. This course became a professorial chair in 1910, which he held until 1940. Perrin was the moving force behind the establishment of the Centre National de la Recherche Scientifique, the principal French organization for scientific research. He also helped found the Parisian science museum, the Palais de la Découverte. During the 1936–1938

Popular Front cabinets of Léon Blum, he served as undersecretary of state for scientific research, and he worked with the Free French in New York during the early years of the Second World War.

Perrin's scientific research focused on atomism, thermodynamics, and electromagnetism in the tradition of Ludwig Boltzmann. Perrin was a preeminent experimentalist, whereas Boltzmann was a theorist. Perrin's first published research presented evidence, in 1895, that cathode rays are negatively charged particles, and his 1897 thesis developed ionization theory in connection with cathode rays and X rays. A 1903 textbook on physical chemistry defended the usefulness and validity of molecular-kinetic ideas against the thermodynamic or energetic approach of Ostwald, Mach, and Duhem. In the early 1900s Perrin studied ion transport and electrolytes, and, from 1906, he turned his attention to Brownian motion in colloids. Here, as in his later work on fluorescence and the chemical activity of light, Perrin interested himself in the minute, statistical fluctuations which evidence the discontinuous structure of matter and light. For this work, especially on Brownian motion, he received the Nobel Prize in physics in 1926.

Perrin's work on Brownian motion is summed up masterfully in the 1909 paper "Brownian Movement and Molecular Reality." In his researches he sought to confirm the explanation given by kinetic theory that microscopic displacements of colloid particles are due to their collisions with particles of the fluid in which they are suspended. After demonstrating a height distribution for microscopic colloid particles and deducing a value for Avogadro's number N, he learned from his friend Paul Langevin that these experimental results were in agreement with theoretical predictions by Einstein and by Maryan Smoluchowski. During the next few years Perrin devoted his attention to measuring Brownian motions of translation and rotation, which allowed him to calculate molecular diameters. At the conclusion of the 1909 paper he linked his experimental values for N to values calculated from the data of radioactivity, diffusion, quantum theory, and other sources, demonstrating the role that molecular-kinetic theory might play in connecting the old and new mechanics, and in unifying chemistry and physics.

BROWNIAN MOVEMENT AND MOLECULAR REALITY.

By M. JEAN PERRIN
(Professeur de Chimie Physique, Faculté des Sciences,
Université de Paris.)

TRANSLATED FROM THE

ANNALES DE CHIMIE ET DE PHYSIQUE, 8me SERIES,
September 1909,

BY F. SODDY, M.A., F.R.S.

I.

1. The first indication of the phenomenon.—When we consider a fluid mass in equilibrium, for example some water in a glass, all the parts of the mass appear completely motionless to us. If we put into it an object of greater density it falls and, if it is spherical, it falls exactly vertically. The fall, it is true, is the slower the smaller the object; but, so long as it is visible, it falls and always ends by reaching the bottom of the vessel. When at the bottom, as is well known, it does not tend again to rise, and this is one way of enunciating Carnot's principle (impossibility of perpetual motion of the second sort).

These familiar ideas, however, only hold good for the scale of size to which our organism is accustomed, and the simple use of the microscope suffices to impress on us new ones which substitute a kinetic for the old static conception of the fluid state.

Indeed it would be difficult to examine for long preparations in a liquid medium without observing that all the particles situated in the liquid instead of assuming a regular movement of fall or ascent, according to their density, are, on the contrary, animated with a perfectly irregular movement. They go and come, stop, start again, *mount*, descend, *remount again*, without in the least tending toward immobility. This is the *Brownian movement*, so named in memory of the naturalist Brown, who described it in 1827 (very shortly after the discovery of the achromatic objective), then proved that the movement was not due to living animalculæ, and recognised that the particles in suspension are agitated the more briskly the smaller they are.

2. Projection of the Brownian movement—This phenomenon can be made visible to a whole audience by projection, but this is difficult, and it may be useful to detail the precautions which have enabled me to arrive at a satisfactory result. The image of an electric arc (or better, of the sun) is formed in the preparation, the greater part of the non-luminous heat rays being stopped by means of a cell full of water. The rays, reflected by the particles in suspension, traverse, as for direct observation, an immersion objective and an eyepiece of high magnification, and are then turned horizontally by a total-reflection prism so as to form the image of the granules on a screen of ground glass (ruled in squares by preference, so as to have reference marks), on the farther side of which the audience is. The light is thus better utilised than with an ordinary screen which would diffuse a large part of it in directions where there were no observers. The magnification can be usefully raised to 8,000 or 10,000 diameters.

But it is necessary above all to procure an appropriate emulsion. In the few trials of projection which have been made up till now, the diameter of the granules employed was of the order of a micron, and their image is visible only with difficulty beyond 3 metres (at least with the light of the arc) whether immersion or lateral illumination is used. Smaller granules are still less visible, and one is led to this, at first sight, paradoxical conclusion, that it is better to project large granules than small ones. It is true that their movement is less, but it is still quite sufficient for its essential characteristics to be easily recognised.

It is still necessary to know how to prepare particles

having a diameter of several microns, and we shall see soon that this is equally desirable in regard to certain points in the experimental study proper of the Brownian movement. I shall indicate later (No. 32) how I have succeeded in obtaining large, perfectly spherical granules of gamboge and mastic. With such granules the Brownian movement can still be perceived at a distance of 8 or 10 metres from the screen in a hall which has been made absolutely dark.

3. Persistance of the phenomenon in absence of all causes external to the fluid. Its explanation by the movements of molecules.—The singular phenomenon discovered by Brown did not attract much attention. It remained, moreover, for a long time ignored by the majority of physicists, and it may be supposed that those who had heard of it thought it analogous to the movement of the dust particles, which can be seen dancing in a ray of sunlight, under the influence of feeble currents of air which set up small differences of pressure or temperature. When we reflect that this apparent explanation was able to satisfy even thoughtful minds, we ought the more to admire the acuteness of those physicists, who have recognised in this, supposed insignificant, phenomenon a fundamental property of matter.

Besides, as happens most frequently when it is sought to unravel the genesis of a great directing idea, it is difficult to fix precisely how the hypothesis, which ascribes the Brownian movement to molecular agitation, first appeared and how it was developed.

The first name which calls for reference in this respect is, perhaps, that of Wiener, who declared at the conclusion of his observations, that the movement could not be due to convection currents, that it was necessary to seek for the cause of it in the liquid itself, and who, finally, almost at the commencement of the development of the kinetic theory of heat, divined that molecular movements were able to give the explanation of the phenomenon *.

Some years later Fathers Delsaulx and Carbonnelle

* *Erklärung des atomistischen Wesens des flüssigen Körperzustandes und Bestätigung desselben durch die sogenannten Molekularbewegungen* (*Pogg. Ann.* 1863, cxviii. 79).

published in the *Royal Microscopical Society* and in the *Revue des Questions scientifiques*, from 1877 to 1880, various Notes on the *Thermodynamical Origin of the Brownian Movement* *. In a note by Father Delsaulx, for example, one may read : "the agitation of small corpuscles in suspension in liquids truly constitutes a general phenomenon," that it is "henceforth natural to ascribe a phenomenon having this universality to some general property of matter," and that "in this train of ideas, the internal movements of translation which constitute the calorific state of gases, vapours and liquids, can very well account for the facts established by experiment."

In another Note, by Father Carbonnelle, one, again, may read this : "In the case of a surface having a certain area, the molecular collisions of the liquid which cause the pressure, would not produce any perturbation of the suspended particles, because these, as a whole, urge the particles equally in all directions. But if the surface is of area less than is necessary to ensure the compensation of irregularities, there is no longer any ground for considering the mean pressure ; the inequal pressures, continually varying from place to place, must be recognised, as the law of large numbers no longer leads to uniformity ; and the resultant will not now be zero but will change continually in intensity and direction. Further, the inequalities will become more and more apparent the smaller the body is supposed to be, and in consequence the oscillations will at the same time become more and more brisk . . . . . ."

These remarkable reflections unfortunately remained as little known as those of Wiener. Besides it does not appear that they were accompanied by an experimental trial sufficient to dispel the superficial explanation indicated a moment ago ; in consequence, the proposed theory did not impress itself on those who had become acquainted with it.

On the contrary, it was established by the work of M. *Gouy* (1888), not only that the hypothesis of molecular agitation gave an admissible explanation of the Brownian movement, but that no other cause of the movement could

* *See* for this bibliography an article which appeared in the *Revue des Questions scientifiques*, January 1909, where M. Thirion very properly calls attention to the ideas of these *savants*, with whom he collaborated.

be imagined, which especially increased the significance of the hypothesis *. This work immediately evoked a considerable response, and it is only from this time that the Brownian movement took a place among the important problems of general physics.

In the first place, M. Gouy observed that the Brownian movement is not due to vibrations transmitted to the liquid under examination, since it persists equally, for example, at night on a sub-soil in the country as during the day near a populous street where heavy vehicles pass. Neither is it due to the convection currents existing in fluids where thermal equilibrium has not been attained, for it does not appreciably change when plenty of time is given for equilibrium to be reached. Any comparison between Brownian movement and the agitation of dust-particles dancing in the sunlight must therefore be set aside. In addition, in the latter case, it is easy to see that the neighbouring dust-particles move in general in the same sense, roughly tracing out the form of the common current which bears them along, whereas the most striking feature of the Brownian movement is the absolute independence of the displacements of neighbouring particles, so near together that they pass by one another. Lastly, neither can the unavoidable illumination of the preparation be suspected, for M. Gouy was able abruptly to reduce it a thousand times, or to change its colour considerably, without at all modifying the phenomenon observed. All the other causes from time to time imagined have as little influence; even the nature of the particles does not appear to be of any importance, and henceforward it was difficult not to believe that these particles simply serve to reveal an internal agitation of the fluid, the better the smaller they are, much as a cork follows better than a large ship the movements of the waves of the sea.

Thus comes into evidence, in what is termed a *fluid in equilibrium*, a property eternal and profound. This equilibrium only exists as an average and for large masses; it is a statistical equilibrium. In reality the whole fluid is

* *Journal de Physique*, 1888, 2nd Series, vii. 561; *Comptes rendus*, 1889, cix. 102; *Revue générale des Sciences*, 1895, 1.

agitated indefinitely and *spontaneously* by motions the more violent and rapid the smaller the portion taken into account; the statical notion of equilibrium is completely illusory.

4. Brownian movement and Carnot's principle.—There is therefore an agitation maintained indefinitely without external cause. It is clear that this agitation is not contradictory to the principle of the conservation of energy. It is sufficient that every increase in the speed of a granule is accompanied by a cooling of the liquid in its immediate neighbourhood, and likewise every decrease of speed by a local heating, without loss or gain of energy. We perceive that thermal equilibrium itself is also simply a statistical equilibrium. But it should be noticed, and this very important idea is again due to M. Gouy, that the Brownian movement is not reconcilable with the rigid enunciations too frequently given to Carnot's principle ; the particular enunciation chosen can be shown to be of no importance. For example, in water in equilibrium it is sufficient to follow with the eyes a particle denser than water to see it at certain moments rise spontaneously, absorbing, necessarily, work at the expense of the heat of the surrounding medium. So it must not any longer be said that perpetual motion of the second sort is impossible, but one must say : "On the scale of size which interests us practically, perpetual motion of the second sort is in general so insignificant that it would be absurd to take it into account." Besides such restrictions have long been laid down : the point of view that Carnot's principle expresses simply a law approximated to has been upheld by Clausius, Maxwell, Helmholtz, Boltzmann, and Gibbs, and in particular may be recalled the *demon*, imagined by Maxwell, which, being sufficiently quick to discern the molecules individually, made heat pass at will from a cold to a hot region without work. But since one is limited to the intervention of invisible molecules, it remained possible, by denying their existence, to believe in the perfect rigidity of Carnot's principle. But this would no longer be admissible, for this rigidity is now in opposition to a *palpable reality*.

On the other hand, the practical importance of Carnot's principle is not attacked, and I hardly need state at length that it would be imprudent to count upon the Brownian movement to lift the stones intended for the building of a house. But the comprehension of this important principle becomes in consequence more profound : its connection with the structure of matter is better understood, and the conception is gained that it can be enunciated by saying that spontaneous co-ordination of molecular movements becomes the more improbable the greater the number of molecules and the greater the duration of time under consideration *.

5. The kinetic molecular hypothesis.—I have said that the Brownian movement is explained, in the theory of M. Gouy and his predecessors, by the incessant movements of the molecules of the fluid, which striking unceasingly the observed particles, drive about these particles irregularly through the fluid, except in the case where these impacts exactly counterbalance one another. It has, to be sure, been long recognised, especially in explanation of the facts of diffusion, and of the transformation of motion into heat, not only that substances in spite of their homogeneous appearance, have a discontinuous structure and are composed of separate *molecules*, but also that these molecules are in incessant agitation, which increases with the temperature and only ceases at absolute zero.

Instead of taking this hypothesis ready made and seeing how it renders account of the Brownian movement, it appears preferable to me to show that, possibly, it is logically suggested by this phenomenon alone, and this is what I propose to try.

What is really strange and *new* in the Brownian movement is, precisely, that it never stops. At first that seems in contradiction to our every-day experience of friction. If for example, we pour a bucket of water into a tub, it seems natural that, after a short time, the motion possessed by the liquid mass disappears. Let us analyse further how this apparent equilibrium is arrived at : all the particles had at

* With regard to the general significance of the principle I should refer to the very interesting considerations developed by J. H. Rosny, Senior, in his book on Pluralism, pp. 85-91 (F. Alcan, 1909).

first velocities almost equal and parallel ; this co-ordination is disturbed as soon as certain of the particles, striking the walls of the tub, recoil in different directions with changed speeds, to be soon deviated anew by their impacts with other portions of the liquid. So that, some instants after the fall, all parts of the water will be still in motion, but it is now necessary to consider quite a small portion of it, in order that the speeds of its different points may have about the same direction and value. It is easy to see this by mixing coloured powders into a liquid, which will take on more and more irregular relative motions.

What we observe, in consequence, so long as we can distinguish anything, is not a cessation of the movements, but that they become more and more chaotic, that they distribute themselves in a fashion the more irregular the smaller the parts.

Does this de-co-ordination proceed indefinitely ?

To have information on this point and to follow this de-co-ordination as far as possible after having ceased to observe it with the naked eye, a microscope will be of assistance, and microscopic powders will be taken as indicators of the movement. Now these are precisely the conditions under which the Brownian movement is perceived: we are therefore *assured* that the de-co-ordination of motion, so evident on the ordinary scale of our observations, does not proceed indefinitely, and, on the scale of microscopic observation, we *establish* an equilibrium between the co-ordination and the de-co-ordination. If, that is to say, at each instant, certain of the indicating granules stop, there are some in other regions at the same instant, the movement of which is re-co-ordinated automatically by their being given the speed of the granules which have come to rest. So that it does not seem possible to escape the following conclusion :

Since the distribution of motion in a fluid does not progress indefinitely, and is limited by a spontaneous re-co-ordination, it follows that the fluids are themselves composed of granules or *molecules*, which can assume all possible motions relative to one another, but in the interior of which dissemination of motion is impossible. If such molecules had no existence it is not apparent how there would be any limit to the de-co-ordination of motion.

On the contrary if they exist, there would be, unceasingly, partial re-co-ordination ; by the passage of one near another, influencing it (it may be by *impact* or in any other manner), the speeds of these molecules will be continuously modified, in magnitude and direction, and from these same chances it will come about sometimes that neighbouring molecules will have concordant motions. In addition, even without this absolute concordance being necessary, it will at least come about frequently that the molecules in the region of an indicating particle will assume in a certain direction an excess of motion sufficient to drive the particle in that direction.

The Brownian movement is permanent at constant temperature : that is an experimental fact. The motion of the molecules which it leads us to imagine is thus itself also permanent. If these molecules come into collision like billiard balls, it is necessary to add that they are perfectly elastic, and this expression can, indeed, be used to indicate that in the molecular collisions of a thermally isolated system the sum of the energies of motion remains definitely constant.

In brief the examination of Brownian movement alone suffices to suggest that every fluid is formed of elastic molecules, animated by a perpetual motion.

6. The atoms. Avogadro's constant.—From this, as is well known, diverse considerations of chemistry, and particularly the study of substitution, lead to the idea of the existence of atoms. When, for example, calcium is dissolved in water, only one half of the hydrogen contained in the latter is displaced. The hydrogen of this water, and in consequence the hydrogen of each molecule, is therefore composed of two distinct parts. No experiments lead to any further differentiation, and it is reasonable to regard these two parts as indivisible, by all chemical methods, or in a word, they are *atoms*. On the other hand, every mass of water, and in consequence each molecule of water, weighs 9 times the hydrogen it contains : the molecule of water, which contains

2 atoms of hydrogen, weighs therefore 18 times the atom of hydrogen. In a similar manner, it may be established that the molecule of methane, for example, weighs 16 times more than the atom of hydrogen. Thus, by a purely chemical method, through the conception of the atom, the ratio 16/18, of the weight of a molecule of methane to a molecule of water, can be reached.

Now this same ratio, precisely, is arrived at by comparison of the masses of similar volumes of methane and water vapour in the gaseous state under similar conditions of temperature and pressure. Thus these two masses, which have the same ratio as the two kinds of molecules, must contain as many molecules the one as the other. This result is general for the different gases, so that in consequence we arrive, in an experimental manner, at the celebrated proposition enunciated in the form of an hypothesis by Avogadro, about a century ago, and taken up again a little later by Ampère:

"Any two gases, taken under the same conditions of temperature and pressure, contain in the same volume the same number of molecules."

It has become customary to name as the gram-molecule of a substance, the mass of the substance which in the gaseous state occupies the same volume as 2 grams of hydrogen measured at the same temperature and pressure. Avogadro's proposition is then equivalent to the following:

"*Any two gram-molecules contain the same number of molecules.*"

This invariable number N is a universal constant, which may appropriately be designated *Avogadro's Constant.* If this constant be known, the mass of any molecule is known: even the mass of any atom will be known, since we can learn by the different methods which lead to chemical formulæ, how many atoms of each sort there are in each molecule. The weight of a molecule of water, for example, is $\frac{18}{N}$; that of a molecule of oxygen is $\frac{32}{N}$, and so on for each molecule. Similarly the weight of the oxygen atom, obtained by dividing the gram-atom of oxygen by N, is $\frac{16}{N}$; that of the atom of hydrogen is $\frac{1{\cdot}008}{N}$, and so on for each atom.

7. The constant of molecular energy.—It is easy to see that if we know Avogadro's constant we can calculate the mean kinetic energy of translation of different molecules, and conversely, the value of this energy will give us N. Let us elaborate this important point a little.

If fluids are composed of molecules in motion, the pressure which they exert on the boundaries which limit their expansion is accounted for by the impacts of their molecules against these boundaries, and, in the case of gases (the molecules of which are very remote, relatively, one from another), it has been established, thanks to the successive arguments, created or modified by Joule, Clausius, and Maxwell, that this conception, at first somewhat vague, contains the precise relation

$$pv = \frac{2}{3}nw$$

where p is the pressure which n molecules of mean kinetic energy w develop in the volume v.

If the mass of gas under consideration is one gram-molecule, n becomes equal to N and pv to RT, T being the absolute temperature and R the constant of a perfect gas (equal in C.G.S. units to $83{\cdot}2 \times 10^6$); the preceding equation may then be written

$$\frac{2}{3}\mathrm{N}w = \mathrm{RT}$$

or

$$w = \frac{3\mathrm{R}}{2\mathrm{N}}\mathrm{T}.$$

Now the constant N is the same for all substances. The molecular kinetic energy of translation has thus for all gases the same mean value, proportional to the absolute temperature

$$w = \alpha\mathrm{T}.$$

The constant α, which may be named the *constant of molecular energy*, equal to $\frac{3R}{2N}$, is, like N, a universal constant.

It is evident that both of these constants will be known as soon as one is.

8. The atom of electricity.—A third universal constant is also reached at the same time as N or α, and this is encountered in the study of the phenomena of electrolysis. It is known that the *decomposition* by the current of the gram-molecule of a given electrolyte is accompanied always by the passage of the same quantity of electricity: as is well known, this is explained by the conception that in all electrolytes a part at least of the molecules are dissociated into *ions* carrying fixed electric charges, and in consequence sensitive to the electric field; lastly, if the name *faraday* is given to the quantity F of electricity (96,550 coulombs) which passes in the decomposition of 1 gram-molecule of hydrochloric acid, it is known that the decomposition of any other gram-molecule is accompanied by the passage of a whole number of faradays, and, in consequence, that any ion carries a whole number of times the charge on the hydrogen ion. This charge e thus also appears as indivisible, and constitutes the atom of electricity or the electron (Helmholtz).

It is easy to obtain this universal constant if either of the constants, N or α, is known. Since the gram-atom of hydrogen in the ionic state, that is to say N atoms of hydrogen, carries one faraday, then necessarily,

$$Ne = F,$$

which is, in C.G.S. electrostatic units,

$$Ne = 96{,}550 \times 3.10^9 = 29.10^{13};$$

thus, in the same step, the three universal constants N, e, α will be found. Can this be accomplished?

9. Molecular speeds. Maxwell's law of irregularities. Mean free path.—The commencement of the answer to this question, and, at the same time, the approximate determination of the order of molecular magnitude, is due to the admirable efforts of Clausius, Maxwell, and Van der Waals. Without entering into detail I think it useful to summarise the line they have followed.

First, for each gas the mean square, U^2, of the molecular speed is easily calculated from the equation just written

$$\frac{2}{3}Nw = RT.$$

It is sufficient to notice that $2Nw$ can be replaced by MU^2, M representing the gram-molecule of the gas under consideration. Thus it is found that U is of the order of some hundreds of metres per second (435 metres at 0° for oxygen).

As well understood, the molecular speeds are very variable and unequal ; but in a steady state the proportion of molecules which have any definite speed remains fixed. On the hypothesis that the probability of a component x is independent of the values of the components y and z, Maxwell succeeded in determining the law of distribution of molecular speeds. His reasoning demonstrated that, on this hypothesis, the probability of any molecule possessing, along the axis Ox, a component between x and $x+dx$ had the value

$$\frac{1}{U}\sqrt{\frac{3}{2\pi}}\,e^{-\frac{3}{2}\frac{x^2}{U^2}}\,dx.$$

This expression represents the irregularities of molecular motion. It is obtained just the same on the hypothesis that the components along the Ox axis are distributed around the zero value according to the so-called law of *chance* enunciated by Laplace and Gauss.

This *law of the distribution of velocities* permits the calculation of the mean speed Ω, which is not equal to U (any more than $\frac{a+b}{2}$ is equal to the square root of $\frac{a^2+b^2}{2}$), but which, as a matter of fact, differs but little from it

$$\left(\Omega = U\sqrt{\frac{8}{3\pi}}\right).$$

On the other hand, this same law of distribution can be used to test by calculation the hypothesis that the *internal friction* between two parallel layers of gas, moving at different speeds, results from the continuous arrival, in each layer, of molecules coming from the other layer. Maxwell in this way found that the coefficient ζ of internal friction, or viscosity, which is experimentally measurable, should be very nearly equal to one-third of the product of the following three quantities : the absolute density δ of the gas (given by the balance), the mean molecular speed Ω (which we know how to calculate), and the mean free path L which a molecule traverses in a straight line between two successive impacts. More exactly, he found

$$\zeta = 0{\cdot}31\, \delta\, \Omega\, L.$$

The value of the mean free path is thus obtained : for example, for oxygen or nitrogen at ordinary temperature and under atmospheric pressure it is approximately equal to 0·1 micron. At the low pressure of a Crookes' tube it can reach many centimetres.

10. The relation of the mean free path of the molecule to its diameter.—In addition, a line of reasoning, due to Clausius, shows that this same mean free path can be calculated in another manner as a function of the nearness of approach of the molecules and of their dimensions *. It is easy to understand that the smaller the molecules are the nearer their approach, and the larger they are the more they act as obstructions.

But there are certainly other considerations to be taken into account, as, for example, that a molecule in the form of a rod (as in the case, possibly, of certain molecules of the fatty series) will not obstruct in the same way as if it had the form of a sphere. In default of any knowledge of the exact shape of molecules, it has been thought that no great error is likely to result in likening them to spherical balls, having a diameter equal to the mean distance apart of the centres of two molecules on impact. This hypothesis, possibly exact in the case of monatomic molecules (mercury, argon, etc.), is certainly false for other molecules, but it is

* *Pogg. Ann.* 1858.

still possible that it may lead to approximate results in the case of the less complicated molecules such as those of oxygen and nitrogen.

Let us then liken the molecules to spheres. The approximate calculation of Clausius, subsequently modified by Maxwell, showed that the following relation should hold approximately,

$$L = \frac{1}{\pi\sqrt{2}} \frac{1}{nD^2},$$

where D represents the molecular diameter, and n the number of molecules contained in each cubic centimetre. Since L can be calculated, a second relation between n and D will give us the diameter of the molecules and the number n per cubic centimetre. In this case, multiplying the number n by the known volume of the gram-molecule under the conditions of temperature and pressure chosen in the calculation, we shall have the number N of molecules in a gram-molecule, *that is to say the required three universal constants.*

But this second relation between n and D has not been very easy to obtain.

11. First determinations of Avogadro's constant.—To begin with, the molecules in the liquid state cannot be more closely packed than bullets are in a pile of bullets *. Now it is easily established that the volume of bullets is only equal to 73 per cent. of the volume of the pile. So, we have

$$\frac{1}{6}\pi n D^3 < 0{\cdot}73\phi,$$

where ϕ signifies the known volume occupied by the mass of a cubic centimetre of the gas considered in the liquid state at low temperature. This inequality combined with the preceding equation, which gives the product nD^2, leads to a value certainly too great for the molecular diameter, and therefore to values certainly too small for n and N.

The calculation is usually made for oxygen (which gives

* In reality, the original reasoning, due to Loschmidt, was limited to the statement that the volume of the molecules is inferior to that of the liquid and perhaps not more than ten times smaller.

$N > 9.10^{22}$): it is better to make it for a monatomic gas, for which the molecules may really be spherical, and recommencing the calculation for mercury, the mean free path of which at 370° is 21.10^{-6} (Landolt's Tables), I find for the inferior limit of N a higher and therefore more useful value, namely

$$N > 45.10^{22}.$$

As for the molecular diameter, for all the gases considered it is found to be less than the millionth of a millimetre (for the special case of mercury $D < 3{\cdot}5 \times 10^{-8}$).

This indication only puts us in the same position as that of an astronomer who, desiring to know the distance of a star from the Sun, finds at first only that it is farther off from it than Neptune. Failing a precise measurement, at least it is desirable to close in this star between two limits and to know for example whether it is nearer than Sirius.

This second limit can be fixed from a theory of dielectrics due to Clausius and Mossotti: on this theory the dielectric power of a gas depends upon the polarisation of each molecule by influence by the displacement of interior electric charges. Developing this hypothesis, we shall write that the true volume of n molecules is not equal (as is sometimes stated) but certainly greater than the volume u of n perfectly conducting spheres which could be put in the place of the molecules without modifying the dielectric constant K of the medium. An electrostatic calculation gives to u the value $\frac{K-1}{K+2}$; one can thus write

$$\frac{1}{6}\pi n D^3 > \frac{K-1}{K+2}.$$

The constant K, being practically equal to the square of the refractive index (Maxwell), can also be measured directly.

Applying to the case of argon and obtaining nD^2 from Clausius's equation, we obtain

$$N < 200.10^{22}.$$

As for the molecular diameter, it is found, for all the gases so considered, to be greater than a ten-millionth of a millimetre (for the special case of argon $D > 1{\cdot}6 \times 10^{-8}$).

Here are, therefore, the various molecular magnitudes confined between two limits, which as regards the weight of each molecule are to one another as 45 is to 200. That we have no better estimate is mainly because we only know how to evaluate roughly the true volume of n molecules which occupy the unit volume of gas. A more delicate analysis is due to Van der Waals, who appears to have obtained as much in connection with molecular magnitudes as the calculation from the mean free path is able to yield *. The gas equation was obtained by supposing the molecules sufficiently separated from one another for their true volume to be small compared with that occupied by their trajectories and that each molecule suffers no sensible influence, similar to cohesion, attracting it towards the whole of the others. Van der Waals was successful in allowing for these two neglected complications and obtained the celebrated equation

$$\left(p + \frac{a}{v^2}\right)(v-b) = \mathrm{RT},$$

approximately true for the whole fluid state, in which the particular nature of the substance studied comes into evidence through the two constants a and b, of which a represents the influence of cohesion, and b exactly four times the true volume of the molecules of the given mass. Hence once b is known, the equation

$$\frac{1}{6}\pi n \mathrm{D}^3 = \frac{b}{4}$$

in conjunction with the equation of Clausius and Maxwell

$$\pi n \mathrm{D}^2 = \frac{1}{\mathrm{L}\sqrt{2}},$$

enables the unknown n and D to be calculated.

This calculation has been made for oxygen and nitrogen and a value for N nearly equal to 45.10^{22} has been obtained: this choice of substances is not a very suitable one, since it necessitates the consideration of the *molecular diameter* for molecules assuredly not spherical. By utilising for the

* *Continuité des états liquides et gazeux*, 1873.

calculation the values recently given for argon, near its critical point, I find

$$N = 62.10^{22}.$$

I should add that it is not easy to estimate the error possibly affecting this number, because of the lack of rigour of the equation of Clausius-Maxwell and of that of Van der Waals. Unquestionably an uncertainty of 30 per cent. will not be a matter for surprise.

With the determination of Van der Waals we reach the end of the first series of efforts. By methods completely different we proceed to consider similar results for which the determination can be made with greater accuracy.

12. The equipartition of energy.—We have seen that the mean molecular energy is, at the same temperature, the same for all gases. This result remains valid when the gases are mixed. It is indeed known that each gas presses upon the enclosure *as if it alone were present*, that is to say that n molecules of this gas develop in the volume v the same partial pressure as if they were alone, in such a way that $\frac{3}{2}\frac{pv}{n}$ preserves the same value. On the other hand, when we try to repeat the reasoning which led to the relation

$$pv = \frac{2}{3} nw,$$

it is found that this reasoning remains applicable. Thus w must preserve the same value. For example, the molecules of carbon dioxide and water vapour, present in the air, must have the same mean kinetic energy in spite of the difference of their natures and their masses.

This invariability of molecular energy is not confined to the gaseous state, and the beautiful work of Van't Hoff has established that it extends to the molecules of all dilute solutions. Let us imagine that a dilute solution is contained in a *semi-permeable* enclosure, which separates it from the pure solvent: we suppose this enclosure allows free passage to the molecules of the solvent, in consequence of which these molecules cannot develop any pressure, but that it stops the dissolved molecules. The impacts of these molecules against the enclosure will then develop an *osmotic pressure* P, and it is seen, if the reasoning is considered in detail, that the pressure produced by these impacts can be calculated as in the case of a gas, so that in consequence we write

$$Pv = \frac{2}{3} n\mathrm{W},$$

W signifying the mean kinetic energy of translation of n molecules contained in the volume v of the enclosure.

Now Van't Hoff has observed that the experiments of Pfeffer give for the osmotic pressure a value equal to the pressure which would be exerted by the same mass of dissolved substance if it alone occupied in the gaseous state the volume of the enclosure. W is thus equal to w: the molecules of a dissolved substance have the same mean energy as in the gaseous state.

I wish to make a remark on this matter which appears to me to render intuitive an important proposition which the kinetic theory of fluids establishes in a somewhat laborious manner. Van't Hoff's law tells us that a molecule of ethyl alcohol in solution in water has the same energy as one of the molecules of vapour over the solution ; it would still have this energy if it were present in chloroform (that is to say if it were surrounded by chloroform molecules), or even if it were in methyl or propyl alcohol : this indifference to the nature of the molecules of the liquid in which it moves makes it almost impossible to believe that it would not still have the same energy if it were in ethyl alcohol ; that is if it forms one of the molecules of pure ethyl alcohol. It therefore follows that the molecular energy is the same in a liquid as in a gas, and we can now say :

At the same temperature all the molecules of all fluids have the same mean kinetic energy, which is proportional to the absolute temperature.

But this proposition, already so general, can be still further enlarged. According to what we have just seen, the heavy molecules of sugar which move in a sugar solution have the same mean energy as the agile molecules of water. These sugar molecules contain 35 atoms; the molecules of sulphate of quinine contain more than 100 atoms, and the most complicated and heaviest molecules to which the laws of Van't Hoff (or of Raoult which are deduced from them) can be extended may be cited. The mass of the molecule appears absolutely unlimited.

Let us now consider a particle a little larger still, itself formed of several molecules, in a word a *dust*. Will it proceed to react towards the impact of the molecules encompassing it according to a new law? Will it not comport itself simply as a very large molecule, in the sense that its mean energy has still the same value as that of an isolated molecule? This cannot be averred without hesitation, but the hypothesis at least is sufficiently plausible to make it worth while to discuss its consequences.

Here we are then taken back again to the observation of the particles of an emulsion and to the study of this wonderful movement which most directly suggests the molecular hypothesis. But at the same time we are led to render the theory precise by saying, not only that each particle owes its movement to the impacts of the molecules of the liquid, but further that the energy maintained by the impacts is on the average equal to that of any one of these molecules.

The propositions, of which I have just shown the probability, could be looked upon as special cases of the famous theorem of the *equipart.tion of energy* which, with Maxwell's *law of distribution of velocities*, forms the central point of the mathematical theory of molecular motion. This theorem, solved in successive steps, thanks to very numerous attempts, among which may be cited those of Maxwell, Gibbs, Boltzmann, Jeans, Langevin and Einstein, leads to the statement of the mean equality, as regards each *degree of freedom*, of the kinetic energies of translation or of rotation which are assumed in the interior of a fluid consisting of any assemblage of molecules. This theorem has had great importance even beyond the matters here broached, and, for example, it has been the means of predicting, according to the number of atoms in a molecule of gas, the ratio between the specific heats of a gas (at constant pressure and constant volume respectively). But its *proof* calls for complicated calculations, and a more simple, even if less rigorous method, has appeared to me desirable. Besides the word *theorem* should not cause illusion, for hypotheses are introduced or implied in the calculations, as in almost all theories of mathematical physics.

I need hardly say that this is not a criticism, and I think, on the contrary, that the great strength of mathematical physics, in its useful application to research and invention, consists in the bringing to light, according to correct logic (conscious or intuitive) of probabilities which qualitative reasoning would not have disclosed. As is well understood, it cannot be maintained that the theories are sufficient to establish the results they indicate without submitting them to the test of experiment.

In brief, whatever the path pursued, we are led to regard the mean energy of translation of a molecule as equal to that possessed by the granules of an emulsion. So that if we find a means of calculating this granular energy in terms of measurable magnitudes, we shall have at the same time a means of proving our theory. The experiments once made, two cases in general can present themselves. Either the numbers found will be greatly different from those given by the kinetic reasoning referred to above, and in this case, above all, if these numbers change according to the granules studied, the credit of the kinetic theories will be weakened and the origin of the Brownian movement remains undiscovered; or, the numbers will be of the order of magnitude predicted, and, in this case, not only shall we have the right to regard the molecular theory of this movement as established, but further we can seek from our experiments, a means, possibly this time exact, of determining molecular magnitudes. I hope to show that experiment has pronounced decisively in this latter direction *.

* My results were published in the *Comptes rendus* of May 1908 and September 1909.

II.

13. The speed of a granule in suspension is not accessible to measurement.—One method of proceeding appears direct; let us suppose one has the power of measuring the mass of a microscopic particle directly; can we not hope to obtain at least an idea of its mean speed, and in consequence of its energy, by direct readings, possibly by dividing by the time of an observation the distance between the two positions it occupies at the commencement and at the end of the observation (apparent mean speed), possibly by following its trajectory in a *camera lucida* during a given time, and then dividing by this time the total length of its trajectory?

As a matter of fact this was at first essayed *, and values can be found in different papers which are always some microns per second for the mean speed of granules of the order of a micron, and which will assign to these granules a mean energy about 100,000 times less than the kinetic theory indicated for the molecule, and this would completely overthrow the theory of the equipartition of energy.

But such values are *grossly untrue.* The entanglements of the trajectory are so numerous and rapid that it is impossible to follow them, and the trajectory seen is always infinitely shorter and more simple than the real trajectory. At the same time the apparent mean speed of a granule during a given time (the quotient of the displacement by the time) varies *absurdly* in magnitude and direction without in the least tending toward a limit when the time of observation is decreased, as can be seen in a simple manner by noting the position of a granule in a *camera lucida* from minute to minute, then, for example, every five seconds, and better still by photographing them every twentieth of a second, as Victor Henri has done, by cinematographing the movement. As is well understood, the tangent at any point of the trajectory cannot be fixed even in the roughest manner. It is one of the cases where one cannot help recalling those continuous functions which do not allow of derivation, which are regarded simply as mathematical

* Wiener, Ramsay, Exner, Zsigmondy and myself.

curiosities, wrongly since nature can suggest them equally as well as the derived functions.

In brief a direct method is impossible. Here is the path I have followed:

14. Extension of the laws of gases to dilute emulsions.— Let us suppose that it is possible to obtain an emulsion, with the granules all identical, an emulsion which I shall call, for shortness, *uniform*. It appeared to me at first intuitively, that the granules of such an emulsion should distribute themselves as a function of the height in the same manner as the molecules of a gas under the influence of gravity. Just as the air is more dense at sea-level than on a mountain-top, so the granules of an emulsion, whatever may be their initial distribution, will attain a permanent state where the concentration will go on diminishing as a function of the height from the lower layers, and the law of rarefaction will be the same as for the air.

A closer examination confirms this conception and gives the law of rarefaction by precise reasoning, very similar to that which enabled Laplace to correlate the altitude and the barometric pressure.

Let us imagine a uniform emulsion in equilibrium, which fills a vertical cylinder of cross section s. The state of a horizontal slice contained between the levels h and $h+dh$ would not be changed if it were enclosed between two pistons, permeable to the molecules of water, but impermeable to the granules (membranes of parchment-paper or of collodion could effectively play this part). Each of these semi-permeable pistons is subjected by the impact of the granules which it stops to an osmotic pressure. If the emulsion is dilute, this pressure can be calculated by the same reasoning as for a gas or a dilute solution, in the sense that, if at level h there are n granules per unit volume, the osmotic pressure P will be equal to $2/3n\mathrm{W}$, if W signifies the mean granular energy; it will be $2/3\,(n+dn)\,\mathrm{W}$ at the level $h+dh$. Now the slice of granules under consideration does not fall: for this it is necessary that there should be equilibrium between the difference of the osmotic pressures, which urges it upward, and the total weight of the granules, diminished by the buoyancy of the liquid, which urges them downwards. Hence, calling ϕ the volume of each granule, Δ its density, and δ that of the intergranular liquid, we see that

$$-\frac{2}{3} s W dn = ns\, dh\, \phi(\Delta-\delta)g,$$

or

$$-\frac{2}{3} W \frac{dn}{n} = \phi(\Delta-\delta)g \,.\, dh,$$

which, by an obvious integration, involves the following relation between the concentrations n_0 and n at two points for which the difference of level is h :

$$\frac{2}{3} W \log \frac{n_0}{n} = \phi(\Delta-\delta)gh,$$

a relation which may be termed *the equation of distribution* of the emulsion. It shows clearly that *the concentration of the granules of a uniform emulsion decreases in an exponential manner as a function of the height*, in the same way as the barometric pressure does as a function of the altitude *.

If it is possible to measure the magnitudes other than W which enter into this equation, one can see whether it is verified and whether the value it indicates for W is the same as that which has been approximately assigned to the molecular energy. In the event of an affirmative answer, the origin of the Brownian movement will be established, and the laws of gases, already extended by Van't Hoff to solutions, can be regarded as still valid even for emulsions with visible granules.

15. Emulsions suitable for the researches.—Previous observations do not afford any information as to the equilibrium

* I indicated this equation at the time of my first experiments (*Comptes rendus*, May 1908). I have since learnt that Einstein and Smoluchowski, independently, at the time of their beautiful theoretical researches of which I shall speak later, had already seen that the exponential distribution is a necessary consequence of the equipartition of energy. Beyond this it does not seem to have occurred to them that in this sense an *experimentum crucis* could be obtained, deciding for or against the molecular theory of the Brownian movement.

distribution of the granules of an emulsion. It is only known that a large number of colloidal solutions will clarify in their upper part when they are left undisturbed for several weeks or months.

I have made some trials without result upon these colloidal solutions (sulphate of arsenic, ferric hydroxide, collargol, etc.). On the other hand, after some trials, I have been enabled to carry out measurements on emulsions of gamboge, then (with the assistance of M. Dabrowski) on emulsions of mastic.

The *gamboge*, which is used for a water-colour, comes from the desiccation of the latex secreted by *Garcinia morella* (*guttier* of Indo-China). A piece of this substance rubbed with the hand under a thin film of distilled water (as *soap-suds* can be made from a piece of soap) dissolves little by little, giving a beautiful opaque emulsion of a bright yellow colour, in which the microscope shows a swarm of yellow granules of various sizes, all perfectly spherical. These yellow granules can be separated from the liquid in which they are contained by energetic centrifuging, in the same manner as the red corpuscles may be separated from blood serum. They then collect at the bottom of the vessel centrifuged as a yellow mud, above which is a cloudy liquid which is decanted away. The yellow mud diluted anew (by shaking) with distilled water gives the mother emulsion which will serve for the preparation of the uniform emulsions intended for the measurements.

Instead of so using the natural granules the gamboge may be treated with methyl alcohol which entirely dissolves the yellow material (about four-fifths of the raw material) leaving a mucilaginous residue, to the properties of which I shall perhaps have to revert. This alcoholic solution, which is quite transparent and very similar to a solution of bichromate, changes suddenly, on the addition of much water, into a yellow opaque emulsion of the same appearance as the natural emulsion, and like it, composed of spherical granules. They can be separated again by centrifuging from the weak alcoholic liquid which contains them, then diluted with pure water, which gives, as in the preceding case, a mother emulsion which consists of granules of very different sizes, but of which the diameter is usually less than 1μ when the alcoholic and aqueous solutions are mixed without precautions.

I presume that the material so precipitated by water is a definite chemical compound and not a mixture * ; but that does not concern the end here pursued, and it is enough that the granules of the mother emulsion should have the same density, in order that it may serve for the preparation, in the manner about to be described, of uniform emulsions suitable for the measurements. Incidentally it may be said that an emulsion is *pure* when the granules forming it have the same composition (and in consequence the same density).

As regards *mastic*, which is used in the preparation of varnish, it is obtained by making incisions in the bark of the *Pistacia lentiscus* (Chios Island). It does not give an emulsion by direct manipulation with water ; but on leaving

* A rapid physico-chemical study has given me the following results :—

The yellow material, which is soluble in alcohol and equally very soluble in sulphide of carbon and acetic acid, is an acid for it dissolves in alkalies, even when very dilute. The acidity can be titrated by determining what quantity of soda renders a given emulsion just transparent. On the other hand, it is known that methyl alcohol and sulphide of carbon, being incompletely miscible, can give two superimposed liquid phases. The dissolved yellow colouring matter distributes itself very unequally between these two phases, and, if it were a mixture, in all probability its constituents would divide themselves unequally. Now the emulsions obtained from these two phases by addition of water require practically the same quantity of soda per gram of dissolved material to clarify them; more exactly a gram-molecule of soda dissolves 537 grams of yellow material in one emulsion and 542 grams in the other. Finally, the determination of the molecular weight by the cryoscopic method in acetic acid gives the number 555. *It is thus fairly certain that the yellow constituent of gamboge is a pure substance, with a molecular weight in the neighbourhood of* 540, *which may be termed* GUTTIC ACID (*acide guttique*).

I have further observed that this guttic acid, which expels carbon dioxide from carbonates on boiling, is displaced at ordinary temperature by a current of carbon dioxide in a solution of guttate of sodium, and then reforms an emulsion composed of spherical granules. It is thus sufficient to breathe into a clear solution of guttate to cloud it. It may be that the formation of the granules in the plant is to be explained in an analogous manner.

it in contact with methyl alcohol, there is obtained, above a completely insoluble tarry residue, a solution which is probably pure, which gives on dilution with much water an emulsion as white as milk. The granules of this emulsion are spherical and of very diverse size. The substance forming the granules has, according to Johnson, a molecular weight equal to 606, corresponding to the formula $C_{40}H_{62}O_4$.

Here are, therefore, two materials yielding spherical granules (and doubtless this would be the case for all resinous emulsions); for all such the equation of distribution of granules of radius a will be

$$\frac{2}{3}\mathrm{W}\log\frac{n_0}{n}=\frac{4}{3}\pi a^3(\Delta-\delta)gh,$$

or, since the Neperian logarithm is equal to the ordinary logarithm multiplied by 2·303,

$$\mathrm{W}\,2{\cdot}303\log_{10}\frac{n_0}{n}=2\pi a^3(\Delta-\delta)gh.$$

I have successively measured all the quantities which enter into this equation.

16. Fractional centrifuging. Realisation of a uniform emulsion.—It is necessary first to know how to prepare an emulsion in which all the granules have, at least approximately, the same diameter. The procedure which I have employed can be compared to the fractionation of a liquid mixture by distillation. Just as, during distillation, the parts first vaporised are relatively richer in the more volatile constituents, so during centrifuging of a pure emulsion * the parts first deposited are relatively richer in large granules, and it was thought that this might furnish a means of separating the granules according to their size. Here is the technique which seemed the most simple to me:

The vessel of the centrifuge is filled to a definite height, 10 cm. for example, with a pure emulsion; the machine is

* If the primary emulsion contains granules of different densities, fractionation will always separate the granules falling from the same height in the same time, which will no longer be equal.

started at a definite angular speed, for example 30 turns per second (which gives, at 15 cm. from the axis, a centrifugal force about 500 times greater than gravity) : the motive power is cut off from the machine after a definite time (60 minutes for example), and it is allowed to come to rest of itself, which should require some minutes; the vessel is then cautiously removed.

A fairly stiff mud occupies the bottom of the vessel to a well-marked level, of a height generally negligible compared with the height of the liquid ; it contains all the granules which have reached the bottom of the vessel during centrifuging, as closely packed the one on the other, as the granules filling a bag of sand.

Let a_1 indicate the radius which a granule, initially placed at the surface, ought to have to reach the bottom of the vessel just at the moment centrifuging is stopped ; all granules larger will necessarily reach the deposited mud, but the mud will contain in addition many smaller granules, which have had time to reach the bottom because at first they were present in the lower layers of the emulsion.

By means of a siphon, the supernatant liquid is cautiously decanted off, and the vessel is refilled to the original height with distilled water ; the mud is stirred up to separate all the granules, and the preceding operation is repeated with the same angular speed and for the same time of centrifuging. All the granules, of radius above a_1, again have time to reach the bottom, but such of the smaller granules as before were able to reach the bottom because they started near it, do not do so this time if they chance to be initially near the surface. Briefly, the second sediment contains like the first all the granules of radius greater than a_1, and contains far fewer smaller granules.

The supernatant emulsion, which already is paler than the first decanted fraction, is decanted, and the same operations are repeated on the sediment until the supernatant liquid at the end of each centrifuging becomes almost as clear as water. Then this sediment contains all the granules of the primary emulsion, the radius of which exceeds a_1, and does not contain any other ; all the smaller granules have been eliminated.

Let us recommence the same operations upon this latter sediment, but with a time of centrifuging a little shorter than previously. Let a_2 indicate the radius which a granule in the surface ought to have to reach the bottom of the vessel at the end of this operation. The supernatant liquid can only contain granules of radius smaller than a_2, and owing to its origin it cannot contain granules larger than a_1; so that if a_2 is near to a_1, *this liquid is a practically uniform emulsion* which it only remains to decant off.

I do not think it is necessary to explain how, in an analogous manner, from the first fractions can be obtained at will a uniform emulsion of granules, distinctly smaller, or from the sediment a uniform emulsion of granules distinctly larger.

17. Determination of the density of the granules.—The density of the granules of the uniform emulsion upon which we wish to work must be known. I have employed two methods which give concordant results. Both make use of the fact that the mass of resin present in a given sample of emulsion can be estimated with precision by simple drying on the stove. A limiting weight is very quickly reached as soon as the temperature reaches a little above 100° and does not sensibly alter when the temperature is raised to 130° or even 140° however long the treatment lasts.

But at this temperature the resin becomes a very viscous liquid, giving on cooling a transparent glass, probably of the same density as that which formed the granules of the emulsion. We are thus led to investigate the density of this glass. This is done conveniently and exactly, by putting some fragments of it into water to which is added, little by little, potassium bromide, until the fragments remain suspended in the solution, neither rising nor falling. The required density is then equal to the density of this solution which is determined without difficulty (method of Retgers, applicable even to extremely small fragments).

The second method, while perhaps more certain, has the disadvantage of requiring much of the emulsion. At a given temperature, in the neighbourhood of 20°, the masses, m of water and m' of emulsion, are measured, which fill the same specific gravity flask. The mass μ of resin contained in tnis mass m' of emulsion is determined by drying, which gives the mass $(m'-\mu)$ of the intergranular water. If d is the absolute density of water the volume of the flask is $\frac{m}{d}$, that of the intergranular water is $\frac{m'-\mu}{d}$, their difference

$$\left[\frac{m}{d}-\frac{m'-\mu}{d}\right]$$

is the volume of the granules, and the quotient of their mass μ by this volume gives the density sought.

As I have said, the two methods are concordant: to take an example, the density of the granules of a certain emulsion of gamboge was found equal to 1·205 by the first method and 1·207 by the second. Also, the density of the granules of an emulsion of mastic, equal to 1·063 by the first method, was equal to 1·064 by the second.

Care must be taken that such results refer to the case where the intergranular liquid is practically pure water.

Indeed I have observed that, when this liquid contains salts, the density of the granules seems to increase, which is at once explained by the phenomenon of adsorption of the salt at the surface of the granule.

Incidentally *there is here a new method of studying adsorption and of determining the thickness of the transition layer.*

Also, if the intergranular liquid contains a colloid with invisible granules, these latter can coat the large granules of resin and change their apparent density.

This is what occurs in the natural emulsions of gamboge, where there is present (No. **15**), about in the proportion of one-fifth, a colourless colloid, invisible in the ultramicroscope, which one can separate from the yellow granules by centrifuging and washing, so that they give the density obtained for the granules made from alcoholic solutions. Thus badly washed natural granules are heavier than after thorough washing. This source of error, soon recognised, slightly falsified my first determinations *.

* *Comptes rendus*, May 1908.

18. Arrangement of the observations.—It is not, as may well be understood, upon a height of some centimetres or even of some millimetres that the equilibrium distribution of the emulsions I have used can be studied, but upon the small height of a preparation arranged for microscopic observation, in the manner indicated roughly in the diagram (Fig. 1).

Fig. 1.

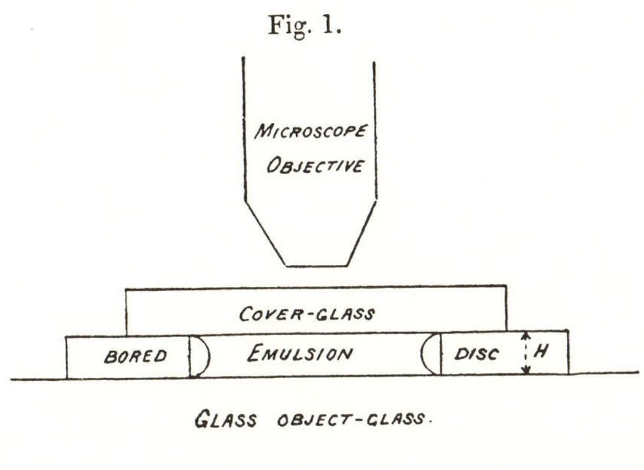

Let us suppose that a very thin glass plate bored with a large hole has been cemented in a fixed position upon a glass slide. Thus will be formed a shallow cylindrical vessel of which the height H will be, for example, about $100\,\mu$ (0·1 mm.) *.

At the centre of this vessel is placed a drop of the emulsion, which is immediately flattened by the cover-glass, and the latter, sticking to the upper face of the perforated glass plate, completely closes the cell. In addition, to prevent all evaporation, theedges of the cover-glass are covered with paraffin or varnish, which admits of a preparation being kept under observation during several days or even weeks.

The preparation is then put on to the stage of a good microscope, which has been carefully levelled. The objective used, being of very high magnifying power, has a small depth of focus, and only those granules can be seen clearly at the

* These requirements are quite satisfied by the *cells for enumeration of the blood corpuscles* (Zeiss), which I have employed.

same time which are present in a very thin horizontal layer, the thickness of which is of the order of a micron. By raising or lowering the microscope the granules in another layer can be seen.

The vertical distance between these two layers corresponds to the height h which enters into the equation of distribution, and this must be exactly known. We obtain it by multiplying the displacement h' of the microscope by the relative refractive index of the two media which the cover-glass separates. As the intergranular liquid is water, h will be equal to $\frac{4}{3}h'$, if a dry objective is employed, and simply equal to h' if, as I have most frequently done, a water immersion is used. As for the displacement h', it is read off directly on the graduated disc, fixed to the micrometer screw actuating the motion of the microscope (the screw of the Zeiss instrument reads to at least the quarter of a micron).

19. Counting the granules.—It is now necessary that we should be able to determine the ratio $\frac{n_0}{n}$ of the concentration of the granules at two different levels. This ratio is obviously equal to the mean ratio of the number of granules visible in the microscope at these two levels. It remains to find these numbers.

That does not at first sight appear to be easy : it is not a question of counting fixed objects, and when the eye is placed to the microscope and some hundreds of granules are seen moving in every direction, besides disappearing unceasingly while at the same time new granules make their appearance, one is soon convinced of the uselessness of attempts to estimate even roughly the mean number of granules present in the layer under observation.

The simplest course appears to be to take instantaneous photographs of this layer, to obtain the number of sharp images of granules there, and, if the emulsion is so dilute that the number is small, to repeat the process until the mean number of granules obtained on the plate can be considered known to the desired degree of approximation, for example, 1 per cent. I have, indeed, employed this proceedure for the relatively large granules, as will appear later. For granules of diameter less than 0·5 μ I have not been able to obtain good images, and I have had recourse to the following device : I placed in the focal plane of the eyepiece an opaque screen of foil pierced with a very small round hole by means of a dissecting-needle. The field of vision is thus very much diminished, and the eye can take in at a glance the exact number of granules visible at a definite instant, determined by a short signal, or during the very short period of illumination which can be obtained by means of a photographic shutter. It is necessary for this that the number does not exceed 5 or 6.

Operating thus at regular intervals, every 15 seconds for example, a series of numbers is noted down of which the mean value approaches more and more nearly a limit which gives the mean frequency of granules at the level studied, in the small cylindrical layer upon which the microscope is set. Recommencing at another level, the mean frequency is there redetermined for the same volume, and the quotient of these two numbers gives the ratio of the concentrations sought. As well understood, instead of making all the readings relating to one level continuously it is better to alternate the readings, making for example 100 at one level, then 100 at another level, then again 100 at the first level, and so on.

Some thousands of readings are required if some degree of accuracy is aimed at. To take an example, I have copied below the numbers given by 50 consecutive readings at two levels 30 μ apart in one of the emulsions I have used :—

3	2	0	3	2	2	5	3	1	2
3	1	1	0	3	3	4	3	4	4
0	3	1	3	1	4	2	2	1	3
1	1	2	2	3	0	1	3	4	3
0	2	2	1	0	2	1	3	2	4

for the lower level, and

2	1	0	0	1	1	3	1	0	0
0	2	0	0	0	0	1	2	2	0
2	1	3	3	1	0	0	0	3	0
1	0	2	1	0	0	1	0	1	0
1	1	0	2	4	1	0	1	0	1

for the upper level.

20. Determination of the radius of the granules.—To be in a position to apply the equation of distribution, we only need a single measurement, namely, that of the radius of the granules of the uniform emulsion studied. I have obtained this radius in three different ways :—

First method : At first, following the example of Sir J. J. Thomson, Langevin, and all those who in recent years have had occasion to determine the dimensions of droplets or dust-particles present in a gas, I have assumed the accuracy of the calculation of Stokes having reference to the movement of a sphere in a viscous medium. According to this calculation, the force of friction which resists the movement of a sphere is at each instant measured by $6\pi\zeta av$, if ζ indicated the viscosity of the liquid, a the radius of the sphere and v its velocity. When the sphere falls with uniform velocity under the sole influence of gravity, the force of friction must be equal to the apparent weight of the sphere in the fluid :

$$6\pi\zeta av = \frac{4}{3}\pi a^3(\Delta - \delta)g$$

an equation which gives a once the speed of fall is measured.

Let us suppose, on the other hand, that one has an extremely high vertical column of the uniform emulsion studied. It will be so far removed from the equilibrium distribution that the granules of the upper layers will fall like the droplets of a cloud, without our having practically to take into consideration the recoil due to the accumulation of granules in the lower layers. The liquid will then become clear in its upper part, and the thickness of the clear zone divided by the time during which the emulsion has been left to itself, will give the speed of fall to which the law of Stokes applies.

This phenomenon takes place in fact in the emulsions I have studied. It is sufficient to fill a *capillary* tube with the emulsion to a height of some centimetres, to seal it at its two ends and to install it vertically in a thermostat, to observe the emulsion gradually leave the upper layers of the liquip, falling like a cloud with a fairly sharp surface, and descending each day by the same amount. Fig. 2 shows the appearance observed. It is necessary to employ a capillary tube to avoid convective movements which confuse the surface of the cloud and which are produced with extreme facility in large tubes.

Fig. 2.

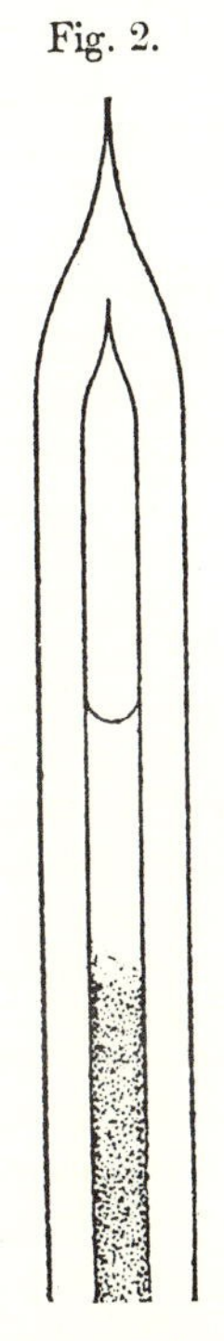

The determination of the radius of the granules is thus possible by an application of the law of Stokes. But this application to such small spheres, although definitely legitimate, gives rise to some objections which I will examine in a moment. It was thus desirable to arrive at the radius of the granules by a different, and, if possible, more direct method.

Second method.—This radius would be obtained in a very certain manner, if it were possible to find how many granules (immediately after shaking) there were in a known *titrated* volume of emulsion. That would give the mass of a granule and in consequence, since its density is known, the radius. It would be sufficient for this to count all the granules present in a cylinder of the emulsion having, as height, the height of the preparation (about 100 μ) and, as base, a surface of known area, engraved previously on the microscope slide, which is done in the *cells for the enumeration of corpuscles*, the bottom of which is divided into squares of 50 μ side. But the counting (or integration), layer by layer, of all the granules present in the height of the preparation carries with it much uncertainty. It is necessary in fact to know exactly the depth of each layer, which is of the order of a micron *, not to speak of other difficulties.

Happily I have had occasion in another connection to notice that in a feebly acid medium (for example 0·01 gram-molecule per litre) the granules of gamboge or of mastic collect on the walls of the glass which holds the preparation. At a perceptible distance from the walls the Brownian movement is in no way modified; but as soon as the chances of the movement bring a granule into contact with the slide or cover-glass, the granule becomes motionless and does not leave the wall. The emulsion is thus progressively impoverished and, after some hours, all the granules it contained are affixed to the walls. Only those, however, can be counted which are fixed in distinct positions and which do not form part of a clotted mass (partial coagulation of the colloid). Without being able to insist upon it here, I am content to say that very minute quantities of a *protecting* colloid, precisely such as is present in the natural latex of gamboge, added to the emulsion studied, prevent the granules from caking together in water acidulated by pure hydrochloric acid. On this account one may operate as follows :—

The uniformemulsion under observation, which has been previously titrated, is shaken, and a known volume of it is mixed with a known volume of feebly acidulated water, and again shaken : a drop of the mixture is taken and arranged on the microscope slide, and at once flattened by a cover-glass, the edges of which are then paraffined, taking care not to displace it, for all parts at first moistened and then

* We do not need to know this thickness when, in order to obtain the ratio of the concentrations at two different levels, we take the ratio of the number of granules visible at these two levels : it is sufficient for our purpose that the depth of the field, whatever it may be, has the same value for these two levels.

abandoned by the liquid carry away the granules. This done, the preparation is left on the stage of the microscope until all the granules have become attached to the walls. A *camera lucida* is then fitted to the microscope and, focussing on the bottom of the preparation, the contour which corresponds to one of the squares engraved upon the slide is drawn : the image of each of the granules fixed inside this square is marked by a point : then, adjusting the microscope until the granules fixed to the upper face are sharply defined, the images of these granules within the same contour are marked in the same way, which correspond in consequence to the same right prism of emulsion. The points on the drawing obtained can be subsequently counted at leisure, and their number is equal to the number of granules sought.

The same work is then recommenced upon another portion of the preparation, and so on until the mean value of the number of granules marked in each square can be considered well known. An obvious calculation then permits the number of granules contained in unit volume of the primary titrated emulsion to be found and gives in consequence the required radius, by a *second method* into which the law of

Fig. 3.

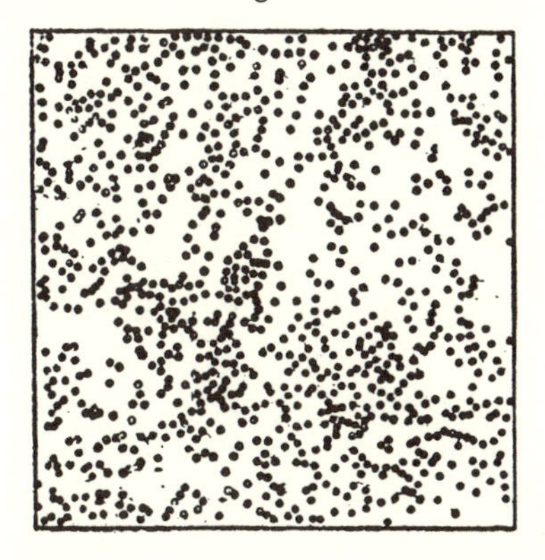

Stokes does not enter. Fig. 3 is a photograph of one of the drawings I have made on an emulsion, the granules of which had a radius equal to 0·212 μ.

The use of the *camera lucida,* fatiguing in other respects, would have been avoided by directly photographing the granules fixed to the walls. But the eye is more sensitive than the photographic plate as regards the visibility of very small clear granules on a bottom almost equally clear (it must not be forgotten that the granules are transparent spheres), and I have only been able to employ photography for granules having a diameter exceeding a demimicron.

Third method.—In this case, without the granules aggregating together in irregular clots under the influence of the acid, it often happens that they arrange themselves in little rectilinear rods composed of 3, 4 or 5 granules, which can be seen moving for an instant before fixing themselves to the bottom. The length of these little rods can be measured easily in a *camera lucida* or on a photographic plate, when the diameter of a single granule (owing to magnification due to diffraction) can only be estimated in a very rough manner. This gives a *third method*, not very exact, but very direct, for the measurement of the diameter of the granules of a uniform emulsion.

For larger grains again, of the order of a micron, the regularity of the deposit becomes greater, and the granules arrange themselves one beside the other (but not one upon the other). Fig. 4 (Pl. I.) gives the photograph of the natural granules of gamboge obtained from a very nearly uniform emulsion : the direct measurement of the radius is then very easy (0·50 μ in the case figured).

21. Extension of the law of Stokes.—The three preceding methods give concordant results. Here is, in hundredths of a micron, the radii which they have indicated for different uniform emulsions. The numbers referring to the same emulsion occupy the same line : the middle column gives the number found by the application of the law of Stokes : the left-hand column the number found by the counting of the granules, of which the total has a known mass ; the right-hand column gives the numbers found by direct measurement of the length of the columns formed by granules in juxtaposition.

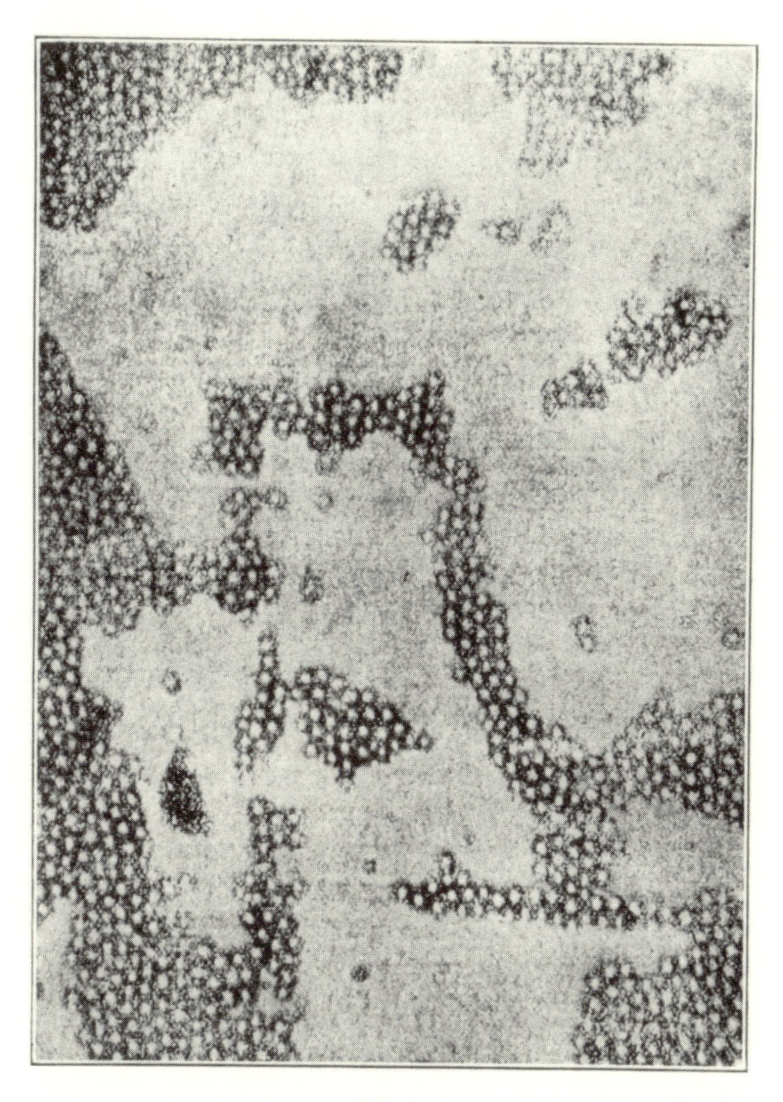

Fig. 4.

	Numbering.	Law of Stokes.	Columns of granules.
Mastic.........	...	52	54
Gamboge	...	49	50
	46	45	45·5
	30	29	30
	21·2	21·3	
	14	15	

The numbers of the fifth line, which have reference to an emulsion more uniform than the others, are those which have been determined with the greatest accuracy. I have counted about 11,000 granules in different regions of different preparations to obtain the figure 21·2 of the first column.

In brief, the three methods employed justify themselves by their concordance. Further, this concordance brings out certain important consequences concerning the law of Stokes.

This law has been established by a calculation postulating conditions of continuity apparently very far from being fulfilled in the case of spheres animated by a very lively Brownian movement. Further, the speed considered is the true speed of the sphere with reference to the fluid. Now this true speed, which, as is always the case for the Brownian movement, changes unceasingly its direction and magnitude, has nothing in common with the constant vertical velocity, so incomparably smaller, with which the cloud, formed of a great number of granules, falls through the liquid. So long as I had determined the radius of these granules only by the formula of Stokes, it was still perfectly legitimate to make reservations on the accuracy of the results obtained by a risky method, as Jacques Duclaux observed in connection with my first publication *. Naturally the same reservations should be extended to all the cases where the formula of Stokes has been applied in the same manner ; in particular to the celebrated results which the school of Sir J. J. Thomson have obtained for the condensation of droplets of water on the ions, droplets of the order of a micron, and *even more strongly* to the researches of Langevin, or his followers, on

* J. Perrin, *Comptes rendus*, 1908, cxlvi. 967, and J. Duclaux, *ibid.* cxlvii. 131.

the *large ions* of the atmosphere, of which the size of the particle is of the order of the hundredth of a micron.

The concordance of the preceding measurements will dispel these doubts; but, precisely because it is not *a priori* evident, it contains something new and gives us the right to regard as experimentally established the following proposition, which is an extrapolation from the law of Stokes:

When a force, constant in magnitude and direction, acts in a fluid on a granule agitated by the Brownian movement, the displacement of the granule, which is perfectly irregular at right angles to the force, takes in the direction of the force a component progressively increasing with the time and in the mean equal to $\frac{\mathrm{F}t}{6\pi\zeta a}$, F *indicating the force*, t *the time*, ζ *the viscosity of the fluid, and* a *the radius of the granule.*

The preceding experiments show that this law is valid in the domain of microscopic quantities, and the verification, pushed even to the threshold of ultramicroscopic magnitudes, scarcely leaves a doubt that the law may be still valid for the far smaller granules of ordinary colloids, or for the *large ions* found in gases.

I presume that it still holds for molecules as large as sulphate of quinine, but I doubt if this extension can remain rigorous for molecules of radius smaller than, or but little larger than, that of the molecules of the solvent. Later on, a reason of an experimental kind will be advanced (No. **36**); but it may be now observed that the formula indicates a zero friction for a zero radius, while the real friction, which depends upon the probability of encounters between the granule under consideration and the molecules of the solvent, can only disappear if the latter also at the same time become infinitely small.

Further than this the extreme limit to which the law of Stokes can be extended does not concern the end here pursued, and, being now in possession of all the means of measurement which are necessary for the verification and utilisation of the equation of distribution of a uniform emulsion, we pass on to see what results these means give and at the same time to settle the question of the origin of the Brownian movement.

22. The progressive rarefaction as a function of the height.— Let us consider a vertical cylinder of the emulsion, arranged for microscopic observation in the manner detailed in No. **18**. At first, after the shaking which necessarily accompanies the manipulation, the granules of this emulsion have an almost uniform distribution. But, if our kinetic theory is exact, this distribution will change from the time the preparation is left at rest, will attain a limiting state, and in this state the concentration will decrease in an exponential manner as a function of the height.

This is just what experiment verifies. At first practically as many granules are visible in the upper layers as in the lower layers of the emulsion. A few minutes suffice for the lower layers to become manifestly richer in granules than the upper layers. If then the counting of the granules (No. **19**) at two different levels is commenced, the ratio $\frac{n_0}{n}$ of the concentrations at these levels is found to have a value gradually increasing for some time, but more and more slowly, and which ends by showing no systematic variation. With the emulsions I have employed three hours is sufficient for the attainment of a well-defined limiting distribution in an emulsion left at rest, for practically the same values are found after 3 hours as after 15 days. Those emulsions which have not been rendered aseptic are occasionally invaded by elongated and very active protozoa, which, by stirring up the emulsion like fishes agitating the mud of a pond, much diminish the inequality of distribution between the upper and lower layers. But if one has patience to wait until these microbes, through lack of food, die and fall inert to the bottom of the preparation, which takes two or three days, it will be found that the initial limiting redistribution is exactly regained, and this possesses all the characters of the distribution of a permanent regime.

Once this permanent state is attained, it is easy to see whether the concentration decreases in an exponential manner as a function of the height. The following measurements show that it is so.

At first I worked on granules of gamboge of radius approximately equal to 0·14 μ, which were studied in a cell having a height of 110 μ. The concentrations of the granules were determined in five equidistant planes, the lowest plane being taken 5 μ above the bottom of the preparation (to eliminate the possible influence of the boundary), the distance between two consecutive planes being 25 μ, so that the uppermost plane was 5 μ below the surface.

The numbers found were between themselves as

100, 116, 146, 170, 200,

whereas the numbers

100, 119, 142, 169, 201,

which do not differ from the preceding by more than the limits of experimental error, are in geometrical progression. The distribution of the granules is thus quite exponential, as is the case for a gas in equilibrium under the influence of gravity. Only the diminution of the concentration to one-half, which for the atmosphere is produced by a height of about 6 kilometres, is produced here in a height of 0·1 millimetre.

But this fall of concentration is still too feeble for the exponential character of the decrease to be quite manifest. I have therefore tried to secure with larger granules a more rapid fall of concentration.

My most careful series has been done with granules of gamboge having a radius of 0·212 μ. The readings have been made, in a cell having a height of 100 μ, in four equidistant horizontal planes cutting the vessel at the levels

5 μ, 35 μ, 65 μ, 95 μ.

These readings, made by direct counting through a needle-hole (No. **19**), relate to 13,000 granules and give, respectively for these levels, concentrations proportional to the numbers

100, 47, 22·6, 12,

which are practically equal to the numbers

100, 48, 23, 11·1,

which again are exactly in a geometrical progression.

Thus the exponential distribution cannot be doubted, each elevation of 30 μ here decreasing the concentration to about half its value.

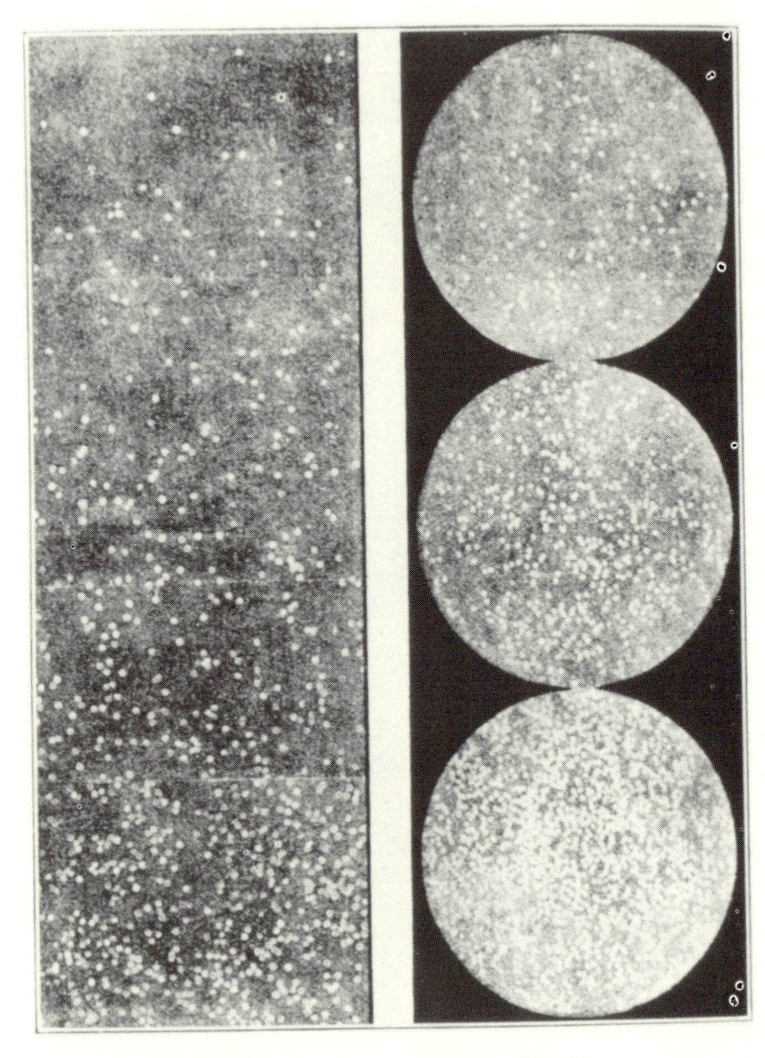

FIG. 5. FIG. 5 A.

FIG. 5. Equilibrium distribution of granules of gamboge (0·6 μ diameter; 4 levels taken every 10 μ).

FIG. 5 A. Equilibrium distribution of an emulsion of mastic (1 μ diameter: 3 levels taken every 12 μ).

A third series, considerably less exact, which refers to 3000 granules of an impure gamboge, more dense than the pure gamboge, has been carried out by a different method (counting of the granules on a photographic plate). The radius of the granules differed little from 0·29 μ. This time an elevation of 30 μ was sufficient to lower the concentration to a tenth of its value : more precisely, the concentrations at the levels

$$5\,\mu, \quad 15\,\mu, \quad 25\,\mu, \quad 35\,\mu,$$

were between themselves as the numbers

$$100, \quad 43, \quad 22, \quad 10,$$

but little different from the numbers

$$100, \quad 45, \quad 21, \quad 9{\cdot}4,$$

which are in geometrical progression.

In fig. 5 (Pl. II.) are shown, one above the other, drawings reproducing the distribution of the granules in four of the photographs from which the preceding numbers were obtained: the progressive rarefaction is evident. This rarefaction is very striking when, keeping the eyes fixed upon the preparation, the microscope is rapidly raised by means of its micrometer screw. The granules are then seen rapidly to become scarcer, just as the atmosphere becomes rarer around an ascending balloon, with this difference, that 10 μ in the emulsion is equivalent to 6 kilometres in the air.

I have studied emulsions of gamboge with still heavier granules. For one of these, in the lower layers, to which the measurements are limited, the concentration fell to one-fourth of its value for an elevation of 6 μ. At 60 μ height, it will therefore be two million times less than at the bottom. Thus for emulsions of this kind once the permanent regime is attained, no granules are ever visible in the upper layers of the vessel employed, of height about 100 μ.

Lastly, with the aid of M. Dabrowski I have established the exponential distribution in the case of emulsions of *mastic*. For example, for an emulsion where the granules had a diameter of about 1 μ ($a=0{\cdot}52\,\mu$), four photographs taken at intervals of 6 μ, the one from the other, showed respectively

$$1880, \quad 940, \quad 530 \quad \text{and} \quad 305$$

images of granules, the numbers being but little different from

1880, 995, 528 and 280,

which decrease in geometrical progression. Fig. 5 *a* gives the distribution of the granules for this emulsion of mastic, in three horizontal layers placed 12 μ one above the other. There again the exponential diminution is manifest.

23. Molecular agitation is indeed the cause of the Brownian movement *.—Since the ratio of the concentrations at two points depends only on the vertical distance between them, the equation of distribution

$$2{\cdot}303\ \mathrm{W}\log_{10}\frac{n_0}{n}=2\pi a^3(\Delta-\delta)gh,$$

established in Nos. **14** and **15**, gives for each emulsion a well defined value of the granular energy W. If our kinetic theory is perfectly exact, this value will not depend upon the emulsion chosen, and will be equal to the mean energy *w* of any molecule at the same temperature. Or, what comes to the same thing, the value N′ of the expression $\frac{3}{2}\frac{\mathrm{RT}}{\mathrm{W}}$ will be independent of the radius and density of the granules studied and will be equal to the expression $\frac{3}{2}\frac{\mathrm{RT}}{w}$, that is (No. **7**) to Avogadro's constant N, which we already know in an approximate manner (No. **11**). This is all, therefore, tantamount to seeing whether, with different emulsions, N′ is placed in the neighbourhood of 60.10^{22}, the number indicated by Van der Waals' equation. The series of measurements which have enabled the exponential law of rarefaction to be established, together with others not yet described, give the answer to this question. The first series, already dealt with, refers to granules of gamboge of radius approximately equal to 0·14 μ, only moderately purified and washed, as I discovered too late, without being able since to improve the measurements as no specimen of this emulsion has been preserved. The observations, which were too difficult or impossible with

* Jean Perrin, *Comptes rendus*, 1908, cxlvi. 167, and cxlvii. 530.

immersion illumination, were made with lateral illumination which is appropriate for ultramicroscopic magnitudes, and referred to about 3000 granules (direct enumeration through a needle-hole). They led, allowing for the margin through the various sources of error, suitable for this series, to a value of N′ lying between

$$50.10^{22} \quad \text{and} \quad 80.10^{22}.$$

A second series, made with well-washed granules

$$(\Delta-\delta=0{\cdot}21),$$

of radius about double (equal approximately to $0{\cdot}30\,\mu$) but only moderately uniform, also referred to about 3000 granules and gave for N′ the value

$$75.10^{22}.$$

A third series, of comparable accuracy, has been done on granules about as large as the preceding ($a=0{\cdot}29\,\mu$) of an impure gamboge ($\Delta-\delta=0{\cdot}30$), by a different process (dotting the granules upon a photographic plate) ; it gave

$$65.10^{22}.$$

A fourth series refers to granules considerably larger ($a=0{\cdot}45\,\mu$), nearly 30 times heavier than the first granules, which were well purified ($\Delta-\delta=0{\cdot}21$). These granules were so heavy that their concentration fell to one-quarter of its value for a height of 6 μ. Dotting of about 4000 granules (direct counting) gave for N′

$$72.10^{22}.$$

Lastly, it was desirable to work with a material other than gamboge. A fifth series has been done, with the aid of M. Dabrowski, on the granules of mastic. For equal volumes, these granules have an apparent weight in water three times less than the granules of pure gamboge

$$(\Delta-\delta=0{\cdot}063).$$

We have worked on granules which have a diameter a little greater than 1 micron ($a=0{\cdot}52\,\mu$) and are easily photographed.

Two plates have been taken, the one during ascent, the other during descent, at 6 equidistant horizontal layers, at intervals of 6 μ. The twelve plates so obtained, on which a total of about 7500 granules are visible, gave for N′ the value

$$70.10^{22}.$$

Thus the values of the granular energy deduced from the preceding series are concordant within the limits of accuracy of the experiments, although the mass of the granules has varied as 1 : 40, the difference of density between the granules and the medium has varied as 1 : 4·7, and the rapidity of rarefaction as a function of the height has varied as 1 : 30. Incidentally, there is here, in a province which will soon be enlarged, *an experimental verification of the equal distribution of energy between different masses.*

But, further, it is manifest that these values agree with that which we have foreseen for the molecular energy. The mean departure does not exceed 15 per cent., and the number given by the equation of Van der Waals does not allow of this degree of accuracy.

I do not think this agreement can leave any doubt as to the origin of the Brownian movement. To understand how striking this result is, it is necessary to reflect that, before this experiment, no one would have dared to assert that the fall of concentration would not be negligible in the minute height of some microns, or that, on the contrary, no one would have dared to assert that all the granules would not finally arrive at the immediate vicinity of the bottom of the vessel. The first hypothesis would lead to the value *zero* for N′, while the second would lead to the value *infinity*. That, in the immense interval which *a priori* seems possible for N′, the number found should fall precisely on a value so near to the value predicted, certainly cannot be considered as the result of chance.

Thus the molecular theory of the Brownian movement can be regarded as experimentally established, and, at the same time, *it becomes very difficult to deny the objective reality of molecules.* At the same time we see the law of gases, already extended by Van't Hoff to dilute solutions, extended to uniform emulsions. The Brownian movement offers us, on a different scale, the faithful picture of the movements possessed, for example, by the molecules of oxygen dissolved in the water of a lake, which, encountering one another only rarely, change their direction and speed by virtue of their impacts with the molecules of the solvent.

It may be interesting to observe that the largest of the granules, for which I have found the laws of perfect gases followed, are already visible in sunlight under a strong lens. They behave as the molecules of a perfect gas, of which the gram-molecule would weigh *200,000 tons.*

I add lastly, that all the measurements detailed in this paragraph have been made on dilute emulsions, which in the parts richest in granules only contain a thousandth part of resin, and where the osmotic pressure does not reach a thousand millionth of an atmosphere.

This last figure shows to what a point I was removed from the conditions under which it has been possible to reveal (Malfitano), and then to measure (J. Duclaux) the osmotic pressure of colloidal solutions with very fine granules closely crowded together. It may be that a generalisation, more or less analogous to that of Van der Waals, will give, one day, by means of a reasoning from the kinetic theory, the osmotic pressure of such solutions.

24. Precise determination of Avogadro's constant.—Recapitulating, equal granules distribute themselves in a dilute emulsion as heavy molecules obeying the laws of perfect gases, and the equation of their distribution, since W may now be replaced by $\frac{3}{2}\frac{RT}{N}$, can be written

$$2{\cdot}303\frac{RT}{N}\log_{10}\frac{n_0}{n}=\frac{4}{3}\pi a^3 g(\Delta-\delta)h.$$

Once this has been well established, this same equation affords a means for determining the constant N, and the constants depending upon it, which is, it appears, *capable of an unlimited precision.* The preparation of a uniform emulsion and the determination of the magnitudes other than N which enter into the equation can in reality be pushed to whatever degree of perfection desired. It is simply a question of patience and time; nothing limits *a priori* the accuracy of the results, and the mass of the atom can be obtained, if desired, with the same precision as the mass of the Earth. I scarcely need observe, on the other hand, that even perfect measurements of compressibility might not be able to prevent an uncertainty of perhaps 40 per cent. in the value of N, deduced from the equation of Van der Waals, by means of hypotheses which we know are certainly not completely exact *.

The values found for N by the five series of experiments detailed give a rough mean of $69 . 10^{22}$; the most careful of the series is the one made with *mastic* (dotting upon photographic plates) which gives $70 . 10^{22}$.

I have made, with *gamboge*, a sixth series already mentioned above on various occasions, which I consider considerably more accurate still. The mean radius of the granules of the emulsion employed was found equal to 0·212 μ, by counting 11,000 granules of a titriated emulsion, and to 0·213 μ by application of the law of Stokes. The difference of density between the material of the granules and the intergranular water was 0·2067 at 20°, the temperature to which the measurements refer. 13,000 granules were counted at different heights (direct observation through a needle-hole), and it was verified that the distribution was quite exponential, each elevation of 30 μ lowering the concentration to about half of its value (exact figures are given in No. **22**). The value resulting from these measurements is $70{\cdot}5 \times 10^{22}$.

25. Numerical values of Avogadro's constant and of the constants depending upon it.—Thus, then, one is led to adopt for Avogadro's constant the value

$$N = 70{\cdot}5 \times 10^{22}.$$

The number n of the molecules per cubic centimetre of gas in normal conditions of temperature and pressure,

* Sphericity of the molecules, and various simplifications in the reasoning which lead to the expression for the mean free path, make it impossible to specify exactly what uncertainty exists in the numerical coefficients of the approximate equations which connect the viscosity, the mean free path, and the molecular diameter (Nos. 9 and 10).

obtained by dividing the preceding by the volume, 22,400, of the gram-molecule, is thus

$$n = 3{\cdot}15 \times 10^{19}.$$

The constant of molecular energy (No. 6), equal to $\frac{3R}{2N}$, is, consequently, in C.G.S. units

$$\alpha = 1{\cdot}77 \times 10^{-16},$$

and in consequence the mean kinetic energy of any molecule at 0°, equal to 273 α, is, in ergs,

$$w = 0{\cdot}48 \times 10^{-13}.$$

Lastly, the charge of the electron, or atom of electricity, is obtained by dividing the faraday by N (No. 8), and is therefore in C.G.S. electrostatic units,

$$e = 4{\cdot}11 \times 10^{-10}.$$

26. Weight and dimensions of molecules or atoms.—The mass of any molecule or atom is obtained in an obvious manner with the same precision. For example, since there are in 32 grams of oxygen N molecules, each molecule of oxygen will have the mass

$$O_2 = \frac{32}{N} = 45{\cdot}4 \times 10^{-24},$$

and the atom of oxygen will be

$$O = \frac{16}{N} = 22{\cdot}7 \times 10^{-24}.$$

In the same way each molecule will be obtained by dividing by N the gram-molecule of the corresponding compound, and each atom by dividing by N the gram-atom of the corresponding element. The lightest of all the atoms, that is to say, the atom of hydrogen, has therefore the mass

$$H = \frac{1{\cdot}008}{N} = 1{\cdot}43 \times 10^{-24}.$$

Lastly, the mass of one of the identical *corpuscles* which carry the negative electricity of the cathode-rays or of the β-rays is itself obtained accurately, since it is known that it is 1775 times smaller than that of the atom of hydrogen (Classen). This corpuscular mass, the latest element of matter revealed to man, is thus

$$e = 0{\cdot}805 \times 10^{-27}.$$

As for the *dimensions* of the molecules, we can, now that we know n, deduce them from the equation (No. **10**) of Clausius-Maxwell,

$$L = \frac{1}{\pi \sqrt{2}} \frac{1}{nD^2},$$

for all gases for which the mean free path L (that is to say, definitively, the viscosity) is known.

For example, at 370°, the mean free path of the molecule of mercury under atmospheric pressure is deduced from the viscosity 6.10^{-4} of the gas by Maxwell's equation

$$\eta = 0{\cdot}31 \rho \Omega L,$$

which gives for L the value $2{\cdot}1 \times 10^{-5}$. On the other hand, at 370°, n is equal to $3{\cdot}15 \times 10^{19} \frac{273}{273 + 370}$. The diameter sought is therefore equal to the square root of

$$\frac{1}{\pi \sqrt{2}} \cdot \frac{1}{3{\cdot}15 \times 10^{19}} \cdot \frac{643}{273} \cdot \frac{10^5}{2{\cdot}1},$$

that is to say, practically $2{\cdot}8 \times 10^{-8}$.

It is in this way I have calculated some of the molecular diameters, as follows :—

Helium	$1{\cdot}7 \times 10^{-8}$.
Argon	$2{\cdot}7 \times 10^{-8}$.
Mercury	$2{\cdot}8 \times 10^{-8}$.
....	....
Hydrogen	2×10^{-8}.
Oxygen	$2{\cdot}6 \times 10^{-8}$.
Nitrogen	$2{\cdot}7 \times 10^{-8}$.
Chlorine..............	4×10^{-8}.
....	....
Ether	6×10^{-8}.

It is clear that, in the case of polyatomic molecules, one can only deal with a badly defined diameter, of which the determination, although but little affected by variations of the mass, cannot from its nature have the certainty possible for the masses.

Incidentally it may be observed that a molecule of hydrogen is insignificant compared to our own body to the same degree as that in turn is insignificant compared to the Sun.

Lastly, even the diameter of the corpuscle can be arrived at, if it is supposed, with Sir J. J. Thomson, that all its inertia is of electromagnetic origin, in which case its diameter is given by the equation *

$$D = \frac{4}{3} \frac{e^2}{mV^2},$$

where V signifies the velocity of light, m the mass of the corpuscle and e its charge, that is to say $4 \cdot 1 \times 10^{-10}$. From this there results for D the value $0 \cdot 33 \times 10^{-12}$, a value enormously smaller than the diameter of the smallest atoms.

III.

27. The formulæ of Einstein.—The preceding experiments allow, as we have seen, the origin of the Brownian movement to be established and the various molecular magnitudes to be determined. But another experimental advance was possible, and has been suggested by Einstein † at the conclusion of the very beautiful theoretical investigations of which I must now speak ‡.

We have not, so far, given a precise characterisation to the activity of the Brownian movement which agitates a definite granule, and we have only observed that its true speed is not directly measurable. Without further troubling about the infinitely tangled trajectory which the granule describes in a given time, Einstein considered simply its

* *See* LANGEVIN, *Thesis*, p. 70.

† "May soon an investigator succeed in deciding the important question in the theory of heat here proposed!"

‡ *Ann. der Physik*, 1905, 549, and 1906, 371.

displacement during this time, the displacement being defined as the length of the rectilinear segment which separates the point of departure from the point of arrival. The mean of the displacements suffered by the granule (or by a large number of identical granules) during periods of time of the same duration is the *mean displacement* appertaining to this duration.

Let us consider, provisionally, granules having the same density as the intergranular liquid; then their movement is *perfectly irregular* not only at right angles to the vertical (as under ordinary conditions) but in all senses.

In virtue of this perfect irregularity, the successive displacements distribute themselves around the mean displacement ω exactly according to the law indicated by Maxwell for the distribution of the molecular speeds around the mean speed (No. **9**), and in the same way the mean square, E^2, of the displacement will be equal to $\frac{3\pi}{8}\omega^2$.

Let us suppose that the granules are unequally distributed in the liquid : they will diffuse toward the regions of lower concentration and naturally the more rapidly the more lively their movement is, that is to say, according as their mean displacement in a given time is the greater. The mathematical analysis of this idea is not very difficult * ; it implies no new hypothesis and leads to the very simple equation

$$E^2 = 6D\tau,$$

where τ indicates the duration of time considered, and D the coefficient of diffusion ; this equation may be written, dividing each side by 3,

$$\xi^2 = 2D\tau,$$

ξ^2 signifying the mean square of the projection of the displacement on the axis Ox.

Let us now suppose that the granules are subjected to a force constant in magnitude and direction. Their movement, modified in the direction of this force, will not be changed at right angles to this direction and the preceding equation will still remain applicable in all that concerns the projection

* *See* EINSTEIN, *Ann. der Physik*, 1906, 557.

of the displacements upon a horizontal plane when the granules have not the same density as the intergranular liquid.

It remains to express the coefficient of diffusion as a function of the quantities experimentally accessible. In the case of spherical granules of radius *a*, Einstein arrived at this easily by considering the condition of permanent regime which is realised when a constant force *, acting upon the granules, maintains the concentrations different, in spite of diffusion, in layers perpendicular to the direction of the force. Then, by writing that as many granules pass in each instant across all planes perpendicular to the force in one sense, under the influence of the force, as pass in the other sense, under the influence of diffusion †, Einstein obtained the equation

$$D=\frac{RT}{N}\frac{1}{6\pi a\zeta};$$

but this time by the aid of hypotheses which are not necessarily implied by the irregularity of the Brownian movement.

One of these hypotheses, by which the viscosity of the liquid ζ is introduced, consisted in supposing that the law of Stokes applied in the case of granules animated by the Brownian movement. I have shown previously (No. **21**) that this extension, then disputable, can be experimentally established, which relieves us from seeking a theoretical justification.

The other hypothesis, which is already familiar to us, by which Avogadro's constant N was introduced, consists in supposing that the mean energy of a granule is equal to the molecular energy. It is precisely on that account that the theory of Einstein suggests a verification of the hypothesis which attributes the origin of the Brownian movement to molecular agitation.

This verification theoretically is easy : it is sufficient to compare the two preceding equations to obtain the equation

$$\xi^2=\tau\frac{RT}{N}\frac{1}{3\pi a\zeta},$$

* Which is not necessarily gravity.

† *Ann. der Physik*, 1906, 554.

in which, excepting N, only magnitudes which are directly measurable appear. It only remains to be seen whether the values of N given by this formula agree with the values otherwise found *.

Making a further step, and employing the deduction that, if the equipartition of energy is verified, the mean energy of rotation around an axis is equal to the mean energy of translation parallel to an axis, Einstein † has even succeeded in obtaining an equation which gives in a given time the mean square, α^2, of the rotation of a granule around an axis

$$\alpha^2 = \tau \frac{RT}{N} \frac{1}{4\pi\zeta a^3},$$

which also n serve as a starting point of an experimental verification, more difficult but not impossible, as I shall show further on (No. **32**). But first I wish to deal with those formulæ of Einstein which have reference to movements of translation.

28. Experimental proof of the theory of Einstein. First attempts.—It may first be remarked that it is the volume of the granules and not their mass which enters into this formula. Particles of dense metallic dust, droplets of oil, and even air bubbles should have thus, for equal volumes, exactly the same agitation. This is actually what good observers have already for a long time affirmed †. True these are only *impressions* not supported by any exact determinations, but they suffice at least to show that, contrary to what might have been thought, a heavy and a light granule of the same size show nearly the same agitation.

* It appears right to recall that Smoluchowski, almost at the same time as Einstein and by another method, arrived at a formula, but little different, in his remarkable Memoir upon a *kinetic theory of Brownian movement* (*Bulletin de l'Académie des Sciences de Cracovie*, July 1906, translated into French), where is to be found, in addition to some very nteresting reflections, an excellent historical summary of work previous to 1905.

† *Ann. der Physik*, 1906, 371-380.

† Jevons, *Proc. Manchester Phil. Soc.*, 1869, 78. Carbonelle and Thirion, *Revue des Questions scientifiques*, 1880, 5. Gouy, *Comptes rendus*, 1889, cix. 102.

Beyond the size of the granule, the theory of Einstein does not take into account the electrification which in general occurs at the surface of contact of a liquid. The contrary has been supposed by various authors who affirm, without otherwise assigning the reason, that the electrification of the granules was the condition necessary for their agitation. The error of this hypothesis has been shown by Svedberg *, who, by gradually adding traces of sulphate of aluminium to a colloidal solution of silver, reversed the sign of electrification of the granules and passed through a zero value of this electrification without perceiving at any instant the least abatement in the activity of the Brownian movement.

As regards the influence of temperature and viscosity some interesting observations of Exner may first be cited, although their significance indeed may be uncertain, but being anterior to the theory of Einstein, they could not in any case have been influenced by this theory †.

Exner worked on granules of gamboge, of which he estimated the radius (correct to about 25 per cent. ?) according to the appearance of the image (enlarged as is known by diffraction). He followed as well as possible in the *camera lucida* the trajectory during a given time, and dividing by this time the total curvilinear path so obtained, hoped to arrive at the true speed of the granule, at least approximately. We have seen (No. **13**) that such estimates are altogether false and that the true speed is enormously greater than the apparent speed so obtained. But the ratio, at two temperatures, of the lengths of trajectories delineated during a given time may not be (?) very different from the ratio of the lengths of chords joining the extremities of the trajectories. In other words, the ratio of the supposed *speeds* observed by Exner at two temperatures may be approximately equal to the ratio of the displacements during the same time for these two temperatures.

Now Exner states that the *speed* of a granule is multiplied by about 1·6, when the temperature is raised from 20° to 71°. This number is nearly equal to 1·7, the square root of the

* *Nova Acta Soc. Sc.*, ii. Upsala, 1906.

† E. Exner, *Ann. der Physik,* 1900, ii. 843.

ratio $\frac{273+71}{273+20} \cdot \frac{0 \cdot 010}{0 \cdot 004}$, which, according to Einstein, ought to be the square of the ratio of the mean displacements during the same time at the two temperatures considered. It can be seen that there is here at least a presumption of a partial verification of the formula in question.

Incidentally, it can be seen from this example that the activity of the Brownian movement which accompanies a variation of temperature depends above all on the correlative variation of viscosity. Exner, who hoped to arrive at the true speeds, considered that what he believed to be the kinetic energy of the granule was very far from varying proportionally to the absolute temperature, and concluded wrongly that the granules could not be regarded as analogous to the molecules of a fluid.

Some years later, and this time in possession of the formula of Einstein, Svedberg attempted at once an experimental control and thought he had obtained a satisfactory verification *. But I must say that this part of his work, very interesting in other respects †, does not appear to me to justify the optimistic conclusions he draws from them, and leaves the question proposed without answer. The displacements which he has observed are from 4 to 6 times larger than what according to his calculations should verify the formula ; at first sight, taking into account the experimental difficulties, there is a temptation to see there at least a rough agreement, but a careful examination discloses a discrepancy actually enormous. Actually a displacement 4 to 6 times too large requires, if the formula is exact, a radius 25 times smaller than the radius assumed by Svedberg, that is to say, for the same weight of substance about 12,000 times more granules in a given volume. Now one of the methods employed by Svedberg to find the radius consisted precisely in the counting of the number of granules in a known volume

* *Studien zur Lehre von den kolloidalen Losungen* (*Nova Acta Reg. Soc. Sc. Upsaliensis*, 4th series, ii., Upsala 1907), and "Ion," 1909.

† The absence, referred to above, of all relation between the contact electrification and the Brownian movement, and the discovery of colloidal metallic solutions in non-ionised liquids.

of titrated emulsion, and it is quite impossible that he should have seen in this volume 12,000 times fewer granules than it contained. But further : Svedberg, to make his calculations, attributed to Avogadro's constant N the value 4.10^{23}, which was admissible at that time, but which is certainly too little by almost one-half. Giving N a more exact value it will be seen that the mean displacements he indicates are more than 7 times too great, and it is necessary that he should have found in the volume explored 125,000 times too few granules. The obvious conclusion from the experiments of Svedberg would thus be, contrary to what he says, that the formula of Einstein is certainly false.

Fortunately there is probably little in common between the magnitudes ξ and τ which enter into the formula and the ill-defined magnitudes introduced in their place by Svedberg. By the aid of a suitable flow he impressed upon the emulsion a uniform movement so rapid that, on account of the persistence of luminous impressions, each granule would give to the eye a brilliant trajectory. By virtue of the Brownian movement this curve is jagged at right angles to the displacement of the whole. But from what is known of the absolute irregularity of the Brownian movement, it would appear quite impossible that these trajectories can be, as Svedberg, evidently the victim of an illusion, describes them, " lines regularly undulated, of well-defined amplitude and wave-length ! " By comparison with a micrometer eyepiece, Svedberg *estimated* the magnitude of the quantities (in reality non-existent) which he calls *the wave-length* and the *amplitude of oscillation*. Taking into account the speed of flow of the liquid he calculated then the time *of duration of the oscillation*, which will be the time during which a granule suffers, at right angles to the displacement of the whole, a displacement equal to double the amplitude. I do not think it necessary to insist upon the uncertainty, to my mind *complete*, which results from a method so questionable and from estimations so vague. On the contrary, it is *a priori* completely correct to obtain the mean displacement of a granule in a fixed time by dotting the successive positions of the image of this granule on photographs taken at equal intervals of time. Victor Henri has made in this sense a cinematographic study of the Brownian movement of the natural granules of the latex of Caoutchouc. He worked with relatively large granules, the diameter of which was estimated, from the size of the images, to be about 1 μ *.

Except that the mean displacement in a given time varied very nearly as the square root of the time, the whole of these measurements appeared unfavourable to the theory of Einstein. In neutral water the mean displacement, being nearly three times greater than the formula would indicate, can only agree with it if the diameter of the granule evaluated at 1 μ, was in reality 8 times smaller; now this is not admissible (at this diameter the granule would not even be visible, the illumination being direct).

But, more serious still, it seemed that traces of acid or alkali, which do not appreciably change the viscosity, and besides are insufficient to coagulate the caoutchouc granules, abate their movement very markedly. For example, in feebly acidulated water, the mean displacement became about 9 times less than in neutral water, which requires for granules of the same external appearance a diameter 80 times greater than in neutral water. This enormous variation is completely irreconcilable with the theory of Einstein, and, in general, with all theories which neglect the nature of the intergranular fluid and only make it intervene by its viscosity.

As far as I can judge by conversation, this then produced, among the French physicists who closely follow these questions, a current of opinion which struck me very forcibly as proving how limited, at bottom, is the belief we accord to theories, and to what a point we see in them instruments of discovery rather than of veritable *demonstrations*.

Without hesitation it was supposed that the theory of Einstein was incomplete or inexact. On the other hand, molecular agitation could not be abandoned as the origin of the Brownian movement, since I had shown by experiment † that a dilute emulsion behaves as a very dense perfect gas, of which the molecules have a weight equal to that of the

* *Comptes rendus*, 1908, 18th May and 6th July.

† *Comptes rendus*, 1908, cxlvi. 967.

granules of the emulsion. It had therefore to be supposed that some unjustifiable complementary hypothesis had entered unperceived into the reasoning of Einstein.

29. Experimental confirmation of Einstein's theory.—However, as Victor Henri had only estimated the diameter of his granules, and as he himself had made some reservations as to the generality of his results, I thought it would still be of use to measure the mean displacement of the granules of exactly known diameter which I knew how to prepare. A student who worked in my laboratory, M. Chaudesaigues, agreed to take charge of these measurements *. In default of a chronophotographic apparatus, he dotted the position of a granule in the *camera lucida* from half-minute to half minute, recommencing with another granule, and so on, making in general four readings on each granule.

Right from the first measurements it became manifest, contrary to what might have been expected, that the displacements verified at least approximately the equation of Einstein. At the same time I satisfied myself that the addition of traces of acid did not appreciably alter the movement of the granules, provided that these granules were remote from the walls †. In brief, it was necessary to suppose that some unknown complication, or some source of systematic error, had vitiated the results of Victor Henri, for the measurements, of which I am about to give a summary, cannot leave any doubt of the rigorous exactitude of the formula proposed by Einstein.

As I said, the granules which I had prepared were dotted in a *camera lucida*, the microscope being vertical, which gives the horizontal projection of the displacements. Working on squared paper, the projection on two rectangular axes of the different segments are so obtained directly, but it is useless to measure them, for the sum of the squares of these projections is equal to the sum of the squares of the segments,

* *Comptes rendus*, 1908, cxlvii. 1044, and *diplome d'Etudes*, Paris, 1909.

† Although this addition annulled and then changed the sign of the electrification, which these granules assume by contact with water.

whence it follows that, to obtain the mean square of the projection upon an axis, it is sufficient to measure these segments one by one, to calculate their squares, and to take the half of the mean of these squares. It only remains then to see whether the value given for N by the equation of Einstein,

$$\xi^2 = \tau \frac{RT}{N} \frac{1}{3\pi a \zeta},$$

agrees, within the error of experiment, with the value already determined.

In a preliminary trial, M. Chaudesaigues studied some relatively large granules of gamboge, of radius about 0·45 μ, which were moderately uniform. He noted the displacement of 40 of these granules during one minute and of 25 during two minutes; these positions gave for N the value 94.10^{22}. On the other hand, 30 granules practically identical, of a slightly greater radius, equal to 0·50 μ, gave me 66.10^{22}, which makes a mean of 80.10^{22} for this group of granules.

M. Chaudesaigues then studied the granules of radius equal to 0·212 μ, which had served for my most exact determination of N (No. **24**). The two tables following summarise the measurements made with two series of 50 granules, following each from 30 seconds to 30 seconds for two minutes, the viscosity being 0·011 for the first series (water at 17°) and 0·012 for the second:—

First Series.

Time in seconds.	Mean Horizontal Displacement (in μ).	$\xi^2 \times 10^{-8}$.	$N \times 10^{-22}$.	N (mean).
30	8·9	50·2	66	
60	13·4	113·5	59	73×10^{22}
90	14·2	1 8	78	
120	15·2	144	89	

Second Series.

Time in seconds.	Mean Horizontal Displacement (in μ).	$\xi^2 \times 10^{-8}$.	$N \times 10^{-22}$.	N (mean).
30	8·4	45	68	
60	11·6	86·5	70·5	68×10^{22}
90	14·8	140	71	
120	17·5	195	62	

Lastly, in a third series, always with granules of the same radius, the intergranular liquid was water containing much sugar, nearly five times more viscous than pure water. The mean displacement in 30 seconds, now equal to 4·7 μ, is so reduced nearly in the predicted ratio (within about 10 per cent.) and gives for N the value 56×10^{22}, rather lower than the preceding, without, however, the divergence exceeding what is possible by reason of the irregularities of statistical values, and of sources of error generally, which are a little augmented by the greater complication of the experiment.

The rough mean of these four series of measurements, which is practically equal to the mean of the two best series taken alone, is exactly

$$70 \,.\, 10^{22},$$

and is practically identical with that found by the completely different method, based upon the distribution of the granules in permanent regime. The arrangement is as perfect as possible and, to reiterate, no doubt can remain.

If, to the different values indicated for N in the preceding Tables, is attributed a *weight* proportional to the number of determinations which gave them (more numerous, for example, for the interval of 30 seconds than for that of 120 seconds) a slightly different general mean will be obtained, namely 68·7 (not 64 as erroneously published in the *Comptes rendus*) instead of 70. I do not think it necessary to make this small correction, because of a source of error, the explanation of which is of some interest, which is of greater importance for the short intervals of time than for the long. Each time a granule is dotted, a small error is really made, analogous to that in target-shooting, which obeys the laws of chance and has the same effect on the readings as if a second Brownian movement were superimposed upon that which it is desired to observe. The corresponding error, which of necessity causes an increase in the calculated mean square ξ^2, while insignificant for large intervals of time and for small viscosities, becomes of the greater importance as the time-interval is reduced and the viscosity increased. It will always have the effect of diminishing slightly the value which would be given for N by rigorously exact dotting.

30. Second confirmation of Einstein's formula.—It was desirable to check these results by changing the substance employed and, as for the distribution in height, with the collaboration of M. Dabrowski, I have repeated the measurements, substituting mastic for gamboge.

The granules of the uniform emulsion studied had a radius equal to 0·52 μ. The illumination was provided by an Auer light and, as in the preceding experiments, the beam traversed a thick cell full of water, which stopped almost all the rays capable of warming the water of the preparation. This was immersed in water, so that the objective was used with immersion, and from time to time the temperature (which it is important to know correctly, because of its great influence on the viscosity) was measured by introducing a thermometer into the tube of the microscope in contact with the objective.

We first made two series of measurements, taking turns at the microscope, each dotting the granules every 30 seconds at the call of the other. In each series this interval of 30 seconds corresponded to about 200 points, the interval of 60 seconds to 100 points, and so on. The results are summarised in the following Table :—

Time in seconds.	$N \times 10^{-22}$.	
	First Series.	Second Series.
30	57	69
60	64	65
120	67	64
240	70	88

Lastly, in a third series, we have measured for 200 distinct granules the magnitude of the displacement in 2 minutes. These 200 measurements have given for N the value

$$77 \,.\, 10^{22}.$$

These different measurements indicate for N a mean value comprised, according to the widest conventions that can be made as to the relative importance of the numbers dotted, between $72 \,.\, 10^{22}$ and $74 \,.\, 10^{22}$, and so is about

$$73 \,.\, 10^{22}.$$

Taking into account the measurements already made with gamboge, it can be said that the 3000 displacements together indicate for $N \,.\, 10^{-22}$ the value

$$71{\cdot}5,$$

which agrees well with the value 70·5 (probably a little nearer) obtained by the fundamentally different method I used first.

The figure here reproduced (fig. 6, p. 64) shows three drawings obtained by tracing the segments which join the consecutive positions of the same granules of mastic at intervals of 30 seconds. It is the half of the mean square of such segments which verifies the formula of Einstein. One of these drawings shows 50 consecutive positions of the same granule. They only give a very feeble idea of the prodigiously entangled character of the real trajectory. If the positions were indicated from second to second, each of these rectilinear segments would be replaced by a polygonal contour of 30 sides, relatively as complicated as the drawing here reproduced, and so on. One realises from such examples how near the mathematicians are to the truth in refusing, by a logical instinct, to admit the pretended

Fig. 6.

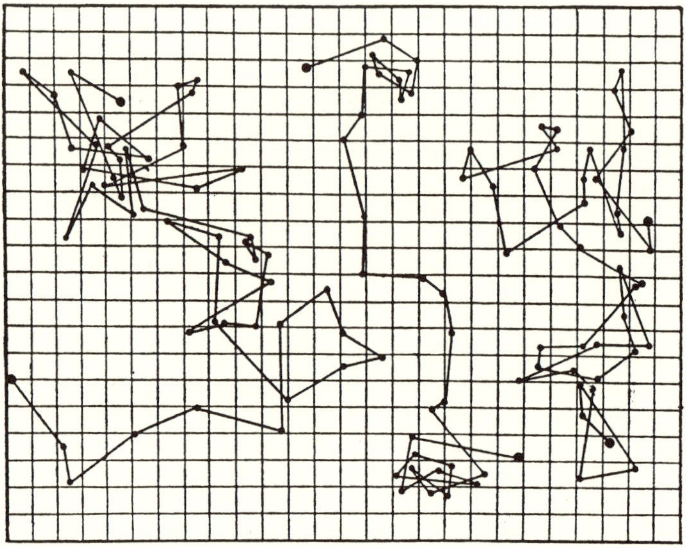

geometrical demonstrations, which are regarded as experimental evidence for the existence of a tangent at each point of a curve.

31. The law of distribution of the displacements.—We have shown (No. **27**) that, in the case of granules having the density of the intergranular liquid, the displacements in a given time ought to distribute themselves around the mean displacement according to the law of irregularity of Maxwell (No. **9**). It is useful to verify directly this important law. This can be done in various ways.

First the probability that the component along the axis Ox should be comprised between x and $x+dx$ should be

$$\frac{1}{\xi}\frac{1}{\sqrt{2\pi}}e^{-\frac{x^2}{2\xi^2}}dx,$$

designating always by ξ^2 the mean square of the component x, a result which will hold valid for all horizontal axes when the granules no longer have the density of the intergranular liquid (No. **27**). Of $\mathfrak{N}$ observations the number which will give components lying between x_1 and x_2 will then be calculable from the expression

$$\mathfrak{N}\frac{1}{\xi}\frac{1}{\sqrt{2\pi}}\int_{x_1}^{x_2} e^{-\frac{x^2}{2\xi^2}}dx.$$

M. Chaudesaigues has made this calculation, relatively to an arbitrary horizontal axis, for the displacements suffered in 30 seconds by the granules of *gamboge* (Tables of No. **29**). The numbers n of displacements having their projection comprised within two given limits (multiples of 1·7 μ which correspond to 5 mm. of the squared paper) are indicated in the following Table :—

Projections (in μ) comprised between	First Series.		Second Series.	
	n (found).	n (calc.).	n (found).	n (calc.).
0 and 1·7	38	48	48	44
1·7 „ 3·4	44	43	38	40
3·4 „ 5·1	33	40	36	35
5·1 „ 6·8	33	30	29	28
6·8 „ 8·5	35	23	16	21
8·5 „ 10·2	11	16	15	15
10·2 „ 11·9	14	11	8	10
11·9 „ 13·6	6	6	7	5
13·6 „ 15·3	5	4	4	4
15·3 „ 17·0	2	2	4	2

Another verification, perhaps still more striking, the idea of which I owe to Langevin, consists in transporting parallel to themselves the observed horizontal displacements, in such a manner as to give them all a common origin. The extremities of the vectors so obtained should distribute themselves around this origin as bullets fired at a target distribute themselves around the bull's-eye. This is well seen in fig. 7, where I have recorded 365 observations relating to granules of mastic, of which I have spoken in the preceding paragraph. Here again the checking of the law of distribution can be

Fig. 7.

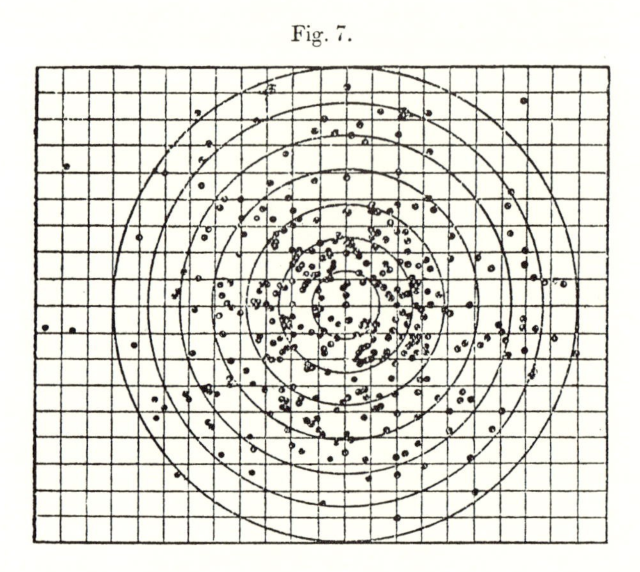

quantitative. For if the law of probability given for a component x is admitted, it is easy to see that the probability of a horizontal displacement having a length comprised between r and $r+dr$ is given by the expression

$$\frac{1}{2\pi\xi^2} e^{-\frac{r^2}{2\xi^2}} 2\pi r\, dr\ ;$$

or, simplifying and replacing $2\xi^2$ by the mean square, ρ^2, of the horizontal displacement,

$$\frac{2}{\rho^2} e^{-\frac{r^2}{\rho^2}} r\, .\, dr$$

of which the integral is simply $-e^{-\frac{r^2}{\rho^2}}$, so that it follows that the number of displacements comprised between r_1 and r_2 can be immediately calculated.

In the case of the preceding figure, ρ is equal to $7 \cdot 16\,\mu$, and I find for the number of displacements comprised between two fixed limits :—

Displacements (in μ) comprised between	n observed.	n calculated.
0 and 2	24	27
2 and 4	76	71
4 and 6	90	84
6 and 8	67	76
8 and 10	45	54
10 and 12	34	30
12 and 14	20	14
14 and 16	4	5
16 and ∞	5	4

I find, lastly, a third form of verification in accord with the value actually found for the mean square of the horizontal displacement ρ^2 and with that which can be assigned to it as soon as the mean value ϵ of the horizontal displacements is known.

The reasoning is very analogous to that which permits it to be shown, assuming the law of Maxwell, that the mean square of the speed U^2 is obtained by multiplying by $\frac{3\pi}{8}$ the square Ω^2 of the mean speed (No. **9**). Let us indicate this reasoning.

We see that out of $\mathfrak{N}$ displacements, there are between r and $r+dr$

$$\mathfrak{N}\frac{2}{\rho^2}e^{-\frac{r^2}{\rho^2}}r\,.\,dr,$$

which have for the sum of their lengths

$$\mathfrak{N}\frac{2}{\rho^2}e^{-\frac{r^2}{\rho^2}}r^2\,.\,dr.$$

The sum of all the lengths is thus

$$\mathfrak{N}\frac{2}{\rho^2}\int_0^\infty \epsilon^{-\frac{r^2}{\rho^2}}r^2\,dr,$$

which is

$$\mathfrak{N}\frac{2}{\rho^2}\frac{\sqrt{\pi}}{4}\rho^3=\mathfrak{N}\rho\sqrt{\frac{\pi}{4}},$$

and the mean length ϵ of the displacement which is obtained by dividing this sum by the total number $\mathfrak{N}$ of the displacements is thus

$$\epsilon=\rho\sqrt{\frac{\pi}{4}},$$

that is to say, very nearly,

$$\rho=\frac{9}{8}\epsilon.$$

This result is well verified. For example, the displacements which have served for the preceding figure have a mean equal to 6·4 μ. The value predicted for ρ by this calculation is hence

$$7{\cdot}21\ \mu,$$

which is in good agreement with the value

$$7{\cdot}16\ \mu$$

actually found.

In brief, Maxwell's law of irregularity is verified indisputably in its application to the displacement of the granules of an emulsion.

That shows that the probability of a certain value x of the projection of a displacement on the axis Ox does not depend upon the values of the components y and z (No. **9**). After this it is difficult to doubt the independence of the three components of the speed. It comes to the same thing to say that it is now difficult to doubt Maxwell's law of distribution of velocities, although the completely direct verification, realised here for the displacements, is still wanting for the velocities.

32. Special study of very large granules.—The preceding reasoning does not contain any restriction as to the size of the granules, and, as regards this, seems to assume simply that the mass of the fluid can be considered very large relatively to the mass of the granules, so enabling the influence of the boundaries to be neglected. So that if no hypothesis has escaped us, objects of large size have still a perceptible Brownian movement, since according to the formula of Einstein a ball of 1 mm. in diameter should have in water at 20° an agitation corresponding to a mean displacement of 1 μ each minute.

Without being able to experiment upon such large particles, I have, however, considerably extended the region in which certain verification of the preceding laws can be obtained.

It was first necessary *to prepare* at will spherical granules considerably larger than those of the emulsions so far studied. I have succeeded in doing this in a manner which the following reasoning will elucidate.

When, without precautions, some water is poured into an alcoholic solution of resin, an aqueous solution is suddenly formed which is very strongly supersaturated with resin at every point. If what is known in other directions of the spontaneous separation of an unstable phase into two phases (formation of small drops, precipitation of crystals) be borne in mind, it is not surprising that the *germs* around which the insoluble phase will grow, appear to be in extremely great numbers. Each of these can only exhaust a very small space and will give therefore a very small granule, so that actually, when an alcoholic solution is violently mixed with water, the granules produced have a diameter in general far below 1 μ. But if, instead of using pure water, the alcoholic solution is mixed with water containing much alcohol, in which the solubility of the resin, although small, is still appreciable, there will be a chance of the germs being produced in much smaller numbers, and in consequence the spheres of resin, which can only form around such germs, may be much larger than in the preceding case.

This, in reality, I have effected by allowing pure water to flow slowly out of a funnel with a narrow orifice under an alcoholic solution of about 5 per cent. of gamboge or mastic. A zone of continuous passage then, necessarily, is established between the two liquids : as soon as a layer contains sufficient water for the supersaturation to be considerable there, germs form, which grow by intercepting the resin arriving from the upper layers, without the supersaturation ever rising to such a degree as will bring about the appearance of very numerous germs. Finally, by reason of this growth the granules soon become so heavy that they fall, in spite of their Brownian movement, passing through the lower layers of pure water, where they are washed, to the bottom of the vessel, and it only remains to collect them by decantation. I have thus precipitated all the resin out of alcoholic solutions of gamboge or mastic in the form of spheres, of which the diameter, practically never less than 2 μ, is generally in the neighbourhood of 10 μ and can attain 50 μ. These large spheres have the appearance of glass balls, yellow for gamboge, colourless for mastic, which are easily broken into irregular fragments: they frequently appear perfect and give, in the same way as lenses, a recognisable real image of the luminous source which illuminates the preparation (for example an Auer mantle) upon which it is easy to focus the microscope. But nearly as often they contain inclusions of slightly different refractivity. I have not been able to settle completely the origin and nature of these inclusions *: thanks to them it is easy to perceive the irregular movements of rotation of the spheres.

Lastly, in exceptional fashion it happens that a granule may be formed of two spheres joined together all round a little circle, resulting evidently from the welding of two spheres while they were in progress of growth around their respective germs. From the double point of view of the origin of *germs* and of their speed of growth, these various appearances present some interest beyond the special end here pursued.

* These inclusions cannot have a composition very different from that of the rest of the granules, for they scarcely modify its density, as proved by the method of Retgers of floating them in an aqueous solution of urea. I think that they are formed by a very viscous mixture enclosing a little alcohol, similar to that which slowly separates when an alcoholic solution already containing a little water is diluted with a very little water (the next day there is found a thin, very viscous layer of almost pure resin at the bottom of the vessel). Some drops of this nature, coming by Brownian movement into a layer where the vitreous resin, properly so-called, is separating can be united in the more rapid growth of the spheres of this pure resin (the reason of this rapidity being unknown to me).

However it may be with regard to these new problems, we now know how to prepare large spherical granules. They can be separated according to size by a fractionation similar to that successfully used for the small granules (No. **16**), but in this case centrifuging is not necessary, and the separation is simply accomplished by making use of the fact that the largest granules fall the most rapidly. Let us see how the fundamental laws of the Brownian movement can be followed for such granules having, for example, a diameter of 10 to 12 μ.

One can scarcely hope to study in water their progressive rarefaction as a function of the height. It suffices to apply the formula of rarefaction (No. **24**) to perceive that in the case of mastic each elevation of only 1 μ suffices to divide the concentration of the granules by about 60,000 (the rarefaction will be still more rapid with gamboge). This comes to saying that all the granules are assembled together in the immediate neigbourhood of the bottom, which can actually be verified, but which does not permit of measurement.

On the contrary, the granules distribute themselves throughout the depths of the preparation if a density practically equal to that of the granules is given to the intergranular liquid by dissolving therein a suitable substance. Even then, the quantitative verification of the law of distribution still remains practically impossible, for to make measurements approximate to a hundredth part, it would be necessary to determine the densities to a millionth.

But the measurement of the mean displacement does not appear, at least *a priori*, as if it would present any serious difficulty, and one can try to see if the formula of Einstein still applies.

I have therefore added different substances to the intergranular liquid, so as to give it the same density as the granules. A complication soon became manifest, for the greater number of these substances *coagulate* the granules, showing in addition in the most happy manner in what the phenomenon of **coagulation** consists, which is less easy to comprehend for ordinary colloidal solutions with ultra-microscopic granules. Under the influence of the coagulant the large granules studied are seen to arrange themselves in clusters of conjoined granules, rather like bunches of grapes, or even like regular pilings of bullets.

This simple and direct result divests of probability the complicated hypothesis, permissible so long as we cannot *see* the phenomenon of coagulation in detail, according to which the granules of a coagulum can be connected the one to the other without being in contact.

In quantity necessary to cause the granules to float in the middle of the liquid, the salts ordinarily considered as feeble coagulants, and even sugar, all coagulated my large granules of mastic. Urea alone had a coagulating power weak enough to allow the movements of single granules to be followed.

The intergranular liquid of suitable density contained about 27 per cent. of urea and its viscosity was 1·28 times that of pure water. A portion of the granules then floated between two layers of liquid and could be usefully observed: their number was always very small, since very feeble differences of density, it is easy to see, are sufficient to bring all the granules close to the bottom or to the surface. Certain measurements were made in a cell 1 mm. high, as high relatively to the large granules as the ones first used were to the small. But on examination the movements appeared the same in this high cell as in cells of only 100 μ; and it is to be supposed that the distribution of molecular movements around the granule which determines its movement, attains its normal regime at a distance from the walls which has nothing to do with the size of the granules. I have followed in the *camera lucida* two of these granules which were practically equal (of diameter 11·5 μ), and have measured about 100 displacements. They gave for N by the application of Einstein's formula $78 \cdot 10^{22}$. In other words the mean horizontal displacement per minute at 25° was found equal to

$$2{\cdot}35\ \mu,$$

whereas the calculated value ($N = 70{\cdot}5 \times 10^{22}$) would be

$$2{\cdot}50\ \mu.$$

Bearing in mind all the difficulties encountered and the

small number of points, the agreement is almost unexpected, and there is no doubt that Einstein's theory remains valid.

Now, the fundamental principle of this theory is the *theorem* of the equipartition of kinetic energy. It is therefore established by the preceding experiments that a granule of mastic of 11·5 μ diameter has the same mean kinetic energy as the smallest granules studied in my other experiments *which weigh about* 60,000 *times less. In this, I think, will be found much the most extended verification up to this day of the equipartition of the kinetic energy of translation.*

33. The Brownian movements of rotation and the energy of rotation.—Lastly, thanks to the inclusions which render manifest a spontaneous irregular rotation of large spherical granules, I have been able to establish by experiment one of the most important propositions of the kinetic theory, that is to say: the mean equality of the energies of rotation and translation. This proposition enables the equation given by Einstein for the rotations to be established:

$$\alpha^2 = \tau \frac{RT}{N} \frac{1}{4\pi\zeta a^3},$$

and it is sufficient to see whether this equation is verified.

In fact, although the existence of a lively Brownian movement of rotation has often been recorded, no one has ever tried to measure it, which can be understood when it is observed that if the formula is exact, a mean rotation of 100° per second is indicated for granules of 1 μ in diameter. But for my granules of 10 μ to 15 μ the predicted rotation is no more than some degrees per minute and should be easily measurable. I have, indeed, succeeded in fixing from minute to minute the orientation of spheres of mastic having a diameter of about 13 μ in suspension in a solution of urea: it is sufficient for this purpose to dot the successive positions of small inclusions, the distances of which from the centre have been measured. Then the necessary elements for calculating the component of the rotation around any axis are known. The numerical calculations, of which the details are of no interest, give for N, after making use of about 200 angular measurements, the value $65 \cdot 10^{22}$. In other words, these measurements indicate for $\sqrt{\alpha^2}$ per minute the value

$$14^{\circ}{\cdot}5,$$

whilst the calculation predicts a rotation of

$$14^{\circ}.$$

The agreement is remarkable, if one thinks of the difficulties of measurement and the complete uncertainty which *a priori* surrounded even the order of magnitude of the rotation. The granules utilised for these measurements were about 100,000 times heavier than the granules of gamboge first studied.

So *the equipartition of energy* is established throughout this great interval. Incidentally, its verification for the rotations is an experimental confirmation of the reasoning from the kinetic theory which has enabled the ratio $\frac{C}{c}$ of the specific heats of a perfect gas to be predicted.

34. Recapitulating, the molecular kinetic theory of Brownian movement has been verified to such a point in all its consequences that, whatever prepossession may exist against Atomism, it becomes difficult to reject the theory. In the second place, the quantitative study of the law of distribution of the granules of an emulsion on the one hand, and, on the other, of the activity of Brownian movement, leads, in two different ways, to *exactly the same value* for Avogadro's constant, which is essentially invariable on the kinetic theory.

It is interesting to compare this value with those obtained in other ways. Although, for the most part, they do not yet allow of any precision, their agreement assumes great importance, as demonstrating the extreme diversity of the methods which have furnished them.

Without being able to explain these methods in detail, I wish at least to enumerate them, so as to facilitate a just perspective of the whole of the questions, in which molecular reality imposes itself on the attention most forcibly.

IV.

35. Indications given by diffusion.—As this point of view is immediately related to the theory of Einstein, I will first say a few words in general as to the somewhat vague information which can be obtained from the measurement of the coefficient of diffusion.

According to one of the formulæ of Einstein the coefficient of diffusion of spherical granules is given by the expression

$$\frac{RT}{N}\frac{1}{6\pi\zeta a}.$$

It may be hoped that this formula still applies roughly to the case of molecules as small as sugar or phenol, and thus it may be seen whether the formula, so assumed, leads to acceptable values of N. This is naturally what Einstein * tried as soon as he was in possession of this formula.

In the case of sugar at 18° one should have approximately

$$\frac{0{\cdot}33}{86{,}400}=\frac{83{\cdot}2\times 10^{6}\times 291}{6\pi\times 0{\cdot}0105}\frac{1}{Na},$$

or

$$Na = 3.10^{-16}.$$

It remains to find the radius (?) of the sugar molecule. The simplest plan is to consider it as approximately given by the specific volume of solid sugar (Langevin), or, in a manner a little more precise still, to observe that in the solid the molecules cannot be packed closer than in a pile of bullets (No. **11**), and are probably less scattered than in an ordinary liquid (where, according to Van der Waals, the apparent volume is four times the real volume of the molecules).

There results for N a value comprised between

$$85.10^{22} \quad \text{and} \quad 150.10^{22}.$$

The same calculation applied to phenol, for which the structural formula indicates a very compact molecule, gives for N a better value, comprised within

$$60.10^{22} \quad \text{and} \quad 100.10^{22}.$$

* *Ann. der Physik.* 1906, xix. 289.

In reality Einstein obtained the radii by a more complicated and uncertain process, according to the difference between the viscosities of pure water and the solution (*Ann. der Physik,* 1905). He so found in the case of sugar a value of N equal to

$$40 \,.\, 10^{22}.$$

All these values are roughly concordant, and better can scarcely be hoped from a line of reasoning which supposes the molecules of saccharose to be spherical (it is much more probable that they resemble long cylinders). For the cyclic molecule of phenol the result is already much better and practically as near as the result obtained by application of the theory of Van der Waals (No. **11**).

Now the reasoning of Einstein supposes the *law of Stokes* to be valid. It is therefore probable that this law, the exactitude of which I have proved directly as far as dimensions of the order of a tenth of a micron (No. **21**), *still remains exactly verified for large molecules, the diameter of which does not reach the thousandth of a micron.* This without doubt is the most interesting result which we owe to this consideration of the coefficients of diffusion. It will permit us presently to apply the law of Stokes with safety to the case of ions in movement through a gas (No. **38**).

36. Indications given by the mobility of ions (in liquids).— A still more daring extension of the law of Stokes is at the bottom of a very ingenious idea developed by M. Pellat *. Let v be the mean speed of electric transport, shown by a monovalent ion of radius a and charge e in an electric field H. We shall have, if the law of Stokes is applicable,

$$6\pi\zeta av = \mathrm{H}e,$$

an equation which finds at least a partial verification in the well-known fact that the speed v is proportional to the field. Let us multiply the two sides of this equation by Avogadro's

* *Traité d'Électricité*, vol. iii. p. 56.

constant N : we shall have, remembering that Ne has the value 29×10^{13} electrostatic units,

$$6\pi\zeta\frac{v}{H}aN = 29 \cdot 10^{13}.$$

If, on the other hand, we suppose that the volume of the *charged* ion may be approximately calculated, starting from the volume Φ of the gram-atom in the solid state, in the same way as that of a neutral molecule may be calculated from the gram-molecule of the solid (an hypothesis which will appear possible but not necessary, if the manner in which the radius of an atom is defined by the impacts (No. **10**) is borne in mind) we shall have, approximately, as indicated in the preceding paragraph,

$$0{\cdot}25\ \Phi < \tfrac{4}{3}\pi a^3 N < 0{\cdot}73\ \Phi,$$

and these two relations will furnish an order of magnitude for a and N.

Let us apply this, not to mercury as M. Pellat did (for the mobility of the mercury ion is unknown and can only be fixed in a hypothetical manner), but to the monovalent ions which are the best studied.

For silver, the atomic volume of which is low, we have

$$63 \cdot 10^{22} < N < 108 \cdot 10^{22},$$

giving, let us say, a mean value of $85 \cdot 10^{22}$, practically as near as the values given by Van der Waals.

But the alkali-metals, and especially cæsium, which, while possessing in the state of ions a mobility not greatly different from that of silver, have a much higher atomic volume (which, besides, is very well known), give much less satisfactory values for N. Thus, to 30 per cent. more or less, the potassium ion will give the value $30 \cdot 10^{22}$ and the cæsium ion $15 \cdot 10^{22}$, almost three times too low. Such a disagreement would not have seemed very large only a few years ago, and no one would have dared to affirm even that there was a certain disagreement. Now we have the right to say that this result shows, either that the law of Stokes begins decidedly no longer to apply at this extreme degree of smallness (without, however, the deviation becoming yet very great), or that the atom of potassium, for example, has a radius two and a half times smaller when it is in water, in the form of an ion, than when it is in the solid metal.

It would be necessary to obtain a second relation, got without reference to the law of Stokes, to elucidate this question.

37. Indications drawn from the blue colour of the sky.—A very curious and completely different method, due to Lord Rayleigh, brings in the diffraction of the light which comes from the sun by the molecules of the atmosphere.

When a pencil of white light penetrates a medium where fine dust-particles are present, the trajectory of the ray is rendered visible laterally, owing to the light diffused or diffracted by these particles. The phenomenon persists when the particles become more and more fine (and it is this which makes possible *ultramicroscopic* observation), but the opalescent light diffracted tends to become blue, the light of short wave-length thus suffering the more pronounced diffraction. Further, the light so scattered is found to be polarised in the plane passing through the incident ray and the eye of the observer.

No limit of minuteness is *a priori* assigned to the diffracting particles. Lord Rayleigh supposes that even the molecules act like the particles still visible in the microscope, and that this is the origin of the blue light which comes to us from the sky during the day. In accord with this hypothesis the blue light of the sky, observed in a direction perpendicular to the solar rays, is strongly polarised. It is, in addition, difficult to suppose that this is due to a diffraction by dust-particles properly so-called, for the *blue of the sky* is scarcely enfeebled when we ascend 2000 to 3000 metres in the purest atmosphere, well above the greater part of the dust-particles which sully the air in the immediate neighbourhood of the soil.

Without resting content with this qualitative conception, Lord Rayleigh, developing the elastic theory of light, has calculated the ratio which should exist, according to his hypothesis, between the intensity of the direct solar radiation and that of the blue light. In a precise manner, let us suppose that the sky is observed in a direction making an angle ϕ with the vertical and an angle β with the solar rays; the illuminations e and E obtained at the focus of an objective, successively directed towards this region of the sky and towards the sun, are for each wave-length λ in the ratio

$$\frac{e}{\mathrm{E}}=\left(9\pi^3\omega^2\frac{1+\cos^2\beta}{2\cos\phi}\right)\frac{p}{\mathrm{M}g}\frac{\mathscr{R}^2}{\lambda^4}\frac{1}{\mathrm{N}},$$

ω indicating the apparent semi-diameter of the sun, p the atmospheric pressure at the place of observation, g the acceleration of gravity, M the mass of the gram-molecule of air, $\mathscr{R}$ the molecular refractive power of air

$$\left(\text{that is }\frac{\mathrm{M}}{d}\frac{n^2-1}{n^2+2}\right)$$

and N Avogadro's constant, which can thus be fixed by this equation, supposing it to be exact. The probability of this exactitude is, besides, increased by the fact that Langevin, starting from the electromagnetic theory, obtained exactly the same equation (n^2 being replaced by the dielectric constant K). It will be seen that the extreme violet of the spectrum suffers a diffraction about sixteen times greater than the extreme red, which well explains the observed colour.

To be exact a test of this theory should be carried out at a height sufficient to avoid disturbances due to *dust-particles* (fumes, mists, *large ions*, etc.). Further, the measurements ought to be spectro-photometric. This last condition is unfortunately not realised in the only data so far available, due to M. Sella, who compared at the same instant, from the summit of Monte Rosa, the brightness of the sun at a height of 40° above the horizon and the brightness of the sky at the zenith. The ratio was found equal to 5 million. Putting this into the formula, and leaving for λ an uncertainty which appears suitable, a value is found for N comprised within

$$30.10^{22} \text{ and } 150.10^{22}.$$

So, in so far as the order of magnitude is concerned, this very interesting theory of Lord Rayleigh is verified, and it is permissible to think that more complete experiments will yield, by this means, a precise determination of N.

38. Direct measurement of the charge of an ion in a gas.— Instead of making an attempt to determine Avogadro's constant or the molecular energy, we may exert ourselves to determine directly the atom of electricity, which, as we have seen, is simply related to them. This is what the physicists of the Cambridge School have succeeded in doing, by determining the charge carried by the ion in gases.

It is not possible to know *a priori* whether, for example, the charge e' of an ion developed in a gas by the passage of X rays bears a simple relation to the charge e which a monovalent ion transports in electrolysis. Naturally, precise measurements of e and e' would decide the question ; but they are not necessary, and to Townsend we owe the establishment, since 1900, by an experimental and theoretical research of extraordinary ingenuity, of the fact that, to about one-hundredth part, the two sorts of ions carry the same charge, the common value e of which it remains to determine otherwise *.

Let us consider ions of the same sign, supposed identical, present in a gas after exposure to X-rays : whatever be their size they will have the same mean kinetic energy as the gas molecules and *will diffuse* in the gas in consequence of this molecular agitation.

Let D be their coefficient of diffusion. On the other hand, let u be the uniform velocity which ions of charge e' assume in this same gas under the influence of an electric field H. N designating always Avogadro's constant, the following equation can be established :

$$Ne' = \frac{RT}{D} \frac{u}{H}.$$

Without reproducing Townsend's reasoning, which makes appeal to some propositions of the kinetic theory not referred to in the course of this Memoir, I will observe that this

* *Phil. Trans. of the Royal Soc.* 1900, 129, translated into French in *Ions, Électrons, Corpuscles*, vol. ii. 920 (Gauthier-Villar, publisher).

equation can be very simply obtained by applying to the ions under consideration the formula given by Einstein for the coefficient of diffusion, namely,

$$D = \frac{RT}{N} \frac{1}{6\pi\zeta a}.$$

In reality, applying the law of Stokes to the movement in the electric field, we can write

$$He' = 6\pi a\zeta v,$$

and by multiplication, the precise equation of Townsend is obtained.

It is therefore sufficient to know the ratio $\frac{u}{H}$ (or the mobility of the ion), and the coefficient of diffusion D, to determine the product Ne'. Townsend himself has measured this coefficient of diffusion in various gases (air, oxygen, hydrogen, carbon dioxide); using then the measurements of the mobility of ions previously made for these same gases, he has found for Ne' values of which the mean agrees, to less than 1 per cent., with the value 29.10^{13}, fixed by electrolysis for the product Ne. This is a result of primary importance which notably enlarges the ideas, to which electrolysis gives rise, of the existence of an atom of electricity.

But although the first exact demonstration of the invariability of the atomic charge is due to Townsend, Sir J. J. Thomson had already succeeded in showing that the two charges are at least of the same order of magnitude, by attacking directly the measurement in absolute units of the charge e' *. For this he made use of the fact, established by C. T. R. Wilson, that in a moist gas, freed from dust-particles and suddenly supersaturated by the cooling produced by an expansion, the droplets of water form around the ions present in the gas. The method can be summed up as follows:—

* *Phil. Mag.* 1898, xlvi. 528; translated into French in *Ions, Électrons, Corpuscles,* vol. ii. 802.

By one of the usual methods the charge E, present in the form of ions per cubic centimetre of the gas, maintained in a constant state of ionisation, is measured, which gives the product ne' of the number of ions present in this volume by the charge e' sought. Then, on suddenly expanding the volume by a known amount, the condensation of a mass of water is brought about, which can be calculated from the known laws of adiabatic expansion. Let m be this mass of water per cubic centimetre of the original gas. If each ion has acted as a germ, this mass is divided between n droplets and, if a is the radius of each droplet, we have

$$m = n\frac{4}{3}\pi a^3.$$

Now the radius a can be obtained from the law of Stokes by measuring the velocity of fall of the cloud under the action of gravity. Since the product ne' is already known, n and in consequence e' can be calculated.

Sir J. J. Thomson has so found for e', in the case of the ions given by X-rays, values between $6{\cdot}5 \times 10^{-10}$ and $3{\cdot}4 \times 10^{-10}$, the latter seeming to him the more probable (1903). In the case of the negative ions which are produced by ultraviolet light at the surface of zinc, he found for e' the practically double value $6{\cdot}8 \times 10^{-10}$. There results from this that the constant of Avogadro ought to be comprised between

$$42.10^{22} \quad \text{and} \quad 85.10^{22},$$

the uncertainty of the mean value being at least 30 per cent. This is the degree of precision of the determination of Van der Waals.

Interesting and instructive as this method is, it contains large sources of error ; it is supposed, in particular, that each ion acts as a germ, that each germ only consists of one ion, and that the whole quantity of water calculated has been in reality condensed. The uncertainties are eliminated by an improvement, due to H. A. Wilson *, who measured the

* *Phil. Mag.* 1903; translated into French in *Ions, Électrons, Corpuscles*, vol. ii. 1107.

ratio of the velocity of fall of the drops, under the influence of gravity alone, and under the influence of gravity assisted or opposed by a vertical electric field H. We have obviously

$$\frac{v_1}{v_2} = \frac{\frac{4}{3}\pi a^3 g}{\frac{4}{3}\pi a^3 g + He'},$$

a always being given by the law of Stokes :

$$\frac{4}{3}\pi a^3 g = 6\pi\zeta a v_1.$$

H. A. Wilson so found that under the influence of the electric field the charged cloud divided itself into two or even three clouds of different velocities, corresponding to charges which are between themselves as 1, 2, and 3. A drop can therefore absorb many ions, at least unless there are polyvalent ions present in the gas. Further, the value found for e' with the least charged cloud varied notably from one experiment to another, jumping suddenly for example, under conditions apparently identical, from $2{\cdot}7 \times 10^{-10}$ to $4{\cdot}4 \times 10^{-10}$. The experiments as a whole indicated, to about 30 per cent. more or less, the value $3{\cdot}2 \times 10^{-10}$ for e, and in consequence for N the value

$$90.10^{22}.$$

In spite of the ingenuity of the improvements realised by H. A. Wilson, a large uncertainty, therefore, still remains, due perhaps (Rutherford) to the evaporation of the drops during their fall.

However, more recently, some new experiments, made according to the same method by Millikan and Begemann *, appear to have permitted more accuracy, and give for e the value $4{\cdot}05 \times 10^{-10}$, and in consequence for N the value

$$72.10^{22},$$

* *Physical Review*, February 1908, 197.

the uncertainty being possibly only a few per cent. more or less *.

39. Charge of "large ions" present in gases.—The preceding measurements may now be compared with some attempts which have been made to determine the charge, which, according to a process elucidated by Langevin, an ultramicroscopic dust-particle assumes in an ionised gas. Without being able to give here the details of his analysis, it is known that every ion, brought by molecular agitation near to such a dust-particle; is attracted by *the electric image* which it develops in the medium of greater dielectric power, and in consequence sticks to the dust-particle, which remains neutral so long as it receives equal numbers of ions of both signs, but becomes charged when it absorbs an excess of the ions of the one sign. This total charge will therefore be a whole number of electrons, rarely much greater than unity, since when one charge is fixed, ions of the same sign are repelled.

M. Ehrenhaft and M. de Broglie have independently verified these conceptions by most beautiful experiments, no longer following the total displacement of a cloud of particles, but by measuring the individual displacement of these particles †. In their experiments, the air charged

* *Note by the Translator.*—Still more recently Prof. Millikan discusses some further results (*Phil. Mag.* Feb. 1910, 209). In the first place, two corrections have to be introduced into the above value for e ($=4{\cdot}05\times10^{-10}$), raising it to $4{\cdot}57\times10^{-10}$. An important improvement was the use of radium as ionising agent instead of the X-rays, as in Wilson's experiments, which are very variable. Secondly, it has been found possible, by means of an electric field, to hold up *individual* drops carrying multiple charges (2, 3, 4, 5, and 6 respectively) practically stationary in the field of vision for a considerable part of a minute. In this way it was found possible to compare the fall under gravity with the movement, if any, under the electric field *for the same drop.* The mean adopted for all the closely agreeing results is $e=4{\cdot}65\times10^{-10}$, and the uncertainty is reckoned as only about 2 per cent. This makes the value of N

$$62.10^{22}.$$

† Ehrenhaft, *Akad. der Wiss. in Wien*, March 1909, and *Physikal. Zeit.* 1909, 308; de Broglie, *Comptes rendus,* May 1909, and *Le Radium*, 1909, 203.

with dust-particles (tobacco-smoke, for example) is drawn through a small transparent box, maintained at constant temperature, where the luminous rays from a powerful source converge. The microscope is placed at right angles to the path of the rays which enables these dust-particles to be seen as very brilliant points animated by a very brisk Brownian movement. When an electric field is made to act at right angles to gravity and to the axis of the microscope, three groups of granules are instantly distinguished: one moving in the direction of the field showing that their charge is positive; another group moving in the inverse direction, which are therefore negatively charged; and, finally, a third group continuing to dance about on the spot and which are therefore neutral.

Precise measurements would be possible if the granules had not, as unfortunately is the case, very varied sizes (and, no doubt, forms). Nevertheless, judging by their brightness, it is possible that they do not depart much from a certain mean diameter deduced, by application of the law of Stokes, from their velocity of fall in a vertical direction *. It only remains to measure the mean speed of the displacement at right angles, in the direction of the electric field, to determine the electric charge, by a second application of the law of Stokes, according to the equation

$$6\pi\zeta a v = \mathrm{H}e.$$

So it is found that the exact value of e cannot differ much from $4{\cdot}6 \times 10^{-10}$ (Ehrenhaft) or $4{\cdot}5 \times 10^{-10}$ (de Broglie), which makes N

$$64.10^{22}.$$

In spite of the uncertainty pointed out, I am inclined to consider this method more precise and more easy to perfect than that which depends upon the condensation of drops of water by expansion.

* M. de Broglie satisfied himself at the same time, by photographic measurements of the displacements, that Einstein's formula remains, at least approximately, applicable. This confirmation is interesting, in spite of its inferior accuracy compared with that previously obtained with gamboge, because of the great difference of the conditions, especially as regards the viscosity (fifty times feebler in air than in water).

40. Values deduced from radioactive phenomena.—Lastly, an admirable investigation of Rutherford, enlarging still further the idea of the atom of electricity, enables this magnitude to be obtained in many different ways, starting from observations having reference to radioactive bodies *.

It is known that the α-rays, given off from radioactive bodies, carry charges of positive electricity ; further, that when they strike sulphide of zinc, they develop there small stars of light (*scintillations*) which disappear instantly. These two phenomena have given Rutherford two completely different means of counting the number p of positive projectiles radiated in a second by 1 gram of radium, for the projected particles can reveal their existence *individually*, either by an impulse in an electrometer or by a scintillation. These two methods agree to about 1 or 2 per cent., and give for the required value

$$p = 3{\cdot}4 \times 10^{10}.$$

If, on the other hand, the total charge of positive electricity radiated per second by a given quantity of radium can be measured, a simple division will give the charge e_0 of the α-particle. The measurement is, in fact, difficult, and Rutherford has found for e_0 values comprised between

$$8{\cdot}3 \times 10^{-10} \quad \text{and} \quad 10 \times 10^{-10},$$

that is, about double the atom of electricity. The α-particle is therefore a bivalent ion (in a more precise manner Rutherford has demonstrated that it is a bivalent atom of helium).

The elementary charge e will therefore be one-half of e_0, and taking the mean of the values found

$$4{\cdot}65 \times 10^{-10},$$

the value of N will be

$$62.10^{22}.$$

This value is a little less than the value I have found from the study of the Brownian movement.

But Rutherford himself quotes other radioactive facts, which involve equally his fundamental determination of the

* RUTHERFORD and GEIGER, *Proc. Roy. Soc.* June 1908; translated into French in *Le Radium*, 1908, vi. 257.

loss p of α-particles, radiated by 1 gram of radium, and which lead to numbers practically identical with mine.

One of these relates to an investigation of Boltwood from which it results that the period of transformation of radium can be simply measured and that the transformation is half accomplished in 2000 years. N signifying always Avogadro's constant, and the gram-atom of radium being 226·5 grams, there results from this that the number of atoms of radium which break up per gram during a second, which is probably equal to the number $3{\cdot}4 \times 10^{10}$ of α-particles radiated during the same time, is also equal to

$$N \frac{1{\cdot}09 \times 10^{-11}}{226{\cdot}5},$$

whence it results that the value of N is

$$\mathbf{70{\cdot}6} \times 10^{22},$$

which is precisely the value I have found.

On the other hand, accepting always for p the number $3{\cdot}4 \times 10^{10}$ according to Rutherford, it may be supposed that the number of helium atoms produced in one second by 1 gram of radium in radioactive equilibrium is four times $3{\cdot}4 \times 10^{10}$, since in this radium there are four products, each emitting per second the same number of α-particles, that is to say, atoms of helium. If, therefore, the volume of helium disengaged per second is known, the number of atoms contained in this volume will be known, and in consequence the number of atoms N contained in 1 gram-atom of helium is directly obtained. Now some very careful measurements have been made by Sir James Dewar of the volume of helium disengaged in a day by 1 gram of radium (0·37 cubic mm.) *. As M. Moulin † has observed, this leads to the value of N

$$\mathbf{71}.10^{22},$$

which again is practically the value which the study of the

* This should have been 0·499 cubic mm. (see Proc. Roy. Soc. 1910, A, lxxxiii. 404). A later result is 0·463. (Trans.)

† *La valeur la plus probable de la charge atomique* (*Le Radium*, 1909, vi. 164).

Brownian movement has given me. The extraordinary coincidence of the results obtained, by means so profoundly different, is all the more striking because they cannot have exerted any influence, even unconsciously, upon one another; for example, the calculation of M. Moulin was only made after the accomplishment and publication of my researches.

41. Values deduced from the laws of dark radiation.—Lastly, it will not be found less surprising that almost the same numbers have once again been found, starting from relative measurements of the infra-red part of the spectrum of *black* bodies, by the reasoning developed by Lorentz and Planck.

The kinetic theory of metals, such as Sir J. J. Thomson and Drude * have conceived it, has for its fundamental hypothesis the existence in the metals of electric corpuscles, probably identical with those which constitute the cathode-rays, which move in all directions in the metal like the molecules of a gas. Every movement of electricity in a conductor is a movement of these corpuscles as a whole ; but, further, their mean energy of movement increases with the temperature ; being in brisker motion in the hotter regions of the metal, they transmit from place to place this brisker movement by their impacts, and this is what the thermal conductivity of metals consists in. Developing this idea more exactly, Drude supposes that the mean corpuscular energy is equal to the mean molecular energy, and shows that the thermal conductivity should then be proportional to the electric conductivity, which is known as the law of Wiedemann and Franz. Further, the coefficient of proportionality can be calculated *a priori*, and the predicted values agree well with the values which the different metals give.

This remarkable quantitative agreement justifies Drude's hypothesis (and extends, this time into the region of the infinitely small, what we know of the equipartition of energy).

This admitted, Lorentz observes that according to a known

* Their memoirs are translated into French in *Ions, Electrons, Corpuscles* (Gauthier-Villars).

law of electromagnetism, these corpuscles which go and come in all directions radiate energy each time their speed changes in direction and magnitude, and according to him, this radiation is precisely and definitively the light which the metal emits at the temperature considered *.

He calculates this radiation, developed in a Fourier series after analysing the emission into rays of different wave-lengths, and confines himself after that to waves of which the period is very large in comparison to the mean free path of the corpuscles. For these long waves he calculates similarly the absorbing power of the metal, and obtains (according to Kirchhoff's law) the expression for the dark radiation by dividing the emissive power by the absorbing power †. The result is that, per unit volume, the energy of radiation dA corresponding to wave-lengths comprised between λ and $\lambda + d\lambda$ is

$$\frac{16\pi}{3} w \frac{d\lambda}{\lambda^4},$$

w indicating the corpuscular (or molecular) energy at the temperature considered. This expression may also be written

$$8\pi \frac{RT}{N} \frac{d\lambda}{\lambda^4},$$

and to find N it only remains to measure this energy (which has been done, at least approximately, in the numerous measurements which refer to the distribution of energy in the spectrum of dark bodies). From measurements by Lummer and Pringsheim, Lorentz thus deduces for N the value

$$77.10^{22}.$$

Independently of Lorentz, by a more complicated theory, Max Planck had already arrived at the same formula. The

* I may observe that, in this conception, there is really nothing of periodicity in the radiation received in each instant at a point of an isothermal space: *this radiation in equilibrium is a kind of Brownian movement in the ether.*

† His calculations (simplified by Langevin) will be found in the first volume of *Ions, Electrons, Corpuscles*, p. 500.

discussion of the experimental results has led him to a value of N a little different,

$$61.10^{22}.$$

The mean of these two values is **69**.10^{22}, which is very near the preceding values.

There is no doubt that as soon as more certain values of the constant of radiation are available an exact measurement of N will be possible in this direction.

42. Comparison of all the values obtained.—A table will serve to recapitulate usefully the various phenomena which enable N to be calculated, which taken altogether form what may be termed the proof of *molecular reality*.

Phenomena studied.		$N.10^{-22}$.
Viscosity of gases taking into account	the volume of the liquid state ..	>45
	the dielectric power of the gas ..	<200
	the exact law of compressibility .	60·
Brownian Movement.	Distribution of uniform emulsion	**70·5**
	Mean displacement in a given time	71·5
	Mean rotation in a given time	65
Diffusion of dissolved substances		40 to 90
Mobility of ions in water		60 to 150
Brightness of the blue of the sky		30 to 150
Direct measurement of the atomic charge.	Droplets condensed on the ions	60 to 90*
	Ions attached to fine dust-particles .	64
Emission of α-projectiles.	Total charge radiated	62
	Period of change of radium	70·5
	Helium produced by radium	71
Energy of the infra-red spectrum..................		60 to 80

The most probable value always appears to me **70·5** × 10^{22}. The corresponding values of the other molecular magnitudes are given in paragraphs **26** and **27**.

* **62**, See Translator's Note, p. 84.

43. Molecular reality.—I think it impossible that a mind, free from all preconception, can reflect upc• the extreme diversity of the phenomena which thus converge to the same result, without experiencing a very strong impression, and I think that it will henceforth be difficult to defend by rational arguments a hostile attitude to molecular hypotheses, which, one after another, carry conviction, and to which at least as much confidence will be accorded as to the principles of energetics. As is well understood, there is no need to oppose these two great principles, the one against the other, and the union of Atomistics and Energetics will perpetuate their dua triumph.

Lastly, although with the existence of molecules or atoms the various realities of number, mass, or charge, of which we have been able to fix the magnitude, obtrude themselves forcibly, it is manifest that we ought always to be in a position to express all the visible realities without making any appeal to elements still invisible. But it is very easy to show how this may be done for all the phenomena referred to in the course of this Memoir.

Firstly, so far as concerns each special law, the constant N is simply a completely known numerical factor, figuring in the enunciation of the law. For example, the law of agitation, predicted by Einstein and established in the course of this work, is expressed by the equation

$$\xi^2 = \frac{RT}{7.10^{23}} \frac{1}{3\pi \zeta a},$$

where all the terms are measurable.

But what is perhaps more interesting, and what brings out in some ways what is now tangible in molecular reality, is to compare two laws in which Avogadro's constant enters. The one expresses this constant in terms of certain variables, $a, a', a'', \ldots,$

$$N = f[a, a', a'' \ldots .];$$

the other expresses it in terms of other variables, $b, b', b'', \ldots,$

$$N = g[b, b', b'' \ldots .].$$

Equating these two expressions we have a relation

$$f[a, a', a'', \ldots] \equiv g[b, b', b'', \ldots],$$

where only evident realities enter, and which expresses a profound connection between two phenomena at first sight completely independent, such as the transmutation of radium and the Brownian movement. For example, if we compare the law of the distribution of the energy A of dark radiation as a function of the wave-length (No. **41**) and the law of rarefaction of a uniform emulsion as a function of gravity (No. **14**), we perceive that these two laws are not independent and that the one is connected to the other by the equation

$$\frac{1}{dA}\frac{d\lambda}{\lambda^4} = \frac{1}{8\pi}\log\frac{n_0}{n}\frac{1}{(\Delta-\delta)\phi gh},$$

an equation in which all the terms are measurable.

The discovery of such relationships marks the point where the underlying reality of molecules becomes a part of our scientific consciousness.

44. Conclusion.—I think I have given in this Memoir the present state of our knowledge of the Brownian movement and of molecular magnitudes. The personal contributions which I have attempted to bring to this knowledge, both by theory and experiment, will I hope elucidate it, and will show that the observation of emulsions gives a solid experimental basis to molecular theory. The principal results established in the course of this work are in summary :—

The preparation of emulsions with equal spherical granules, of an exactly measured radius, chosen at will ;

The extension of Stokes's law to the domain of microscopic magnitudes ;

The demonstration that the laws of perfect gases apply to uniform emulsions ;

The *exact* determination, from this experimental fact, *of the various molecular magnitudes, and of the charge of the electron* ;

The experimental confirmation, for the rotations as well as for the translations, *of the equipartition of energy*, and of the beautiful theoretical investigations of Einstein ;

Lastly, arising out of this confirmation, a second *exact* determination, agreeing with the first of the various molecular magnitudes.

As we have seen, I was aided in this last part by M. Chaudesaigues, who made with much skill the greater part of the measurements of displacement for gamboge, and who thought of verifying their good agreement with the *law of chance*. On the other hand, I owe to the friendly insistence of M. Dabrowski the repetition, upon a second substance *mastic*, of the first experiments made with gamboge, which has increased their certainty. In these new experiments his able and devoted collaboration has been very useful to me, and I offer him my affectionate thanks. Lastly, I have still to thank M. Dastre, who has been so kind as to put at my disposal the powerful centrifugal machine, which was indispensable for my fractionations.

RUTHERFORD, ERNEST. (b. near Nelson, New Zealand, 30 August 1871; d. Cambridge, England, 19 October 1937)

Rutherford studied at Nelson College and Canterbury College, Christchurch in New Zealand before coming to Cambridge University's Cavendish Laboratory on a scholarship in 1895. By this time, he had already obtained B.A., B.Sc., and postgraduate M.A. degrees, and he had devised a wireless wave detector. After collaborating from 1895 to 1898 with J. J. Thomson, Rutherford taught and directed the physics laboratory at McGill University in Montreal during the period from 1898 to 1907; he then succeeded Arthur Schuster in the physics chair and laboratory at Manchester University in England. When J. J. Thomson resigned from the Cavendish Laboratory in 1919, Rutherford became its director. Like Thomson he was Cambridge professor of physics and professor of natural philosophy at the Royal Institution. In 1908 Rutherford received the Nobel Prize in physics for his work in radioactivity.

It was Rutherford's interest in electromagnetism that led him to the Cavendish and to his first collaboration with J. J. Thomson on the creation by X rays of "ions" in a gas discharge tube. He began studying uranium emissions at Cambridge, distinguishing two types which he called alpha and beta, and he continued this train of research at McGill with Frederick Soddy, and later with Bertram Boltwood, Hans Geiger, Ernest Marsden, and others. Alone, and with his collaborators, Rutherford elucidated the properties and the character of the alpha-radiations as doubly charged helium ions; he worked out decay and recovery curves for the radioelements beginning with thorium; developed the theory of the transmutation of radioelements to account for their activity; and worked out a model for the structure of the atom. Rutherford and Geiger developed techniques for counting alpha particles, and Rutherford's student H. G. J. Moseley discovered that the elements' emission of X rays could be used to identify them, according to atomic "numbers" corresponding to the positive charge of the nucleus.

That each chemical atom has a densely packed, positively charged nucleus was the revolutionary conclusion made by Rutherford in his 1911 paper "The Scattering of Alpha and Beta Particles by Matter and the Structure of the Atom." Rutherford wrote this paper after pondering the 1909 experimental results of Hans Geiger and Ernest Marsden measuring the occasional large deflections of alpha-particles by a sheet of thin metal foil. Speculating on a solar-system model for the atom, Rutherford found that his calculations from this model coincided with experimental results for the scattering of alpha-particles.

The 1911 paper appeared a few months before the 1911 Solvay Congress and was received favorably by a number of participants, including Jean Perrin who, earlier in 1901, had suggested a similar model without firm evidence for it. Perrin's work on Brownian motion now provided experimental confirmation of the real dimensions of molecular and atomic particles. Rutherford's work on radioactivity suggested that atoms, like molecules, have a substructure. Much of the discussion at the 1911 Solvay Congress was to focus on the quantum theory, and it was Niels Bohr's later (1913) incorporation into Rutherford's solar-system model of a quantum explanation of the behavior of the moving orbital electrons that put physical and chemical atomism on a new footing. The new path opened into the twentieth-century world of statistical and quantum mechanics.

LXXIX. *The Scattering of α and β Particles by Matter and the Structure of the Atom.* By Professor E. RUTHERFORD, *F.R.S., University of Manchester* *.

§ 1. IT is well known that the α and β particles suffer deflexions from their rectilinear paths by encounters with atoms of matter. This scattering is far more marked for the β than for the α particle on account of the much smaller momentum and energy of the former particle. There seems to be no doubt that such swiftly moving particles pass through the atoms in their path, and that the deflexions observed are due to the strong electric field traversed within the atomic system. It has generally been supposed that the scattering of a pencil of α or β rays in passing through a thin plate of matter is the result of a multitude of small scatterings by the atoms of matter traversed. The observations, however, of Geiger and Marsden † on the scattering of α rays indicate that some of the α particles must suffer a deflexion of more than a right angle at a single encounter. They found, for example, that a small fraction of the incident α particles, about 1 in 20,000, were turned through an average angle of 90° in passing through a layer of gold-foil about ·00004 cm. thick, which was equivalent in stopping-power of the α particle to 1·6 millimetres of air. Geiger ‡ showed later that the most probable angle of deflexion for a pencil of α particles traversing a gold-foil of this thickness was about 0°·87. A simple calculation based on the theory of probability shows that the chance of an α particle being deflected through 90° is vanishingly small. In addition, it will be seen later that the distribution of the α particles for various angles of large deflexion does not follow the probability law to be expected if such large deflexions are made up of a large number of small deviations. It seems reasonable to suppose that the deflexion through a large angle is due to a single atomic encounter, for the chance of a second encounter of a kind to produce a large deflexion must in most cases be exceedingly small. A simple calculation shows that the atom must be a seat of an intense electric field in order to produce such a large deflexion at a single encounter.

Recently Sir J. J. Thomson § has put forward a theory to

* Communicated by the Author. A brief account of this paper was communicated to the Manchester Literary and Philosophical Society in February, 1911.

† Proc. Roy. Soc. lxxxii. p. 495 (1909).

‡ Proc. Roy. Soc. lxxxiii. p. 492 (1910).

§ Camb. Lit. & Phil. Soc. xv. pt. 5 (1910).

explain the scattering of electrified particles in passing through small thicknesses of matter. The atom is supposed to consist of a number N of negatively charged corpuscles, accompanied by an equal quantity of positive electricity uniformly distributed throughout a sphere. The deflexion of a negatively electrified particle in passing through the atom is ascribed to two causes—(1) the repulsion of the corpuscles distributed through the atom, and (2) the attraction of the positive electricity in the atom. The deflexion of the particle in passing through the atom is supposed to be small, while the average deflexion after a large number m of encounters was taken as $\sqrt{m} \, . \, \theta$, where θ is the average deflexion due to a single atom. It was shown that the number N of the electrons within the atom could be deduced from observations of the scattering of electrified particles. The accuracy of this theory of compound scattering was examined experimentally by Crowther * in a later paper. His results apparently confirmed the main conclusions of the theory, and he deduced, on the assumption that the positive electricity was continuous, that the number of electrons in an atom was about three times its atomic weight.

The theory of Sir J. J. Thomson is based on the assumption that the scattering due to a single atomic encounter is small, and the particular structure assumed for the atom does not admit of a very large deflexion of an α particle in traversing a single atom, unless it be supposed that the diameter of the sphere of positive electricity is minute compared with the diameter of the sphere of influence of the atom.

Since the α and β particles traverse the atom, it should be possible from a close study of the nature of the deflexion to form some idea of the constitution of the atom to produce the effects observed. In fact, the scattering of high-speed charged particles by the atoms of matter is one of the most promising methods of attack of this problem. The development of the scintillation method of counting single α particles affords unusual advantages of investigation, and the researches of H. Geiger by this method have already added much to our knowledge of the scattering of α rays by matter.

§ 2. We shall first examine theoretically the single encounters † with an atom of simple structure, which is able to

* Crowther, Proc. Roy. Soc. lxxxiv. p. 226 (1910).

† The deviation of a particle throughout a considerable angle from an encounter with a single atom will in this paper be called "single" scattering. The deviation of a particle resulting from a multitude of small deviations will be termed "compound" scattering.

produce large deflexions of an α particle, and then compare the deductions from the theory with the experimental data available.

Consider an atom which contains a charge $\pm Ne$ at its centre surrounded by a sphere of electrification containing a charge $\mp Ne$ supposed uniformly distributed throughout a sphere of radius R. e is the fundamental unit of charge, which in this paper is taken as $4{\cdot}65 \times 10^{-10}$ E.S. unit. We shall suppose that for distances less than 10^{-12} cm. the central charge and also the charge on the α particle may be supposed to be concentrated at a point. It will be shown that the main deductions from the theory are independent of whether the central charge is supposed to be positive or negative. For convenience, the sign will be assumed to be positive. The question of the stability of the atom proposed need not be considered at this stage, for this will obviously depend upon the minute structure of the atom, and on the motion of the constituent charged parts.

In order to form some idea of the forces required to deflect an α particle through a large angle, consider an atom containing a positive charge Ne at its centre, and surrounded by a distribution of negative electricity Ne uniformly distributed within a sphere of radius R. The electric force X and the potential V at a distance r from the centre of an atom for a point inside the atom, are given by

$$X = Ne\left(\frac{1}{r^2} - \frac{r}{R^3}\right)$$

$$V = Ne\left(\frac{1}{r} - \frac{3}{2R} + \frac{r^2}{2R^3}\right).$$

Suppose an α particle of mass m and velocity u and charge E shot directly towards the centre of the atom. It will be brought to rest at a distance b from the centre given by

$$\tfrac{1}{2}mu^2 = NeE\left(\frac{1}{b} - \frac{3}{2R} + \frac{b^2}{2R^3}\right).$$

It will be seen that b is an important quantity in later calculations. Assuming that the central charge is $100\,e$, it can be calculated that the value of b for an α particle of velocity $2{\cdot}09 \times 10^9$ cms. per second is about $3{\cdot}4 \times 10^{-12}$ cm. In this calculation b is supposed to be very small compared with R. Since R is supposed to be of the order of the radius of the atom, viz. 10^{-8} cm., it is obvious that the α particle before being turned back penetrates so close to the central charge, that the field due to the uniform distribution of negative electricity may be neglected. In general, a simple calculation shows that for all deflexions greater than a degree, we may without sensible error suppose the deflexion due to the field of the central charge alone. Possible single deviations due to the negative electricity, if distributed in the form of corpuscles, are not taken into account at this stage of the theory. It will be shown later that its effect is in general small compared with that due to the central field.

Consider the passage of a positive electrified particle close to the centre of an atom. Supposing that the velocity of the particle is not appreciably changed by its passage through the atom, the path of the particle under the influence of a repulsive force varying inversely as the square of the distance will be an hyperbola with the centre of the atom S as the external focus. Suppose the particle to enter the atom in the direction PO (fig. 1), and that the direction of motio

Fig. 1.

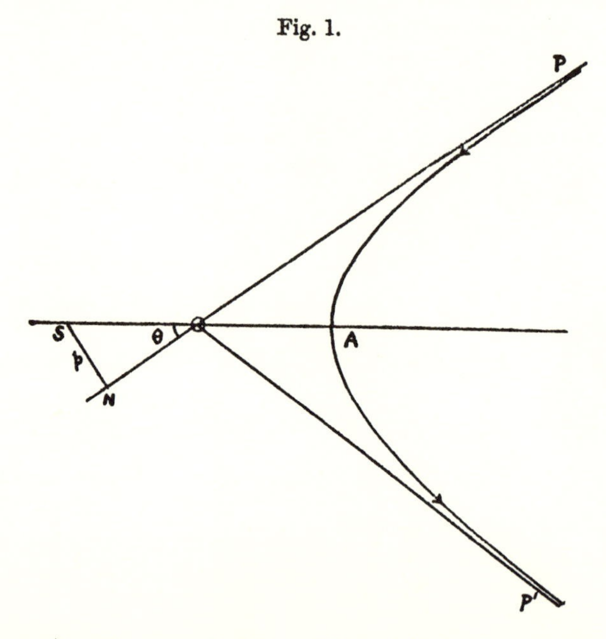

on escaping the atom is OP′. OP and OP′ make equal angles with the line SA, where A is the apse of the hyperbola. p=SN=perpendicular distance from centre on direction of initial motion of particle.

Let angle $POA=\theta$.

Let V=velocity of particle on entering the atom, v its velocity at A, then from consideration of angular momentum

$$pV=SA\,.\,v.$$

From conservation of energy

$$\tfrac{1}{2}mV^2=\tfrac{1}{2}mv^2-\frac{NeE}{SA},$$

$$v^2=V^2\left(1-\frac{b}{SA}\right).$$

Since the eccentricity is $\sec\theta$,

$$SA=SO+OA=p\operatorname{cosec}\theta(1+\cos\theta)$$

$$=p\cot\theta/2,$$

$$p^2=SA(SA-b)=p\cot\theta/2(p\cot\theta/2-b),$$

$$\therefore\quad b=2p\cot\theta.$$

The angle of deviation ϕ of the particle is $\pi-2\theta$ and

$$\cot\phi/2=\frac{2p}{b}^{*}\quad\ldots\ldots\ldots\quad(1)$$

This gives the angle of deviation of the particle in terms of b, and the perpendicular distance of the direction of projection from the centre of the atom.

For illustration, the angle of deviation ϕ for different values of p/b are shown in the following table:—

p/b....	10	5	2	1	·5	·25	·125
ϕ......	$5^\circ\!\cdot\!7$	$11^\circ\!\cdot\!4$	28°	53°	90°	127°	152°

§ 3. *Probability of single deflexion through any angle.*

Suppose a pencil of electrified particles to fall normally on a thin screen of matter of thickness t. With the exception of the few particles which are scattered through a large angle, the particles are supposed to pass nearly normally through the plate with only a small change of velocity. Let n=number of atoms in unit volume of material. Then the number of collisions of the particle with the atom of radius R is πR^2nt in the thickness t.

* A simple consideration shows that the deflexion is unaltered if the forces are attractive instead of repulsive.

The probabilty m of entering an atom within a distance p of its centre is given by

$$m=\pi p^2 nt.$$

Chance dm of striking within radii p and $p+dp$ is given by

$$dm=2\pi pnt\,.\,dp=\frac{\pi}{4}ntb^2\cot\phi/2\operatorname{cosec}^2\phi/2\,d\phi, \quad (2)$$

since $$\cot\phi/2=2p/b.$$

The value of dm gives the *fraction* of the total number of particles which are deviated between the angles ϕ and $\phi+d\phi$.

The fraction ρ of the total number of particles which are deflected through an angle greater than ϕ is given by

$$\rho=\frac{\pi}{4}ntb^2\cot^2\phi/2. \quad \ldots\ldots \quad (3)$$

The fraction ρ which is deflected between the angles ϕ_1 and ϕ_2 is given by

$$\rho=\frac{\pi}{4}ntb^2\left(\cot^2\frac{\phi_1}{2}-\cot^2\frac{\phi_2}{2}\right). \quad \ldots \quad (4)$$

It is convenient to express the equation (2) in another form for comparison with experiment. In the case of the α rays, the number of scintillations appearing on a *constant* area of a zinc sulphide screen are counted for different angles with the direction of incidence of the particles. Let r = distance from point of incidence of α rays on scattering material, then if Q be the total number of particles falling on the scattering material, the number y of α particles falling on unit area which are deflected through an angle ϕ is given by

$$y=\frac{Q\,dm}{2\pi r^2\sin\phi\,.\,d\phi}=\frac{ntb^2\,.\,Q\,.\operatorname{cosec}^4\phi/2}{16r^2}. \quad \ldots \quad (5)$$

Since $b=\frac{2NeE}{mu^2}$, we see from this equation that the number of α particles (scintillations) per unit area of zinc sulphide screen at a given distance r from the point of incidence of the rays is proportional to

(1) $\operatorname{cosec}^4 \phi/2$ or $1/\phi^4$ if ϕ be small;

(2) thickness of scattering material t provided this is small;

(3) magnitude of central charge Ne;

(4) and is inversely proportional to $(mu^2)^2$, or to the fourth power of the velocity if m be constant.

In these calculations, it is assumed that the α particles scattered through a large angle suffer only one large deflexion. For this to hold, it is essential that the thickness of the scattering material should be so small that the chance of a second encounter involving another large deflexion is very small. If, for example, the probability of a single deflexion ϕ in passing through a thickness t is 1/1000, the probability of two successive deflexions each of value ϕ is $1/10^6$, and is negligibly small.

The angular distribution of the α particles scattered from a thin metal sheet affords one of the simplest methods of testing the general correctness of this theory of single scattering. This has been done recently for α rays by Dr. Geiger *, who found that the distribution for particles deflected between 30° and 150° from a thin gold-foil was in substantial agreement with the theory. A more detailed account of these and other experiments to test the validity of the theory will be published later.

§ 4. *Alteration of velocity in an atomic encounter.*

It has so far been assumed that an α or β particle does not suffer an appreciable change of velocity as the result of a single atomic encounter resulting in a large deflexion of the particle. The effect of such an encounter in altering the velocity of the particle can be calculated on certain assumptions. It is supposed that only two systems are involved, viz., the swiftly moving particle and the atom which it traverses supposed initially at rest. It is supposed that the principle of conservation of momentum and of energy applies, and that there is no appreciable loss of energy or momentum by radiation.

* Manch. Lit. & Phil. Soc. 1910.

Let m be mass of the particle,

v_1 = velocity of approach,

v_2 = velocity of recession,

M = mass of atom,

V = velocity communicated to atom as result of encounter.

Let OA (fig. 2) represent in magnitude and direction the momentum mv_1 of the entering particle, and OB the momentum of the receding particle which has been turned through an angle AOB $=\phi$. Then BA represents in magnitude and direction the momentum MV of the recoiling atom.

Fig. 2.

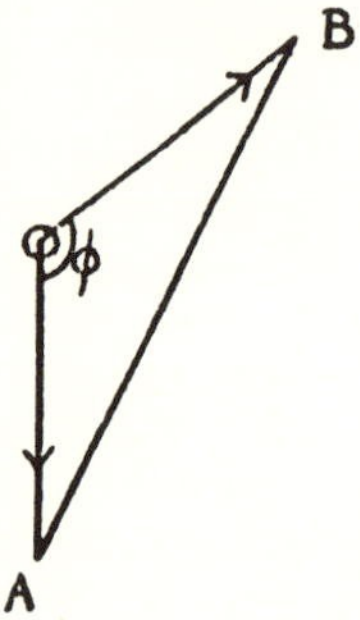

$$(MV)^2=(mv_1)^2+(mv_2)^2-2m^2v_1v_2\cos\phi. \quad (1)$$

By the conservation of energy

$$MV^2=mv_1^2-mv_2^2. \quad . \quad . \quad (2)$$

Suppose $M/m=K$ and $v_2=\rho v_1$, where ρ is <1.

From (1) and (2),

$$(K+1)\rho^2-2\rho\cos\phi=K-1,$$

or $$\rho=\frac{\cos\phi}{K+1}+\frac{1}{K+1}\sqrt{K^2-\sin^2\phi}.$$

Consider the case of an α particle of atomic weight 4, deflected through an angle of 90° by an encounter with an atom of gold of atomic weight 197.

Since $K=49$ nearly,

$$\rho=\sqrt{\frac{K-1}{K+1}}=\cdot 979,$$

or the velocity of the particle is reduced only about 2 per cent. by the encounter.

In the case of aluminium $K=27/4$ and for $\phi=90°$ $\rho=\cdot 86$.

It is seen that the reduction of velocity of the α particle becomes marked on this theory for encounters with the lighter atoms. Since the range of an α particle in air or other matter is approximately proportional to the cube of the velocity, it follows that an α particle of range 7 cms. has its range reduced to 4·5 cms. after incurring a single deviation of 90° in traversing an aluminium atom. This is of a magnitude to be easily detected experimentally. Since the value of K is very large for an encounter of a β particle with an atom, the reduction of velocity on this formula is very small.

Some very interesting cases of the theory arise in considering the changes of velocity and the distribution of scattered particles when the α particle encounters a light atom, for example a hydrogen or helium atom. A discussion of these and similar cases is reserved until the question has been examined experimentally.

§ 5. *Comparison of single and compound scattering.*

Before comparing the results of theory with experiment, it is desirable to consider the relative importance of single and compound scattering in determining the distribution of the scattered particles. Since the atom is supposed to consist of a central charge surrounded by a uniform distribution of the opposite sign through a sphere of radius R, the chance of encounters with the atom involving small deflexions is very great compared with the chance of a single large deflexion.

This question of compound scattering has been examined by Sir J. J. Thomson in the paper previously discussed (§ 1). In the notation of this paper, the average deflexion ϕ_1 due to the field of the sphere of positive electricity of radius R and quantity Ne was found by him to be

$$\phi_1 = \frac{\pi}{4} \cdot \frac{\mathrm{N}e\mathrm{E}}{mu^2} \cdot \frac{1}{\mathrm{R}}.$$

The average deflexion ϕ_2 due to the N negative corpuscles supposed distributed uniformly throughout the sphere was found to be

$$\phi_2 = \frac{16}{5} \frac{e\mathrm{E}}{mu^2} \cdot \frac{1}{\mathrm{R}} \sqrt{\frac{3\mathrm{N}}{2}}.$$

The mean deflexion due to both positive and negative electricity was taken as

$$(\phi_1^2 + \phi_2^2)^{1/2}.$$

In a similar way, it is not difficult to calculate the average deflexion due to the atom with a central charge discussed in this paper.

Since the radial electric field X at any distance r from the centre is given by

$$X=Ne\left(\frac{1}{r^2}-\frac{r}{R^3}\right),$$

it is not difficult to show that the deflexion (supposed small) of an electrified particle due to this field is given by

$$\theta=\frac{b}{p}\left(1-\frac{p^2}{R^2}\right)^{3/2},$$

where p is the perpendicular from the centre on the path of the particle and b has the same value as before. It is seen that the value of θ increases with diminution of p and becomes great for small values of ϕ.

Since we have already seen that the deflexions become very large for a particle passing near the centre of the atom, it is obviously not correct to find the average value by assuming θ is small.

Taking R of the order 10^{-8} cm., the value of p for a large deflexion is for α and β particles of the order 10^{-11} cm. Since the chance of an encounter involving a large deflexion is small compared with the chance of small deflexions, a simple consideration shows that the average small deflexion is practically unaltered if the large deflexions are omitted. This is equivalent to integrating over that part of the cross section of the atom where the deflexions are small and neglecting the small central area. It can in this way be simply shown that the average small deflexion is given by

$$\phi_1=\frac{3\pi}{8}\frac{b}{R}.$$

This value of ϕ_1 for the atom with a concentrated central charge is three times the magnitude of the average deflexion for the same value of Ne in the type of atom examined by Sir J. J. Thomson. Combining the deflexions due to the electric field and to the corpuscles, the average deflexion is

$$(\phi_1^2+\phi_2^2)^2 \quad \text{or} \quad \frac{b}{2R}\left(5\cdot 54+\frac{15\cdot 4}{N}\right)^{1/2}.$$

It will be seen later that the value of N is nearly proportional to the atomic weight, and is about 100 for gold. The effect due to scattering of the individual corpuscles expressed by the second term of the equation is consequently small for heavy atoms compared with that due to the distributed electric field.

Neglecting the second term, the average deflexion per atom is $\frac{3\pi b}{8R}$. We are now in a position to consider the relative effects on the distribution of particles due to single and to compound scattering. Following J. J. Thomson's argument, the average deflexion θ_t after passing through a thickness t of matter is proportional to the square root of the number of encounters and is given by

$$\theta_t = \frac{3\pi b}{8R}\sqrt{\pi R^2 . n . t} = \frac{3\pi b}{8}\sqrt{\pi n t},$$

where n as before is equal to the number of atoms per unit volume.

The probability p_1 for compound scattering that the deflexion of the particle is greater than ϕ is equal to $e^{-\phi^2/\theta_t^2}$.

Consequently $$\phi^2 = -\frac{9\pi^3}{64} b^2 nt \log p_1.$$

Next suppose that single scattering alone is operative. We have seen (§ 3) that the probability p_2 of a deflexion greater than ϕ is given by

$$p_2 = \frac{\pi}{4} b^2 . n . t \cot^2 \phi/2.$$

By comparing these two equations

$$p_2 \log p_1 = -{\cdot}181 \phi^2 \cot^2 \phi/2,$$

ϕ is sufficiently small that

$$\tan \phi/2 = \phi/2,$$

$$p_2 \log p_1 = -{\cdot}72.$$

If we suppose $p_2 = {\cdot}5$, then $p_1 = {\cdot}24$.

If $p_2 = {\cdot}1$, $p_1 = {\cdot}0004$.

It is evident from this comparison, that the probability for any given deflexion is always greater for single than for compound scattering. The difference is especially marked when only a small fraction of the particles are scattered through any given angle. It follows from this result that the distribution of particles due to encounters with the atoms is for small thicknesses mainly governed by single scattering. No doubt compound scattering produces some effect in equalizing the distribution of the scattered particles; but its effect becomes relatively smaller, the smaller the fraction of the particles scattered through a given angle.

§ 6. *Comparison of Theory with Experiments.*

On the present theory, the value of the central charge Ne is an important constant, and it is desirable to determine its value for different atoms. This can be most simply done by determining the small fraction of α or β particles of known velocity falling on a thin metal screen, which are scattered between ϕ and $\phi+d\phi$ where ϕ is the angle of deflexion. The influence of compound scattering should be small when this fraction is small.

Experiments in these directions are in progress, but it is desirable at this stage to discuss in the light of the present theory the data already published on scattering of α and β particles.

The following points will be discussed :—

(*a*) The "diffuse reflexion" of α particles, *i. e.* the scattering of α particles through large angles (Geiger and Marsden).

(*b*) The variation of diffuse reflexion with atomic weight of the radiator (Geiger and Marsden).

(*c*) The average scattering of a pencil of α rays transmitted through a thin metal plate (Geiger).

(*d*) The experiments of Crowther on the scattering of β rays of different velocities by various metals.

(*a*) In the paper of Geiger and Marsden (*loc. cit.*) on the diffuse reflexion of α particles falling on various substances it was shown that about 1/8000 of the α particles from radium C falling on a thick plate of platinum are scattered back in the direction of the incidence. This fraction is deduced on the assumption that the α particles are uniformly scattered in all directions, the observations being made for a deflexion of about 90°. The form of experiment is not very suited for accurate calculation, but from the data available it can be shown that the scattering observed is about that to be expected on the theory if the atom of platinum has a central charge of about 100 e.

(*b*) In their experiments on this subject, Geiger and Marsden gave the relative number of α particles diffusely reflected from thick layers of different metals, under similar conditions. The numbers obtained by them are given in the table below, where z represents the relative number of scattered particles, measured by the number of scintillations per minute on a zinc sulphide screen.

Metal.	Atomic weight.	z.	$z/A^{3/2}$.
Lead	207	62	208
Gold	197	67	242
Platinum	195	63	232
Tin	119	34	226
Silver	108	27	241
Copper	64	14·5	225
Iron	56	10·2	250
Aluminium ...	27	3·4	243
		Average	233

On the theory of single scattering, the fraction of the total number of α particles scattered through any given angle in passing through a thickness t is proportional to $n.A^2t$, assuming that the central charge is proportional to the atomic weight A. In the present case, the thickness of matter from which the scattered α particles are able to emerge and affect the zinc sulphide screen depends on the metal. Since Bragg has shown that the stopping power of an atom for an α particle is proportional to the square root of its atomic weight, the value of nt for different elements is proportional to $1/\sqrt{A}$. In this case t represents the greatest depth from which the scattered α particles emerge. The number z of α particles scattered back from a thick layer is consequently proportional to $A^{3/2}$ or $z/A^{3/2}$ should be a constant.

To compare this deduction with experiment, the relative values of the latter quotient are given in the last column. Considering the difficulty of the experiments, the agreement between theory and experiment is reasonably good *.

The single large scattering of α particles will obviously affect to some extent the shape of the Bragg ionization curve for a pencil of α rays. This effect of large scattering should be marked when the α rays have traversed screens of metals of high atomic weight, but should be small for atoms of light atomic weight.

(*c*) Geiger made a careful determination of the scattering of α particles passing through thin metal foils, by the scintillation method, and deduced the most probable angle

* The effect of change of velocity in an atomic encounter is neglected in this calculation.

through which the α particles are deflected in passing through known thicknesses of different kinds of matter.

A narrow pencil of homogeneous α rays was used as a source. After passing through the scattering foil, the total number of α particles deflected through different angles was directly measured. The angle for which the number of scattered particles was a maximum was taken as the most probable angle. The variation of the most probable angle with thickness of matter was determined, but calculation from these data is somewhat complicated by the variation of velocity of the α particles in their passage through the scattering material. A consideration of the curve of distribution of the α particles given in the paper (*loc. cit.* p. 496) shows that the angle through which half the particles are scattered is about 20 per cent greater than the most probable angle.

We have already seen that compound scattering may become important when about half the particles are scattered through a given angle, and it is difficult to disentangle in such cases the relative effects due to the two kinds of scattering. An approximate estimate can be made in the following way :—From (§ 5) the relation between the probabilities p_1 and p_2 for compound and single scattering respectively is given by

$$p_2 \log p_1 = -\cdot 721.$$

The probability q of the combined effects may as a first approximation be taken as

$$q = (p_1^2 + p_2^2)^{1/2}.$$

If $q = \cdot 5$, it follows that

$$p_1 = \cdot 2 \quad \text{and} \quad p_2 = \cdot 46.$$

We have seen that the probability p_2 of a single deflexion greater than ϕ is given by

$$p_2 = \frac{\pi}{4} n \, . \, t \, . \, b^2 \cot^2 \phi/2.$$

Since in the experiments considered ϕ is comparatively small

$$\frac{\phi \sqrt{p_2}}{\sqrt{\pi n t}} = b = \frac{2NeE}{mu^2}.$$

Geiger found that the most probable angle of scattering of the α rays in passing through a thickness of gold equivalent in stopping power to about ·76 cm. of air was 1° 40′. The angle ϕ through which half the α particles are turned thus corresponds to 2° nearly.

$$t = \cdot 00017 \text{ cm.} \, ; \; n = 6\cdot 07 \times 10^{22} \, ;$$

$$u \text{ (average value)} = 1\cdot 8 \times 10^9.$$

$$E/m = 1\cdot 5 \times 10^{14} \text{ . E.S. units} \, ; \; e = 4\cdot 65 \times 10^{-10}.$$

Taking the probability of single scattering =·46 and substituting the above values in the formula, the value of N for gold comes out to be 97.

For a thickness of gold equivalent in stopping power to 2·12 cms. of air, Geiger found the most probable angle to be 3° 40′. In this case $t=\cdot00047$, $\phi=4^{\circ}\cdot4$, and average $u=1\cdot7\times10^9$, and N comes out to be 114.

Geiger showed that the most probable angle of deflexion for an atom was nearly proportional to its atomic weight. It consequently follows that the value of N for different atoms should be nearly proportional to their atomic weights, at any rate for atomic weights between gold and aluminium.

Since the atomic weight of platinum is nearly equal to that of gold, it follows from these considerations that the magnitude of the diffuse reflexion of α particles through more than 90° from gold and the magnitude of the average small angle scattering of a pencil of rays in passing through gold-foil are both explained on the hypothesis of single scattering by supposing the atom of gold has a central charge of about 100 *e*.

(*d*) *Experiments of Crowther on scattering of β rays.*—We shall now consider how far the experimental results of Crowther on scattering of β particles of different velocities by various materials can be explained on the general theory of single scattering. On this theory, the fraction of β particles p turned through an angle greater than ϕ is given by

$$p=\frac{\pi}{4}n\,.\,t\,.\,b^2\cot^2\phi/2.$$

In most of Crowther's experiments ϕ is sufficiently small that $\tan\phi/2$ may be put equal to $\phi/2$ without much error. Consequently

$$\phi^2=2\pi\,n\,.\,t\,.\,b^2 \quad \text{if } p=1/2.$$

On the theory of compound scattering, we have already seen that the chance p_1 that the deflexion of the particles is greater than ϕ is given by

$$\phi^2/\log p_1=-\frac{9\pi^3}{64}n\,.\,t\,.\,b^2.$$

Since in the experiments of Crowther the thickness t of matter was determined for which $p_1=1/2$,

$$\phi^2=\cdot96\pi\,n\,t\,b^2.$$

For a probability of 1/2, the theories of single and compound

2 Y 2

scattering are thus identical in general form, but differ by a numerical constant. It is thus clear that the main relations on the theory of compound scattering of Sir J. J. Thomson, which were verified experimentally by Crowther, hold equally well on the theory of single scattering.

For example, if t_m be the thickness for which half the particles are scattered through an angle ϕ, Crowther showed that $\phi/\sqrt{t_m}$ and also $\frac{mu^2}{E}\cdot\sqrt{t_m}$ were constants for a given material when ϕ was fixed. These relations hold also on the theory of single scattering. Notwithstanding this apparent similarity in form, the two theories are fundamentally different. In one case, the effects observed are due to cumulative effects of small deflexions, while in the other the large deflexions are supposed to result from a single encounter. The distribution of scattered particles is entirely different on the two theories when the probability of deflexion greater than ϕ is small.

We have already seen that the distribution of scattered α particles at various angles has been found by Geiger to be in substantial agreement with the theory of single scattering, but cannot be explained on the theory of compound scattering alone. Since there is every reason to believe that the laws of scattering of α and β particles are very similar, the law of distribution of scattered β particles should be the same as for α particles for small thicknesses of matter. Since the value of mu^2/E for the β particles is in most cases much smaller than the corresponding value for the α particles, the chance of large single deflexions for β particles in passing through a given thickness of matter is much greater than for α particles. Since on the theory of single scattering the fraction of the number of particles which are deflected through a given angle is proportional to kt, where t is the thickness supposed small and k a constant, the number of particles which are undeflected through this angle is proportional to $1-kt$. From considerations based on the theory of compound scattering, Sir J. J. Thomson deduced that the probability of deflexion less than ϕ is proportional to $1-e^{-\mu/t}$ where μ is a constant for any given value of ϕ.

The correctness of this latter formula was tested by Crowther by measuring electrically the fraction I/I_0 of the scattered β particles which passed through a circular opening subtending an angle of 36° with the scattering material. If

$$I/I_0=1-e^{-\mu/t},$$

the value of I should decrease very slowly at first with

increase of t. Crowther, using aluminium as scattering material, states that the variation of I/I_0 was in good accord with this theory for small values of t. On the other hand, if single scattering be present, as it undoubtedly is for α rays, the curve showing the relation between I/I_0 and t should be nearly linear in the initial stages. The experiments of Madsen * on scattering of β rays, although not made with quite so small a thickness of aluminium as that used by Crowther, certainly support such a conclusion. Considering the importance of the point at issue, further experiments on this question are desirable.

From the table given by Crowther of the value $\phi/\sqrt{t_m}$ for different elements for β rays of velocity $2 \cdot 68 \times 10^{10}$ cms. per second, the values of the central charge Ne can be calculated on the theory of single scattering. It is supposed, as in the case of the α rays, that for the given value of $\phi/\sqrt{t_m}$ the fraction of the β particles deflected by single scattering through an angle greater than ϕ is ·46 instead of ·5.

The values of N calculated from Crowther's data are given below.

Element.	Atomic weight.	$\phi/\sqrt{t_m}$.	N.
Aluminium	27	4·25	22
Copper	63·2	10·0	42
Silver	108	15·4	78
Platinum	194	29·0	138

It will be remembered that the values of N for gold deduced from scattering of the α rays were in two calculations 97 and 114. These numbers are somewhat smaller than the values given above for platinum (viz. 138), whose atomic weight is not very different from gold. Taking into account the uncertainties involved in the calculation from the experimental data, the agreement is sufficiently close to indicate that the same general laws of scattering hold for the α and β particles, notwithstanding the wide differences in the relative velocity and mass of these particles.

As in the case of the α rays, the value of N should be most simply determined for any given element by measuring

* Phil. Mag. xviii. p. 909 (1909).

the small fraction of the incident β particles scattered through a large angle. In this way, possible errors due to small scattering will be avoided.

The scattering data for the β rays, as well as for the α rays, indicate that the central charge in an atom is approximately proportional to its atomic weight. This falls in with the experimental deductions of Schmidt *. In his theory of absorption of β rays, he supposed that in traversing a thin sheet of matter, a small fraction α of the particles are stopped, and a small fraction β are reflected or scattered back in the direction of incidence. From comparison of the absorption curves of different elements, he deduced that the value of the constant β for different elements is proportional to nA^2 where n is the number of atoms per unit volume and A the atomic weight of the element. This is exactly the relation to be expected on the theory of single scattering if the central charge on an atom is proportional to its atomic weight.

§ 7. *General Considerations.*

In comparing the theory outlined in this paper with the experimental results, it has been supposed that the atom consists of a central charge supposed concentrated at a point, and that the large single deflexions of the α and β particles are mainly due to their passage through the strong central field. The effect of the equal and opposite compensating charge supposed distributed uniformly throughout a sphere has been neglected. Some of the evidence in support of these assumptions will now be briefly considered. For concreteness, consider the passage of a high speed α particle through an atom having a positive central charge Ne, and surrounded by a compensating charge of N electrons. Remembering that the mass, momentum, and kinetic energy of the α particle are very large compared with the corresponding values for an electron in rapid motion, it does not seem possible from dynamic considerations that an α particle can be deflected through a large angle by a close approach to an electron, even if the latter be in rapid motion and constrained by strong electrical forces. It seems reasonable to suppose that the chance of single deflexions through a large angle due to this cause, if not zero, must be exceedingly small compared with that due to the central charge.

It is of interest to examine how far the experimental evidence throws light on the question of the extent of the

* *Annal. d. Phys.* iv. 23. p. 671 (1907).

distribution of the central charge. Suppose, for example, the central charge to be composed of N unit charges distributed over such a volume that the large single deflexions are mainly due to the constituent charges and not to the external field produced by the distribution. It has been shown (§ 3) that the fraction of the α particles scattered through a large angle is proportional to $(N e E)^2$, where Ne is the central charge concentrated at a point and E the charge on the deflected particle. If, however, this charge is distributed in single units, the fraction of the α particles scattered through a given angle is proportional to Ne^2 instead of N^2e^2. In this calculation, the influence of mass of the constituent particle has been neglected, and account has only been taken of its electric field. Since it has been shown that the value of the central point charge for gold must be about 100, the value of the distributed charge required to produce the same proportion of single deflexions through a large angle should be at least 10,000. Under these conditions the mass of the constituent particle would be small compared with that of the α particle, and the difficulty arises of the production of large single deflexions at all. In addition, with such a large distributed charge, the effect of compound scattering is relatively more important than that of single scattering. For example, the probable small angle of deflexion of a pencil of α particles passing through a thin gold foil would be much greater than that experimentally observed by Geiger (§ *b–c*). The large and small angle scattering could not then be explained by the assumption of a central charge of the same value. Considering the evidence as a whole, it seems simplest to suppose that the atom contains a central charge distributed through a very small volume, and that the large single deflexions are due to the central charge as a whole, and not to its constituents. At the same time, the experimental evidence is not precise enough to negative the possibility that a small fraction of the positive charge may be carried by satellites extending some distance from the centre. Evidence on this point could be obtained by examining whether the same central charge is required to explain the large single deflexions of α and β particles; for the α particle must approach much closer to the centre of the atom than the β particle of average speed to suffer the same large deflexion.

The general data available indicate that the value of this central charge for different atoms is approximately proportional to their atomic weights, at any rate for atoms heavier than aluminium. It will be of great interest to examine experimentally whether such a simple relation holds also for the lighter atoms. In cases where the mass of the deflecting atom (for example, hydrogen, helium, lithium) is not very different from that of the α particle, the general theory of single scattering will require modification, for it is necessary to take into account the movements of the atom itself (see § 4).

It is of interest to note that Nagaoka * has mathematically considered the properties of a "Saturnian" atom which he supposed to consist of a central attracting mass surrounded by rings of rotating electrons. He showed that such a system was stable if the attractive force was large. From the point of view considered in this paper, the chance of large deflexion would practically be unaltered, whether the atom is considered to be a disk or a sphere. It may be remarked that the approximate value found for the central charge of the atom of gold ($100\,e$) is about that to be expected if the atom of gold consisted of 49 atoms of helium, each carrying a charge $2\,e$. This may be only a coincidence, but it is certainly suggestive in view of the expulsion of helium atoms carrying two unit charges from radioactive matter.

The deductions from the theory so far considered are independent of the sign of the central charge, and it has not so far been found possible to obtain definite evidence to determine whether it be positive or negative. It may be possible to settle the question of sign by consideration of the difference of the laws of absorption of the β particle to be expected on the two hypotheses, for the effect of radiation in reducing the velocity of the β particle should be far more marked with a positive than with a negative centre. If the central charge be positive, it is easily seen that a positively charged mass if released from the centre of a heavy atom, would acquire a great velocity in moving through the electric field. It may be possible in this way to account for the high velocity of expulsion of α particles without supposing that they are initially in rapid motion within the atom.

Further consideration of the application of this theory to these and other questions will be reserved for a later paper, when the main deductions of the theory have been tested experimentally. Experiments in this direction are already in progress by Geiger and Marsden.

University of Manchester,
April 1911.

* Nagaoka, Phil. Mag. vii. p. 445 (1904).

First Solvay Conference, Brussels, Belgium, 1911

Appendices

Primary Sources

1. CARLSRUHE COMPTE RENDU. [1860] "Compte rendu des séances du Congrès international des chimistes réuni à Carlsruhe le 3, 4 et 5 septembre 1860," Anlage VIII (pp. 671-688) in Volume I of Richard Anschütz, *August Kekulé*, 2 vols. (Berlin: Verlag Chemie, G.M.B.H., 1929). The official account was written by Adolphe Wurtz. In German, "Vier Jahrzehnte chemischer Forschung unter besonderer Rücksicht auf Baden als Heimstätte der Chemie," pp. 346-355 in *Festgabe zum Jubiläum der Vierzigjahrigen Regierung seiner Königlichen Hoheit des Grossherzogs Friedrich von Baden* (Karlsruhe, 1892).
2. CANNIZZARO, STANISLAO. [1858] "Sketch of A Course of Chemical Philosophy," Alembic Club Reprint No. 18 (Edinburgh: E. and S. Livingstone Ltd., 1947). Copies of this paper were distributed at the 1860 Karlsruhe Congress. Originally published as "Sunto di un corso di filosofia chimica fatto nella Reale Università di Genova," *Nuova Cimento, 7 (1858), 321-366.* In German translation, in Wilhelm Ostwald's *Klassiker der Exacten Wissenschaften,* no. 30 (Leipzig, 1891).
3. THOMSON, WILLIAM. LATER LORD KELVIN. [1867] "On Vortex Atoms," *Philosophical Magazine*, [4], 34, (1867), 15-24. Originally appeared in *Proceedings of the Royal Society of Edinburgh*, 6 (1867), 94-105.
4. WILLIAMSON, A. W. [1869] "On the Atomic Theory," *Journal of the Chemical Society (London), 22 (1869), 328-365.* Followed by "Discussion on Dr. Williamson's Lecture on the Atomic Theory," *Journal of the Chemical Society (London)*, 22 (1869), 433-441.
5. MACH, ERNST. [1872] "Mechanical Physics," Chapter III (pp. 42-58 and notes 3-5 on pp. 85-88), in Ernst Mach, *History and Root of the Principle of the Conservation of Energy*, trans. Philip E. Jourdain (Chicago: Open Court, 1911). Appeared originally as "Die mechanische Physik," pp. 20-33 and 54-56 in *Die Geschichte und Die Wurzel des Satzes der Erhaltung der Arbeit* (Prague, 1872).
6. LOCKYER, J. NORMAN. [1874] "Atoms and Molecules Spectroscopically Considered," Chapter IV (pp. 113-144), in Norman Lockyer, *Studies in Spectrum Analysis*, 2d ed. (London: Kegan Paul and Co., 1878). By the same title in *Nature*, 10 (1874), 69-71 and 89-90.

7. MAXWELL, JAMES CLERK. [1875] "On the Dynamical Evidence of the Molecular Constitution of Bodies," *Journal of the Chemical Society (London)*, 28 (1875), 493-508. Also appeared in *Nature*, 11 (1875), 357-359 and 374-377.
8. MARIGNAC, CHARLES. [1877] "Chemical Equivalents and Atomic Weights Considered as Bases of a System of Notation," trans. P. Casamajor, *American Journal of Science*, [3], 115 (1878), 89-98, and "Answer of M. Marignac," *American Journal of* Science, [3], 115 (1878), 187-189. Originally appeared in *Moniteur Scientifique*, 19 (September, 1877), 920-926 and (December, 1877), 1256-1257.
9. BERTHELOT, MARCELLIN. [1877] "On Systems of Chemical Notation," trans. P. Casamajor, *American Journal of Science*, [3], 115 (1878), 184-187. Originally appeared in *Moniteur Scientifique*, 19 (December, 1877), 1254-1256.
10. HELMHOLTZ, HERMANN VON. [1881] "On the Modern Development of Faraday's Conception of Electricity," pp. 132-159 in C. S. Gibson and A. J. Greenaway, eds. *Faraday Lectures 1869-1928* (London: The Chemical Society: Burlington House, 1928). Appeared originally in *Journal of the Chemical Society (London)*, 39 (1881), 277-304.
11. ARRHENIUS, SVANTE.[1887] "On the Dissociation of Substances in Aqueous Solution," Part II (pp. 43-67) in *The Foundations of the Theory of Dilute Solutions*, Alembic Club Reprint No. 19 (Edinburgh: The Alembic Club, 1929). Appeared as "Ueber die Dissociation der im Wasser gelösten Stoffe," *Zeitschrift für Physikalische Chemie*, 1 (1887), 631-648.
12. MENDELEEV, DMITRI. [1889] "The Periodic Law of the Chemical Elements," pp. 160-182 in C. S. Gibson and A. J. Greenaway, ed. *Faraday Lectures 1869-1928* (London: The Chemical Society: Burlington House, 1928). Appeared originally in *Journal of the Chemical Society* (London), 55 (1889), 634-656.
13. OSTWALD, WILHELM [1895] "Emancipation from Scientific Materialism," trans. F.G. Donnan and F. B. Kenrick, *Science Progress: A Quarterly Review of Current Scientific Investigation.* (London: The Scientific Press, 1894-1898), 4 (February, 1896), 419-436. Appeared originally as "Die Ueberwindung des wissenschaftlichen Materialismus," *Verhandlungen der Gesellschaft Deutscher Naturforscher und Aertze*, 67, I (1895), 155-168. In French translation, "La Déroute de l'atomisme contemporaine," *Revue Générale des Sciences*, 6 (1895), 953-958.

14. BOLTZMANN, LUDWIG. [1897] "On the Necessity of Atomic Theories in Physics," *The Monist*, 12 (October, 1901), 65–79. Translated from *Annalen der Physik*, 60 (1897), 231-247. Also in Ludwig Boltzmann, *Populäre Schriften* (Leipzig: J. A. Barth, 1905), 140–161.
15. THOMSON, J. J. [1897] "Cathode Rays," *Philosophical Magazine*, [5], 44 (1897), 293-316.
16. POINCARÉ, HENRI. [1900] "Relations between Experimental Physics and Mathematical Physics," trans. George K. Burgess, *The Monist*, 12 (July, 1902), 516-543. Originally appeared in *Rapports présentés au Congrès International de Physique réuni à Paris en 1900,* 4 vols. (Paris: Gauthier-Villars, 1900), Vol. I, 1-29. Also as Chapter IX, "Hypotheses in Physics," pp. 140–159 in *Science and Hypothesis* (London: Walter Scott, 1905).
17. RUTHERFORD, ERNEST and FREDERICK SODDY. [1902] "The Cause and Nature of Radioactivity," *Philosophical Magazine*, [6], 4 (1902), 370-396.
18. EINSTEIN, ALBERT. [1905] "On a Heuristic Point of View about the Creation and Conversion of Light," pp. 91–107 in D. Ter Haar, *The Old Quantum Theory* (Oxford, Pergamon Press, 1967). Also trans. in H. A. Boorse and L. Motz, *The World of the Atom* (New York: Basic Books, 1966), Vol. I, 544–557. Appeared originally as "Ueber einen die Erzeugung und Verwandlung des Lichtes betreffenden heuristischen Gesichtspunkt," *Annalen der Physik*, 17 (1905), 132–148.
19. DUHEM, PIERRE [1906] "Physical Theory and Natural Classification," Part I, Chapter II. (pp. 19-30) and "Primary Qualities," Part II, Chapter II (pp. 121-131) in *The Aim and Structure of Physical Theory*, trans. Philip P. Wiener (Princeton University Press, 1954). Published originally in *La Theorie Physique: Son Objet, Sa Structure* (Paris, 1906).
20. PERRIN, JEAN [1909] *Brownian Movement and Molecular Reality*, trans. F. Soddy (London: Taylor and Francis, 1910), 93 pp. Originally appeared as "Mouvement brownien et réalité moléculaire," *Annales de Chimie et de Physique*, [8], 18 (1909), 1-114. In German, in the *Kolloidchemische Beihefte*, 1 (1910), 221 ff.
21. RUTHERFORD, ERNEST. [1911] "The Scattering of α and β particles by Matter and the Structure of the Atom," *Philosophical Magazine*, [6] 21 (1911), 669-688. Also in D. ter Haar, *the Old Quantum Theory* (Oxford: Pergamon Press, 1967), 108–131.

Anlage 8.

Compte rendu des séances du Congrès international des chimistes réuni à Carlsruhe le 3, 4 et 5 Septembre 1860[1]).

L'idée de provoquer une réunion internationale de chimistes appartient à M. Kekulé. C'est pendant l'automne de 1859 qu'il a eu occasion de faire les premières ouvertures à cet égard à M. Weltzien d'abord et puis à M. Wurtz. A la fin du mois de Mars 1860, ces trois savants se trouvant réunis à Paris ont concerté les premières mesures à prendre pour réaliser le projet en question. Dans le but de recueillir les adhésions des hommes les plus marquants dans la science, une première circulaire à été redigée. Elle mentionnait, en généraux termes, les divergences qui se sont manifestées dans les vues théoriques des chimistes, et l'urgence d'y mettre un terme par une entente commune, au moins sur certaines questions.

Le premier appel ayant été favorablement accueilli, on s'est entendu sur le lieu et sur l'époque de la réunion et on a arrêté la rédaction d'une circulaire qui, adressée à tous les chimistes de l'Europe, leur exposait l'objet et le but d'un Congrès international dans les termes suivants[2]):

Paris, 15 Juin 1860.

»Monsieur et très-honoré Confrère.

»Le grand développement qu'a pris la chimie dans ces dernières années et les »divergences qui se sont manifestées dans les opinions théoriques, rendent oppor- »tun et utile un Congrès ayant pour but la discussion de quelques questions »importantes au point de vue des progrès futurs de la science.

»Les soussignés convient à cette réunion tous les chimistes autorisés par »leurs travaux ou leur position à émettre un avis dans un débat scientifique.

---

1) Für die Durchsicht des französischen Textes bin ich meinem Kollegen Herrn Prof. Dr. Eugen Gaufinez in Bonn zu bestem Dank verpflichtet.

2) Das Rundschreiben wurde in deutscher, französischer und englischer Sprache versendet. Der deutsche Text ist: „Carlsruhe, den 10. Juli 1860", der englische: „London, 1st July, 1860" datiert.

In die französische Niederschrift der Kongreßverhandlungen nahm ich den französischen Text des Rundschreibens auf, der verglichen mit dem deutschen und englischen unter den Unterschriften den Namen „Regnault" enthält, der unter dem deutschen und englischen Text fehlt; unter dem letzteren fehlt auch der Name: „Mitscherlich".

»Une telle assemblée ne saurait prendre des délibérations ou des résolutions »obligatoires pour tous; mais par une discussion libre et approfondie elle pour- »rait faire disparaître certains malentendus et faciliter une entente commune »sur quelques-uns des points suivants:

»Définition de notions chimiques importantes, comme celles qui sont exprimées »par les mots: atome, molécule, équivalent, atomique, basique.

»Examen de la question des équivalents et des formules chimiques.

»Etablissement d'une notation et d'une nomenclature uniforme.

»Sans espérer que les déliberations de l'assemblée soient de nature à concilier »toutes les opinions et à faire disparaître immédiatement toutes les dissidences, »les soussignés pensent néanmoins que de tels travaux pourront préparer, dans »l'avenir, un accord si désirable entre les chimistes, au moins en ce qui concerne »les questions les plus importantes. Une commission pourrait être chargée de »poursuivre l'étude de ces questions et d'y intéresser les Académies ou Sociétés »savantes disposant des moyens matériels nécessaires pour les résoudre.

»Le Congrès se réunira à Carlsruhe le 3 septembre 1860.

»Notre collègue M. Weltzien, professeur à l'Ecole polytechnique de cette »ville, veut bien se charger des fonctions de commissaire général. En cette »qualité, il recevra les adhésions des futurs membres du Congrès et ouvrira »l'assemblée, le jour indiqué, à neuf heures du matin.

»En terminant, et dans le but d'éviter des omissions regrettables, les sous- »signées prient les personnes auxquelles cette circulaire sera adressée de vouloir »bien la communiquer aux savants, leurs amis, dûment autorisés à assister à la »réunion projetée.

Babo de, Freiburg.
Balard, Paris.
Bekétoff, Kasan.
Boussingault, Paris.
Brodie, Oxford.
Bunsen, Heidelberg.
Bussy, Paris.
Cahours, Paris.
Cannizzaro, Genua.
Deville, H., Paris.
Dumas, Paris.
Engelhardt, St. Petersburg.
Erdmann, O. L., Leipzig.
Fehling de, Stuttgart.
Frankland, London.
Fremy, Paris.
Fritzsche, St. Petersburg.
Hofmann, A. W., London.
Kekulé, Gent.
Kopp, H., Gießen.
Hlasiwetz, Innsbruck.
Liebig, J. de, München.
Malaguti, Rennes.
Marignac, Genf.
Mitscherlich, Berlin.
Odling, London.
Pasteur, Paris.
Payen, Paris.
Pebal, Wien.
Peligot, Paris.
Pelouze, Paris.
Piria, Turin.
Regnault, V., Paris.
Roscoe, Manchester.
Schroetter, A., Wien.
Socoloff, St. Petersburg.
Staedeler, Zürich.
Stas, Brüssel.
Strecker, Tübingen.
Weltzien, C., Karlsruhe.
Will, H., Gießen.
Williamson, W., London.
Wöhler, F., Göttingen.
Wurtz, Ad., Paris.
Zinin, St. Petersburg.

NB. Vous voudrez bien faire connaître votre adhésion soit directement à M. Weltzien (Ecole polytechnique, Carlsruhe), soit à M. A. Kekulé, professeur de chimie à l'Université de Gand, qui se chargera de la transmettre à M. Weltzien.

»Le nombre des adhérents a été considérable et le 3 Sept. 1860 140 chimistes [1]) se sont trouvés réunis dans la salle des séances de la seconde chambre des Etats, local que S. A. R. le Grand-Duc de Bade a bien voulu mettre à la disposition du congrès.

»Voici les noms des chimistes présents [2]):

I. Belgien: *Brüssel:* Stas. *Gent:* Donny, A. Kekulé.

II. Deutschland: *Berlin:* Ad. Baeyer, G. Quincke. *Bonn:* Landolt. *Breslau:* Lothar Meyer. *Cassel:* Guckelberger. *Clausthal:* Streng. *Darmstadt:* E. Winkler. *Erlangen:* v. Gorup-Besanez. *Freiburg i. B.:* v. Babo, Schneyder. *Gießen:* Boeckmann, H. Kopp, H. Will. *Göttingen:* F. Beilstein. *Halle a. S.:* W. Heintz. *Hannover:* Heeren. *Heidelberg:* Becker, O. Braun, R. Bunsen, L. Carius, E. Erlenmeyer, O. Mendius, Schiel. *Jena:* Lehmann, H. Ludwig. *Karlsruhe:* A. Clemm, R. Müller, J. Nessler, Petersen, K. Seubert, Weltzien. *Leipzig:* O. L. Erdmann, Hirzel, Knop, Kuhn. *Mannheim:* Gundelach, Schroeder. *Marburg a. L.:* R. Schmidt, Zwenger. *München:* Geiger. *Nürnberg:* v. Bibra. *Offenbach:* Grimm. *Rappenau:* Finck. *Schönberg:* R. Hoffmann. *Speyer:* Keller, Mühlhäuser. *Stuttgart:* v. Fehling, W. Hallwachs. *Tübingen:* Finckh, A. Naumann, A. Strecker. *Wiesbaden:* Casselmann, R. Fresenius, C. Neubauer. *Würzburg:* Scherer, v. Schwarzenbach.

III. England: *Dublin:* Apjohn. *Edinburg:* Al. Crum Brown, Wanklyn, F. Guthrie. *Glasgow:* Anderson. *London:* B. J. Duppa, G. C. Foster, Gladstone, Müller, Noad, A. Normandy, Odling. *Manchester:* Roscoë. *Oxford:* Daubeny, G. Griffeth. F. Schickendantz. *Woolwich:* Abel.

IV. Frankreich: *Montpellier:* A. Béchamp, A. Gautier, C. G. Reischauer. *Mühlhausen i. E.:* Th. Schneider. *Nancy:* J. Nicklès. *Paris:* Boussingault, Dumas, C. Friedel, L. Grandeau, Le Canu, Persoz, Alf. Riche, P. Thénard, Verdét, Wurtz. *Straßburg i. E.:* Jacquemin, Oppermann, F. Schlagdenhauffen, Schützenberger. *Tann:* Ch. Kestner, Scheurer-Kestner.

V. Italien: *Genua:* Cannizzaro. *Pavia:* Pavesi.

VI. Mexiko: Posselt.

VII. Österreich: *Innsbruck:* Hlasiwetz. *Lemberg:* v. Pebal. *Pesth:* Th. Wertheim. *Wien:* V. v. Lang, A. Lieben, Folwarezny, F. Schneider.

VIII. Portugal: *Coïmbra:* Mide Carvalho.

IX. Rußland: *Charkow:* Sawitsch. *St. Petersburg:* Borodin, Mendelejeff, L. Schischkoff, Zinin. *Warschau:* T. Lesinski, J. Natanson.

X. Schweden: *Harpenden:* J. H. Gilbert. *Lund:* Berlin, C. W. Blomstrand. *Stockholm:* Bahr.

XI. Schweiz: *Bern:* C. Brunner, H. Schiff. *Genf:* C. Marignac. *Lausanne:* Bischoff. *Reichenau bei Chur:* A. v. Planta. *Zürich:* J. Wislicenus.

XII. Spanien: *Madrid:* R. de Suna.

---

[1]) Die gedruckte und durch handschriftliche Zusätze ergänzte Liste der Mitglieder enthält 126 Namen. (A.)

[2]) Ich habe die Teilnehmer nach den Ländern geordnet und nach den Städten, in denen sie damals tätig waren. (A.)

Première séance du Congrès.

M. W e l t z i e n, commissaire général, a ouvert la première séance par le discours suivant:

Meine Herren!

Als provisorischer Geschäftsführer habe ich die Ehre eine Versammlung zu eröffnen, wie eine derartige zuvor wohl nie getagt hat.

Zwar traten seit 1822 fast jährlich auf O k e n's Anregung nach dem Vorbilde schweizerischer Versammlungen die deutschen Naturforscher und Aerzte in den verschiedenen Städten ihres Vaterlandes zu wissenschaftlichem Verkehr zusammen; es fanden diese Versammlungen Nachahmung in England, Frankreich, und noch in den letzten Jahren vereinigten sich auch die skandinavischen Naturforscher zu ähnlichen Zusammenkünften.

Es sind dieses aber immer Männer, welche zwar den verschiedenen Theilen der Naturwissenschaften und der Medizin ihre Kräfte zuwenden, welche aber stets denselben Nationalitäten angehören.

Die wissenschaftliche Beschäftigung in diesen Versammlungen ist hauptsächlich durch Vorträge bezeichnet, welche über eigene Arbeiten nach freier Wahl jedes Einzelnen gehalten werden, deren Gegenstand an kein voraus festgestelltes Programm gebunden ist.

Ein reger, freundschaftlicher Verkehr, gewürzt durch eine Reihe von Festen, vereinigt eine Anzahl von Tagen die s t a m m - und s p r a c h v e r w a n d t e n Naturforscher und Aerzte.

Nicht so unsere heutige Versammlung.

Zum ersten Male sind hier die Vertreter einer einzigen Naturwissenschaft, und zwar der jüngsten, versammelt; diese Vertreter gehören aber fast allen Nationalitäten an. Wir sind verschiedenen Stammes und sprechen verschiedene Sprachen, aber wir sind fachverwandt, uns verbindet ein wissenschaftliches Interesse, uns vereinigt dieselbe Absicht.

Wir sind versammelt zu dem bestimmten Zwecke, den Versuch zu machen, in gewissen, für unsere schöne Wissenschaft wichtigen Punkten eine Einigung anzubahnen.

Bei der außerordentlich raschen Entwicklung der Chemie, besonders bei der massenhaften Ansammlung des thatsächlichen Materials, sind die theoretischen Ansichten der Forscher und die Ausdrücke in Wort und Symbol weiter auseinander gegangen, als zur gegenseitigen Verständigung zweckmäßig und besonders für das Lehren ersprießlich ist. Und doch bei der Wichtigkeit der Chemie für die übrigen Naturwissenschaften, bei der Unentbehrlichkeit derselben für die Technik muß es im höchsten Grade wünschenswerth und geboten erscheinen, ihr eine exactere Form zu geben, damit es möglich werde, dieselbe in verhältnißmäßig kurzer Zeit wissenschaftlich zu lehren.

Um dies zu erlangen, sollten wir nicht gezwungen sein, verschiedene Ansichten und Schreibweisen, wobei die Verschiedenheiten wenig Wesentlichkeiten bieten, vorzutragen, nicht mit einer Nomenclatur belastet sein, welcher bei einer Masse von unnöthigen Symbolen meist alle rationelle Basis abgeht, und die zur

Vermehrung des Uebelstandes sich meist von einer Theorie ableitet, welche jetzt kaum mehr Gültigkeit besitzt.

Die zahlreiche Betheiligung an der Versammlung ist wohl ein deutliches Zeichen, daß diese Mißstände allseitig erkannt sind und eine Beseitigung derselben im Wege der Einigung im höchsten Grade wünschenswerth erscheint. Die Erreichung dieses Zieles ist ein so schöner Preis, daß es wohl der Mühe werth ist, den Versuch hierzu zu machen.

Den ersten Gedanken zu einem Chemiker-Congresse sprach unser College Kekulé schon vor längerer Zeit gegen mich aus. In diesem Frühjahr that ich die ersten Schritte zu seiner Verwirklichung. Das Zeitgemäße des Unternehmens wurde vielfach anerkannt, allerseits fand ich zuvorkommende Unterstützung, so daß ich nicht zweifle, diese Versammlung wird berufen sein, in der Geschichte unserer Wissenschaft einen nicht unwichtigen Zeitabschnitt zu begründen.

Die Stadt Carlsruhe, welcher vor zwei Jahren das Glück zu Theil wurde, eine der glänzendsten Versammlungen der deutschen Naturforscher und Aerzte zu beherbergen, hat jetzt die Ehre, die erste internationale Chemiker-Versammlung in ihren Mauern vereinigt zu sehen.

Carlsruhe ist die Hauptstadt eines zwar kleinen, aber gesegneten Landes, in welchem unter einem erhabenen Fürsten, einer liberalen Regierung Wissenschaften und Künste blühen und ihre Vertreter, geachtet und unterstützt, mit Freudigkeit und Liebe ihrem Berufe folgen können.

Indem ich Sie in dieser Stadt herzlich willkommen heiße, zweifle ich nicht, daß dieselbe Freudigkeit auch unsere Versammlungen durchdringen und hoffe, daß die Wissenschaft mit Befriedigung einst auf die Versammlung zurückblicken werde.

Après ce discours Mr. le commissaire général invite M. Bunsen à prendre le fauteuil, celui-ci refuse, et prie l'assemblée d'engager M. Weltzien à diriger la déliberation dans cette première séance. M. Weltzien est designé comme président. MM. Wurtz, Strecker, Kekulé, Odling, Roscoe, Schischkoff sont appelés à remplir les fonctions de secrétaires et prennent place au bureau.

Sur la proposition de M. Kekulé, l'assemblée décide qu'une commission sera chargée de préparer la rédaction des questions qui seront soumises aux déliberations du Congrès.

M. Kekulé prend la parole pour exposer le programme de ces questions.

(Suit l'analyse du discours de M. Kekulé.) [1])

M. Erdmann insiste sur la nécessité de faire porter les délibérations et les résolutions de l'assemblée sur des questions de forme, plutôt que sur des points de doctrine.

Une discussion s'engage sur la question de savoir si la commission tiendra ses séances à huis clos ou en présense de l'assemblée.

Après quelques paroles échangées entre MM. Fresenius, Kekulé, Wurtz, Boussingault, H. Kopp, l'assemblée décide, sur la notion de ce dernier membre, que les séances auront lieu à huis clos.

---

[1]) Vgl. Anlage 9. (A.)

Première Séance de la Commission.

La commission s'est réunie le 3 Sept. à 11 heures sous la présidence de M. H. Kopp.

M. le Président propose d'engager la discussion sur la notion de molécule et d'atome et invite MM. Kekulé et Cannizzaro, dont les études ont spécialement embrassé cette question, à prendre la parole.

M. Kekulé insiste sur la nécessité de distinguer la molécule de l'atome, et, en principe au moins, la molécule physique de la molécule chimique.

M. Cannizzaro ne peut concevoir la notion de la molécule chimique. Pour lui, il n'existe que des molécules physiques, et la loi d'Ampère et d'Avogadro est la base des considérations relatives à la molécule chimique. Celle-ci n'est autre chose: que la molécule gazeuse.

M. Kekulé pense au contraire que ce sont de données chimiques qui doivent servir de base à la définition et à la détermination de la molécule (chimique) et que les considérations physiques ne doivent être invoquées qu'à titre de contrôle.

M. Strecker fait observer que dans certains cas l'atome est identique à la molécule, c'est ainsi pour l'éthylène.

W. Wurtz dit que l'oxygène et les éléments diatomiques en général étant comparables à l'éthylène, on peut éprouver un certain embarras à définir la molécule chimique de ces corps. Les considérations physiques conduisent à les envisager comme des molécules formées de deux atomes [1]), mais jusqu'ici aucune donnée chimique ne paraît militer en faveur de cette duplication.

M. H. Kopp résumant la discussion, dit que la nécessité de séparer la notion de la molécule de la notion de l'atome paraît établie, que la notion de la molécule peut être fixée à l'aide de considérations purement chimiques et qu'il n'est pas nécessaire de faire intervenir uniquement dans cette définition la densité; qu'enfin il paraît naturel de nommer molécule la quantité la plus grande, atome la quantité la plus petite. L'orateur formule, en terminant, la première question à proposer à l'assemblée.

Cette question est ainsi conçue:

»Est-il convenable d'établir une distinction entre les termes molécule et atome, »de nommer molécules les plus petites quantités des corps qui entrent dans une »réaction ou qui en sortent et qui d'ailleurs sont comparables en ce qui concerne »les propriétés physiques, de nommer atomes les plus petites quantités des corps »qui sont contenus dans les molécules?«

M. Fresenius appelle l'attention sur l'expression *atome composé* et dit que ces deux mots impliquent contradiction. L'observation de M. Fresenius motive la rédaction de la seconde question à soumettre à l'assemblée, et qui est ainsi conçue:

»L'expression *atome composé* peut-elle être supprimée et remplacée par les »expressions *radical* ou *résidu?*«

---

[1]) In Kekulé's Manuscript hat Kekulé hier die Randglosse: „nicht immer!" zugefügt. (A.)

M. Kopp reprend la suite du programme exposé par M. Kekulé et appelle l'attention sur la définition du mot *équivalent.* Il lui semble que la notion de l'équivalent est parfaitement claire et se distingue nettement de la notion de la molécule et de celle de l'atome. En conséquence la commission adopte sans discussion la 3me proposition à soumettre à l'assemblée, et qui est ainsi conçue:

»La notion des équivalents est empirique et indépendante de la notion de la »molécule et de l'atome.«

La séance continue sous la présidence de M. Erdmann[1]. La discussion s'engage sur la *notation:* M. Kekulé fait remarquer qu'on peut employer la notation moléculaire et atomistique ou bien la notation en équivalents, mais que dans un système comme dans l'autre il est nécessaire de s'en tenir d'une manière rigoureuse et conséquente à la notation une fois adoptée.

La signification du mot équivalent est l'objet de quelques remarques. M. Béchamps dit qu'on ne peut admettre l'équivalence que dans les cas où les fonctions des corps sont identiques.

M. Schischkoff ne partage pas cette opinion. Il pense que la notation d'équivalence et de quantités équivalentes est indépendante des fonctions chimiques. Tout le monde admet une équivalence entre le chlore et l'hydrogène. Après quelques observations présentées sur ce sujet par différents autres membres, la séance est levée.

2e Séance du Congrès.

Présidence de M. Boussingault.

En prenant place au fauteuil, M. Boussingault remercie le congrès d'avoir appelé à l'honneur de le présider un savant dont les études ont eu pour objet plutôt des questions de chimie appliquée que des points de théorie abstracte. M. le Président voit dans le choix que le congrès a bien voulu faire un gage de l'union entre ce que l'on nomme l'ancienne et la nouvelle chimie. Il s'élève contre cette dénomination et fait remarquer que ce n'est pas la chimie qui vieillit, mais les chimistes.

M. le Président annonce que le travail de la commission n'est pas prêt. que cependant elle est tombée d'accord sur la rédaction de trois questions à soumettre aux délibérations de l'assemblée. Il invite un des secrétaires à lui en donner connaissance.

M. Strecker prend la parole et donne lecture à l'assemblée des questions rédigées par la commission et indiquées plus haut.

M. Kekulé développe les points indiqués dans la première question.

En ce qui concerne l'hypothèse fondamentale qu'on peut faire sur la nature de la matière, l'orateur se demande s'il faut adopter l'hypothèse atomistique ou si l'on peut se contenter d'une hypothèse dynamique. La première lui paraît préférable. L'hypothèse de Dalton est vérifiée par tout ce que l'on sait de la nature des gaz. On est autorisé à admettre dans les

---

[1]) In Kekulé's Manuscript hat Kekulé die Randbemerkung: „Kekulé und Will lehnt ab", hinzugefügt.

gaz de petites unités, de petits individus, et lorsqu'un même corps peut affecter l'état gazeux, l'état solide et l'état crystallin, il est possible que les molécules crystallines soient précisément les petits individus gazeux dont il s'agit, ou que ceux-ci soient une fraction des autres; mais la nature de ces relations ne peut pas être précisée. Ce qui est sûr, c'est que dans les réactions chimiques il existe une quantité qui y entre ou qui en sort en plus petite proportion et jamais dans une fraction de cette proportion. Ces quantités sont les plus petites qui puissent exister à l'état de liberté. Ce sont là les molécules définies chimiquement. Mais ces quantités ne sont pas indivisibles, les réactions chimiques parviennent à les couper et à les résoudre en particules absolument indivisibles. Ces particules sont les atomes. Les éléments eux mêmes, lorsqu'ils sont libres, constituent des molécules formées d'atomes [1]). Ainsi la molécule de chlore libre est formée de deux atomes. On est donc conduit à admettre différentes unités moléculaires et atomistiques:

1) les molécules physiques
2) les molécules chimiques
3) les atomes.

Les molécules physiques gazeuses ne sont pas démontrées identiques avec les molécules physiques solides et liquides. En second lieu, les molécules chimiques ne sont pas démontrées identiques avec les molécules gazeuses. Ainsi il n'est pas avéré que la plus petite quantité d'un corps qui entre dans une réaction soit aussi la plus petite quantité de ce corps, qui joue un rôle dans les phénomènes de chaleur.

Il faut dire pourtant qu'ordinairement la molécule chimique est identique avec la molécule gazeuse. On a même prétendu que la première ne représente jamais autre chose que la seconde. Pour l'orateur il n'en est pas ainsi. La molécule chimique a une existence indépendante, et pour qu'on soit autorisé à admettre la distinction dont il s'agit, il suffit de démontrer qu'elle est bien réelle dans quelques cas. Or cela est facile. N'est il pas prouvé, pour la densité de la vapeur du soufre, que les molécules chimiques ne se séparent pas toujours complètement les uns des autres, mais restent soudées dans certaines conditions (à 500°) pour former des molécules physiques?

L'orateur ajoute que l'existence et la grandeur des molécules chimiques peuvent et doivent être déterminées par des démonstrations chimiques et que les données physiques ne suffisent pas pour atteindre ce résultat [2]). Comment pourrait on démontrer, en effet. à l'aide de considérations physiques, que l'acide chlorhydrique est formé d'un seul atome d'hydrogène et d'un seul atome de chlore? Ne suffirait-il pas de multiplier la formule HCl par un certain coefficient, et de faire de même pour toute autre formule pour établir une concordance parfaite entre les propriétés physiques?

---

[1]) In Kekulé's Manuscript findet sich hier die Randbemerkung: „Das Molekül ist bisweilen 1 meist 2 Atom." (A.)

[2]) In Kekulé's Manuscript hat Kekulé die Randbemerkung „Schlagende Beispiele: NH_4Cl, $SO_3 \cdot OH_2$" hinzugefügt. (A.)

M. Cannizzaro prend la parole pour faire observer que la distinction entre les molécules physiques et chimiques ne lui paraît ni nécessaire ni clairement établie.

M. Wurtz émet l'avis que ce point est secondaire et peut être réservé. Il lui semble au contraire que la question relative à la distinction à établir entre les termes molécule et atome est arrivée près de sa conclusion, et que tout le monde semble reconnaître l'utilité d'une telle distinction. Il s'agit de préciser le sens de mots généralement usités, il s'agit purement d'une définition et l'orateur pense que dans une question de ce genre il y aurait peut être convenance et utilité à ce qu'un avis soit émis par l'assemblée à la suite de la discussion. Cet avis n'engagerait d'ailleurs personne et n'aurait rien d'obligatoire.

La discussion s'établit sur la seconde question relative au mot »radical composé.«

M. Miller pense que le langage scientifique ne saurait se passer du mot atome composé. Il y a des atomes de corps simples, il y a des atomes de corps composés.

MM. Kekulé, Natanson, Strecker, Ramon de Luna, Nicklès, Béchamps et d'autres membres présentent diverses observations dans un sens ou dans un autre; mais la discussion sur cette question, comme sur la précédente, n'aboutit à aucune résolution de la part de l'assemblée.

Deuxième séance de la Commission.

Présidence de M. H. Kopp.

M. Kekulé expose ses idées sur la notation chimique. Il fait remarquer qu'on peut employer ou une notation atomistique moléculaire ou une notation en équivalents. Dans la première, la formule chimique représente la molécule, dans la seconde elle représente l'équivalence. Les exemples suivants font comprendre cette distinction.

notation atomistique moléculaire	notation en équivalents
H Cl	H Cl
$H^2 \Theta$	H O
H^3 Az	H az [1])

Ce qu'il importe, c'est de ne pas entremêler et confondre ces notations, comme on le fait si souvent. On les confond en écrivant l'eau HO = 9 et l'ammoniaque $H^3Az = 17$ etc.

M. Cannizzaro insiste sur l'importance des considérations relatives aux volumes dans la question de la notation. Les arguments développés par l'orateur sont reproduits in extenso dans le compte-rendu de la 3ème Séance du Congrès (voir plus loin).

---

1) $H = 1. \quad C = 8. \quad \Theta = 16. \quad Az = 14. \quad az = \frac{14}{3}.$

M. le Président fait remarquer que la discussion va trop loin, que les questions doivent être indiquées plutôt qu'approfondies dans le sein de la commission. Il estime d'ailleurs qu'on peut laisser de côté la discussion relative à la notation en équivalents, telle qu'elle vient d'être formulée par M. Kekulé. Personne ne s'en sert. L'orateur croit qu'il serait convenable de ne pas trop s'attacher à des questions de théorie, touchant le fond des choses et qu'on doit s'en tenir à des questions de forme.

Plusieurs membres expriment une opinion analogue et M. Erdmann en particulier fait remarquer qu'il est urgent d'adopter une notation telle que les symboles dont on se sert représentant une valeur déterminée et toujours la même.

M. le Président, résumant la discussion, reconnaît que, vu les progrès récents de la science, il est probable que certains poids atomiques doivent être doublés, mais que dans la notation où l'on se sert de ces poids doubles, il serait utile d'avoir égard à la notation généralement employée jusqu'ici et de ne pas employer exactement les symboles de la dernière notation réprésentant des valeurs différentes. Il pense qu'il serait convenable, comme mesure transitoire et pour éviter des confusions, d'adopter certains signes pour marquer les différences dont il s'agit. En conséquence, M. le Président approuve l'habitude qu'ont prise quelques chimistes de barrer les poids atomiques doublés. En terminant, il formule de la manière suivante la question à soumettre au Congrès:

»Est-il désirable de mettre en harmonie la notation chimique avec les progrès »récents de la science en doublant un certain nombre de poids atomiques?«

Troisième Séance de la Commission.

Présidence de M. Dumas.

M. Kekulé résume la discussion de la séance précédente et reproduit sous une forme un peu mitigée la question annoncée par M. Kopp. D'après l'orateur cette question doit être posée de la manière suivante:

»Les progrès récents de la science autorisent-ils un changement dans la notation etc?«

M. Strecker propose d'adopter en principe la notation atomistique.

M. le Président insiste avec force sur les inconvénients qui résultent de la confusion actuelle. Il fait remarquer que cet état de choses, s'il devait se prolonger, serait de nature à porter atteinte non seulement à la bonne direction de l'enseignement et aux progrès de la science, mais encore à la sûreté des travaux industriels. Reportons, dit M. le Président, nos souvenirs en arrière d'une vingtaine d'années. La table des poids atomiques de Berzelius était à la fois le soutien de la science entière et le guide infaillible des opérations de l'industrie. Rien ne remplace aujourd'hui cette autorité universellement reconnue et il faut prendre garde que la Chimie ne vienne à déchoir du rang élevé qu'elle a occupé jusqu'ici parmi les sciences.

M. Wurtz est heureux de reconnaître que M. Dumas a placé la question sur son véritable terrain et pense qu'il faut revenir aux principes des poids atomiques et de la notation de Berzelius. Des changements peu importants dans l'interprétation de quelques faits suffiraient d'après l'orateur pour mettre les principes et cette notation en harmonie avec les exigences de la science moderne. La notation qu'il convient d'adopter aujourd'hui n'est point précisément celle de Gerhardt. Gerhardt a rendu d'immenses services à la science. Il est mort aujourd'hui et son nom, dit l'orateur, ne doit être prononcé qu'avec respect. Mais il semble que ce chimiste ait commis deux fautes. L'une touche à la forme seulement, l'autre est inhérente au fond des choses.

Premièrement, au lieu de présenter sa notation comme fondée sur des principes nouveaux, il eût fait plus sagement de la rattacher aux principes de Berzelius, et d'abriter ainsi son innovation sous l'autorité de ce grand nom. En second lieu, il semble que Gerhardt ait commis une erreur en assimilant tous les oxydes de la chimie minérale à l'oxyde d'argent et à l'oxyde de potassium anhydre et en leur attribuant, comme à ceux-ci, la formule $\left.\begin{matrix}R\\R\end{matrix}\right\}\Theta$. Il doit y avoir en chimie minérale des oxydes correspondant à l'oxyde d'éthylène, comme il y a des oxydes qui sont les représentants de l'oxyde d'éthyle et d'autres qui correspondent à l'oxyde de glycéryle, et si l'hydrate de potasse p. ex. peut être comparé à l'alcool, d'autres hydrates doivent être comparés au glycol et à la glycérine. On comprend que ces considérations sont de nature à provoquer et à rendre légitimes quelques changements dans la notation de Gerhardt et dans les poids atomiques qu'il attribuait à certains métaux.

Après une discussion à laquelle prennent part MM. Cannizzaro, Wurtz, Kekulé, ce dernier membre émet l'opinion que la question est suffisamment préparée pour pouvoir être soumise à la délibération du Congrès et demande que la rédaction de cette question soit confiée au bureau.

Cette proposition est adoptée par la commission.

3e Séance du Congrès.

Présidence de M. Dumas.

En prenant place au fauteuil, M. le Président adresse quelques paroles de remerciement à l'assemblée et exprime l'espoir que l'on arrivera à une entente commune sur quelques unes des questions agitées devant le congrès.

M. le Président propose ensuite à l'assemblée la désignation de deux vice-présidents. MM. Will et Miller prennent place au bureau en cette qualité. M. Odling remplace comme secrétaire M. Roscoe qui a dû s'absenter. MM. les Secrétaires donnent ensuite lecture des questions élaborées par la commission et dont la rédaction a été confiée au bureau. Ces questions sont ainsi conçues:

»Est-il désirable de mettre en harmonie la notation chimique avec les progrès de la science?«

»Est-il convenable d'adopter de nouveau les principes de Berzelius, en ce qui concerne la notation, en apportant quelques modifications à ces principes?«

»Est-il désirable de distinguer à l'aide de signes particuliers les nouveaux symboles chimiques de ceux qui étaient généralement en usage, il y a une quinzaine d'années?«

M. Cannizzaro prend la parole pour combattre la seconde proposition. Il lui paraît très peu convenable et très peu logique de faire reculer la science jusqu'au temps de Berzelius, pour lui faire parcourir de nouveau le chemin qu'elle a déjà fait. En effet, des modifications successives ont déjà été apportées au système de Berzelius, et ces modifications nous ont conduit au système de formules de Gerhardt. Et ce n'est point brusquement et sans transition que ces changements ont été introduits dans la science; ils ont été la conséquence de progrès successifs. Si Gerhardt ne les avait point proposés, ils l'auraient été ou par M. Williamson ou par M. Odling, ou par tout autre chimiste qui aurait pris part au mouvement de la science.

»La source où remonte le Système de Gerhardt est la théorie d'Avogadro et d'Ampère sur la constitution uniforme des corps à l'état gazeux. Cette théorie nous conduit à envisager les molécules de certains corps simples, comme susceptibles d'une division ultérieure. M. Dumas comprit l'importance de la théorie d'Avogadro et toute l'étendue de ses conséquences Il posa cette question: Y-a-t il accord entre les résultats de la théorie d'Avogadro et les résultats qu'on déduit des autres méthodes servant à déterminer les poids relatifs des molécules? S'apercevant que la science était encore pauvre en résultats expérimentaux de ce genre, il voulut en rassembler le plus grand nombre possible avant *de se permettre aucune conclusion générale sur cet objet.* Il se mit donc à l'œuvre et à l'aide de la méthode qu'il appliqua à la détermination des densités de vapeurs, il dota la science de résultats précieux. Pourtant il semble qu'il ne l'ait jamais trouvée suffisamment avancée pour tirer des résultats acquis la conclusion générale à laquelle il visait. Quoiqu'il en soit de cette réserve qu'il a cru devoir observer, on peut dire que c'est lui qui a mis les chimistes sur le chemin de la théorie d'Avogadro; car il a contribué plus que tout autre à introduire l'habitude de choisir pour les corps volatils des formules correspondant au même volume que celui qu'occupent l'acide chlorhydrique et l'ammoniaque.

Cette influence de l'école de M. Dumas se révèle de la manière la plus évidente dans un mémoire d'un de ses élèves M. Gaudin. M. Gaudin accepta sans réserve la théorie d'Avogadro. Il a établi une distinction bien tranchée entre les mots *atome* et *molécule,* au moyen de laquelle il put concilier tous les faits avec la théorie. Cette distinction avait déjà été faite par M. Dumas qui, dans ses leçons de philosophie chimique, avait appelé la molécule: *l'atome physique.* Elle est certainement un des pivots du système de Gerhardt.

M. Gaudin, plus conséquent et plus fidèle à la théorie d'Avogadro que ne le fut plus tard Gerhardt, et profitant des nouvelles données expérimentales sur les densités de vapeur, établit que les atomes ne sont pas toujours la même fraction des molécules des corps simples; c. à. d. que ces molécules ne résultent pas toujours du même nombre d'atomes; que, tandis que les molécules de l'oxygène, de l'hydrogène et des corps halogènes sont formées par 2 atomes, la molécule de mercure est faite d'un seul atome. Il alla jusqu'à comparer la composition de volumes égaux d'alcool et d'éther, pour en déduire la composition relative de leurs molécules. Mais son esprit ne saisit pas toutes les conséquences de cette comparaison et les chimistes ont oublié l'idée qu'il en a eu. Et pourtant cette comparaison et un des points de départ de la réforme proposée par G e r h a r d t.

D'autres chimistes, parmi lesquels on peut citer P r o u s t, acceptèrent aussi la théorie d'Avogadro et arrivèrent aux mêmes conséquences générales que M. G a u d i n.

Dans cet état de la science, qu'a fait G e r h a r d t?

Il accepte la théorie d' A v o g a d r o avec la conséquence de la divisibilité des atomes des corps simples, il applique cette théorie à déduire la constitution relative des molécules de l'hydrogène, de l'oxygène, du chlore, de l'azote, de l'acide chlorhydrique, de l'eau et de l'ammoniaque. S'il s'était arrêté là, il n'aurait pas devancé A v o g a d r o et M. D u m a s. Mais il soumit ensuite à un examen général toutes les formules de la chimie organique et s'aperçut que toutes ces formules correspondant à des volumes égaux d'acide chlorhydrique et d'ammoniaque étaient confirmées par toutes les réactions et par toutes les analogies chimiques. Il songea alors à modifier les formules qui faisaient exception à cette règle dont M. D u m a s avait introduit l'habitude. Il tâcha de démontrer que les raisons par lesquelles, on s'était éloigné de la règle des volumes égaux, n'étaient pas fondées. Réduire les formules de tous les corps volatils de la chimie organique à des volumes égaux, tel a été le point de départ des réformes proposés par G e r h a r d t. Les modifications des poids atomiques de certains corps simples, la découverte des relations qu'offrent les hydrates, soit acides, soit basiques avec l'eau ont été la conséquence de ce premier pas. Qu'est-il arrivé ensuite? Les mémorables expériences de M. W i l l i a m s o n sur l'éthérification, sur les éthers mixtes, sur les acétones, celles de G e r h a r d t sur les acides anhydres, celles de M. W u r t z sur les radicaux alcooliques etc. vinrent successivement confirmer ce que G e r h a r d t avait prévu comme conséquence de son système. Ainsi il est arrivé en chimie quelque chose d'analogue à ce qui est arrivé en optique, quand on y a introduit la théorie des ondulations. Cette théorie a fait prévoir avec une admirable exactitude les faits que l'expérience a plus tard confirmés. Le système de G e r h a r d t en chimie n'a pas été moins fécond en prévisions exactes. Il est intimement mêlé et enchaîné dans l'histoire de la science, à tous les travaux chimiques qui l'ont précédés et aux progrès qui l'ont suivi. Ce n'est pas un saut brusque, un fait isolé. C'est un pas régulier en avant, petit en apparence, mais grand par ses résultats. On ne peut désormais effacer ce système de l'histoire de la science; on peut et on doit le discuter et le modifier; mais c'est lui qu'il faut prendre pour point de départ, quand il s'agit d'introduire dans la science un système de formules en accord avec l'état actuel de nos connaissances. Quelques chimistes seront peut-être tentés de dire: la différence entre les formules de G e r h a r d t et celles de B e r z e l i u s est très petite; car la formule de l'eau par exemple est la même dans les 2 systèmes. Mais qu'on y prenne garde. La différence est très petite en apparence, mais grande au fond. B e r z e l i u s était sous l'influence des idées de D a l t o n. L'idée d'une différence entre l'atome et la molécule des corps n'est jamais entrée dans son esprit. Dans tous ses raisonnements il admet implicitement que les atomes des corps simples sont, vis à vis des forces physiques, des unités du même ordre, des atomes composés. C'est par cette raison qu'il a commencé par admettre que des volumes égaux contiennent un même nombre d'atomes. Bientôt il s'aperçut que cette règle ne pouvait s'appliquer qu'aux corps simples et, pendant tout le cours de sa carrière scientifique, il n'attribuait aucune valeur aux atomes des corps composés pour choisir des formules. Il a été même obligé de limiter la règle du nombre égal d'atomes dans des volumes égaux à un très petit nombre de corps simples c. a. d. à ceux qui sont des gaz permanents; introduisant ainsi dans la constitution des gaz et des vapeurs une différence qu'aucun physicien n'a jamais pu admettre. Berzelius n'admettait pas que les molécules des corps simples puissent se diviser en se combinant; il suppose au contraire, que souvent deux molécules forment la quantité qui entre tout entière dans la combinaison. C'est ce qu'il appelait des *atomes doubles*. Ainsi il admet que l'eau et l'acide chlorhydrique contiennent la même quantité d'hydrogène, quantité égale à deux molécules physiques réunies.

Vous voyez donc, Messieurs, quelle différence profonde existe entre les idées de Berzelius et les idées d'Avogadro, d'Ampère, de M. Dumas et de Gerhardt.

Je m'étonne que M. Kekulé ait accepté la proposition de la commission, lui qui dans son livre a dit que Gerhardt est le premier et le seul qui ait bien compris la théorie atomique.

Je crois avoir démontré, continue M. Cannizzaro, qu'il faut prendre pour point de départ d'une discussion des formules celles de Gerhardt, mais je ne soutiens pas qu'il faut les accepter toutes, telles qu'il les a proposées. Loin de là, j'ai essayé, il y a quelques années, d'y introduire certaines modifications, de manière à éviter les inconséquences qui me paraissent exister dans le système de Gerhardt. En effet, il est curieux de voir comment ce chimiste a renié la théorie d'Avogadro, après s'en être servi comme base de ses réformes. Voici comment il s'exprime lui-même: »Il y a des molécules à 1, 2 et 4 vol. comme il y en a à $^1/_2$, »à $^1/_3$, à $^1/_4$ de volume.« (Comptes rendus des travaux de Chimie. année 1851 p. 146). Et il continue ainsi (ibid. p. 147): »On s'étonne peut-être de me voir »soutenir cette thèse, alors que j'ai recommandé et que je recommande encore »tous les jours de suivre en chimie organique une notation regulière, en représen»tant tous les corps volatils par le même nombre de volumes, par 2 ou par 4. Les »chimistes qui voient en cela deux affirmations contradictoires, oublient que je »n'ai jamais avoué le principe précédent, comme une vérité moléculaire, mais »comme une condition à remplir pour arriver à la connaissance de certaines lois »ou de certains rapports qu'une notation arbitraire ou appropriée à des cas parti»culiers laisserait échapper à l'attention de l'observateur.«

Il y avait certainement des faits qui obligeaient Gerhardt à renier la théorie d'Avogadro, mais il y avait aussi des hypothèses gratuites. Les faits étaient les densités de vapeur de l'acide sulfurique monohydraté, des sels ammoniacaux et du perchlorure de phosphore.

Vous savez déjà, Messieurs, qu'à l'occasion de la publication du Mémoire de

M. Deville sur la dissociation de certains composés par la chaleur j'ai essayé le premier d'interpréter le fait de ces densités anormales, en supposant que les corps dont ils s'agit se dédoublent et qu'en réalité dans la détermination de ces densités on pèse un mélange de vapeurs. Après moi, M. H. Kopp a proposé de son côté la même interprétation.

Je ne répèterai pas ici les arguments que nous avons invoqués en faveur de cette interprétation. J'ajouterai seulement qu'un des membres de ce Congrès vient de me dire que le point d'ébullition de l'acide sulfurique est presque constant à des pressions très différentes, fait qui démontre qu'il s'agit ici, non d'un point d'ébullition, mais d'un point de décomposition. D'autres faits, j'en suis convaincu, viendront confirmer l'interprétation que nous avons donnée des densités anormales et feront ainsi disparaître les doutes que quelques savants paraissent encore conserver au sujet de la théorie d'Avogadro.

Mais indépendamment des faits que je viens de citer, il y avait aussi des hypothèses gratuites qui avaient éloigné Gerhardt de la théorie d'Avogadro. Je vais démontrer qu'il en est ainsi.

Gerhardt admettait comme une vérité démontrée que tous les composés métalliques ont des formules analogues à celles des composés hydrogénés correspondants. Cela admis donne aux chlorures de mercure les formules HgCl. Hg^2Cl, en supposant que la molécule de mercure libre est formée comme celle de l'hydrogène de deux atomes. Remarquons que les densités de vapeur conduisent à un résultat différent. En effet, pour représenter la composition de volumes égaux des cinq corps suivants: hydrogène, acide chlorhydrique, mercure, chlorure mercureux, chlorure mercurique, nous aurons les formules suivantes:

$$H^2 \quad HCl \quad Hg^2 \quad Hg^2Cl \quad Hg^2Cl^2$$

La comparaison de ces formules nous montre que dans les molécules du mercure libre et de ses deux chlorures il existe la même quantité de mercure exprimée par Hg^2. que le chlorure mercureux est analogue à l'acide chlorhydrique, tandis que le chlorure mercurique contient dans sa molécule une quantité double de chlore.

Il en résulte que la même raison qui nous a portés à doubler l'atome de carbone, doit aussi nous engager à doubler l'atome de mercure. Cela revient à dire que la quantité de mercure exprimée par Hg^2 dans les formules précédentes représente un seul atome. On voit que dans ce cas l'atome est égal à la molécule du corps libre, que dans les sels mercureux cet atome est l'équivalent d'un seul atome d'hydrogène, tandis que dans les sels mercuriques il est l'équivalent de 2 atomes d'hydrogène. En d'autres termes et pour employer le langage généralement usité aujourd'hui. dans les sels mercureux le mercure est monoatomique. dans les sels mercuriques il est diatomique comme les radicaux des glycols de M. Wurtz.

Il est important de faire remarquer maintenant qu'en doublant le poids atomique de mercure, comme on a doublé celui du soufre, on arrive à des nombres qui s'accordent avec la loi des chaleurs spécifiques.

Mais si l'on double le poids atomique du mercure on est conduit par analogie à doubler ceux du cuivre, du zinc, du plomb, de l'étain etc., en un mot on retombe dans le système des poids atomiques de M. Regnault qui sont d'accord avec les chaleurs spécifiques, avec l'isomorphisme et avec les analogies chimiques.

C'était chose vraiment regrettable que le désaccord entre le système de Gerhardt et la loi des chaleurs spécifiques ainsi qu'avec l'isomorphisme. Ce désaccord avait produit deux chimies différentes, l'une qui traitait des corps inorganiques accordait une grande valeur à l'isomorphisme; l'autre, qui étudiait les corps organiques, n'en tenait aucun compte, de telle sorte que le même corps pouvait n'avoir pas la même formule dans une chimie que dans l'autre. Le désaccord que je viens de signaler provenait de ce que le système de Gerhardt n'était pas conséquent dans toutes ses parties. Il disparaît dès qu'on supprime les inconséquences.

Les densités de vapeur offrent le moyen de déterminer le poid des molécules des corps, soit simples, soit composés. Les chaleurs spécifiques servent à contrôler les poids des atomes et non ceux des molécules. L'isomorphisme révèle les analogies de constitution moléculaire.

A l'appui de la modification que je viens de proposer des poids atomiques de certains métaux, je citerai les faits suivants: Tous les composés volatils de mercure, de zinc, d'étain, de plomb contiennent des quantités de métal représentées dans la notation ordinaire par Hg^2, Zn^2, Sn^2, Pb^2. Ce fait seul suffisait pour nous indiquer que ces quantités représentent les vrais atomes des métaux en question. On pourrait aussi citer ce fait qu'il existe trois oxalates de potassium et d'ammonium (radicaux monoatomiques) tandis qu'il existe seulement 2 oxalates de barium et de calcium (radicaux diatomiques). Mais pour le moment, je n'insiste pas sur ce point et je ne puis disconvenir d'un autre côté, qu'il est un cas où le poids atomique déduit de la comparaison des compositions moléculaires est en désaccord avec celui que l'on déduirait de la chaleur spécifique. Ce cas est relatif au carbone. Mais il se peut qu'ici la loi des chaleurs atomiques reste voilée par d'autres causes qui interviennent dans la chaleur spécifique.

En résumé, Messieurs, je vous propose d'accepter le système de Gerhardt en prenant en considération les modifications que je propose d'apporter aux poids atomiques de certains métaux et aux formules de leurs sels.

Et si nous ne pouvons tomber d'accord pour accepter franchement la base du nouveau système, évitons du moins de prononcer un jugement contraire qui serait sans résultat, soyez en sûrs. En effet nous ne pourrons empêcher que le système de Gerhardt ne gagne des partisans tous les jours. Déjà aujourd'hui, il est accepté par la majorité des jeunes chimistes, de ceux qui prennent la part la plus active aux progrès de la science.

Bornons-nous, dans ce cas, à faire quelques conventions pour éviter la confusion qui résulte de l'emploi des symboles identiques auxquels on attribue des valeurs différentes. C'est ainsi que, généralisant un usage déjà établi, nous pourrons adopter les lettres barrées pour représenter les poids atomiques doublés.«

M. Strecker donne quelques éclaircissements concernant la rédaction de la seconde proposition soumise au Congrès. Cette rédaction mentionnait primitivement le nom de Gerhardt, mais la majorité du bureau avait été d'avis de substituer à ce nom celui de Berzelius. L'orateur n'a pas partagé l'opinion de la majorité. Il lui a semblé qu'il n'y avait pas lieu de remonter jusqu'à Berzelius, auquel on peut reprocher peut-être un défaut de conséquence dans la question des atomes et des équivalents. Ce qui est utile et urgent, c'est d'améliorer ce qui existe, en tenant compte des progrès de la science depuis Berzelius. M. Strecker ajoute que les doctrines dont le »système de Gerhardt« est l'expression, présentent des avantages réels. Quant à lui, il adoptera dorénavant, dans ses Mémoires, les nouveaux poids atomiques, mais il ne pense pas que le moment soit venu de les introduire dans l'enseignement et dans les livres élémentaires.

M. Kekulé partage entièrement les opinions exprimées par M. Cannizzaro. Il lui parait cependant utile de faire une réserve sur un point de détail. M. Cannizzaro considère le chlorure mercureux comme renfermant HgCl (Hg = 200). Il parait plus rationnel à M. Kekulé de l'envisager comme une combinaison analogue à la liqueur des Hollandais c. a. d. comme renfermant Hg^2Cl^2 (Hg = 200) et d'admettre qu'au moment de la vaporisation la molécule Hg^2Cl^2 se scinde.

M. Will ne veut pas entrer dans les détails des questions soumises au Congrès. Il se borne à faire remarquer qu'il faut marcher directement au but. Ce but consiste à trouver une notation claire, logique, incapable de faire naître la confusion dans l'esprit des personnes peu initiées aux formules et propre à exprimer, non seulement les faits depuis longtemps enrégistrés dans la science, mais encore ceux que les découvertes modernes y ajoutent chaque jour.

M. Erdmann propose de laisser tomber les 2 premières questions et de se borner à la discussion de la dernière. Il lui paraît difficile d'arriver à une entente concernant les questions de principe et surtout d'imposer en quelque sorte une notation par un vote.

M. Wurtz fait remarquer que personne n'a eu l'idée d'imposer une opinion quelconque. On se trouve en présence de deux sortes de questions, les unes touchent au fond même des choses, les autres sont des questions de forme. Si les premières ne sont pas encore assez mûres pour qu'il y ait lieu de les trancher par des votes, rien n'empêche, au contraire, de s'entendre et même de voter sur les pures questions de forme.

M. Hermann Kopp constate que sur beaucoup de points théoriques les opinons des chimistes sont partagées. Ces différences d'opinions sont causées en partie par des malentendus et sont reflétées par la notation même. Une discussion peut être très utile pour faire cesser les malentendus.

M. Erlenmeyer propose de se servir toujours des symboles barrés pour exprimer les poids atomiques qui représentent les anciens équivalents doublés [1]).

M. L. Meyer fait remarquer que ce point semble acquis à la discussion, parceque personne n'a élevé une objection à cet égard.

Un débat s'engage entre plusieurs membres sur l'opportunité d'émettre un vote.

M. Cannizzaro est d'avis qu'il est inutile de voter sur la troisième question.

---

[1]) Vgl. Kekulé Ann. (1857) **104,** 132 Anm. [2]) Ursprünglich von Williamson vorgeschlagen. (A.)

M. Boussingault fait observer qu'il paraît difficile que l'on se trompe sur la signification des votes que peut émettre le congrès au sujet des questions qui lui sont soumises. En votant il ne fait qu'exprimer des vœux et il n'entend imposer à personne l'opinion de la majorité.

M. Will se range au même avis.

M. Normandy fait observer que les savants qui proposent d'introduire certaines réformes dans la science concernant la notation sont ceux qui cultivent principalement la chimie organique. Or on peut constater que sur quelques points les savants ne sont pas même d'accord entre eux. Il paraît donc prématuré d'appliquer à la chimie minérale des principes qui sont encore en discussion.

M. Odling prend la parole au sujet des symboles barrés. Il rappelle que Berzelius les a introduits dans la science pour exprimer des atomes doubles. La barre, dit-il, est donc le signe de la divisibilité et il paraît contraire à la logique de barrer les symboles exprimant les atomes indivisibles de l'oxygène et du carbone.

M. Kekulé, tout en accordant que les atomes doubles de Berzelius avaient une signification différente que les atomes indivisibles dont on propose de barrer les symboles, fait remarquer cependant que ces symboles barrés doivent exprimer, non pas la divisibilité des atomes, mais la divisibilité de la valeur exprimé par ces symboles et qui est double de celle qu'on leur attribuait autrefois.

En réponse aux observations présentées par MM. Erdmann et Normandy, M. Kekulé ajoute que personne ne peut avoir la pensée d'imposer une opinion théorique ou une notation par un vote, mais qu'une discussion sur de tels sujets est nécessaire et utile et ne manquera pas de porter des fruits.

Le Congrès consulté par M. le Président exprime le vœu que l'usage des symboles barrés, représentant des poids atomiques doubles de ceux qu'on admettait autrefois, s'introduise dans la science.

M. Dumas lève la 3e et dernière Séance du Congrès, après s'être rendu l'organe de la reconnaissance qu'a inspirée à l'assemblée l'hospitalité dont elle a été l'objet de la part de S. A. R. le Grand-Duc de Bade.

Suggestions for Further Reading

Boorse, H. A., and Motz, L., eds. *The World of the Atom*. New York: Basic Books, 1966.

Brock, W.H. "Lockyer and the Chemists: The First Dissociation Hypothesis." *Ambix* 16 (July, 1969): 81-99.

Brock, W.H., ed. *The Atomic Debates. Brodie and the Rejection of the Atomic Theory*. Leicester, England: Leicester University Press, 1967.

Brock, W.H., and Knight, D.M. "The Atomic Debates." *Isis* 56 (1965): 5-25.

Broda, Engelbert. "The Philosophical Views of Ludwig Boltzmann." In *The Boltzmann Equation: Theory and Applications*, edited by E. G. D. Cohen and W. Thirring. Vienna: Springer-Verlag, 1973.

Brooke, John Hedley. "Laurent, Gerhardt and the Philosophy of Chemistry." *Historical Studies in the Physical Sciences* 6 (1975): 405-429.

Brooke, John Hedley. "Avogadro's Hypothesis and Its Fate: A Case-Study in the Failure of Case-Studies." *History of Science* 19 (1981): 235-273.

Brush, Stephen. "Mach and Atomism." *Synthese* 18 (1968): 192-215.

Brush, Stephen G. "Randomness and Irreversibility." *Archive for History of Exact Sciences* 12 (1974): 1-88.

Brush, Stephen G. *The Kind of Motion We Call Heat: A History of the Kinetic Theory of Gases in the 19th Century*. 2 vols. Amsterdam: North-Holland Publishing Company, and New York: Elsevier, 1976.

Buchdahl, Gerd. "Sources of Scepticism in Atomic Theory." *British Journal for the Philosophy of Science* 10 (1960): 120-134.

Cardwell, D. S. L., ed. *John Dalton and the Progress of Science*. Manchester and New York: Manchester University Press, 1968.

Clark, Peter. "Atomism vs Thermodynamics." In *Method and Appraisal in the Physical Sciences*, edited by Colin Howson. Cambridge, England: Cambridge University Press, 1976.

Colmant, P. "Querelle à l'Institut entre équivalentistes et atomistes." *Revue des Questions Scientifiques* 143 (1972): 493-519.

Conn, G. K. T., and Turner, H. D. *The Evolution of the Nuclear Atom*. New York: American Elsevier, 1965.

Curd, Martin V. "Ludwig Boltzmann's Philosophy of Science: Theories, Pictures and Analogies." Ph.D. diss., University of Pittsburgh, 1978.

Daub, Edward E. "Atomism and Thermodynamics." *Isis* 58 (1967): 293-303.

De Broglie, Maurice. *Les Premiers Congrès de Physique Solvay et l'orientation de la physique depuis 1911.* Paris: Michel, 1951.

De Kosky, R. K. "Spectrosopy and the Elements in the Late Nineteenth-Century: The Work of Sir William Crookes." *British Journal for the History of Science* 6 (1973): 400-423.

De Milt, Clara. "Carl Weltzein [sic] and the Congress at Karlsruhe." *Chymia* 1 (1948): 153-170.

Fisher, Nicholas. "Avogadro, the Chemists, and Historians of Chemistry, Part I." *History of Science* 20 (1982): 77-102.

Freund, Ida. *The Study of Chemical Composition.* Cambridge: Cambridge University Press, 1904.

Garber, Elizabeth. "Molecular Science in Late-Nineteenth Century Britain." *Historical Studies in the Physical Sciences* 9 (1978): 265-297.

Gardner, Michael. "Realism and Instrumentalism in Atomic Theory." *Philosophy of Science* 46 (1979), 1-34.

Hiebert, Erwin N. "The Conception of Thermodynamics in the Scientific Thought of Mach and Planck." *Ernst Mach Institut. Wissenschaftlicher Bericht* 5 (1968): 106 p.

Hiebert, Erwin N. "The Genesis of Mach's Early Views on Atomism." *Boston Studies in the Philosophy of Science* 6 (1970): 79-106.

Holt, Niles. "A Note on Wilhelm Ostwald's Energism." *Isis* 61 (1970): 386-389.

Holton, Gerald. "Subelectrons, Presuppositions, and the Millikan-Ehrenhaft dispute." In *The Scientific Imagination: Case Studies*, by Gerald Holton. Cambridge: Cambridge University Press, 1978: 25-83.

Ihde, Aaron. *The Development of Modern Chemistry.* New York: Harper and Row, 1964.

Klein, Martin J. "Mechanical Explanation at the End of the Nineteenth Century." *Centaurus* 17 (1972): 58-82.

Knight, David M. *Atoms and Elements. A Study of Theories of Matter in England in the Nineteenth Century.* London: Hutchinson, 1967.

Knight, David M., ed. *Classical Scientific Papers: Chemistry.* 2 vols. London: Elsevier, 1968, 1970.

Laudan, Laurens. "Towards a Reassessment of Comte's 'Méthode Positive.' " *Philosophy of Science* 38 (1971): 35–53.
Laudan, Laurens. "The Methodological Foundations of Mach's Opposition to Atomism." In *Space and Time, Matter and Motion*, edited by P. Machamer and R. Turnbull. Columbia: Ohio State University Press, 1976.
Levere, Trevor. *Affinity and Matter: Elements of Chemical Philosophy, 1800–1865*. Oxford: Clarendon Press, 1971.
McCormmach, Russell. "Henri Poincaré and the Quantum Theory." *Isis* 58 (1967): 37–55.
McCormmach, Russell. "H. A. Lorentz and the Electromagnetic View of Nature." *Isis* 61 (1970): 459–497.
McGucken, William. *Nineteenth-Century Spectroscopy: Development of the Understanding of Spectra, 1802–1897*. Baltimore: Johns Hopkins University Press, 1969.
Mehra, A. J. *The Solvay Conferences on Physics*. Dordrecht: Reidel Publishing Company, 1975.
Mellor, David P. *The Evolution of the Atomic Theory*. Amsterdam and New York: Elsevier, 1971.
Metz. André. "La Notation atomique et le théorie atomique en France à la fin du XIXe siècle." *Revue d'Histoire des Sciences* 16 (1963): 233–239.
Muir, M. M. Pattison. *A History of Chemical Theories and Laws*. New York: John Wiley and Sons, 1906.
Nye, Mary Jo. *Molecular Reality. A Perspective on the Scientific Work of Jean Perrin*. London: Macdonald and New York: American Elsevier, 1972.
Nye, Mary Jo. "The Nineteenth-Century Atomic Debates and the Dilemma of an 'Indifferent Hypothesis.' " *Studies in the History of Philosophy of Science* 7 (1976): 254–268.
Nye, Mary Jo. "Berthelot's Anti-Atomism: A 'Matter of Taste?' " *Annals of Science* 38 (1981): 585–590.
Partington, J. R. *A History of Chemistry*. Vol. IV. London: MacMillan, 1964.
Paul, E. P. "Alexander W. Williamson on the Atomic Theory: A Study of Nineteenth-Century British Atomism." *Annals of Science* 35 (1978): 17–32.
Post, H. R. "Atomism 1900." *Physics Education* 3 (1968): 1–13.
Rocke, Alan J. "Atoms and Equivalents: the Early Development of the Chemical Atomic Theory." *Historical Studies in the Physical Sciences* 9 (1978): 225–263.

Rocke, Alan J. "The Reception of Chemical Atomism in Germany." *Isis* 70 (1979): 519–536.

Rocke, Alan J. *Chemical Atomism in the Nineteenth Century: From Dalton to Cannizzaro*. Columbus: Ohio State University Press. In press.

Rosenfeld, Leon. "Men and Ideas in the History of Atomic Theory." *Archives for History of Exact Sciences* 7 (1971): 69–90.

Russell, Colin A. *The History of Valency*. New York: Humanities Press, 1971.

Schonland, Basil. *The Atomists (1805–1933)*. Oxford: Clarendon Press, 1968.

Scott, Wilson L. *Conflict between Atomism and Conservation Theory. 1644–1860*. London: Macdonald and New York: American Elsevier, 1970.

Shinn, Terry. "Orthodoxy and Innovation in Science: The Atomist Controversy in French Chemistry." *Minerva* 18 no. 4 (Winter 1980): 539–555.

Sugiyama, Shigeo. "The Late Nineteenth-Century Atomic Debates and the Mechanical View of Nature." *Kagakushi Kenkyu* (*Journal of the History of Science, Japan*) 16 (1977): 153–160. In Japanese.

Thorpe, T. E. *Essays in Historical Chemistry*. London: Macmillan, 1894.

Topper, D. R. "Commitment to Mechanism: J. J. Thomson, the Early Years." *Archive for History of Exact Sciences* 7 (1971): 393–410.

Van Melsen, Andrew B. *From Atmos to Atom*. New York: Harper Torchbooks, 1960.

Yoshida, Akira. "Charles Adolphe Wurtz et la théorie atomique." *Japanese Studies in the History of Science* 16 (1977): 129–135.